Metal Nano 3D Superlattices

Synthesis, Properties, and Applications

Marie-Paule Pileni

WILEY-VCH

Author

Prof. Marie-Paule Pileni
Sorbonne Université
Department of Chemistry
4 Place Jussieu
75005 Paris Cedex 05
France

Cover Image: © Marie-Paule Pileni,
Sorbonne Université

Library of Congress Card No.: applied for

British Library Cataloguing-in-Publication Data
A catalogue record for this book is available from the British Library.

Bibliographic information published by the Deutsche Nationalbibliothek
The Deutsche Nationalbibliothek lists this publication in the Deutsche Nationalbibliografie; detailed bibliographic data are available on the Internet at <http://dnb.d-nb.de>.

Typesetting Straive, Chennai, India
Printing and Binding CPI Group (UK) Ltd, Croydon, CR0 4YY

Print ISBN 978-3-527-34477-2
ePDF ISBN 978-3-527-34478-9
ePub ISBN 978-3-527-34480-2
oBook ISBN 978-3-527-34479-6

C085132_030123

To Vannina, Christophe, Alix, and Paul

Contents

Acknowledgments

The experiments for the data presented in this book have been performed by MPP team. The experiments were confirmed by several researchers from the group. This reduces the potential artifacts. Furthermore, long-term collaborations with limited group number are another strength.

MPP would like to thank

* Long-term collaborations: *Dr. P.A. Albouy* from Orsay, France; *Dr. P. Bonville* of CEA-Saclay, France; *Professors N. Cerullo, D. Polli,* and *G. De la Valle* of Politecnico di Milano, Italy; *Professor M. El Sayed* of Georgia Tech, USA; *Professor K. Holmberg* of Chalmers University, *Professor B. W. Ninham* of Canberra University, Australia; Sweden; *Professor F. Willig* of Fritz-Haber Institute der MPG, Germany; and *Professor Z. Yang* of Key Laboratory of Colloid and Interface Chemistry of MOE, School of Chemistry and Chemical Engineering, Shandong, University, China
* Short-term collaborations: *Professor S. Ball* of University of Antwerp, Belgium; *Professor F. Carn* of Université de Paris, *Dr. F. Charra* of CEA-Saclay, France; *Professor P. M. Claesson* of KTH, Sweden; *Dr. I. Dobryden* of KTH, Sweden, *Professors J. Freund and J. Urban* of Fritz-Haber Institut der MPG, Germany; *Dr. F. Gazeau* of Université de Paris, France; *Professor J. Israelashivili* of Santa Barbara, USA; *Professor P. Moriarty* of University of Nottingham, UK; *Dr. A. Nicolas- Boluda* of Université de Paris, and *Professor D. M. Smilgies* of Cornell University, USA and *Dr. N. Winckelmans* of University of Antwerp, Belgium.
* Nothing would be possible without my collaborators. Nice collaborative atmosphere has been the key parameter in the success of this book. When that was not possible, the collaboration was stopped. Most of them were PhD students or postdoctorates. Some of them remain in the group after the thesis.

People involved in this book are: *E. Alphandéry, I. Arfaoui, A. Çolak, A. Courty, M. Gauvin, V. Germain, A. I. Henry, Kh. Khazen, E. KLecha, X. Ling, Y. Lalatonne, I. Lisiecki, S. Mourdikoudis, T. Ngo, D. Parker, H. Portales, J. Richardi, S. Roland, J. Romann, C. Salzeman, N. Schaeffer, Y. Wan, L. Wang, J. Wei, C. Yan, J. Yang, N. Yang, P. Yang,* and *Z. Yang*.

Other collaborators not involved in this book are: *P. André, F. Billoudet*$^{+}$, *A. Brioude, Q. Darugar, D. Ingert, N. Feltin, A. Filamkenbo, T. Jian, X. C. Jiang, J. Legrand, L. Levy, W. Qian, M. Maillard, F. Michel, L. Motte, K. Naoe, C. Petit, N. Pinna, F. Pitré, J. Tanori,* and *H. Taleb*.

Introduction

Self-organization of entities into two-dimensional (2D) or three-dimensional (3D) materials, forming highly ordered structures such as atoms in bulk materials, cations and anions in salts, and organic molecules in biological materials, is rather basic. For instance, the gem opal assembled from silica spheres with a closely packed locally periodic structure under moderate compression in geology; the crystal of the tobacco mosaic virus in biological systems. When referring to science and technology, self-organization with a multitude of tunable parameters, such as the types of building blocks, the assembly driving force as well as potential applications of these assembled architectures, has attracted a broad interest among countless chemists, biologists, and physicists. Hence, self-assembly of objects into ordered functional superstructures is a universal process and a prevalent topic in science.

In this book, we concentrate our efforts on self-assemblies of metal nanocrystals. The nanocrystal size is in the range from 2 to 13 nm. To keep their integrities, these metal nanocrystals are coated with various organic molecules. The size distribution of the nanocrystals used here is rather low. Nanocrystals are dispersed in solution. The produced colloidal solution is optically clear and colored. The difference between nanoparticles and nanocrystals is related to the crystallinity of the core. Hence, the core of nanoparticles is amorphous, whereas that of nanocrystals is crystalline. This latter term does not take into account the ordering mode of atoms in the core (single domain or polycrystal). The crystalline structures of nanocrystals called *nanocrystallinity* permit to differentiate between single domain and polycrystal but not the crystalline phase. As matter of fact, single domain is a generic name and corresponds to various crystalline phases such as face-centered-cubic (fcc), hexagonal-close-packed (hcp), and body-centered-cubic (bcc) structures. For Co nanocrystals, the *ε-phase* exists at the nanoscale. Either by slow solvent evaporation of the colloidal solution or by keeping it under saturated solvent atmosphere (the solvent needs to be a bad solvent for the alkyl chains), nanocrystals tend to self-assemble either as films or as aggregates. A careful characterization shows that both films and aggregates are usually assembled in 3D superlattices called *colloidal crystals*, *supercrystals*, or *supracrystals*. When the distribution of nanocrystals is random without regular order, the films and aggregates are considered as amorphous. All over these chapters, the building blocks of 3D superlattices are in the size range between 5 and 10 nm. Similar crystalline structures are obtained in nature. However, the particle sizes (most of the cases ferrite) are in the range of micrometers. In Chapter 1, syntheses of nanocrystals with low size distribution are described. Chapter 2 shows some properties of nanocrystals differing by their nanocrystallinity. It has not been possible to control the nanocrystallinity of all the metal nanocrystals studied. This is due to the techniques used (soft chemistry) to produce nanocrystals. Here, we concentrate on the final structures of Co nanocrystals subjected to oxygen and differing by size and nanocrystallinity. Such diffusion processes in metal have been largely studied with bulk metal and nowadays at the nanoscale. This process is called nano-Kinkerdall effect. Hence, with ε-*phase*, whatever be the nanocrystal size, CoO hollow nanocrystals are produced. With amorphous, fcc, and hcp Co nanocrystals, various final structures are produced such as Co oxide (CoO), shell or yolk/shell structures with CoO shell, and CoO hollow structures, whereas the Co cores keep the crystalline structure they had before oxygen treatment. Some physical properties of metal nanocrystals differing by their nanocrystallinities are described: the localized surface plasmon resonance of Au nanocrystals and the acoustic breathing mode of Au and Co nanocrystals do not depend on the nanocrystallinity. Conversely, the fundamental quadrupolar mode of single-domain Au nanocrystals split, whereas the polycrystalline counterpart does not. Such splitting will be used all over the book chapters as a tool to identify a large collection of nanocrystals differing by their nanocrystallinities. In Chapter 3, the various assemblies obtained by solvent evaporation of Au colloidal solution are described. It is shown that the coating agents, the nanocrystal size, solvent vapor pressure, and temperature play key roles in the final crystalline structures and in the interparticle distances of the 3D superlattices. Chapter 4 shows that by keeping the colloidal solution under solvent saturation during several days, simultaneous crystal growth process takes place: a 3D superlattice film at the solvent–air interface and shaped 3D superlattices precipitated are produced. Such simultaneous 3D superlatticesgrowth processes are produced if the solvent of the colloidal solution has to have a large surface tension as the air–solvent interface and be a bad solvent for the alkyl chain used as coating agent. Very surprisingly, in Chapter 5, it is shown that crystalline structure of colloidal crystals keeps the memory, after evaporation, of the solvent used to disperse Ag nanocrystals. Hence, by tuning the solvent used to disperse nanocrystals, coating agents of nanocrystals, and temperature of solvent evaporation, i.e. solvent volatility, various crystalline structures such as fcc, bcc, and hcp can also be tuned. Finally, keeping the colloidal solution under solvent saturation during several days, simultaneous crystal growth process takes place: a 3D superlattice film at the solvent–air interface and shaped 3D superlattices precipitated are produced, which will be discussed in Chapter 4. To produce such segregation, the solvent of the colloidal solution has to have a large surface tension as the air–solvent interface and be a bad solvent for the alkyl chain used as coating agent. Chapter 6 shows that, by slow solvent evaporation of a magnetic colloidal solutions under magnetic field, long superimposed cylinders with a very regular structure are produced when the applied field is parallel to the substrate. Such mesoscopic structures are obtained when the distance between nanocrystals is very short. The control of interparticle distance between nanocrystals through the coating agent permits the control of the mesoscopic structure from superimposed cylinders to undulated films. When the applied field is perpendicular to the substrate, dots and labyrinths are produced. The difference between these two last structures is related to the size distribution of nanocrystals, i.e. with nanocrystal ordering: With highly ordered structures, dots are produced, whereas labyrinths are obtained with building blocks having a large size distribution. This is attributed to the fact that the stiffnesses of 3D superlattices are larger than amorphous mesoscopic structure. This was confirmed by the mechanical properties described in Chapter 12. Chapter 7 shows that by using nanocrystals having two well-defined sizes, their assemblies, governed by hard sphere interactions, favor the formation of a very large variety of 3D superlattice crystalline structures such as NaCl, AlB_2, $NaZn_{13}$, and Laves phase ($MgZn_2$, $MgCu_2$, $MgNi_2$) structures. However, either through ligand exchange of the coating agent used to coat the nanocrystals, by using ferromagnetic nanocrystals, or by adding surfactant molecules, quasi-3D superlattices are produced. The driving force involved in the formation of quasi-3D superlattices remains an open question. Furthermore, the nanocrystallinity of nanocrystals involved in binary systems and the relative amount of small and large nanocrystals also play a role in the final structure. In some experimental conditions, 3D superlattices characterized by very stable vicinal surfaces are produced. Very surprisingly, Chapters 8 and 9 show analogies between atomic crystals and 3D superlattices. The atoms are replaced by (uncompressible) nanocrystals and atomic bonds by coating agents (carbon chains) that act like mechanical springs holding together the nanocrystals. These analogies are observed in the crystal growth processes, magnetic responses as well as vibrational processes. Hence, similar crystalline structures are obtained with 3D superlattices and atomic crystals. Furthermore, nanocrystals breath coherently in a 3D superlattice as atoms in a nanocrystal. Longitudinal propagation waves are observed in thin 3D superlattice film deposition on a substrate as well as thin atomic crystals grown on a wafer. Chapter 10 points out the importance of the coating agents on the nanocrystal ability to self-assemble in 3D superlattices and on their stabilities. Chapters 11 and 12 point out many of intrinsic properties due to the nanocrystal ordering in 3D superlattices. Hence, mild annealing of an assembly of nanocrystals induces epitaxial growth of triangular single atomic crystals. Their sizes are directly correlated to the coherence length of the nanocrystal assemblies. Furthermore, electronic transport properties through very thick 3D superlattices are observed by scanning tunneling microscopy and spectroscopy at 5 K. The mechanism of such unexpected process remains an open question. The magnetic properties governed by dipolar interactions are pointed out. The mechanical properties of 3D superlattices depend on their growth processes, the crystalline structure of nanocrystals used as building blocks, and the material used (Co, Au, Ag nanocrystals). In Chapter 13, we describe cracks of nanocrystal films as observed in nature, paintings, etc. It is shown that, in the range of few nanometers to centimeters, the surface of the cracks linearly depends on the thickness of the film. This is probably similar in nature and paintings but difficult to search. Finally, from Chapters 14 to 16, it is demonstrated that assemblies of hydrophobic nanocrystals dispersed in aqueous solution are characterized by specific properties. The optical properties of Au 3D superlattices show both collective modes related to the 3D superlattice size and the fingerprint of nanocrystals used as building blocks. Such latter properties are also described in Chapter 12 where Coulomb blockage of isolated nanocrystals is observed in a 3D superlattice. These new suprastructures of either Au or ferrite nanocrystals behave as universal nanoheaters. Such behavior opens a new research area. In the presence of cancer cells, these suprastructures are internalized keeping the nanocrystal assemblies. This also opens a new research area. As a matter of fact, we can expect collective properties in cancer cells. The first example is shown by studying the photothermal properties of suprastructures internalized in cancer cells. Such property is not limited to cells. Veal tendon can be burnt by laser irradiation of 3D superlattices deposited on it.

Obviously, these metallic 3D superlattices are a new generation of materials in their infancy and many new intrinsic properties will be discovered in the future. The analogy between 3D superlattices and atomic crystals (Chapters 8 and 9) allows us to claim that other specific properties will be discovered in a large number of research are as involving energy release, catalysis, properties of transport, biomedicine, etc. As already observed in some cases, the scaling laws in 3D superlattices with respect to atomic crystals will have to be revisited. This claim is strongly supported by some results presented here such as (i) The possibility of identifying, as in Nature, a bipyramidal morphology as an intermediate in the growth process of truncated to triangular tetrahedral assemblies (Chapter 8). (ii) The speed of sound through 3D superlattices is very low compared to that measured in atomic crystals (Chapter 9). (iii) The frequency of the respiration process of nanocrystals in 3D superlattices is greater than that of atoms in nanocrystals. (iv) Very thick 3D superlattices (up to 5 μm) can be surprisingly imaged by STM and their electronic properties studied (Chapter 11). (v) Large single crystals of triangular atoms are produced by soft annealing of 3D superlattices obtained under ultra high vacuum (UHV) (Chapter 11), etc.

1

Syntheses of Metal Nanocrystals

Metal nanocrystals have been extensively developed over this last two decades. Here, we will focus on spherical metal nanocrystals. We study the most important factors [1] involved to control size, size distribution, composition, and crystalline structure called nanocrystallinity. However, other parameters could emerge in the future.

1.1 Nanocrystal Growth Processes and Control of Size and Distribution

The nanocrystal growth processes with control of size and size distribution have been largely studied over these past decade [1–18]. LaMer developed a theoretical model of hydrosols nucleation and growth processes [19, 20]. Ostwald observed that the atomic mobility by transfer matter from small to larger particles reduces the free energy associated with the particle/matrix interfacial area. This is called Ostwald ripening. These models have been experimented for nanoparticles/nanocrystals. The nucleation process is where nuclei (seeds) acts as templates for crystal growth. Homogeneous nucleation occurs when nuclei form uniformly throughout the parent phase, whereas heterogeneous nucleation forms at structural inhomogeneities (container surfaces, impurities, grain boundaries, dislocations). In liquid phase, heterogeneities occur much easier since a stable nucleating surface is already present. The growth of nanoparticles depends on the surface reaction and the monomer's diffusion to the surface. In many cases, the same synthesis does not produce the same final size and shape (see below). This phenomenon may be attributed to either the presence of some impurities, additives present in a very low concentration in the reactants, or sequence of reactants inducing heterogeneous nucleation. The latter is due to the changes on the growth mechanism that markedly differs with seed size and/or internal crystallinity. Consequently, various chemical methods have been developed to control the size and shape of these nanocrystals by changing the preparative conditions such as the kind and/or the amount of protecting agent.

Coating agents are needed to keep the integrity of nanocrystals. However, in reality as shown below, ligands have a vital role not only in influencing the size and size distribution of the nanocrystals but also, more importantly, in defining their interaction/interface with the environment. The ligand-mediated nanocrystal synthesis is still an open question. Actually, the role of nanocrystal/ligand differs with the metal ion and the organic ligand used. Hence, the ligand and metal nature are key parameters. At this point, we can mention the strength of ligand–nanocrystal surface interaction and the labile nature of the ligands either in the free or complex forms with the formation of clusters that it is produced with the nanocrystal constituents. The most used coating agents are surfactants. Here, we will concentrate to surfactant molecules. For most of the studies, the head polar groups are either thiol (SH), amine (NH_2), or carboxylic (COOH) groups, whereas the alkyl chains are either alkane or oleic groups. In most of the cases, size selection processes are needed to produce nanocrystals with low size distribution. Carboxyl acid is covalently bound with Co and Fe atoms [21, 22]; thiol derivatives are "quasi" covalently bound to Au atoms, whereas it is lesser than Ag ones [23]. Conversely, amine groups [24] are weakly bound to most of the metallic atoms (Figure 1.1).

To control the size distribution, the digestive ripening permits to convert polydispersed to uniform nanocrystals: the first step is to break the nanocrystal large size distribution into ligand/metal complexes by ligand addition, then isolate the ligand-stabilized nanocrystals, and finally heat, at high boiling point, the solvent of the dispersed nanocrystals in the presence of ligand [25–27]. Then, the control in size and size distribution, based on van der Waals or dipolar forces, can be reached by size selection procedure. By mixing bad and good solvents for the coating agents, as obtained with polymers [28], the bad solvent for the alkyl chains induces precipitate of the largest nanocrystals, leaving the smaller nanocrystals in solution. After centrifugation, the precipitate is dissolved with a good solvent for the alkyl chains,

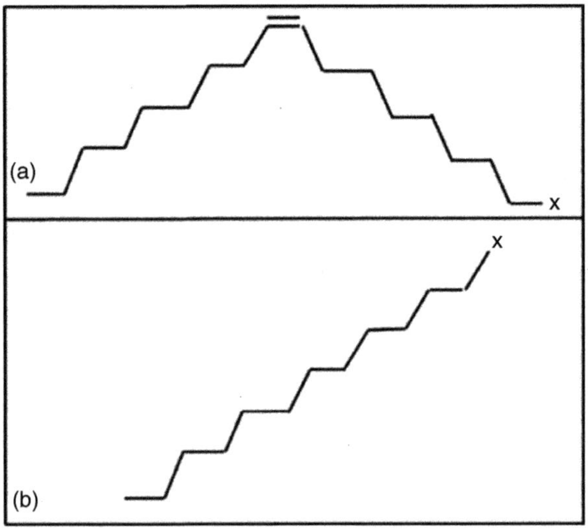

Figure 1.1 Various coating agents used to coat nanocrystals: (a) oley group ($C_{18}H_{36}X$) and (b) alkyl chains ($C_{12}H_{25}X$). X is the head polar group: X = SH thiol derivatives, X = COOH acid derivatives, and X = NH_2 amine derivatives.

leading to a homogeneous clear solution, and again a bad solvent is added inducing a new precipitate. Similarly, the supernatant solvent is evaporated, and the same procedure as the precipitate is proceeded. Such process is repeated several times before reaching a sufficient low size distribution (≤10%) and consequently to be able to self-assemble in 2D or 3D superlattices. A rather large number of mixed solvents is used to size select nanocrystals: the solvent/nonsolvent pairs for the coating agent, e.g. hexane/ethanol, heptane/ethanol, chloroform/ethanol, chloroform/methanol, and pyridine/hexane, are usually used. Ligands, considered as capping/passivating agents, play a major role in the control of nanocrystal size and is far from only providing stability to the nanocrystal against aggregation.

With the most used syntheses to produce well-defined metal nanocrystals with a very low size distribution, several procedures are used: reverse micelles, inorganic chemical reaction, hot injection processes, and thermal decomposition. They are based on chemical reduction of the metal ion by various reducing agents such as hydrazine, sodium borohydride ($NaBH_4$), citrate, and glycol. The control of nanocrystal size depends on (i) the relative concentrations of the reactants, (ii) the coating agent concentration, (iii) the structure of the coating agent, (iv) the relative amount of reactant and that of the coating agent (the higher the stabilizer agent compared with reactants, the smaller nuclei formed and consequently smaller nanocrystal size), and (v) the mixture of two coating agents (one binds tightly to the nanocrystal surface hindering growth, and the other less tightly bound favors the rapid growth).

1.2 Crystalline Structure of Metal Nanocrystals

Generally, the nanocrystal size is determined by the number of atoms produced and consequently by the number of the nanoparticles formed depending on the kinetics of nucleation and growth processes. The nanocrystal crystalline structure called nanocrystallinity is determined by the competition between the kinetics of structural transformation of the nanoparticles and that of delivering separation between nucleation/growth processes (e.g. a burst of nucleation) and diffusion-controlled growth [29–34].

1.2.1 Co Nanocrystals

Cobalt has three nearly isoenergetic crystal structures (Figure 1.2): hexagonal close-packed (hcp), face-centered cubic (fcc), and epsilon (ε). Bulk cobalt displays two stable phases of hcp and fcc at ambient pressure. At the nanoscale, it can possess the epsilon phase with a complex cubic primitive structure, which is similar to that of β-manganese, with a unit cell parameter of 6.097 Å [35]. This metastable crystal structure (ε phase) can be transformed to hcp or fcc structure by annealing the sample at a suitable temperature [36]. Due to its relatively smaller magnetocrystalline anisotropy constant ($K_c \approx 1.5 \times 10^5\ \mathrm{J\,m^{-3}}$), ε-structure is widely known as a softer magnetic material, compared with the hcp and fcc phases.

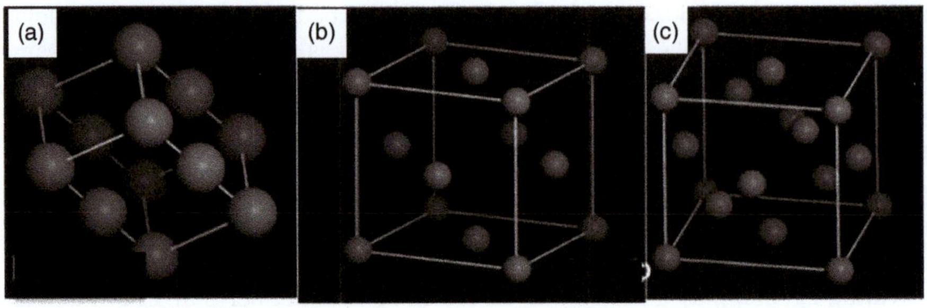

Figure 1.2 Crystalline structures of Co nanocrystals (a) hcp, (b) fcc, and (c) ε-phase.

1.2.2 Au, Ag, and Cu Nanocrystals

At the atomic level, the densest possible packing of four spheres is a tetrahedron [29, 37–39]. At the earliest stages of growth of a solid, the atoms reorganized into a completely new structure each time an atom is added. For fcc materials, the nanocrystals considered as compact packing of spheres are cuboctahedra, decahedra, and icosahedra (Figure 1.3). The surface of a cuboctahedral nanocrystals consisted of eight (111) planes and six (100) planes with closed atomic shells and represents the most stable form. The defined number of atoms resulted exclusively from the geometric arrangements and cannot be explained by the electronic shell model. The decahedra and icosahedra are structures with fivefold symmetry and are called multiple-twinned particles (MTPs). In fact, decahedral nanocrystals consisted of five deformed tetrahedral subunits twinned by their {111} planes, whereas icosahedral clusters originated from the twinning of 20, (111) faces, deformed tetrahedral subunits.

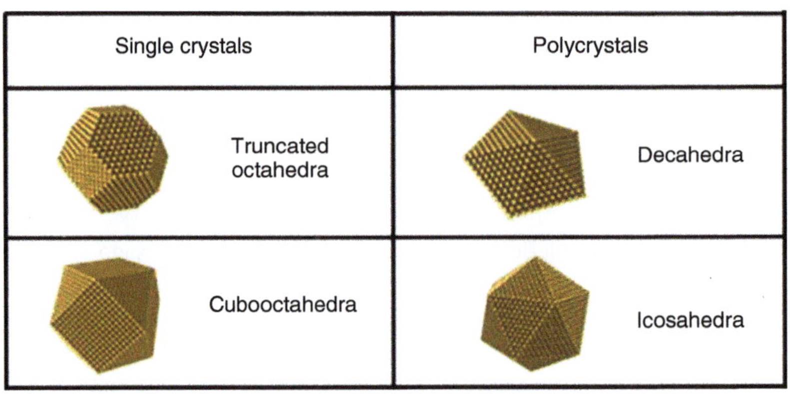

Figure 1.3 Various shapes of fcc nanocrystal arrangements.

1.3 Various Techniques Used to Produce Metal Nanocrystals and Control Their Sizes and Distribution

A very large number of techniques have been used to produce nanocrystals, with very low size distribution. Significant advantages of methodologies using high-boiling-point solvents and organometallic precursors in the presence or absence of surfactants favor the control of nanoparticles size and improve the crystalline structure of the nanocrystals [11, 33, 40].

1.3.1 Reverse Micelles

Reverse micelles [2, 17, 41–48] are well known water-in-oil droplets stabilized by surfactants [49]. The best surfactant providing uniform droplets is sodium bis(2-ethylhexyl)sulfosuccinate, Na(AOT), called Aerosol OT. The amount of water controls the droplet size (Figure 1.4a). The average droplet diameter increases linearly with the water content, w, defined as $w = [H_2O]/[AOT]$. By collisions these droplets exchange their water contents and form again two independent droplets (Figure 1.4b). This process is used to produce nanosized material by either chemical reduction of metal ions or coprecipitation reactions. At low water content, the number of water molecules per surfactant is too small to hydrate the counterions and the head polar groups. This induces strong interactions between water molecules and the head polar groups. The water molecules are then considered as "bound water."

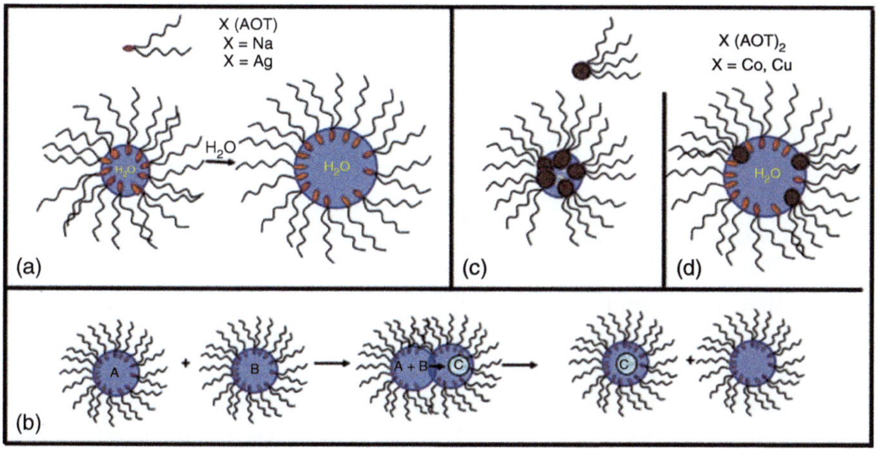

Figure 1.4 (a) Control the size of the droplet with the amount of water with Na bis(2-ethylhexyl)sulfosuccinate, Na(AOT). Similar behavior is observed with Ag(AOT). (b) Exchange process by collision between two droplets to favor chemical reaction. (c) Reverse micelles from functionalized surfactant Co and Cu (tetra (2-ethylhexyl) sulfosuccinate), $X(AOT)_2$. (d) Mixed micelles.

An increase of the water content induces a progressive appearance of "free" water molecules in the center of the droplets, whereas some of water molecules remain bound to the interface. With Na(AOT), the bulk water phase is reached around $w = 10$. To produce metal nanocrystals without any oxide formation, the metal ions have to be in environment of bound water molecules and not into the bulk phase, i.e. they have to be located at the oil–water interface of the droplets. This explains why we synthesized functionalized surfactants $X_n(AOT)_n$ with $n = 1$ or 2, i.e. the sodium ions of AOT are replaced by metallic ones (Figure 1.4c). With $X_n(AOT)_n$, reverse micelles are formed to produce reverse micelles, but the amount of water inside reverse micelles is limited ($w \neq 5$) [42]. To provide a control of the droplet size by the water content, we used mixed micelles, Na(AOT), $X_n(AOT)_n$. In such condition, the size of the droplet remains controlled by the amount of water molecules involved in the formation of reverse micelles that can then play the role of template (Figure 1.4d).

To synthesize amorphous Co nanoparticles, mixed micelles [$Co(AOT)_2$/Na(AOT)] are used. Sodium borohydride ($NaBH_4$) added to the micellar solution reduces the cobalt ions. Immediately after $NaBH_4$ addition, the color of the micellar solution changes from pink to black, indicating the formation of colloidal Co nanoparticles. Coating agents with carboxylic as head group dodecanoic acid molecules added to the colloidal solution induce a chemical bond between the oxygen of carboxyl group and the Co atoms located at the interface. The coated Co nanoparticles are then washed and centrifuged several times with ethanol to remove all the AOT surfactant. Moreover, the black powder obtained is dispersed in hexane, and the colloidal solution is isotropic. Size selection process as described above takes place to produce Co nanoparticles with a ~10% size distribution. The entire synthesis is carried out in an N_2 glove box using de-oxygenated solvents to prevent nanoparticle oxidation. The ratios of [$NaBH_4$]/[$Co(AOT)_2$] and [Na(AOT)]/[$Co(AOT)_2$] control the nanoparticle size. In contrary to others, the nanoparticles are amorphous. The average diameter and size distribution evolve from 4 to 9 nm and 9% to 13%, respectively (Figure 1.5a). The relative ratio between Na(AOT) and $Co(AOT)_2$ controls the nanoparticle size. By annealing process, under an inert atmosphere, the crystalline structure of Co nanoparticles is improved, and a transition from amorphous to single-domain hcp structure takes place [47, 48]. The annealing process does not significantly affect the nanocrystal diameter and its size distribution.

Amorphous Co nanoparticles deposited on the TEM grid are annealed under inert gas, 60 minutes at 250 °C. During such process, as shown in Figure 1.6, a transition of the crystalline Co nanoparticles from amorphous to single-domain hcp phase takes place. Note the hcp nanocrystal size increases compared with the corresponding amorphous counterparts. This is due to hcp lattice fringes. Hence, the crystallinity of Co nanoparticles is controlled by keeping the other parameters constant as size and surface coating. The control of amorphous Co nanoparticle size is obtained by mixing Na(AOT) with $Co(AOT)_2$ [50].

To produce Ag nanocrystals by using reverse micelles, a colloidal solution of Na(AOT) and Ag(AOT) is mixed with Na(AOT) reverse micelles containing hydrazine [51]. The nanocrystals are extracted from the micellar solution and then are

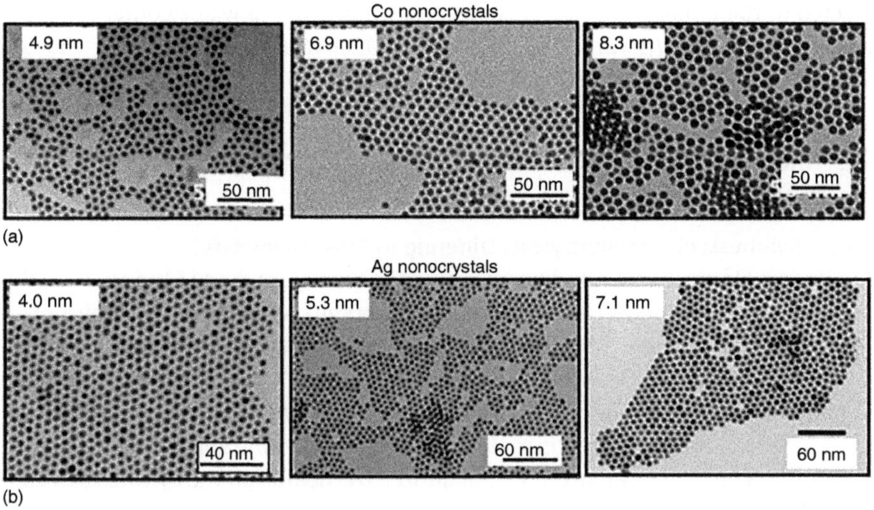

Figure 1.5 (a) Increase in the size of Co nanoparticles with controlling *R*, *w*, and the relative amount of $Co(AOT)_2$. (b) Reduction of mixed reverse micelles of Ag(AOT)/Na(AOT) (6/4) by hydrazine with $w = 2$. By controlling the ratio R = [Ag(AOT)]/[hydrazine] from 4 to 0.7, the Ag nanocrystal sizes are (a) 4, (b) 5, (c) 6, and (d) 7 nm, respectively.

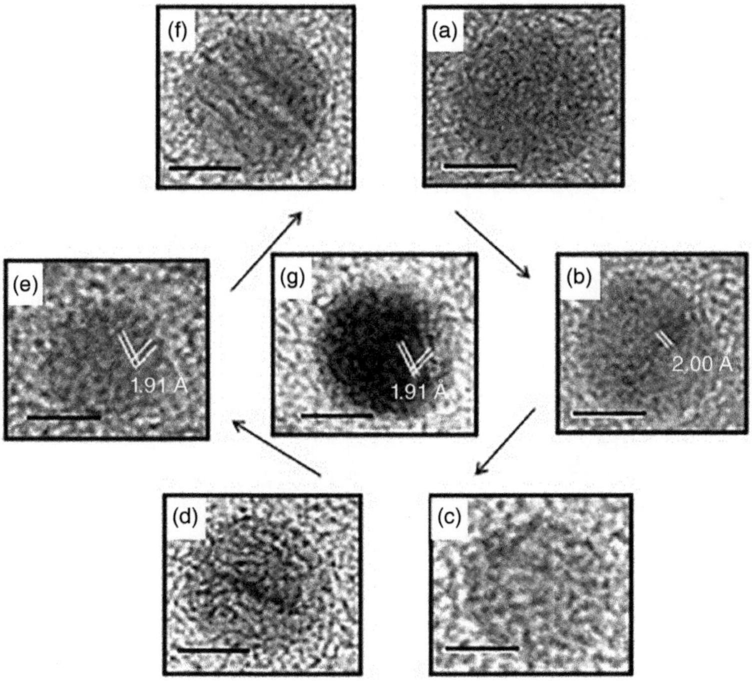

Figure 1.6 Schematic illustration with high-resolution TEM (HRTEM) images of the evolution process of 7.2-nm Co nanocrystals: (a) amorphous Co nanoparticles, (b) hcp Co with large domains, (c) and (d) very large domains, (e) small hcp single crystals, (f) hcp single crystals with stacking faults, and (g) hcp single crystals. The scale bar is 3 nm.Source: Yang et al. [48]/American Chemical Society.

subjected to the ethanol/hexane wash cycles. Changing the water content and the ratios of [Ag(AOT)]/[hydrazine] and [Ag(AOT)]/[Na(AOT)] controls the Ag nanocrystal from 4 to 7.3 nm (Figure 1.5b).

1.3.2 Inorganic Chemical Reaction to Produce Au and Ag Nanocrystals

1.3.2.1 Synthesis of Au Nanocrystals Differing by Their Diameters

Au nanocrystals are synthesized by revisiting the Stucky method [51–54]: two solutions (A) and (B) are mixed under nitrogen protection. (A) is a preheated solution of chlorotriphenylphosphine Au(I) dissolved in toluene containing dodecanethiol, and (B) is amine–borane complex dissolved in toluene. The ratio of the volume of A/B controls the size of the nanocrystals from 7.8 to 5.8 nm. To decrease the size from 5.1 to 4 nm, the solution B is replaced by *tert*-butylamine and ammonia borane complexes, respectively (Figure 1.7). For simplicity, details on the synthesis to control the nanocrystal size are given in Table 1.1.

Let us consider 5-nm Au nanocrystals: at the end of an organometallic synthesis, a drop of Au nanocrystals dispersed in toluene is deposited on TEM grid. The dark-field TEM image (Figure 1.8) shows homogeneous or inhomogeneous contrasts corresponding to single and polycrystalline nanocrystals, respectively. After analyzing

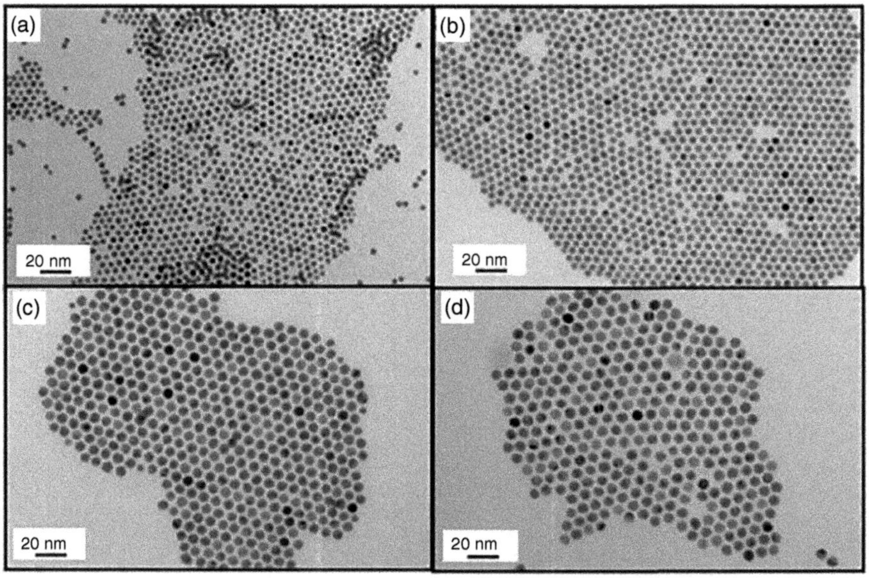

Figure 1.7 TEM image and corresponding histograms of 4 (a), 5 (b), 7 (c), and 8 nm (d) obtained by deposition of a drop of colloidal solution of Au nanocrystals coated with dodecanethiol and dispersed in toluene .Source: Goubet et al. [55]/JOHN WILEY & SONS, INC.

Table 1.1 Experimental conditions to produce Au nanocrystals differing by their average diameters.

Diameters (nm)	**4.3**	**5.1**	**5.8**	**7.2**	**7.8**
σ (%)	9	7	8	6	6
V_{doc} (μl)	500	500	500	125	125
N_{bo}	2.5	5	5	5	10
S	Et/TO	TO	TO	TO	TO

Solutions A and B are 0.25 mmol of chlorotriphenylphosphine Au(I) and *tert*-butylamine borane complex dissolved in 25 ml toluene, respectively. D, s, V_{doc}, N_{bo}, S, Eth, and To are the Au nanocrystal diameter, size distribution, the volume of dodecanethiol in 25 ml solution, the amount in mmol in 2 ml of solution, solvent, ethanol, and toluene, respectively.

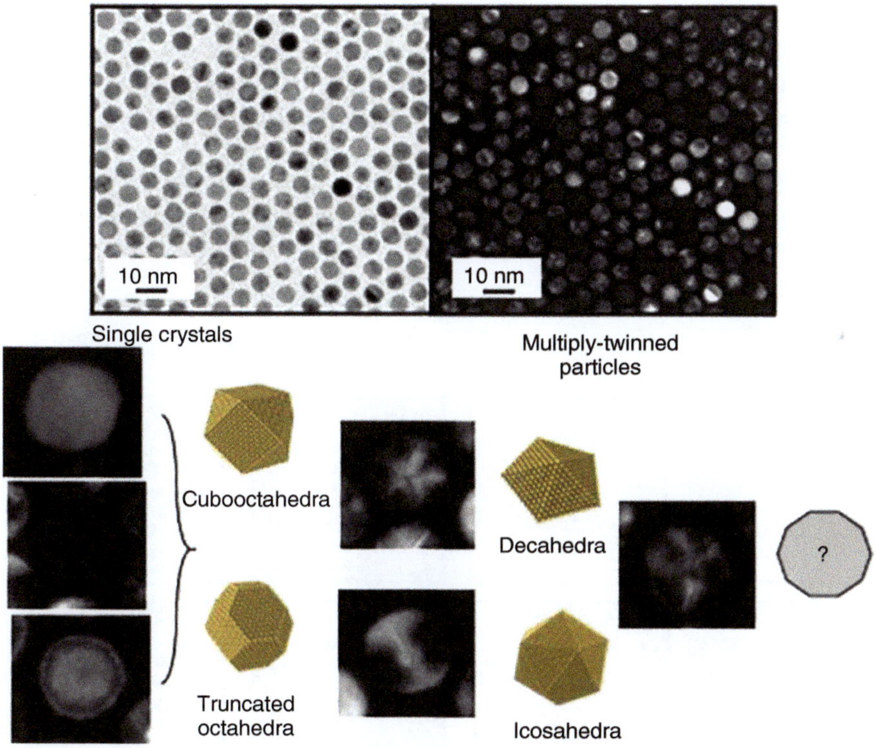

Figure 1.8 Bright- and dark-field TEM image of 5.1-nm Au nanocrystals with homogeneous (single-domain) and heterogeneous (polycrystal) contrasts. Source: Portalès et al. [56]/American Chemical Society.

the conical dark-field TEM images over a total population of 2000 nanocrystals, the amount of single-crystalline nanocrystals is found to be approximately 30% of the full population. The major portion (65%) is composed of MTPs and/or polycrystals, while 5% is poorly twinned nanocrystals. On increasing the Au nanocrystals size, the amount of single domain markedly decreases.

In Chapter 4, we describe nanocrystal segregation processes with the formation of interfacial and precipitate 3D superlattices by subjecting colloidal solution to toluene-saturated atmosphere after keeping during seven days. Both interfacial and precipitate 3D superlattices are characterized by single-domain 5-nm Au nanocrystals, whereas the corresponding polycrystalline phase remains in solution [54]. This crystalline selection process makes possible to select 5-nm single-domain and polycrystalline nanocrystals. The growth of the seeds is induced in a solution of Au–oleylamine complex, produced by mixing $HAuCl_4$ in pure oleylamine [57]. After dissolution, the solution is yellow colored and freshly used. Then, colloidal solutions of either single-domain (Figure 1.9a) or polycrystalline (Figure 1.9f) 5-nm nanocrystals are mixed with the Au–oleylamine complex solution, and the volume is completed with toluene. The seeded growth takes place at fixed temperature. During a few hours, Au–oleylamine complex is slowly reduced on the seeds. Adjusting the gold complex concentration can control the resulting nanocrystal size. After reaction, the nanocrystals are washed with ethanol and redispersed in chloroform.

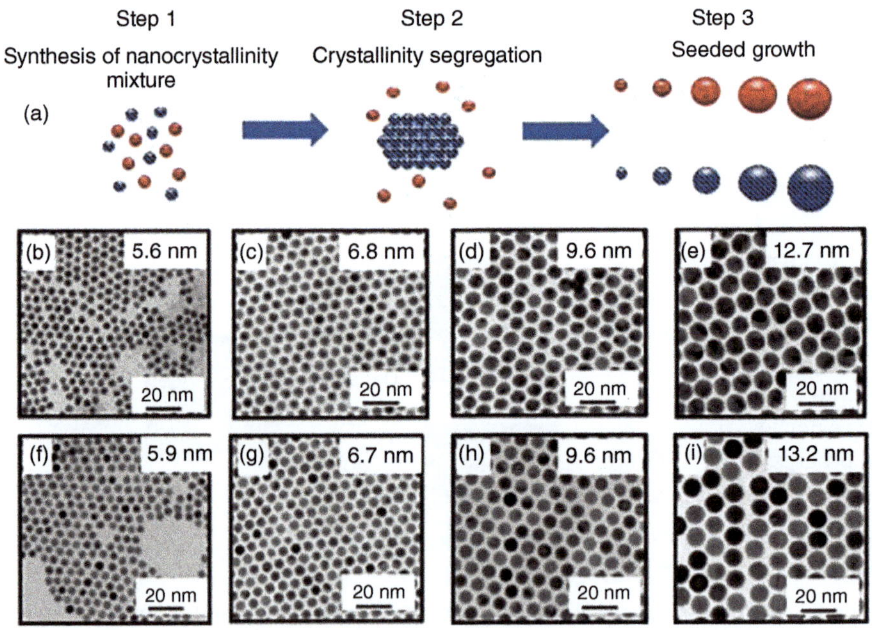

Figure 1.9 (a) A schematic image showing the formation of Au nanocrystals with tunable size and nanocrystallinity. The polycrystalline nanocrystals are represented in red, and single-domain nanocrystals in blue with straight lines. (b–e) TEM images of NPs synthesized using polycrystalline seeds: (b) 5.6, (c) 6.8, (d) 9.6, and (e) 12.7 nm. (f–i) TEM images of NPs synthesized using single-domain seeds: (f) 5.9, (g) 6.7, (h) 9.6, and (i) 13.2 nm. The scale bars are 20 nm. Source: Goubet et al. [57]/Royal Society of Chemistry.

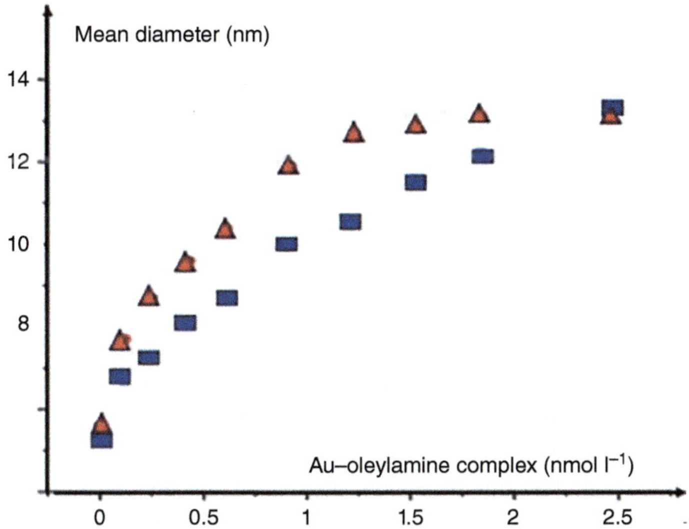

Figure 1.10 Typical mean diameter of single-domain (blue squares) and polycrystalline (red circles) nanocrystals as a function of gold complex concentration during the seeded growth. Source: Goubet et al. [57]/Royal Society of Chemistry.

Hence, the control of nanocrystal size from 5 to 13 nm is obtained with either single-domain (Figure 1.9b–e) and polycrystalline (Figure 1.9f–i) nanocrystals.

Figure 1.10 shows the increase in the seed growth that is rather fast for polycrystalline nanocrystals compared with the corresponding single nanocrystals. By tuning the Au–oleylamine concentration, the control of mean diameter can be reduced to 0.5 nm.

1.3.2.2 Synthesis of 5-nm Polycrystalline Silver Nanocrystals

Same procedure as described above is developed [51]. The precursor chlorotriphenylphosphine gold(I) is replaced by its Ag counterparts [chlorotris(triphenylphosphine) silver]. A solution of this precursor is mixed with dodecanethiol. A solution of *tert*-butylamine borane complex at 0.17 M is then added for the reduction. After few hours, some small particles (<2 nm) start to appear in the solution and progressively. After leaving the solution to stir overnight, nanocrystals with the diameter of 5.1 nm having very narrow size distribution are obtained (Figure 1.11).

1.3.3 Thermal Decomposition

The thermal decomposition process has been well developed to produce a rather large number of metal and metal oxide nanoparticles [18, 36, 59]. Hence, cobalt carbonyl ($Co_2(CO)_8$) is dissolved in dioctylamine in the presence of oleic acid to produce fcc Co nanocrystals (Figure 1.12). By adjusting the ligand concentration and decomposition temperature, the multidomain fcc Co nanocrystals having a diameter of 5–10 nm with a rather low size distribution (from 7% to 10%) are produced.

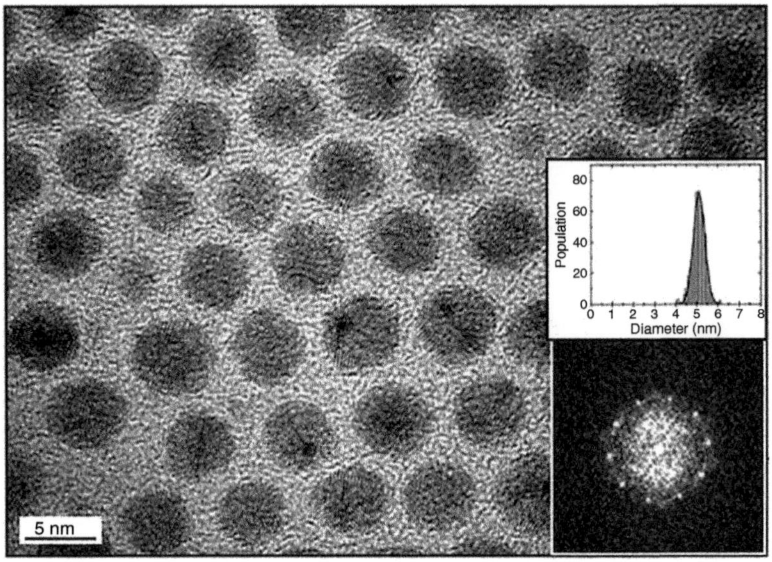

Figure 1.11 HRTEM characterization of silver MTPs. HRTEM image and size distribution histogram of multiply twinned 5.1 nm Ag nanocrystals. Inserts: The size distribution histogram is determined from over 500 silver nanocrystals from the colloidal solution used and power spectrum as typically obtained from a silver icosahedron oriented along the fivefold axis. Source: Portalès et al. [58].

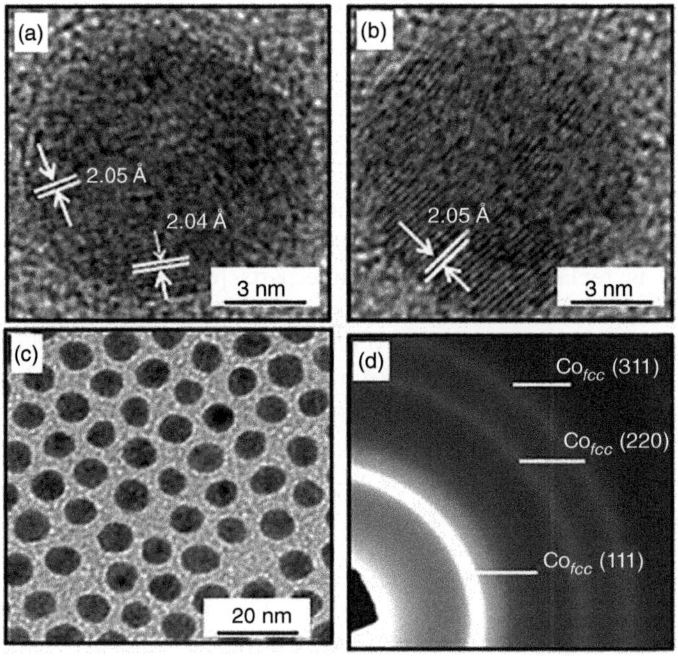

Figure 1.12 HRTEM (a, b) and TEM (c) images and electron diffraction pattern (d) of fcc Co nanocrystals produced as described above. Engineering the magnetic dipolar interactions in three-dimensional binary supracrystals via mesoscale alloying. Source: Yang et al. [60]/John Wiley & Sons.

1.3.4 Hot Injection

This method is based in injection of a cool solution of precursor molecules into hot liquid [4, 35, 61–63]. The injection leads to the instantaneous formation of seeds. The decrease of solvent temperature prevents the formation of new nuclei. The reactant container is a suspension of reasonably rather homogeneous nuclei together with considerable amounts of precursors. To slow the growth of the existing nuclei without any new nucleation, the solution is heated below the corresponding injection temperature. The omnipresent ligand molecules slow down the growth considerably by coordinating the surface atoms, thus forming a steric barrier for reactants. The slow growth at relatively high temperatures allows the nanocrystals to anneal and to form nearly single-domain lattices at the same distance as those of the bulk phase. At the end of the syntheses, size selection takes place by the addition of a non-solvent to reach low size distribution. The ligands remain attached to the surface atoms, and the suspensions are sterically stabilized. The nanocrystal size increases with temperature to which the nuclei grow. Note that another approach is a rapid injection in hot water of the reactants [64].

Single-domain ε-phase Co nanocrystals are produced by hot injection: a mixture of trioctylphosphine oxide (TOPO), oleic acid, and (1,2-dichlorobenzene) DCB is heated to under flowing nitrogen. $Co_2(CO)_8$ dissolved in DCB is quickly injected in the solution under vigorously stirring. The rapid injection of the cobalt complex into hot organic solution can minimize the nucleation stage, inducing burst nucleation of Co monomers. The solution is held at high temperature and then cooled down to room temperature. The as-prepared nanocrystals are isolated by adding anhydrous ethanol. Comparison of bright- and dark-field images of homogeneous dark or bright nanocrystals (Figure 1.13a,b) indicates the formation of single-domain nanocrystals (Figure 1.13). The high-resolution TEM image and the diffraction pattern (Figure 1.13c,d) confirms the formation of single domain of ε-phase Co nanocrystals. The number of nuclei formed in the hot injection synthesis of ε-Co nanocrystals depends more on temperature kinetics after the injection than on the injection itself [65]. This indicates a faster recovery of temperature after injection leading to smaller-sized nanocrystals. The size of Co nanocrystals is tuned, from 5 to 10 nm, by adjusting both concentration of oleic acid and temperature.

This hot injection method is used to produce Ag nanocrystals coated with oleylamine (C_{18}–NH_2) and dodecylamine (C_{12}–NH_2). Briefly, $AgNO_3$ and dodecane diol are separately dissolved in *o*-dichlorobenzene (DCB). These solutions are called A and B, respectively, for simplicity. A is rapidly injected into a hot B solution. The mixture reacts in few minutes. The solution is then cooled down to room temperature. With oleylamine (C_{18}–NH_2), this procedure produces the Ag nanocrystal size from 2.2 to 7 nm, (Figure 1.14). The change of the nanocrystals is obtained by changing the relative ratio of $AgNO_3$ and oleylamine. For larger nanocrystal size (from 8 to 13 nm), oleylamine-coated nanocrystals are prepared following a modified one-pot method as described elsewhere [8].

Hence, this procedure permits to control the nanocrystal size from 2 to 13 nm. The TEM images in Figure 1.14 clearly show that the nanocrystals, regardless of their size, are well arranged in a two-dimensional hexagonal close-packed array. This is typical of

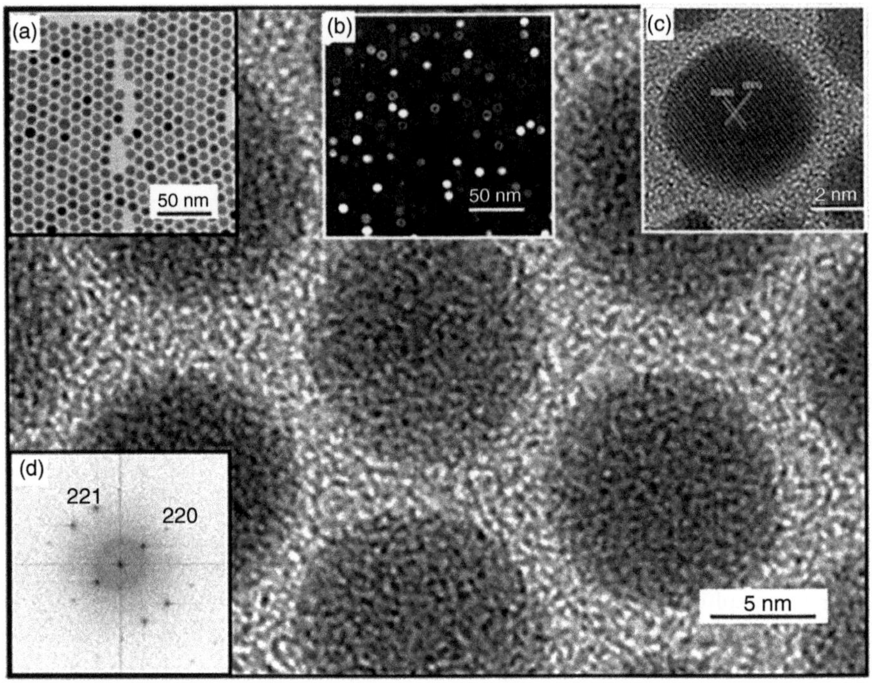

Figure 1.13 High-resolution of TEM images of 7.6-nm ε-phase Co nanocrystals, long range TEM images bright (b) and conical dark field (c), high-resolution TEM image of isolatednanocrystal (d), and corresponding electron diffraction pattern (e). The ε-phase Co nanocrystals are produced with OA/TOPO= 2.4.

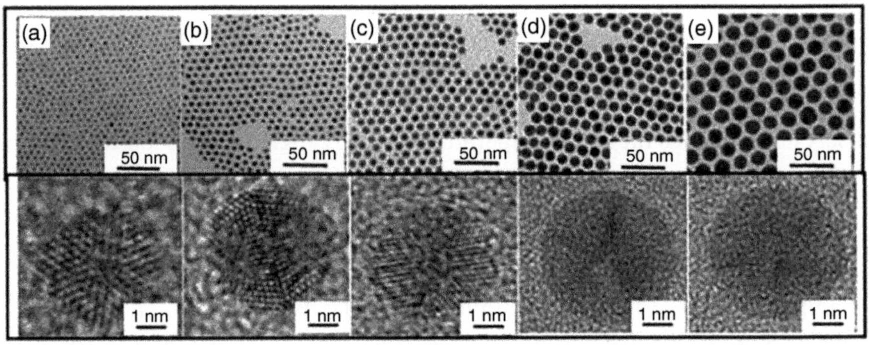

Figure 1.14 TEM images and high-resolution TEM of oleylamine-coated silver NCs produced by either hot injection (a–c) or one-pot method (d, e): (a) 2.2, (b) 4.1, (c) 5.9, (d) 8.7, and (e) 12.9 nm.

nanocrystals with low size distribution (typically below 10%). Note that these nanocrystals are obtained without any post-synthesis size selection process or size-focusing ripening process. This is, from the best of our knowledge, the first report of Ag nanocrystal preparation with such a low size distribution. High-resolution TEM images of the nanocrystals reveal that all nanocrystals are MTPs with icosahedral morphologies. Replacing oleylamine (C_{18}–NH_2) by dodecylamine (C_{12}–NH_2), which is a similar

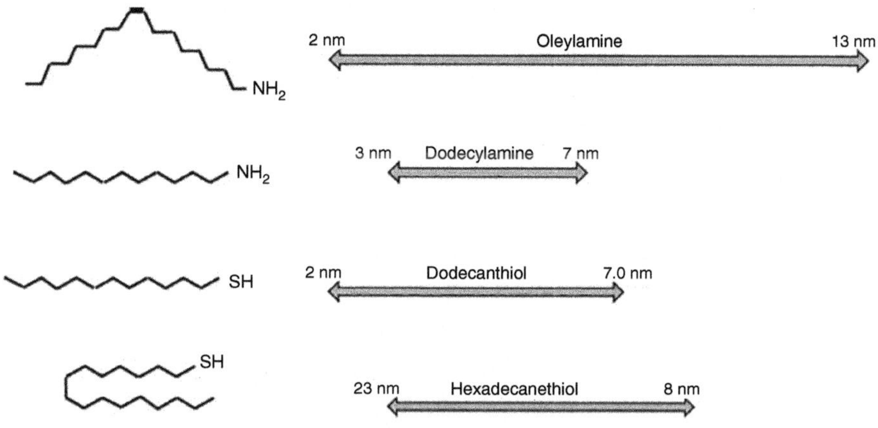

Figure 1.15 Average size of Ag nanocrystals produced by the same procedure and coated with various coating agents.

$-NH_2$-terminated coating agent with a shorter alkyl chain, stable Ag nanocrystals are produced. In this case, the size of the dodecanethiol-coated Ag nanocrystals is controlled from 3.0 to 6.3 nm. Again, the size distribution of these nanocrystals is kept below 10% without the need of size segregation procedures. Such hot injection process cannot be used for the preparation of alkanethiol-coated Ag nanocrystals since $AgNO_3$ cannot be dissolved in DCB in the presence of –SH-terminated coating agent. When a mixture of $AgNO_3$ and dodecanethiol (C_{12}–SH) is injected into hot solutions containing polyols, small nanocrystals with large size distribution are produced. Hence, to obtain Ag nanocrystals having similar sizes with a new coating agent such as alkanethiol, the ligand exchange process from nanocrystals coated with oleylamine is used. The colloidal solution of Ag nanocrystals coated with oleylamine is mixed with a small amount of C_{12}–SH; the mixture is kept under vigorous stirring. Hence, this method is highly powerful to control the size and size distribution (<10%) without any other complementary processes to reduce the size distribution. However, the size range of nanocrystals depends on the coating agent (Figure 1.15).

1.4 An Example to Show the Importance of the Reactant Sequence to Produce Nanocrystals

Here, we will show how the sequence of reactants plays a key factor in the final size of Au nanocrystals [13]. By revisited Brust method [66, 67] to produce Au nanocrystals coated with dodecanethiol, the coating agent is added to the reactant ($HAuCl_4 \cdot 3H_2O$) either before or after the chemical reduction. The addition of C_{12}–SH before the chemical reduction induces the formation of polymeric Au(I) thiolate species. At the end of the synthesis, rather small single-domain nanocrystals with an average diameter of 1.5 nm and a size distribution of 25% are produced (Figure 1.16a). When C_{12}–SH is added at the end of the synthesis, 3.5-nm MTP nanocrystals with 20% size distribution are obtained (Figure 1.16b). Hence, the size of the nanocrystals

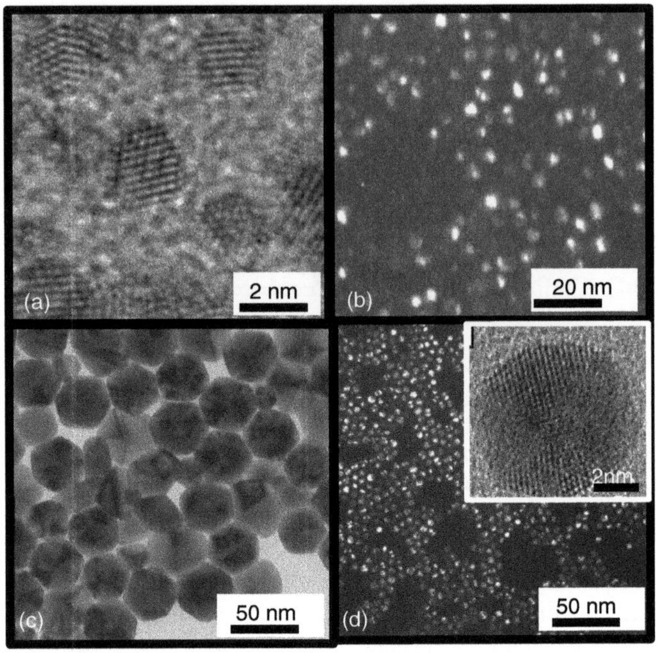

Figure 1.16 TEM image of nanocrystals produce by dodecanethiol addition. Before (a) and after (b) addition of the reducing agent. By thermal treatments of (a) and (b) dispersed in toluene, the final products are (c) and (d), respectively. Source: Goubet et al. [13]/ American Chemical Society.

is controlled by the sequence of reaction additions. These two nanocrystals (1.5 and 3.5 nm) considered as seeds (I and II, respectively) are dispersed in toluene and subjected to the same thermal treatment at 463 K during the same period of time [68]. At the end of the process, seed I produced very well-faceted hexagonal particles (85%) and triangular (9%) and pentagonal (6%) Au nanocrystals (Figure 1.16c). Thus, thermal treatment of 1.5-nm cuboctahedral Au nanocrystals induces the formation of large icosahedral particles. With 3.5 nm as seed (II), 5.8-nm Au nanocrystals with 9% size distribution are produced (Figure 1.16d). This is explained as follows: small cuboctahedra (1.5 nm) subjected to thermal treatment melt and reach a liquidlike phase. This leads by slow cooling rate to the formation of mainly large icosahedral particles that are the most stable Au fcc structures and the presence of small amounts of other structures. By simulation [69, 70], it is found that crystallization of the interior and external atoms leads to an icosahedral structure. The discovery of, at equilibrium, the Au nanoparticles consisting of 85% icosahedral particles is explained by the low surface energy of Au(111) surface [69]. This confirms that the kinetics of nucleation and growth in clusters play a key role in determining the structure of the nanocrystals. The stochastic nature of nucleation explains why other shapes are produced as well. Conversely, when the seed size is larger, they no longer melt, and migration of atoms/adatoms between nanocrystals becomes the predominant growth and shaping mechanism [69, 71].

1.5 N-Heterocyclic Carbene Ligands for Au Nanocrystals Stabilization

N-Heterocyclic carbenes (NHCs) [72–80] have attracted increasing attention as neutral C-ligands for nanomaterial stabilization and functionalization [81–103]. The NHC high flexibility allows fine-tuning of steric and electronic properties and facilitates the introduction of functional groups at different positions [104]. The N-substituents of the NHCs point out from both sides and are oriented toward the metallic surface, inducing significant interactions with the metallic surface and the neighboring ligands. Therefore, NHC structure modulations can dramatically affect the NHC bonding mode, the surface covering and ligand packing density, and the number of free metallic sites available for catalysis. Furthermore, the strong σ-donating properties of NHCs are expected [105], as in metal complexes [106], to provide strong NHC–metal bonds at the surfaces. The most commonly used method for obtaining NHCs involves deprotonation of the precursor azolium salts [105]. In contrast to many metal–NHC complexes, free NHCs are highly sensitive to oxygen, water, and carbon dioxide and should be handled with exclusion of air and moisture. In many experimental procedures, NHCs and metal-ligated NHCs are generated *in situ* under inert atmosphere [72, 73]. Various synthetic methods have been tested for the formation of metal (M) NCs stabilized by NHC ligands (M = Ir, Ag, Au, Ru, Pd, Pt) [81–99]. Some NHC-based metal NCs have been applied in metal-catalyzed hydrogenation reactions (M = Ir, Ru, Pd, Pt) [82, 88–90, 92, 99]. The enhanced stability of NHC-based SAMs and NCs to oxidative treatments was reported [100, 101].

Two new methods are proposed to produce Au nanocrystals coated with various NHC ligands as described in Figure 1.17 [76–80]:

(i) *One-phase reduction method*: the control of nanocrystal size and stability is the first major challenge when new ligands are introduced. Previous study shows one-phase procedure to produce Au nanocrystals with triphenylphosphine gold chloride in the presence of an excess of dodecanethiol as described in Section 1.3.2.1 [54, 107]. Let us consider similar approach by using [(NHC)AuCl] complexes. Ammonia–borane complex (NH_3BH_3) in toluene at 100 °C is used as reducing agent (Figure 1.17). The best data are obtained with **5** ligand as precursor. Au nanocrystals with an average size of 9.8 nm and a narrow size distribution of ±6.9% are produced. However, a limited long-term stability was observed for **L5@Au** nanocrystal that precipitates from the toluene solution after 12 hours. Hence, NHC can play a dual role both as a ligand for the initial Au^{I} species and as a stabilizing agent for the metallic NCs formed (Figure 1.18).

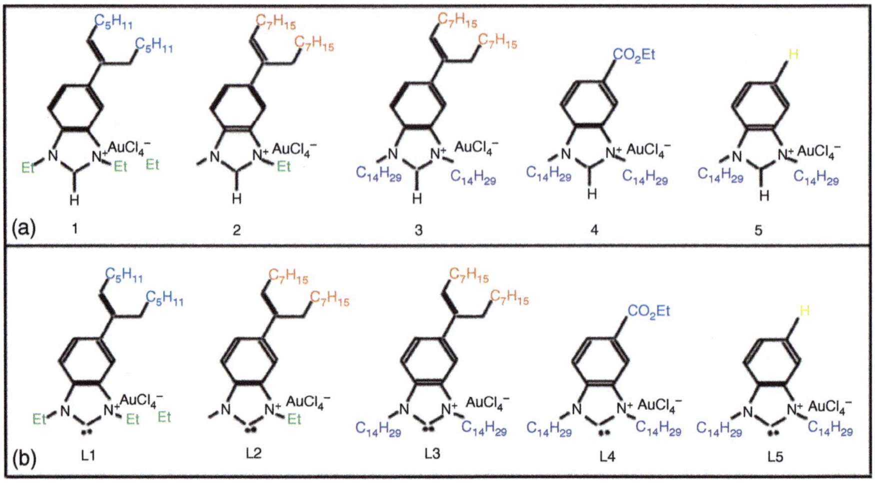

Figure 1.17 Structures of NHC ligands **1–5** and deprotonated ligands (**L1–L5**) for Au nanocrystals. Source: Based on Refs. [76–80].

Figure 1.18 Strategy for synthesizing NHC-coated Au nanocrystals by one-phase reduction of well-defined [(NHC)AuCl] complexes.

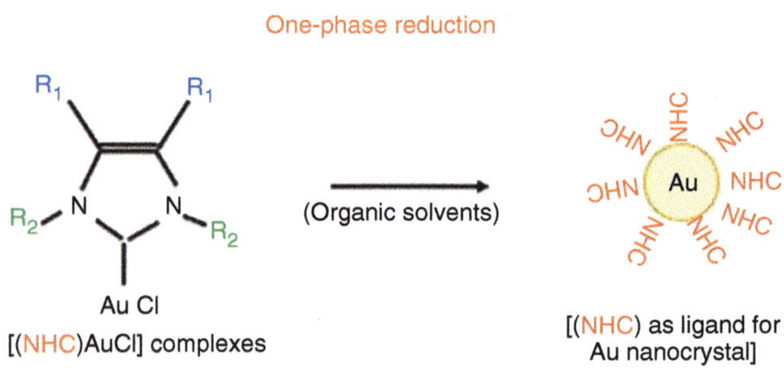

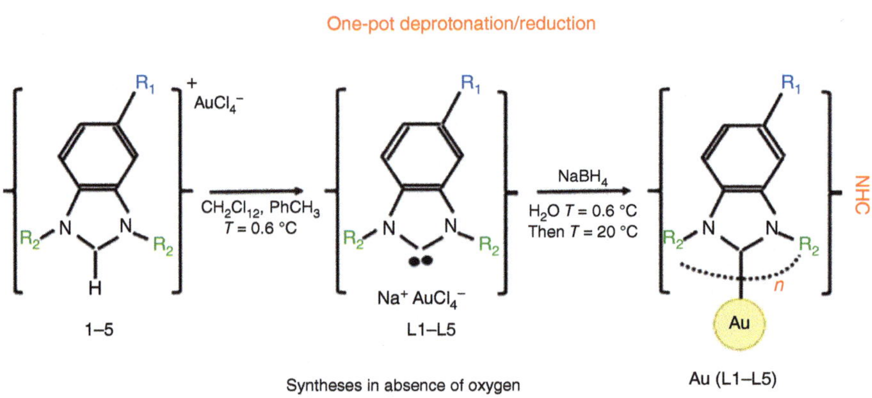

Figure 1.19 Typical procedure for the preparation of NHC-coated Au nanocrystals (**L1–L5@AuNC**) from benzimidazolium chloroaurate salts (in a glove box).

(ii) *One-pot deprotonation/reduction method*: to produce Au nanocrystals coated with various NHC ligands, a method based on deprotonation/reduction reaction of chloroaurate(III) imidazolium salts as precursors is used. In a previous study [91], well-defined metal–NHC complexes and air-sensitive NHC ligands are generated *in situ* and not isolated. This sequence gives large Au nanocrystals (16.6 nm) with a wide size distribution of ±49% [91]. Here, benzimidazolium chloroaurate salts in which the heterocycle is functionalized by C_7–C_{14} long alkyl chains are used (Figure 1.17).

All the five salts, although differing in the length and position of the alkyl chains, are found to generate stabilized Au nanocrystals (**L1–L5@Au** nanocrystals) after treatment by NaH as the base and reduction by $NaBH_4$ at 0–1 °C. In brief, the synthetic route is shown in Figure 1.19: the chemical reaction takes place in a glove box. A suspension of sodium hydride in a 1 : 1 mixture of CH_2Cl_2/toluene is added to a solution of chloroaurate benzimidazolium salt in CH_2Cl_2. The mixture is stirred. A solution of $NaBH_4$ in water is added dropwise while maintaining the ice bath temperature. After addition, the mixture is stirred at room temperature. The solvents are removed under a nitrogen flow to give a dark powder. The powder is washed with ethanol and the suspension is centrifuged. The supernatant (EtOH) is removed. The residue is dried under a nitrogen stream to give Au nanocrystals as a dark powder. Table 1.2 shows that the size distribution, before size selection, is very wide. To reduce such distribution, size selection is based on a mixture of good and bad solvents. Table 1.2 shows that nanocrystals with narrower size distributions of 9–12% could successfully. The final **L1–L5@Au** nanocrystals are stable for weeks as clear purple-red colloidal solutions in toluene. By deposition of the colloidal solution, nanocrystals self-assemble in compact hexagonal network (Figure 1.20). The nanocrystal size changes with the ligand used and evolves from 3.5 to 6 nm (Table 1.2).

Table 1.2 Summary of NHC-coated Au NC **L1–L5@Au** nanocrystal properties: average diameter (*d*) and size distribution (*s* %).

Au nanocrystal	Before size selection		After size selection	
	d (nm)	*σ* (%)	*d* (nm)	*σ* (%)
L1 Au nanocrystals	5.9	13.3	6.0	9.2
L2 Au nanocrystals	5.7	20.4	5.7	11.2
L3 Au nanocrystals	3.5	19.9	4.0	11.9
L4 Au nanocrystals	5.4	16.4	5.2	10.2
L5 Au nanocrystals	4.7	15.5	5.0	10.1

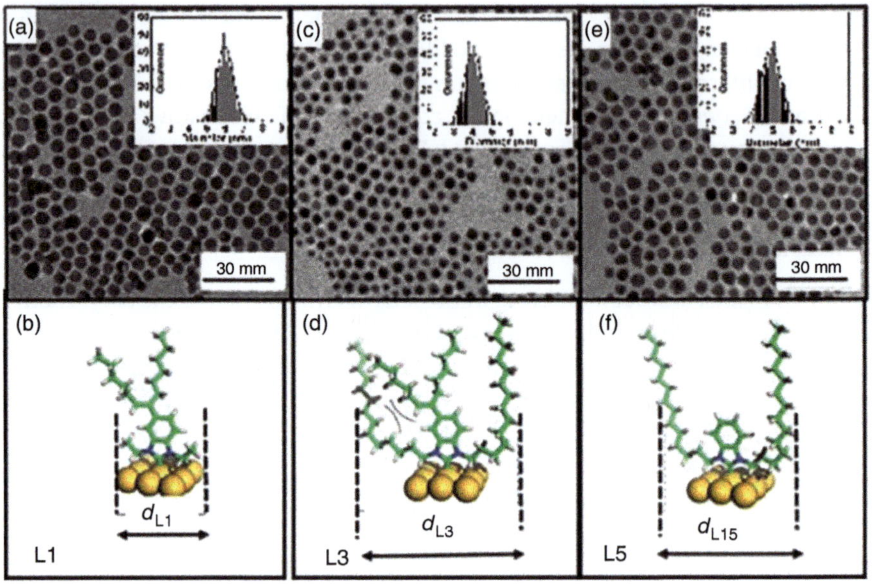

Figure 1.20 TEM images and insetcorresponding size distribution of Aunanocrystals coated with various N-heterocyclic carbenes. (a) nanocrystalscoated with L1 (b), (c) nanocrystals coated with L3 (d), (e) nanocrystalscoated with L5 (f). Source: Ling et al. [78]/American Chemical Society. Source: (a, c, e) Ling et al. [78]/American Chemical Society.

1.6 Conclusion

In this chapter, we described a very large number of methods to control metal nanocrystal sizes with a rather low size distribution. This favors to self-assemble nanocrystals in 2D and 3D superlattices. Furthermore, we find that, in some specific cases, it has been possible to control the crystalline structures of nanoparticles called nanocrystallinity. In the other chapters, we will discuss the importance of the nanocrystallinity on 3D superlattices.

References

1 Pileni, M.P. (2003). Role of soft colloidal templates in the control of size and shape of inorganic nanocrystals. *Nat. Mater.* 2: 145–150.
2 Pileni, M.P. (1993). Reverse micelles as microreactors. *J. Phys. Chem.* 97: 6961–6973.
3 Sugimoto, T. (2001). *Monodispersed Particles*. Amsterdam: Elsevier.
4 Puntes, V.F., Zanchet, D., Erdonmez, C.K., and Alivisatos, A.P. (2002). Synthesis of hcp nanodisks. *J. Am. Chem. Soc.* 124: 12874.
5 Prasad, B.L.V., Stoeva, S.I., Sorensen, C.M., and Klabunde, K.J. (2002). Digestive ripening of thiolated gold nanoparticles: the effect of alkyl chain length. *Langmuir* 18: 7515–7520.

6 Arbouet, A., Voisin, C., Christofilos, D. et al. (2003). Electron-phonon scattering in metal clusters. *Phys. Rev. Lett.* 90: 177401-1–177401-4.

7 Daniel, M.C. and Astruc, D. (2004). Gold nanoparticles: assembly, supramolecular chemistry, quantum size related properties and applications toward biology, catalysis, and nanotechnology. *Chem. Rev.* 104: 293–346.

8 Wang, J., Zhuang, Q., Peng, Y., and Li, A. (2005). A general strategy for nanocrystal synthesis. *Nature* 437: 121–124.

9 Kwon, S.G., Piao, Y., Park, J. et al. (2007). Kinetics of monodisperse iron oxide nanocrystal formation by "heating-up" process. *J. Am. Chem. Soc.* 129: 12571.

10 Kwon, S.G. and Hyeon, T. (2011). Formation mechanisms of uniform nanocrystals via hot-injection and heat-up methods. *Small* 7: 2685.

11 Kwon, S.G. and Hyeon, T. (2008). Colloidal chemical synthesis and formation kinetics of uniformly sized nanocrystals of metals, oxides, and chalcogenides. *Acc. Chem. Res.* 41: 1696–1709.

12 Robinson, I., Zacchini, S., Tung, L.D. et al. (2009). *Chem. Mater.* 21: 3021.

13 Goubet, N., Ding, Y., Brust, M. et al. (2009). A way to control the gold nanocrystals size: using seeds with different sizes and subjecting them to mild annealing. *ACS Nano* 3: 3622–3628.

14 Smith, D.K., Miller, N.R., and Korgel, B.A. (2009). Iodide in CTAB prevents gold nanorod formation. *Langmuir* 25 (16): 9518–9524.

15 Carbone, L. and Cozzoli, P.D. (2010). Colloidal heterostructured nanocrystals synthesis and growth mechanisms. *Nano Today* 5: 449.

16 Zhang, R., Khalizov, A., Wang, L. et al. (2012). Nucleation and growth of nanoparticles in the atmosphere. *Chem. Rev.* 112: 1957.

17 Taleb, A., Petit, C., and Pileni, M.P. (1997). Synthesis of highly monodisperse silver nanoparticles from AOT reverse micelles: a way to 2D and 3D self-organization. *Chem. Mater.* 9: 950–959.

18 Fievet, F., Ammar-Merah, S., Brayner, R. et al. (2018). The polyol process: a unique method for easy access to metal nanoparticles with tailored sizes, shapes and compositions. *Chem. Soc. Rev.* 47: 5187–5233.

19 LaMer, V.K. and Dinegar, R.H. (1950). Theory, production and mechanism of formation of monodispersed hydrosols. *J. Am. Chem. Soc.* 72: 4847.

20 LaMer, V.K. (1952). Nucleation in phase transition. *Ind. Eng. Chem.* 44: 1270.

21 Wu, N., Fu, L., Su, M. et al. (2004). Interaction of fatty acid monolayers with cobalt nanoparticles. *Nano Lett.* 4: 383–386.

22 Legrand, J., Petit, C., Bazin, D., and Pileni, M.P. (2000). Collective effect on magnetic properties of 2D superlattices of nanosized cobalt particles. *Appl. Surf. Sci.* 164: 186–192.

23 Love, J.C., Estroff, L.A., Kriebel, J.K. et al. (2005). Self-assembled monolayers of thiolates on metals as a form of nanotechnology. *Chem. Rev.* 105: 1103–1169.

24 Mourdikoudis, S. and Liz-Marzan, L.M. (2013). Oleylamine in nanoparticle synthesis. *Chem. Mater.* 25 (9): 1465–1476.

25 Shimpi, J.R., Sidhaye, D.S., and Prasad, B.L.V. (2017). Digestive ripening: a fine chemical machining process on the nanoscale. *Langmuir* 33: 9491–9507.

26 Zhang, S., Zhang, L., Liu, K. et al. (2018). Digestive ripening in the formation of monodisperse silver nanospheres. *Mat. Chem. Front.* 2: 1328–1333.

27 Irzhak, T.F. and Irzhak, V.I. (2019). On the digestive ripening mechanism. *Dokl . Phys. Chem.* 486: 77–79.

28 Scott, R.L. and Magat, M. (1945). The thermodynamics of high-polymer solutions: I. The free energy of mixing of solvents and polymers of heterogeneous distribution. *J. Chem. Phys.* 13: 172.

29 Burda, C., Chen, X.B., Narayanan, R., and El-Sayed, M.A. (2005). Chemistry and properties of nanocrystals of different shapes. *Chem. Rev.* 105: 1025–1102.

30 Baletto, F. and Ferrando, R. (2005). Structural properties of nanoclusters: energetic, thermodynamic and kinetic effects. *Rev. Mod. Phys.* 77: 371–423.

31 Baletto, F., Mottet, C., and Ferrando, R. (2001). Microscopic mechanisms of the growth of metastable silver icosahedra. *Phys. Rev. B* 63: 155408.

32 Baletto, F., Mottet, C., and Ferrando, R. (2000). Reentrant morphology transition in the growth of free silver nanoclusters. *Phys. Rev. Lett.* 84: 5544–5547.

33 Park, J., Joo, J., Kwon, S.G. et al. (2007). Synthesis of monodisperse spherical nanocrystals. *Angew. Chem. Int. Ed.* 46: 4630–4660.

34 Yin, Y. and Alivisatos, A.P. (2005). Colloidal nanocrystal synthesis and the organic–inorganic interface. *Nature* 437: 664–670.

35 Dinega, D.P. and Bawendi, M.G. (1999). A solution-phase chemical approach to a new crystal structure of cobalt. *Angew. Chem.* 38: 1788–1791.

36 Sun, S.H., Murray, C.B., Weller, D. et al. (2000). Monodisperse FePt nanoparticles and ferromagnetic FePt nanocrystal superlattices. *Science* 287: 1989.

37 Kirkland, A.I., Jefferson, D.A., Duff, D.G. et al. (1993). Structural studies of trigonal lamellar particles of gold and silver. *Proc. R. Soc. Lond. A* 440: 589–609.

38 Schadt, M., Cheung, W., Luo, J., and Zhong, C.J. (2006). Molecularly tuned size selectivity in thermal processing of gold nanoparticles. *Chem. Mater.* 18: 5147–5150.

39 Shimizu, T., Terashini, T., Hasegawa, S., and Miyake, M.J. (2003). Size evolution of alkanethiol-protected gold nanoparticles by heat treatment in the solid state. *J. Phys. Chem. B* 107: 2719–2724.

40 Beck, C.G.S., Souza, T.L., Silva, M. et al. (2011). Formation mechanism via a heterocoagulation approach of FePt nanoparticles using the modified polyol process. *J. Phys. Chem. C* 115: 10475–10482.

41 Petit, C. and Pileni, M.P. (1988). Synthesis in situ of cadmium sulfite synthetized in reverse micelles and hydrocarbon gels. *J. Phys. Chem.* 92: 2282–2286.

42 Petit, C., Lixon, P., and Pileni, M.P. (1990). Synthesis in situ of cadmium sulfate semiconductor: 2. Effect of functionalized surfactant. *J. Phys. Chem.* 94: 1598–1603.

43 Petit, C., Lixon, P., and Pileni, M.P. (1993). Synthesis in situ of silver nanocluster in AOT reverse micelles. *J. Phys. Chem.* 97: 12974–12983.

44 Pileni, M.P. (1997). Nanosized particles made in colloidal assemblies. *Langmuir* 13: 3266–3276.

45 Pileni, M.P. (2001). Mesostructured fluids in oil rich regions: structural and templating approaches. *Langmuir* 17: 7476–7487.

46 Lisiecki, I. and Pileni, M.P. (2003). Synthesis of well-defined and low size distribution cobalt nanocrystals: the limited influence of reverse micelles. *Langmuir* 19: 9486.

47 Cavalier, M., Wall, M., Lisiecki, I., and Pileni, M.P. (2011). How can the nanocrystallinity of 7 nm spherical Co nanoparticles dispersed in solution be improved? *Langmuir* 27: 5014–5020.

48 Yang, Z., Cavalier, M., Walls, M. et al. (2012). A phase-solution annealing strategy to control the cobalt nanocrystal anisotropy: structural and magnetic investigations. *J. Phys. Chem. C* 116: 15723–15730.

49 Pileni, M.P., Zemb, T., and Petit, C. (1985). Solubilization by reverse micelles: solute localization and structure perturbation. *Chem. Phys. Lett.* 118: 414.

50 Yang, N., Yang, J., Yang, J. et al. (2015). Control of the oxygen and cobalt atoms diffusion through Co nanoparticles differing by their crystalline structure and size. *Adv. Funct. Mater.* 25: 891–897.

51 Wei, J., Schaeffer, N., Albouy, P.A., and Pileni, M.P. (2015). Surface plasmon resonance properties of silver nanocrystals differing in size and coating agent ordered in 3D supracrystals. *Chem. Mater.* 27: 5614–5621.

52 Maye, M.M., Zheng, W., Leibowitz, F.L. et al. (2000). Heating-induced evolution of thiolate-encapsulated gold nanoparticles: a strategy for size and shape manipulations. *Langmuir* 16: 490–497.

53 Zheng, N., Fan, J., and Stucky, G.D. (2006). One-step one-phase synthesis of monodisperse noble-metallic nanoparticles and their colloidal crystals. *J. Am. Chem. Soc.* 128: 6550–6551.

54 Goubet, N., Portalès, H., Yan, C. et al. (2012). Simultaneous growths of gold colloidal crystals. *J. Am. Chem. Soc.* 134: 3714–3719.

55 Goubet, N., Richardi, J., Albouy, P.A., and Pileni, M.P. (2011). Which forces do control the supracrystal nucleation in organic media. *Adv. Funct. Mater.* 21: 2693–2704.

56 Portalès, H., Goubet, N., Saviot, L. et al. (2010). Crystallinity dependence of the plasmon resonant Raman scattering by anisotropic gold nanocrystals. *ACS Nano* 4: 3490–3497.

57 Goubet, N., Tempra, I., Yang, J. et al. (2015). Size and nanocrystallinity controlled gold nanocrystals: synthesis, electronic and mechanical properties. *Nanoscale* 7: 3237–3246.

58 Portalès, H., Goubet, N., Saviot, L. et al. (2008). Probing atomic ordering and multiple twinning in metal nanocrystals through their vibrations. *Proc. Natl. Acad. Sci. USA* 105: 14784–14789.

59 Peng, S., Xie, J., and Sun, S. (2008). Synthesis of Co/MFe_2O_4 (M = Fe, Mn) core/shell nanocomposite particles. *J. Solid State Chem.* 181: 1560.

60 Yang, Z., Wei, J., Bonville, P., and Pileni, M.P. (2015). Engineering the magnetic dipolar interactions in three-dimensional binary supracrystals via mesoscale alloying. *Adv. Funct. Mater.* 25: 4908.

61 Yang, Z., Yang, J., Bergström, J. et al. (2014). Crystal polymorphism: dependence of oxygen diffusion through 2D ordered Co nanocrystals. *Phys. Chem. Chem. Phys.* 16: 9791–9796.

62 Yang, J., Khazen, K., and Pileni, M.P. (2014). How nanocrystallinity and order define the magnetic properties of ε-Co supracrystal. *J. Phys. Condens. Matter* 26: 295303.

63 Wei, J., Schaeffer, N., and Pileni, M.P. (2016). Solvent-mediated crystallization of nanocrystal 3D assemblies of silver nanocrystals: unexpected superlattice ripening. *Chem. Mater.* 28: 293–302.

64 Williams, J.V., Kotov, N.A., and Savage, P.E. (2009). A rapid hot-injection method for the improved hydrothermal synthesis of CdSe nanoparticles. *Ind. Eng. Chem. Res.* 48 (9): 4316–4321.

65 Timonen, J.V.I., Seppala, E.T., Ikkala, O., and Ras, R.H.A. (2011). From hot-injection synthesis to heating-up synthesis of cobalt nanoparticles: observation of kinetically controllable nucleation. *Angew. Chem. Int. Ed.* 50: 2080–2084.

66 Brust, M., Walker, M., Bethell, D. et al. (1994). Synthesis of thiol-derivatized gold nanoparticles in a two-phase liquid/liquid system. *J. Chem. Soc., Chem. Commun.* 801–802.

67 Brust, M., Fink, J., Bethell, D. et al. (1995). Synthesis and reactions of functionalised gold nanoparticle. *J. Chem. Soc., Chem. Commun.* 1655–1656.

68 Zhou, M., Chen, S., and Zhao, S. (2006). Synthesis of icosahedral gold nanocrystals: a thermal process strategy. *J. Phys. Chem. B* 110: 4510–4513.

69 Chushak, Y.G. and Bartell, L.S. (2001). Melting and freezing of gold nanoclusters. *J. Phys. Chem. B* 105: 11605.

70 Nam, H.S., Hwang, N.M., Yu, B.D., and Yoon, J.K. (2002). Formation of an icosahedral structure during the freezing of gold nanoclusters: surface-induced mechanism. *Phys. Rev. Lett.* 89: 275502-1–275502-4.

71 Shim, J.H., Lee, B.J., and Cho, Y.W. (2002). Thermal stability of unsupported gold nanoparticle: a molecular dynamics study. *Surf. Sci.* 512: 262–268.

72 Bourissou, D., Guerret, O., Gabbai, F.P., and Bertrand, G. (2000). Stable carbenes. *Chem. Rev.* 100: 39–92.

73 Herrmann, W.A. (2002). N-Heterocyclic carbenes: a new concept in organometallic catalysis. *Angew. Chem. Int. Ed.* 41: 1290–1309.

74 Hahn, F.E. and Jahnke, M.C. (2008). Heterocyclic carbenes: synthesis and coordination chemistry. *Angew. Chem. Int. Ed.* 47: 3122–3172.

75 Hopkinson, M.N., Richter, C., Schedler, M., and Glorius, F. (2014). An overview of N-heterocyclic carbenes. *Nature* 510: 485–496.

76 Ling, X., Schaeffer, N., Roland, S., and Pileni, M.P. (2013). Nanocrystals: why do silver and gold N-heterocyclic carbene precursors behave differently? *Langmuir* 29: 12647–12656.

77 Ling, X., Roland, S., and Pileni, M.P. (2015). Superior oxygen stability of N-heterocyclic carbene-coated Au nanocrystals. *Langmuir* 31: 12873–12882.

78 Ling, X., Roland, S., and Pileni, M.P. (2015). Supracrystals of N-heterocyclic carbene-coated Au nanocrystals. *Chem. Mater.* 27: 414–423.

79 Ling, X. (2015). Unpublished experiments. PhD thesis. UPMC-University, Paris.

80 Roland, S., Ling, X., and Pileni, M.P. (2016). N-Heterocyclic carbene ligands for Au nanocrystal stabilization and three-dimensional self-assembly. *Langmuir* 32: 7683–7696.

81 Ott, L.S., Cline, M.L., Deetlefs, M. et al. (2005). Nanoclusters in ionic liquids: evidence for N-heterocyclic carbene formation from imidazolium-based ionic liquids detected by ^{2}H NMR. *J. Am. Chem. Soc.* 127: 5758–5759.

82 Scholten, J.D., Ebeling, G., and Dupont, J. (2007). On the involvement of NHC carbenes in catalytic reactions by iridium complexes, nanoparticle and bulk metal dispersed in imidazolium ionic liquids. *Dalton Trans.* 5554–5560.

83 Huang, R.T.W., Wang, W.C., Yang, R.Y. et al. (2009). Liquid crystals of gold(I) N-heterocyclic carbene complexes. *Dalton Trans.* 7121–7131.

84 Hurst, E.C., Wilson, K., Fairlamb, I.J.S., and Chechik, V. (2009). N-Heterocyclic carbene coated metal nanoparticles. *New J. Chem.* 33: 1837–1840.

85 Vignolle, J. and Tilley, T.D. (2009). N-Heterocyclic carbene-stabilized gold nanoparticles and their assembly into 3D superlattices. *Chem. Commun.* 7230–7232.

86 Aslanov, L.A., Zakharov, V.N., Zakharov, M.A. et al. (2010). Stabilization of silicon nanoparticles by carbenes. *Russ. J. Coord. Chem.* 36: 330–332.

87 Kang, J.N., Yang, H.Y., Kim, H.I., and Son, S.U. (2010). Colloidal synthesis of cubic-phase copper selenide nanodiscs and their optoelectronic properties. *Chem. Mater.* 22: 3586–3588.

88 Ranganath, K.V.S., Kloesges, J., Schäfer, A.H., and Glorius, F. (2010). Asymmetric nanocatalysis: N-heterocyclic carbenes as chiral modifiers of Fe_3O_4/Pd nanoparticles. *Angew. Chem. Int. Ed.* 49: 7786–7789.

89 Lara, P., Rivada-Wheelaghan, O., Conejero, S. et al. (2011). Ruthenium nanoparticles stabilized by N-heterocyclic carbenes: ligand location and influence on reactivity. *Angew. Chem. Int. Ed.* 50: 12080–12084.

90 Gonzales-Galves, D., Lara, P., Rivada-Wheelaghan, O. et al. (2013). NHC-stabilized ruthenium nanoparticles as new catalysts for the hydrogenation of aromatics. *Catal. Sci. Technol.* 3: 99–105.

91 Serpell, C.J., Cookson, J., Thompson, A.L. et al. (2013). Haloaurate and halopalladate imidazolium salts: structures, properties, and use as precursors for catalytic metal nanoparticles. *Dalton Trans.* 42: 1385–1393.

92 Baquero, E.A., Tricard, S., Flores, J.C. et al. (2014). Highly stable water-soluble platinum nanoparticles stabilized by hydrophilic N-heterocyclic carbenes. *Angew. Chem. Int. Ed.* 53: 13220–13224.

93 Song, S.G., Satheeshkumar, C., Park, J. et al. (2014). N-Heterocyclic carbene-based conducting polymer–gold nanoparticle hybrids and their catalytic application. *Macromolecules* 47: 6566–6571.

94 Liu, H.X., He, X., and Zhao, L. (2014). Metallamacrocycle-modified gold nanoparticles: a new pathway for surface functionalization. *Chem. Commun.* 50: 971–974.

95 Richter, C., Schaepe, K., Glorius, F., and Ravoo, B.J. (2014). Tailor-made N-heterocyclic carbenes for nanoparticle stabilization. *Chem. Commun.* 50: 3204–3207.

96 Rodriguez-Castillo, M., Laurencin, D., Tielens, F. et al. (2014). Reactivity of gold nanoparticles towards N-heterocyclic carbenes. *Dalton Trans.* 43: 5978–5982.

97 Crespo, J., Guari, Y., Ibarra, A. et al. (2014). Ultrasmall NHC-coated gold nanoparticles obtained through solvent free thermolysis of organometallic Au(I) complexes. *Dalton Trans.* 43: 15713–15718.

98 MacLeod, M.J. and Johnson, J.A. (2015). PEGylated N-heterocyclic carbene anchors designed to stabilize gold nanoparticles in biologically relevant media. *J. Am. Chem. Soc.* 137: 7974–7977.

99 Ferry, A., Schaepe, K., Tegeder, P. et al. (2015). Negatively charged N-heterocyclic carbene-stabilized Pd and Au nanoparticles and efficient catalysis in water. *ACS Catal.* 5: 5414–5420.

100 Zhukhovitskiy, A.V., MacLeod, M.J., and Johnson, J.A. (2015). Carbene ligands in surface chemistry: from stabilization of discrete elemental allotropes to modification of nanoscale and bulk substrates. *Chem. Rev.* 115: 11503–11532.

101 Weidner, T., Baio, J.E., Mundstock, A. et al. (2011). NHC-based self-assembled monolayers on solid gold substrates. *Aust. J. Chem.* 64: 1177–1179.

102 Zhukhovitskiy, A.V., Mavros, M.G., Van Voorhis, T., and Johnson, J.A. (2013). Addressable carbene anchors for gold surfaces. *J. Am. Chem. Soc.* 135: 7418–7421.

103 Crudden, C.M., Horton, J.H., Ebralidze, I.I. et al. (2014). Ultra stable self-assembled monolayers of N-heterocyclic carbenes on gold. *Nat. Chem.* 6: 409–414.

104 Benhamou, L., Chardon, E., Lavigne, G. et al. (2011). Synthetic routes to N-heterocyclic carbene precursors. *Chem. Rev.* 111: 2705–2733.

105 Nelson, D.J. and Nolan, S.P. (2013). Quantifying and understanding the electronic properties of N-heterocyclic carbenes. *Chem. Soc. Rev.* 42: 6723–6753.

106 Pyykkö, P. and Runeberg, N. (2006). Comparative theoretical study of N-heterocyclic carbenes and other ligands bound to Au^{I}. *Chem. Asian J.* 1: 623–628.

107 Yang, Z., Lisiecki, I., Walls, M., and Pileni, M.P. (2013). Nanocrystallinity and the ordering of nanoparticles in 2D superlattices: controlled formation of either core/shell (Co/CoO) or hollow CoO nanocrystals. *ACS Nano* 7: 1342–1350.

2

Influence of the Nanoparticle Crystalline Structures Called Nanocrystallinities on Various Properties

The chemical and physical properties related to the crystalline structure of metal nanoparticles called nanocrystallinity have been poorly investigated up to now. This is due to the difficulty of controlling the internal structure of nanoparticles. Some progress in the syntheses has been achieved, enabling studies of specific chemical and physical properties. The nanocrystallinity control received much less attention and remains an open question.

We study properties of Au and Co nanocrystals differing by their nanocrystallinities. It plays a major role in the interdiffusion processes, called nano-Kirkendall effects, of cobalt and oxygen through the nanocrystal interface. It is demonstrated that the nanocrystal size, nanocrystallinity, and nanocrystal environment are key parameters in the final structures. With Au nanocrystals, keeping the same chemical environment, it is shown that the localized surface plasmon resonance is slightly modified between monodomain and polycrystalline Au nanocrystals. The quadrupole vibrational modes of single-domain Au nanocrystals are split as they are degenerated with a polycrystalline counterpart. The elastic moduli of the 3D superlattice are significantly larger with a single domain as building blocks than with a polycrystalline phase. Moreover, with Au and Co nanocrystals, the properties of the vibration dynamics show that the respiratory modes keep the same frequencies, while the damping times change between monodomains and polycrystals.

2.1 Nano-Kirkendall

The nano-Kirkendall effect has been rather well studied [1–7]. Various techniques are used: mild annealing of their amorphous counterparts synthesized in reverse micelles [1, 2] produces single-domain hcp structures (Figure 2.1a,b). The corresponding ε-phase is obtained by hot injection process (Figure 2.1c) [8, 9]. A thermal decomposition [6, 10] procedure produces polycrystalline fcc with well-resolved individual crystalline domains (Figure 2.1d). The control of the nanoparticle size is

Metal Nano 3D Superlattices: Synthesis, Properties, and Applications, First Edition.
Marie-Paule Pileni. Published 2023 by WILEY-VCH GmbH.

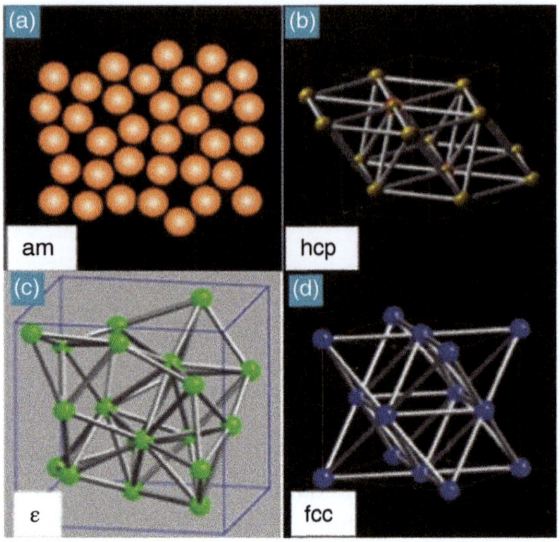

Figure 2.1 Scheme of Co nanoparticles differing by their nanocrystallinities: (a) amorphous nanoparticle, (b) hcp nanocrystals, (c) ε-phase nanocrystals, and (d) large domains of fcc nanocrystals.

obtained by adjusting the surfactant and reactive concentrations, as well as the temperature.

By using the various procedures developed in Chapter 1, amorphous Co nanoparticles, single-domain hcp and ε-phase nanocrystals, and polycrystalline fcc nanocrystals are produced (Figure 2.2). The interplanar spacings are 2.00, 2.07, and 2.05 Å for hcp, ε-phase, and fcc crystalline structures of Co nanocrystals corresponding to the most intense in the diffraction pattern of (002), (221), and (111) crystal planes, respectively. Concerning the oxidized form, the spacings at 2.11 and 2.46 Å are seen that were indexed as (200) and (111) reflections of the cubic CoO lattice.

Numerous chemical reactions or microstructural changes in solids take place through solid-state diffusion [11]. In 1942, Ernest Kirkendall investigated interdiffusion of copper and zinc in brass at elevated temperatures, observing a net mass transport accompanied by a noticeable volume reduction in brass. He concluded that the reduction in volume was evident because of the faster movement for zinc species out of brass than the influx of copper species; thus, vacancy-assisted hopping was proposed as the main fashion of atomic transport [12]. In 1947, Smigelskas and Kirkendall reported the experiment under carefully controlled conditions and demonstrated the difference between the diffusion rates of copper and zinc in brass that led to the formation of voids in the species at high temperatures [13]. Although the deterioration of alloyed materials occurs due to the formation of Kirkendall voids, chemists further applied this destructive effect for engineering nanovoids inside nanomaterials.

Over the last decade, inorganic hollow nanostructures with designed fashion, controlled pore volume, and shell thickness have been fascinated by their high surface-to-volume ratio and large pore volume, which would be promising for technological applications including drug delivery system, energy storage, and catalysis [14–16]. Since the first report of Caruso and Mohwald's [17] colloidal-template method on the fabrication of inorganic hollow structures, a wealth of methods has been applied for engineering the hollow nanostructures, among which the template-mediated

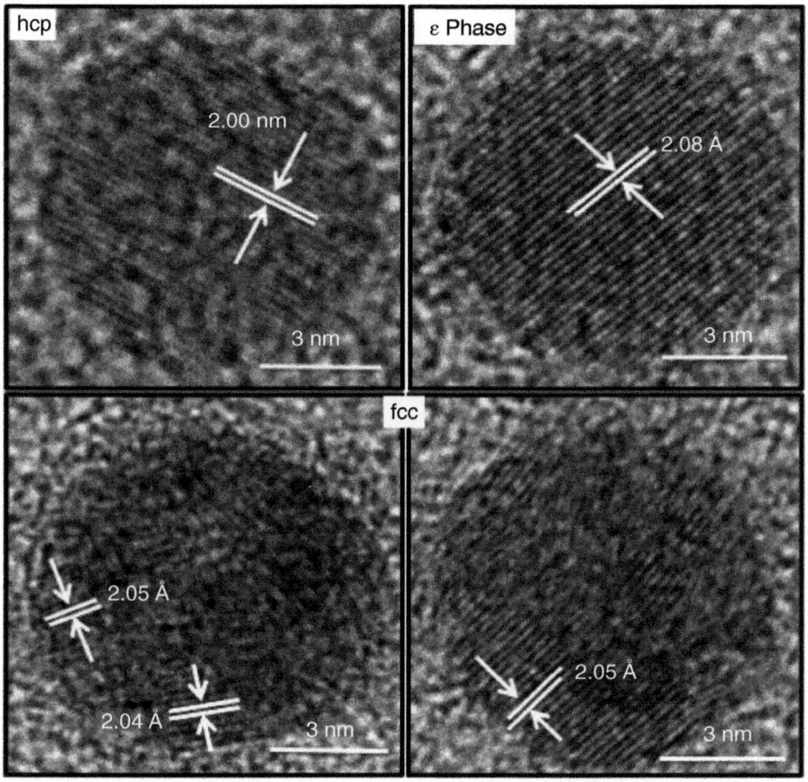

Figure 2.2 Nanocrystals (8 nm) differing by their nanocrystallinities. *Source:* Yang et al. [3–5]/Royal Chemical Society.

synthesis has been demonstrated as a versatile and effective approach. A traditional colloidal-templating method usually requires several procedures, including surface modification of the template, coating the surface of the template colloids with desired materials, and a posttreatment of template removal. The size of hollow interior obtained via such a colloidal-template approach is usually confined by the size of the template. In any case, the sizes of the colloids, such as silica, carbon, or polymer beads, are over hundreds of nanometers, which may veil the intrinsic properties of hollow nanostructures with size in the regime of below 100 nm. In recent years, a new self-templated method, which can overcome the size limitation of colloidal template, has emerged, employing old physical principles, such as Ostwald ripening, oriented attachment, and Kirkendall effect [1, 3–7, 18, 19].

In nanoscale objects, where there is little space for dislocations, and lattice shift in spherical objects may lead to large stresses, Kirkendall voiding takes place, which is not the case for the bulk phase materials. Later on, this principle has been demonstrated to be a general procedure for the production of hollow nanostructures, even an extension to cylindrical nanotubes [20]. Considerable nicely written reviews on this field are available, concerning on the general concept and synthesis of hollow nanostructures by nanoscale Kirkendall effect [21–24].

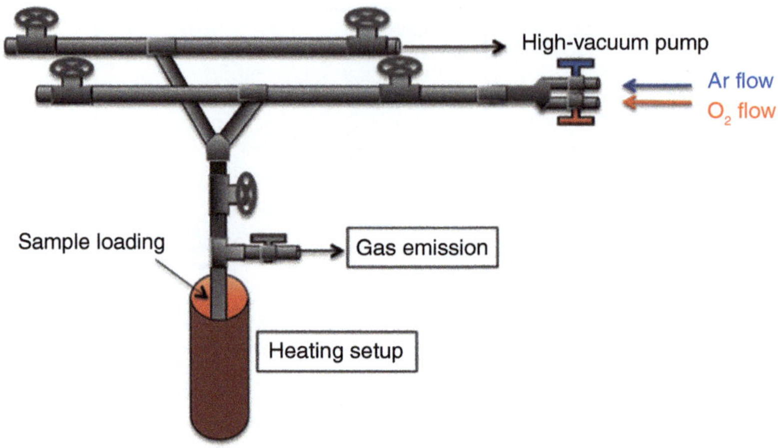

Scheme 2.1 Modified Schlenk line setup applied for the air-free annealing or oxidizing the samples.

To further determine the role of crystal structure in the nanoscale Kirkendall effect, Co nanoparticles with various crystal structures (amorphous, fcc, hcp, and ε-phase) are deposited on a substrate. The nanoparticles with low size distribution self-assemble in hexagonal network. These various substrates are simultaneously placed in a modified Schlenk-line setup. The samples are loaded into the oxidation apparatus (Scheme 2.1). It is pumped to vacuum and then reexposed to Ar three times to remove the air. The sample is then submitted to the pure oxygen flow at 200 °C for 10 minutes, after cooling under Ar flow to room temperature. Note that the atom diffusion processes are reduced when nanoparticles are self-assembled in 2D superlattices compared with isolated nanoparticle on the substrate. The crystallinity of isolated Co nanoparticles deposited on a substrate does not play any role in the oxide formation. However, when they are self-assembled in 2D superlattices, the oxidation process is slowed and produces various types of nanoparticles (hollow, core/shell, yolk/shell structures). The setup applied oxidizing after or not annealing the sample is shown below.

2.1.1 Influence of the Atom Diffusion Processes on Rather Large Nanoparticles (7 nm/8 nm) Differing by Their Nanocrystallinities

2.1.1.1 Amorphous Nanoparticles

Let us first consider the 7.2-nm Co nanocrystals self-ordered in 2D hexagonal network (Figure 2.3) showing that after oxidation the 2D array remains ordered, whereas the nanoparticles' contrasts are no longer homogeneous and exhibit a core/shell structure. This core/shell structure is further revealed by high-resolution high-angle annular dark field–scanning TEM (HAADF-STEM) image (Figure 2.3a). The local superposition chemical map of the sample from STEM–electron energy-loss spectrum (EELS) measurement in Figure 2.3b reveals a blue center and a purple edge corresponding to the Co atom (blue) and oxygen atom (red). This clearly indicates formation of a

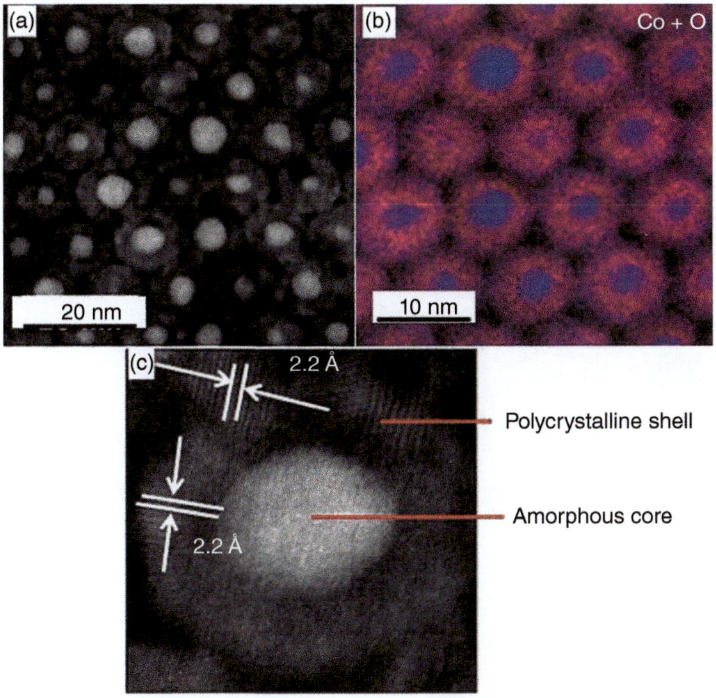

Figure 2.3 STEM-HAADF images of 2D ordered assemblies of the 7.2-nm *amorphous* nanoparticles after oxidation at 200 °C for 10 minutes: (a) low magnification, (b) STEM-EELS spectrum image, and (c) nanocrystal, high resolution. *Source:* Yang et al. [6, 7]/American Chemical Society..

Co/CoO core/shell. Figure 2.3c shows that the inner part of the nanoparticles is in amorphous state, whereas the outer part shows some degree of crystallinity, and the lattice fringe spacing is determined to be 2.2 Å, which is in good agreement with the (200) plane of the CoO cubic phase. Similar data are produced on slightly increase Co nanoparticle size from 7.2 to 8.2 nm. With 8.3-nm Co nanoparticles, as observed above with 7.2 nm, the TEM contrast is not homogeneous and displays a core/shell structure.

2.1.1.2 hcp Nanocrystals

With 7.1-nm hcp Co nanocrystals, the STEM-HAADF image in Figure 2.3a shows hollow structure inside the nanocrystal. Figure 2.3b shows the Co and O projected distributions extracted from an EELS image, and the Co (blue) and O (red) composite signal demonstrates that the hollow nanocrystals are homogeneous in their chemical distribution. The high-resolution image (Figure 2.4c) confirms the formation of hollow nanoparticles with the formation of single-domain CoO shell. Note that throughout the nanocrystal, CoO shell is multi-crystal (Figure 2.3c) with amorphous nanoparticles, whereas it is single domain with hcp nanocrystals (Figure 2.3c).

With 8.3-nm hcp Co nanocrystals exhibit a core/shell structure (Figure 2.5), whereas hollow structure is observed with 7.2-nm hcp nanocrystals (Figure 2.4c).

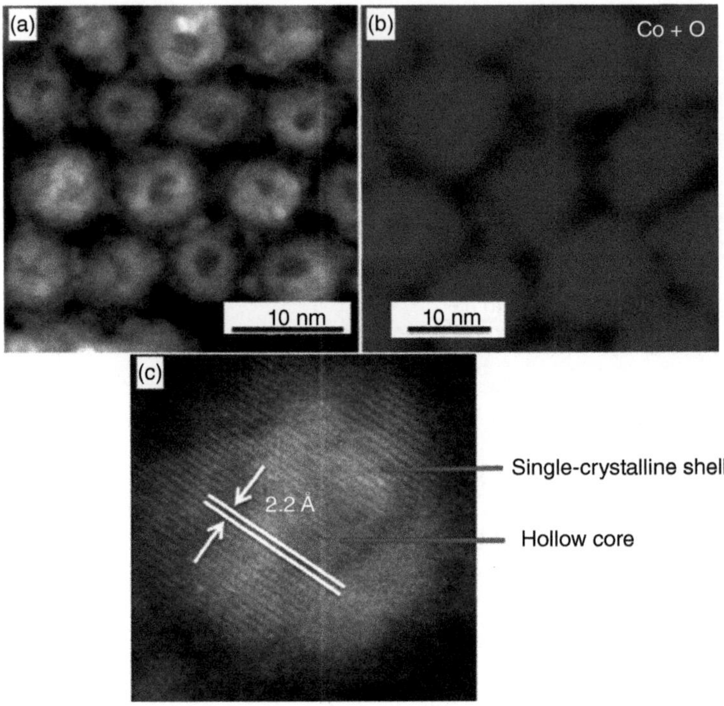

Figure 2.4 STEM-HAADF images of 2D ordered assemblies of the7.2-nm hcp nanocrystals after oxidation at 200 °C for 10 minutes: (a) low magnification; (b) STEM-EELS spectrum image, and (c) high resolution. *Source:* Yang et al. [6, 7]/American Chemical Society.

Figure 2.5 HRTEM images of 8.2-nm hcp Co nanocrystals after oxidation. *Source:* Yang et al. [3–5]/Royal Chemical Society.

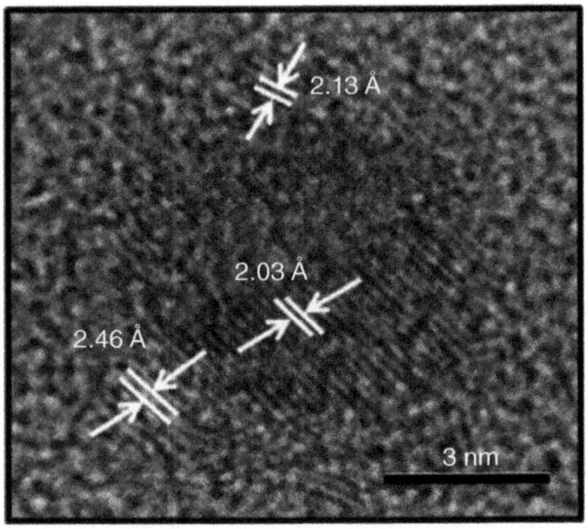

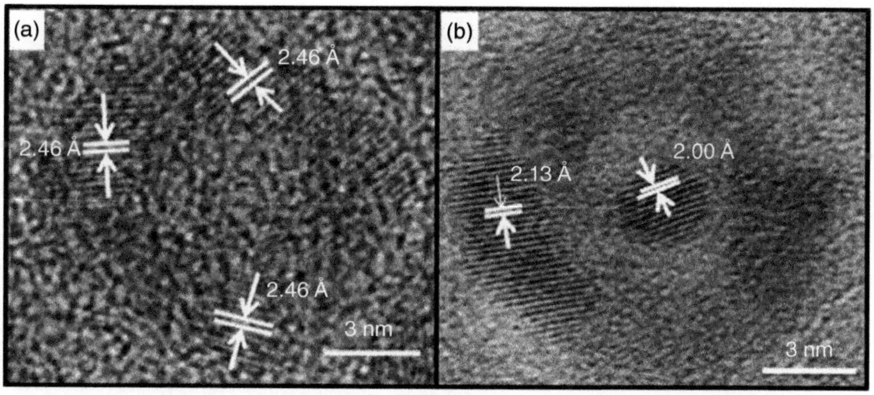

Figure 2.6 HRTEM images of 8.2-nm fcc Co nanoparticles after oxidation. (a) Hollow nanoparticles and (b) yolk/shell structure. *Source:* Yang et al. [3–5]/Royal Chemical Society.

2.1.1.3 fcc Nanocrystals

For 8.1-nm fcc polycrystalline Co nanoparticles, a hollowing process indicative of Kirkendall effect takes place and results in hollow nanoparticles (Figure 2.6a). In addition, a small fraction of the nanoparticles still has separated core materials inside the shell, forming a yolk/shell structure (Figure 2.6b). Such a yolk/shell structure is frequently observed during oxidizing metal nanoparticles, which is a typical intermediate toward ultimate hollow structure of nanoscale Kirkendall effect. Here, the presence of both hollow and yolk/shell structures indicates that the oxidation rate varies among different 8.1-nm Co nanoparticles, which could be attributed to the large distribution of crystal domain sizes in this sample. Nevertheless, the effect of size distribution of the Co nanocrystals cannot be ruled out because the nanocrystal size can also play an important role during the Kirkendall process as discussed below.

2.1.1.4 Epsilon-Phase Nanocrystals

With single-domain ε-phase 8.1-nm Co nanocrystals, the TEM image shows that hollow nanoparticles are produced (Figure 2.7), while the shell of the hollow structure is composed of several large CoO domains with their characteristic lattice fringe, indicating a polycrystalline structure induced by the surface strain. The above explicit examples may allow one to make a conclusion that atomic ordering of the Co nanoparticles is indispensable for the creation of single Kirkendall void inside the nanocrystals.

Here, we demonstrated that Co nanocrystal nanocrystallinity plays an important role during the Co and O atoms diffusion within the Co nanocrystals, as a result of whether or not Kirkendall voids can be generated inside the nanocrystals. Furthermore, hollow structures cannot be produced from amorphous-phase Co nanoparticles in which Co atoms are randomly distributed. This makes the atoms diffuse chaotically and disorderedly; hence, the oxidation process is very difficult after the formation of an initial oxide layer and the core/shell structure forms.

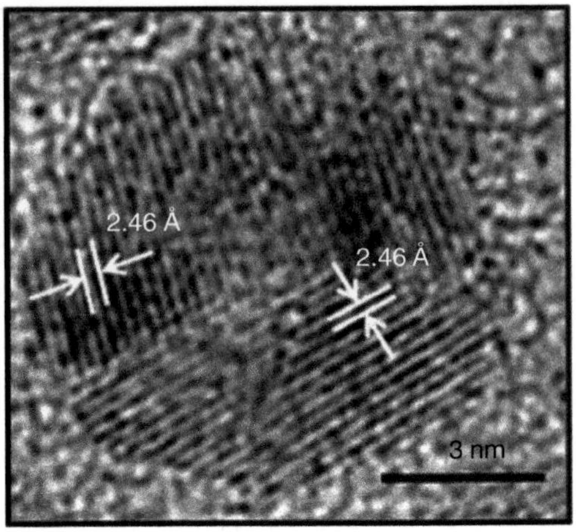

Figure 2.7 HRTEM images of hollow nanoparticle produced by oxidation of ε-phase 8.1-nm Co nanocrystals. *Source:* Yang et al. [6, 7]/American Chemical Society.

In other words, atoms diffuse along the lattices within the crystallized nanoparticles, which leads to the diffusion of oxygen and cobalt atoms directionally. As a result, Kirkendall voids within the nanoparticles generate with the continuous atomic interdiffusion.

2.1.2 Influence of the Nanocrystal Size Related to the Various Crystalline Structures

Different size-dependent oxygen diffusion processes are observed for Co nanoparticles with various nanocrystallinities, including ε and hcp single-domain, fcc polycrystal, and amorphous Co nanoparticles. Based on the comparable size of different crystalline phases and the same experimental conditions, the result is given in Table 2.1. For single-domain ε-phase Co nanocrystals, no significant difference with the nanocrystal size is observed, and consequently, polycrystalline CoO hollow nanoparticles are produced after oxidation. While for single-domain hcp Co nanocrystals, a drastic size effect is observed: the small-sized nanocrystals would result in the *single-domain* CoO hollow nanocrystals, whereas the large-sized nanocrystals lead to the formation of Co/CoO core/shell nanoparticles. By comparison between data obtained from compact crystalline structures (hcp and fcc), because of the presence of crystalline defects in fcc Co nanocrystals (where there are no defects in hcp ones), polycrystalline CoO hollow or yolk/shell structure can be produced with a faster diffusion of cobalt than oxygen atoms. For large nanocrystals, the thick oxide shell will also hamper further diffusion, forming a yolk/shell structure at the large size. Among all the Co nanoparticles investigated, only *amorphous* Co can form a solid nanoparticle without the observation of cavities. The final structure depends on both nanocrystallinity and nanoparticle size (Table 2.1): (i) with amorphous nanoparticles, solid CoO nanoparticles are produced for smaller sizes and are converted to core/shell structures for larger ones. (ii) With single-domain nanocrystals differing by their crystalline structure (ε and hcp phases), marked changes in the final structures are observed upon increasing the nanocrystal size. The ε-phase favors the

Table 2.1 Various final nanoparticle structures subjected to oxygen differing by their nanocrystallinities and their sizes: ε-phase, formation of hollow nanocrystals; hcp Co phase, transition from hollow to core/shell nanocrystals; fcc Co phase, transition from hollow to yolk/shell structure; *amorphous* Co nanocrystals, transition from CoO to core/shell nanocrystals.

<table>
<tr><th>Diameters</th><th rowspan="2">5 nm</th><th rowspan="2" colspan="2">7 nm</th><th rowspan="2" colspan="2">8 nm</th></tr>
<tr><th>Nanocrystallinity</th></tr>
<tr><td>ε-phase

Shell crystalline structure</td><td>
Polycrystal
None</td><td colspan="2">
Polycrystal
None</td><td colspan="2">
Polycrystal
None</td></tr>
<tr><td>Core crystalline structure</td><td></td><td colspan="2"></td><td colspan="2"></td></tr>
<tr><td>hcp

Shell crystalline structure</td><td>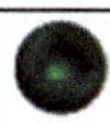
Single domain
None</td><td colspan="2">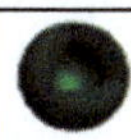
Mixed
None</td><td colspan="2">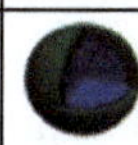
Polycrystal
hcp</td></tr>
<tr><td>Core crystalline structure</td><td></td><td colspan="2"></td><td colspan="2"></td></tr>
<tr><td rowspan="2">fcc

Shell crystalline structure</td><td rowspan="2">
Polycrystal
None</td><td rowspan="2" colspan="2">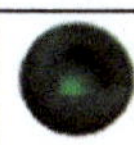
Polycrystal
None</td><td colspan="2">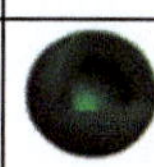 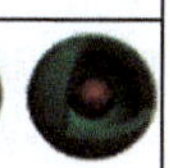
Polycrystal</td></tr>
<tr><td>None</td><td>None</td></tr>
<tr><td>Core crystalline structure</td><td></td><td colspan="2"></td><td colspan="2"></td></tr>
<tr><td rowspan="2">Amorphous

Shell crystalline structure
Core crystalline structure</td><td rowspan="2">
Amorphous
CoO</td><td colspan="2"> </td><td rowspan="2" colspan="2">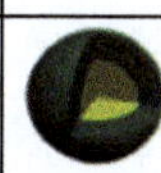
Polycrystal
Amorphous</td></tr>
<tr><td>Amorphous
CoO</td><td>Polycrystal
Amorphous</td></tr>
</table>

formation of a hollow structure, whereas a transition from single-domain hollow to multidomain core/shell structure takes place with the hcp phase. We know that ε-phase Co nanocrystals are a *cubic phase* and consequently are fewer compact than the hcp structure. This could explain the change in the final structure. (iii) With polycrystalline fcc nanocrystals, a transition from hollow to yolk/shell structure is observed. The presence of defects in the fcc crystalline structure could explain the slight differences obtained between hcp single-domain and fcc polycrystalline phases.

From these data, we claim that the nanocrystallinity and the nanoparticle size are key parameters in determining the final structures. This should be taken into account for future applications.

2.1.3 Nanoparticle Environment Effect

2.1.3.1 Isolated Nanocrystals

Nanocrystals are coated with hydrophobic alkyl chains, and consequently the nanoparticle environment includes nanoparticle surface coating and reaction medium. Uniform nanoparticles coated with alkyl chains tends to self-order into periodically ordered array upon evaporation of the carrier solvent. At the end of these monolayers, isolated nanocrystals are detected. This permits to study separately the behavior of isolated and self-assembled nanoparticles differing by their nanocrystallinities.

For 7.2-nm *amorphous* Co nanoparticles, an isolated nanoparticle can be fully oxidized and be transformed to polycrystalline CoO solid nanoparticles (Figure 2.8a–c), whereas core/shell (Co/CoO) nanoparticles take place for 2D ordered nanoparticle array (Figure 2.3). Similarly, for isolated 7-nm hcp nanocrystals, polycrystalline CoO solid nanoparticles would be produced after oxidation treatment (Figure 2.8d–f), whereas hcp Co nanocrystals self-assembled in hexagonal order produce hollow single-crystalline CoO nanocrystals (Figure 2.4). The above result can be explained as follows: when the nanoparticles are self-ordered, the interdigitation of the coating alkyl chains between the adjacent nanoparticles reduces the oxygen diffusion speed, compared with the isolated nanoparticles coated with alkyl chain monolayer. Hence, it is necessary to point out that a faster diffusion of oxygen atom through the

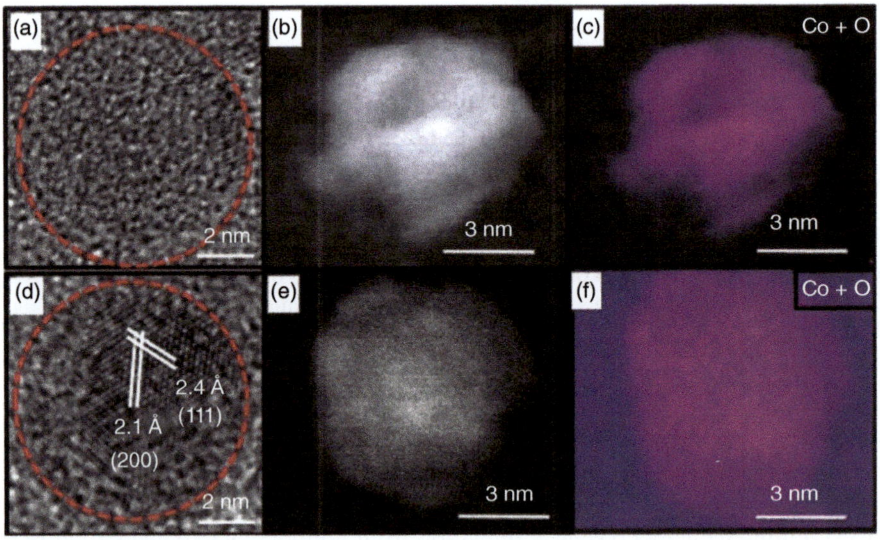

Figure 2.8 HRTEM images (a, d), STEM-HAADF spectrum image (b, e), and STEM-EELS (c, f) of the isolated amorphous Co nanoparticles (a, b, c) and hcp isolated nanocrystals (d, e, f) deposited on TEM grid and subjected to oxidation. *Source:* Yang et al. [6, 7]/ American Chemical Society.

hcp Co nanocrystals can result in the shrinkage of the hollow structure, leaving a solid structure. Thus, the Co and O interdiffusions through isolated hcp Co nanocrystals are not strictly termed as a Kirkendall effect despite the presence of the diffusion couple.

The density of surface protection layer can affect the Kirkendall process. Our result shows that the nanoparticles assembled into compact 2D array effectively increase the density of the protection layer (alkyl chains). Furthermore, the density of the protection layer can also be tuned by controlling the solubility of ligand in the selected solvent [25].

2.1.3.2 Influence of the Electron Beam Irradiation on the Final Nanocrystals

The TEM beam irradiation changes the nanoparticle environment after their exposure to beam irradiation [4].

Let us consider 7.2-nm *amorphous* nanoparticles located on the e-beam unirradiated. After oxidation treatment, the nanoparticles remain self-assembled in hexagonal network, and their average diameters increase from 7.2 to 9.5 nm. A magnified high-angle annular dark-field (HAADF) image (Figure 2.9a) indicates that the shell is composed of various crystallized zones with lattice fringes of 2.13 Å corresponding to the (200) plane of the CoO cubic phase. The core shows similar atomic ordering from which one can observe a few crystalline domains whose typical size does not exceed 1 nm in the core region. This clearly indicates that the nanoparticle core keeps the polycrystalline structure observed before treatment. The local chemical maps show that the Co signal (green) is present throughout the nanoparticles but enhanced in their central region, whereas the O signal (red) is mainly located at the shell region of the nanoparticle. The superposition of the two maps reveals a green center and a yellow edge, corresponding to the Co/CoO core/shell structures with shell thickness of around 3.5 nm (Figure 2.9b). Note that the shell part has a

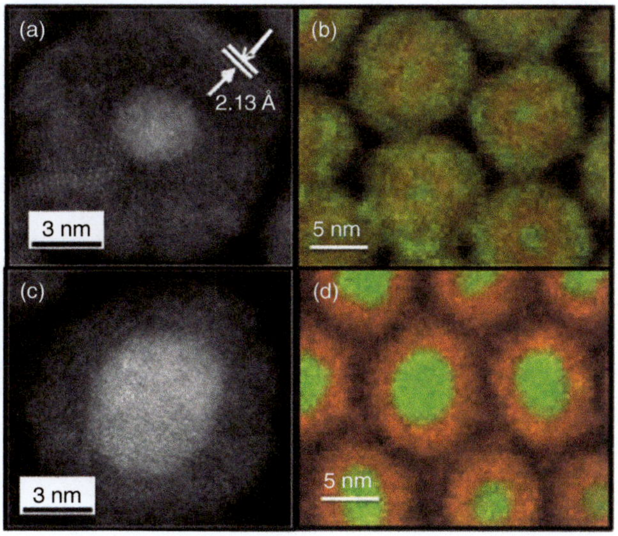

Figure 2.9 (a) and (c) are HAADF-STEM images and superposition of elemental maps Co (green) and oxygen (red) using OK-edge maps of 7.2-nm *amorphous* Co nanoparticles after oxidation corresponding to zone without (a, b) and with (c, d) electron irradiation. *Source:* Yang et al. [4]/Royal Chemical Society.

nonuniform elemental distribution, which can be discerned from the color contrast. The yellow and inner orange contrasts located at the outmost part are attributed to CoO (confirmed by HRTEM) and Co_xO_y ($y > x$) structures, respectively. This indicates the formation a core/double-shell Co/Co_xO_y/CoO structures. Let us consider the same sample subjected to e-beam, as in the unradiated zone the nanoparticle remains self-assembled in 2D superlattices. However, the increase in the nanoparticle size is less than that observed in the unradiated zone. The final nanoparticle size is characterized by an average diameter of 8.5 nm. Figure 2.9c shows an HAADF image of a single-domain nanocrystal with the inner Co core composed of numerous small-crystalline domains, as seen before the oxygen treatment. However, the outer oxide shell exhibits an amorphous structure from which no oxide rings can be identified. From the EELS chemical map (Figure 2.9d), the superposition of Co and O signal shows a thickness of 1.8 nm around a core of 5 nm. Hence, the comparison between the two zones thus shows, for the irradiated nanoparticles, a larger core size, better homogeneity in the chemical analysis, and less well-crystallized oxide.

With hcp Co nanocrystals instead of their *amorphous* counterparts, the unirradiated zone shows the formation of hollow cavities inside the particles. The average diameter of the hollow nanocrystals increases from 7.2 to 8.3 nm. No coalescence is observed. The diameter of the cavity is around 2.1 nm (Figure 2.10a). The HAADF-STEM image (Figure 2.10c) reveals that a cavity is around 2.1 nm and the hollow nanocrystals are well crystallized, with a lattice spacing of 2.18 Å, indicative of (200) planes of cubic CoO. The chemical map shows that the Co (green) and O (red) are both concentrated at the periphery of the nanocrystals with a very homogeneous distribution (Figure 2.10b). When the zone is irradiated in the same experimental

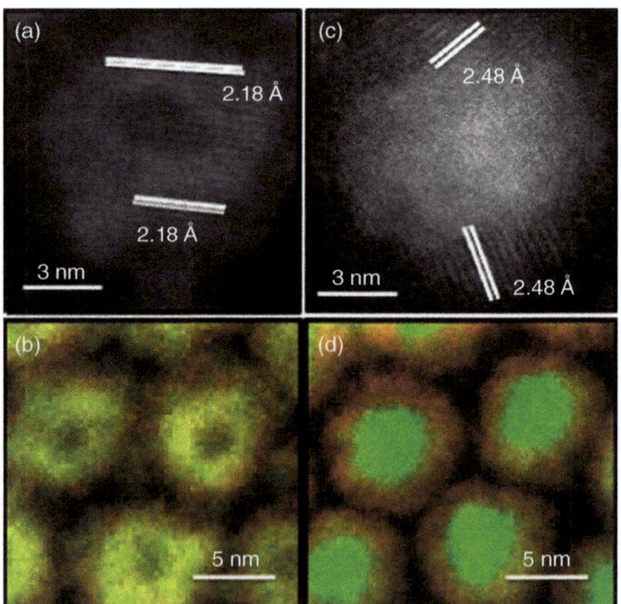

Figure 2.10 (a) and (c) are HAADF-STEM images and superposition of elemental maps Co (green) and oxygen (red) using OK-edge maps of 7.2-nm single-domain hcp Co nanocrystals after oxidation corresponding to zone without (a, b) and with (c, d) electron irradiation. *Source:* Yang et al. [4]/Royal Chemical Society.

condition, core/shell structures are produced (Figure 2.10c). The superimposed chemical maps of Co and O indicate that the chemical distribution is nonuniform. In the Co map, the Co-rich and Co-poor regions are easily distinguished in the center and at the edge, respectively. Conversely, the O-rich and O-poor regions in the oxygen map are at the edge and in the center, respectively, consistent with Co_{hcp}/CoO core/shell structures (Figure 2.10d). Such structures are strikingly different from the hollow CoO nanocrystals generated without pre-irradiation. Note that various sets of lattice fringes corresponding to crystallized CoO domains are observed, whereas single-domain CoO hollow nanocrystals are produced in the non-irradiated zone. Structure cannot be distinguished.

2.1.4 Conclusions

Nanoscale Kirkendall effect, based on atomic interdiffusion through the interface between the heterogeneous solids, is a powerful tool in the fabrication of hollow-based nanostructures. Here, we summarize the strategy for the nanofabrication guided by the nanoscale Kirkendall effect mainly from the oxidation of metal nanocrystals. The overall strategy can be summarized as fabrication of core/shell nanostructures, and the core material has a faster diffusion rate than that of the shell material. After the formation of the core/shell nanostructures, it is essential to know which factor would determine the formation of Kirkendall voids. Several factors, such as nanoparticle size, nanoparticle crystallinity, and nanoparticle environment, are involved in the final structures. It appears that they play an important role during the process of atomic interdiffusion in core/shell nanostructures, despite the core having a faster mobility. The design of hollow nanostructure has unique merits and great potentials for practical applications, such as in catalysis and drug delivery. Nanoscale Kirkendall effect might be a general method to form various hollow nanostructures with controlled conditions. Furthermore, nanoscale Kirkendall effect can be applied to design multicomponent nanostructures, because combining multiple components within individual nanoparticles is an efficient way to control chemical and physical properties to produce advanced biomedicine and efficient catalysts.

2.2 Local Surface Plasmon Resonance (LSPR) of Au Nanocrystals Differing by Their Nanocrystallinity

The optical properties of noble metal nanoparticles related to the dipolar localized surface plasmon resonance (LSPR) band are attributed to the collective oscillation of the conduction electrons of the nanoparticle [26–28]. This is due to their resonant coupling with the electric field of the incident light [29]. Experimentally and from various numerical models, the frequency and strength of these oscillations strongly depend on the nanoparticle size [30] and shape [30–34] and their local dielectric environment [29, 30].

Face-centered cubic (fcc) noble metal nanocrystals exhibit different shapes directly related to their nanocrystallinity. As mentioned in Chapter 1, the single-domain nanocrystals are commonly cuboctahedra or truncated octahedra, while their multiply-twinned counterparts, usually called MTPs, are decahedra and icosahedra [35]. Discrete dipole approximation (DDA) method [36, 37] permits to investigate the LSPR by using the for spherical particles.

Here, we concentrate on Au nanoparticles. Let us first consider the use of DDA method to calculate the 5-nm LSPR spectra differing by their shape, i.e. by their nanocrystallinities. To account for the presence of dodecanethiol molecules at the surface of the nanocrystals, the latter are considered in the simulations to be immersed in a host medium with the same refractive index compared with the coating agent, that is, $n = 1.46$. All the calculated spectra are plotted in Figure 2.11 along with the UV/vis extinction spectrum measured from the colloidal solution [26]. Except for the case of the decahedron, all the others calculated spectra exhibit nearly similar profiles where the intense band centered around 530 nm corresponding to the LSPR of Au nanocrystals is partially superimposed on its high-energy side to the onset of the interband transitions. Furthermore, these spectra are found to be qualitatively in fairly good agreement with the experimental one, in which the LSPR is peaked at around 520 nm. At variance, the spectrum calculated for decahedral

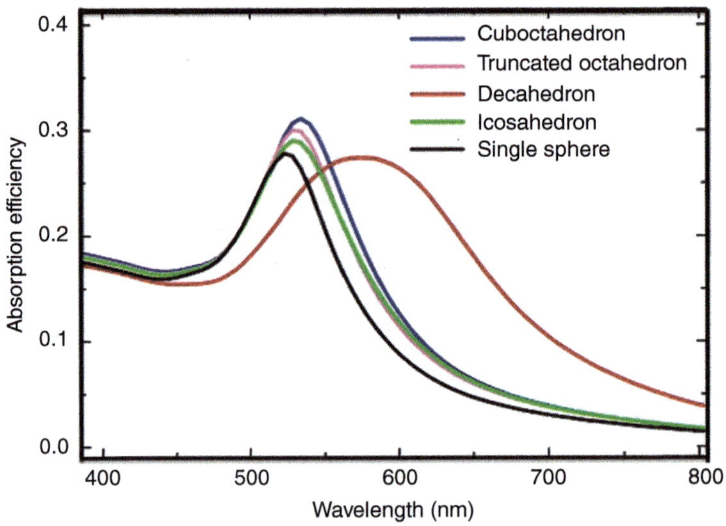

Figure 2.11 Calculated LSPR of Au nanoparticles NPs with cuboctahedral (blue), truncated octahedral (pink), decahedral (red), and icosahedral (green) morphologies. The spectrum calculated for one single sphere (black) is plotted for comparison with the other shapes. The targets of different morphologies are equally sized by taking an effective radius of 2.5 nm for all of them. *Source:* Yang et al. [26]. Copyright 2011 American Physical Society.

nanocrystals shows the LSPR band to be much broader and redshifted with respect to those obtained for the other particle shapes. A study based on the spectroscopic measurements, imaging, and modeling of individual Au decahedra pointed out their specific optical properties [38]. By comparing the various calculated spectra plotted in Figure 2.11, it is worth noting that the cuboctahedron, the truncated octahedron, and the icosahedron almost behave like spheres. The different optical behavior observed for the decahedron is due to its (oblate) shape anisotropy. With regard to this specificity, it should also be noted that the extinction spectrum calculated for the decahedral nanoparticles is orientationally averaged to account for their random orientation in the sample.

Similar behaviors are obtained for Ag and Cu nanoparticles [26].

Experimentally, whatever the nanocrystallinity, the LSPR of 11.6-nm Au nanocrystals coated with dodecanethiol is characterized by an absorption band centered at 527 nm differing by their nanocrystallinity and dispersed in toluene. The LSPR band of polycrystalline nanocrystals is characterized by a weaker intensity and larger linewidth than single-domain nanocrystals having the same average size (Figure 2.12a) [27].

On increasing the nanocrystal size ranging from 6 to 13 nm, the LSPR band remains quite unchanged with a slight shift of the maximum at large nanocrystal size. In the first approximation, Figure 2.13 shows that the LSPR position does not depend on the nanocrystal.

Hence, the optical properties of Au nanocrystals larger than 5 nm can be well represented with a classical approach [39], in which the LSPR linewidth is related to the electronic scattering rate, Γ, given by

$$\Gamma = \Gamma_0 + \frac{A v_f}{R} \tag{2.1}$$

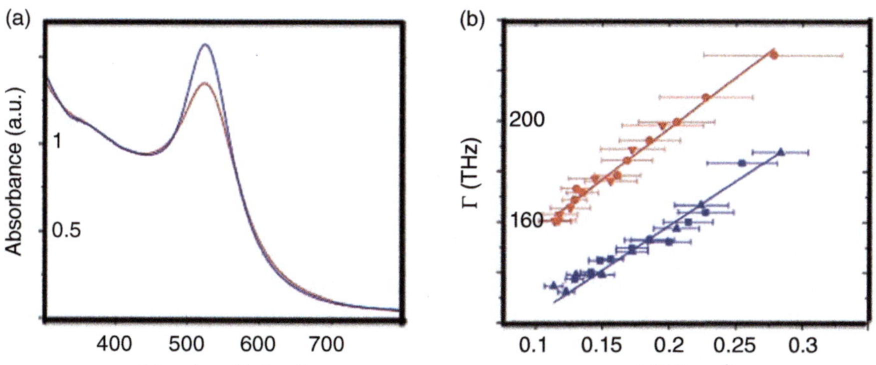

Figure 2.12 (a) Typical localized surface plasmon resonance (LSPR) for an ensemble of single-domain (blue) and polycrystalline (red) nanocrystals with 6.6 nm as diameter and (b) LSPR linewidth versus the inverse of the radius for single-domain (blue squares) and polycrystalline (red circles) nanocrystals. Each symbol corresponds to a different synthesis run. *Source:* Goubet et al. [28]. Copyright Royal Chemical Society.

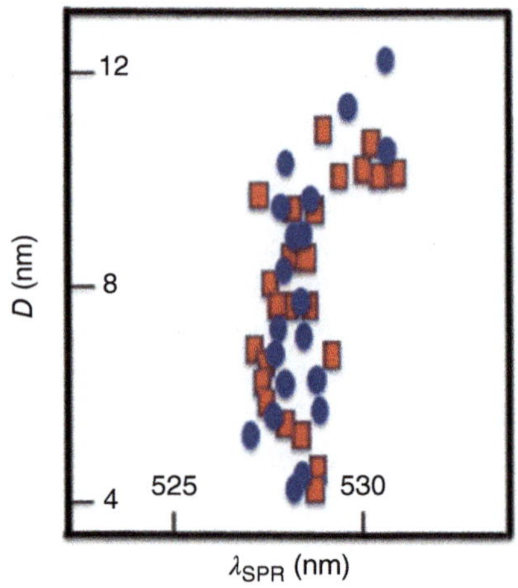

Figure 2.13 Dependence of the LSPR position of nanocrystal differing by their diameter (*D*) and the nanocrystallinity with single-domain (blue) and polycrystalline (red) nanoparticles. *Source:* Based on Goubet et al. [28]. Copyright Royal Chemical Society.

where Γ_0, A, v_f, and R are the intrinsic electron scattering rate, a constant dependent on the dielectric constant of the medium, the Fermi velocity of electrons, and the nanoparticle radius, respectively. By assuming that nanocrystals are spherical, the intrinsic electron damping constant is represented by the different electron collision events that are defined by the Matthiessen law [40] and the absorption due to the interband transition [41]. Indeed, since the LSPR of Au nanocrystals partly overlaps with the interband transition due to its short-wavelength threshold (≈540 nm), one has to take it into account in the following description:

$$\Gamma_0 = \frac{1}{\tau_{e-ph}} + \frac{1}{\tau_{e-e}} + \frac{1}{\tau_{defect}} + \frac{\omega_{LSPR}^3}{\omega_p^2} \varepsilon_2^{ib} \tag{2.2}$$

where $\tau_{e\text{-}ph}$, $\tau_{e\text{-}e}$, and τ_{defect} are the damping times due to the scattering of the electrons on phonons, electrons, and lattice defects (e.g. point defects, dislocations, and grain boundaries), respectively. The last term describes the absorption due to the interband transition where ω_{LSPR} is the LSPR frequency, ω_p is the plasma frequency of the metal, and ε_2^{ib} is the imaginary part of the interband contribution of bound electrons to the dielectric function [41]. Due to the size dependences of $\tau_{e\text{-}ph}$ and $\tau_{e\text{-}e}$, note that due to the size dependences of $\tau_{e\text{-}ph}$ and $\tau_{e\text{-}e}$, Γ_0 differs from the bulk value [41, 42]. However, since the $\tau_{e\text{-}ph}$ and $\tau_{e\text{-}e}$ size dependences mostly concern nanocrystals with diameter of below 5 nm, it can be neglected. The experimental LSPR absorption band is fitted with the superposition of a Lorentzian function and a linear decay that takes into account the absorption due to interband transitions. From such analysis, we extract the LSPR linewidth (Γ) evolving linearly with the inverse of the nanoparticle radius for single domain (blue) and polycrystalline (red) (Figure 2.12b). From such linear fit, the values of the intrinsic scattering rate Γ_0 and of the constant A are deduced. They are $(11.3\,\text{fs})^{-1}$, 0.19, and $(8.6\,\text{fs})^{-1}$, 0.22, for single-domain and polycrystalline nanocrystals, respectively. The difference in the linewidths cannot be attributed to the chemical interface damping, since the nanocrystals are synthesized in the same chemical environment. The similarity of the constant A for both nanocrystallinities confirms that the chemical environment and consequently the dielectric constant of the medium remain unchanged, suggesting comparable electron-surface scattering processes. Since ω_{LSPR} does not depend on the nanocrystal size (Figure 2.13), the absorption due to the interband transition is similar for single-domain and polycrystalline nanocrystals. Then, the observed difference in Γ_0 ($(11.3\,fs)^{-1}$ and $(8.6\,fs)^{-1}$) between single-domain and polycrystalline nanocrystals is attributed to crystal defects increasing the scattering rate, as shown in Eq. (2.2). In particular, it has been shown that the plasmon linewidth of polycrystalline Au nanocrystals is strongly affected by the grain size [43]. We conclude that the collection of polycrystalline nanocrystals has intrinsic dephasing rates of around 20% higher than single-domain nanocrystals due to the electron scattering on their crystal defects. This result reveals the importance and often neglected role of twins in determining the LSPR linewidth. Indeed, the main parameter that is used to calculate the nanoparticle optical response obtained with single-particle spectroscopy is the surface scattering parameter [39, 44]. However, the values of this parameter are often different from one nanoparticle to the other while the crystal defects can also contribute to the LSPR broadening and have to be taken into account.

2.3 Acoustic Vibrational Modes

2.3.1 Breathing Mode

The coherent oscillations for spherical Au nanocrystals differing by size and nanocrystallinity are accurately described as those of an elastically isotropic homogeneous sphere [28, 45, 46]. Due to the isotropy of the thermal excitation, the time-domain experiments mainly probe their fundamental radial breathing mode (electron–phonon scattering) [47, 48]. Following this description, the computed frequencies of the radial breathing mode are expected to be almost identical for single-domain and polycrystalline samples [49], but the experimental results in this direction are still controversial. Strong variations were found between single-domain and polycrystalline structures in the case of silver [50] and hollow Au nanospheres [51]. To clarify this issue, we performed fluence-dependent measurement on nanocrystals having similar size and differing in terms of their nanocrystallinity.

Let us consider first Au nanocrystals, in the size range of 5–12 nm, dissolved in chloroform and differing by their nanocrystallinity. The pump–probe experiments induce a differential transmission ($\Delta T/T$) signal, measured over the entire visible spectral region using a white-light supercontinuum probe pulse entirely covering the LSPR of the different nanospheres [48]. With Au nanocrystals, the 400-nm pump pulse mainly promotes electrons from the 5d to the 6sp Au bands and creates a non-thermal electronic distribution that recovers to the initial equilibrium conditions through a sequence of processes such as electron–electron and electron–phonon

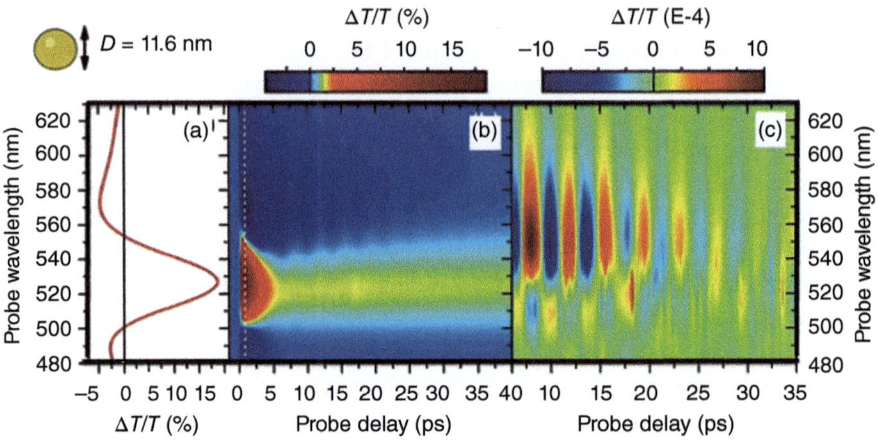

Figure 2.14 Broadband femtosecond pump–probe spectroscopy for a single-crystal sphere of 11.6-nm diameter excited at 400 nm. (a) Spectrum at 800 fs pump–probe delay showing the typical "second derivative" shape due to broadening of the plasmonic resonance. (b) Two-dimensional **Δ***T*/*T* map showing the signal as a function of pump–probe delay and probe wavelength. The vertical cut (dotted line) at 800 fs shows the position of the transient spectrum depicted in panel (a). (c) Map of the coherent oscillations on the two sides of the plasmonic resonance obtained by subtracting the slowly varying exponential decays to the map in panel (b). *Source:* Based on Goubet et al. [28]. Copyright Royal Chemical Society.

scattering and heat dissipation to the surroundings [52]. Figure 2.14 shows typical results for 11.6-nm-diameter single-domain Au nanocrystals. The 2D $\Delta T/T$ map as a function of the probe wavelength and pump–probe delay is presented in panel (b), while panel (a) depicts a vertical cut, namely, a $\Delta T/T$ spectrum for a fixed pump–probe delay of 800 fs. Panel (b) also highlights the typical shape of Au nanocrystals transient spectra after photoexcitation [53]. The increase in the electronic temperature leads to a broadening of the LSPR due to an increase of the imaginary part of the Au dielectric constant. This broadening is responsible for the "second derivative" shape observed in Figure 2.14a and peaked at approximately 525 nm, where the LSPR for small spherical Au nanocrystals (whose extinction coefficient is dominated by absorption rather than scattering) occurs [54]. The electron–phonon relaxation process is also responsible for an energy flow from the electrons to the lattice, and, as a consequence, it drives an impulse versus heating and expansion of the nanocrystals. Because the time scale for heating ($\approx$1 ps) is typically faster than the period of the fundamental radial breathing mode in small nanoparticles, this expansion impulsively launches coherent oscillations that are coupled to the optical response of the sample through a periodical shift in the position of the LSPR peak [52]. In the case of Au spherical nanocrystals, they arise from periodical variations of the volume and, consequently, of the carrier density that directly affects the plasma frequency.

Figure 2.14 highlights the measured coherent oscillations in the 11.6-nm-diameter single-domain Au nanocrystals dispersed in chloroform. The recorded $\Delta T/T$ signal (shown in panel (b) is plotted in panel (c)), for delays longer than 6 ps (to remove the initial fast thermalization process). Subtracting the slowly varying relaxation dynamics for each probe wavelength isolates the oscillatory component of the signal. It is worth noting that the amplitude of the oscillation is larger at longer wavelengths (540–570 nm) with respect to the plasmonic peak (≈525 nm) and a π phase shift occurs for wavelengths at different sides of the LSPR [48], as expected from oscillations arising from a periodic shift of an optical resonance.

Let us first investigate the electron–phonon scattering dynamics occurring during the first ≈5 ps after excitation and then focus on the analysis of the coherent oscillations on the longer time scale. It is known that the time constant for the electron–phonon scattering increases linearly with the excitation fluence. This is explained in the framework of the two-temperature model given that a higher pump intensity creates a higher initial electronic temperature that is directly reflected in the relaxation time constant [54].

Figure 2.12a shows the measured (data points) $\Delta T/T$ dynamics at 525 nm (i.e. the peak of the bleaching signal) at three different pump fluences (normalized to the low-fluence signal) for the single-domain nanocrystals with 11.6-nm diameter. The fits (solid lines) are performed considering the instrumental response function of our system and the contribution of three exponential processes, representing the electron–electron, electron–phonon, and phonon–phonon scattering processes. The electron–electron and phonon–phonon time constants are fixed, while the electron–phonon decay is the only free parameter. Figure 2.15b shows the extracted electron–phonon time constant $\tau_{\text{e-ph}}$ as a function of the pump fluence for single-domain (blue squares) and polycrystalline (red circles) nanocrystals. The two samples are diluted so as to have the same steady-state absorption at 400 nm, to guarantee that the absorbed energy is the same at any given pump fluence. Interestingly, we observe that the nanocrystallinity does not affect the electron–phonon scattering dynamics, as the two datasets are overlapping. By extending the linear fit of the data (black solid line in Figure 2.15b) to the intercept with the vertical axis, we can extrapolate an electron–phonon time constant at zero-power excitation of approximately 625 fs. This value, in good agreement with other measurements on Au nanoparticles [54], corresponds to the value of Au bulk phase [55]. Very similar results are obtained for other couples of samples consisting of single-domain and polycrystalline nanocrystals with a diameter from 5 to 12 nm.

In Figure 2.16a,b, we show the radial breathing mode oscillations detected at 550-nm probe wavelength for the 11.6-nm-diameter spherical nanocrystals with different crystallinity (data points). Similar results are obtained by extending these experiments to other Au nanocrystals differing by the average diameter, D. The amplitude of the oscillations rapidly decays in time, and after six to seven periods, it vanishes below the sensitivity of our measurements. The decay can be attributed mainly to two mechanisms: (i) intrinsic damping of the radial breathing mode caused by energy dissipation, giving rise to an homogeneous broadening and to an exponential decay in time $e^{-t/\tau}$ and (ii) dephasing of the oscillations due to the inhomogeneous broadening related to the size distribution of the nanocrystal, which translates into a superposition of oscillators with different period, giving rise to a e^{-t^2/τ^2} decay profile [48]. Both methods provide very good fits with data shown in

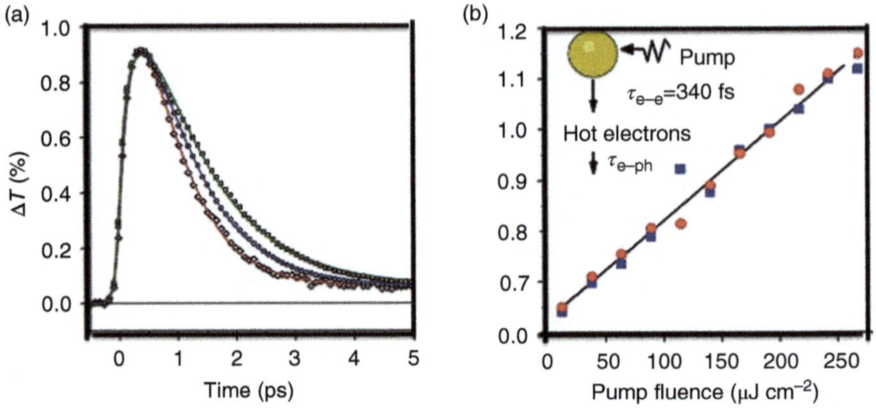

Figure 2.15 Effect of the nanocrystallinity on the electron–phonon scattering. (a) $\Delta T/T$ dynamics (data points) for single-domain NPs with 10.6 nm diameter excited at 400 nm and corresponding fitting (solid lines) obtained considering 180 fs temporal resolution, a fixed electron–electron scattering time constant of 340 fs, and an electron–phonon scattering time constant as a free parameter. Pump fluences: 12.7 μJ cm^{-2} (diamonds and red curve), 140 μJ cm^{-2} (circles and blue curve, scaled down by 8.7 times to normalize the trace to the low-fluence data), and 267 μJ cm^{-2} (squares and green curve, scaled down by 15.9 times). The electron–phonon scattering time constant is kept as a free parameter. (b) Extracted electron–phonon scattering time constant as a function of pump fluence for both single-domain (blue squares) and polycrystalline (red circles) nanocrystals with 10.6-nm diameter. Solid black line is a linear fit to the data (intercept at $\tau_{0e\text{-}ph}$ = 625 fs; slope 1.96 fs cm^2). *Source:* Based on Goubet et al. [28]. Copyright Royal Chemical Society.

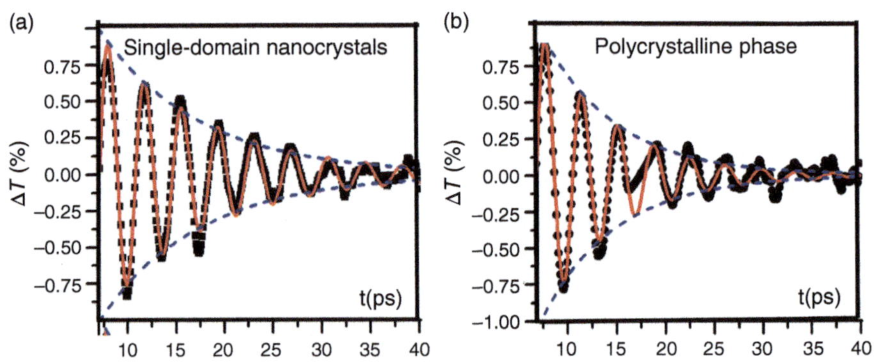

Figure 2.16 Normalized oscillations detected at 550-nm probe wavelength (data points) for single-domain (a) and polycrystalline nanocrystals (b) with 11.6-nm diameter together with a fit following the model described in the text (red solid curve). Blue dashed lines indicate the envelope of the oscillation amplitude, with $e^{-t/\tau}$ exponential decay. *Source:* Based on Goubet et al. [28]. Copyright Royal Chemical Society.

Figure 2.16 with minor differences in the residuals. A pure inhomogeneous broadening related to the size distribution of the nanocrystals is excluded for the following reasons: if the latter is the only cause of the damping of radial breathing mode, the standard deviation, σ_D, of the nanocrystals diameter distribution extracted from the pump–probe data would be larger than that measured from TEM images. In fact, knowing the average diameter $\bar{D}$ of each sample, from the fits of the pump–probe data, the oscillation period T and the time constant τ of the Gaussian decay are extracted so that the standard deviation σ_D can be evaluated from [48]

$$\sigma_D = \frac{\bar{D}T}{\sqrt{2}\grave{A}\tau}$$

By plotting the standard deviation and the size distribution of Au nanocrystals differing by their average diameters, the values deduced from pump–probe measurements and those from TEM images are in the same order of magnitude. However, we have to note that those obtained from the oscillation period are overestimated compared with the TEM ones. This allows us to conclude that the major contribution to the damping of the radial breathing mode oscillation is the intrinsic damping caused by energy dissipation, thanks to the extremely well-defined nanocrystals. In the following, we therefore consider the results obtained from the fits using an exponential decay.

The extracted periods of oscillation period and decay time constants τ are reported in Figure 2.17 as a function of nanocrystal diameter for single-domain (blue squares) and polycrystalline (red circles) nanocrystals. The period of oscillation T increases linearly with the sphere size for both sets of samples [52]. In accordance with

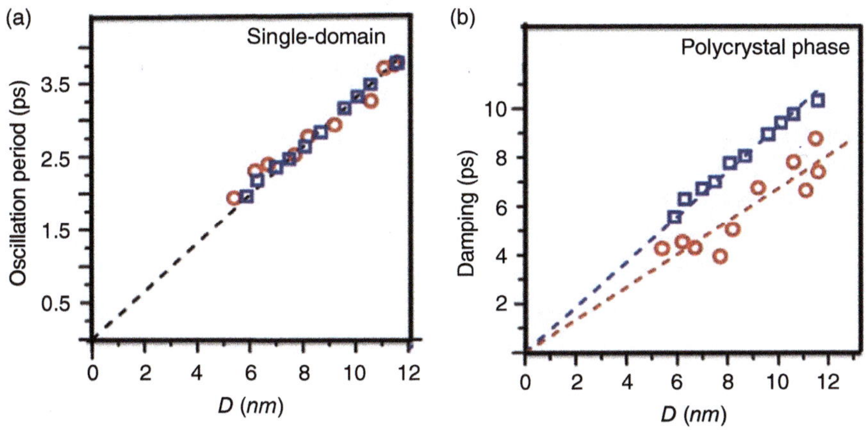

Figure 2.17 Period T (a) and damping τ (b) of the fundamental radial breathing mode as a function of the nanocrystal diameter D for both single-domain (blue squares) and polycrystalline (red circles) samples, together with linear fits (with fixed zero intercept). For the period (panel (a)), the entire dataset of single-domain and polycrystalline NPs was considered, giving T = (330 fs/nm)D. For the damping (panel (b)), the two datasets are aligned in lines with different slopes: τ_S = (930 fs/nm)D for the single-domain NPs (blue squares) and τ_P = (670 fs/nm)D for the polycrystalline ones (red circles). *Source:* Based on Goubet et al. [28]. Copyright Royal Chemical Society.

theoretical predictions [49], Figure 2.17 shows that the radial breathing mode period lies on the same straight line and consequently is not affected by the nanocrystallinity. The linear fit of the entire dataset (with intercept set to zero) gives a slope of $\frac{T}{D} = 330\,\text{fs}/\text{nm}$, which is in good agreement with the inverse of the longitudinal speed of sound in gold $\frac{1}{v_s} = 310\,\text{fs}/\text{nm}$. This suggests that the nanoscale mechanical properties of isolated Au nanocrystals, in particular the Young modulus, which determines the speed of sound, are not significantly affected by nanocrystallinity. From the analysis of Figure 2.17b, we can conclude that the two datasets lie onto straight lines with different slope: $\frac{\tau_S}{D} = 930\,\text{fs}/\text{nm}$ for the single-domain NPs (blue squares) and $\frac{\tau_P}{D} = 670\,\text{fs}/\text{nm}$ for the polycrystalline ones (red circles). The quality factors associated with the two crystallinities are then $Q_S = \pi\frac{\tau_S}{T} \approx 9$ for the single-domain and $Q_P = \pi\frac{\tau_P}{T} \approx 6$ for the polycrystalline nanocrystals.

Let us consider 7.3-nm *amorphous* and hcp Co nanoparticles self-assembled in 2D superlattices on highly ordered pyrolytic graphite (HOPG) substrate. As described above, by extracting the background signal, the resulting maps, in which the oscillatory component of the residual signal is solely represented, are shown in Figure 2.18a,b for the poly- and single-crystalline Co nanocrystals, respectively. Looking at these maps now allows for the clear observation of the oscillating pattern for both samples (single and polycrystal nanocrystals) at each probe wavelength. The oscillations are visible for about five periods, and their damping time is consistent with the low nanocrystal size distribution (10%). The presence of the two maps of similar parallel bands with regular thicknesses, which successively appear in blue and red/yellow colors, indicates that the periodicity of the oscillations is not significantly affected either by the probe wavelength or by a change in the particle nanocrystallinity. Figure 2.18c,d shows the time traces of the oscillating component of the $\Delta T/T$ signal measured at three selected probe wavelengths as extracted from the 2D maps displayed in Figure 2.18a,b. All the traces have been fitted (see gray thick lines in Figure 2.18c,d) with a damped sinusoidal function. By averaging the results obtained at several probe wavelengths and for a number different positions on the sample, we obtain $\nu^{[\text{poly}]} = 606 \pm 8\,\text{GHz}$ and $\nu^{[\text{single}]} = 566 \pm 10\,\text{GHz}$ for the poly- and single-crystalline samples, respectively.

The experimental values of the frequencies obtained with 7.3-nm Co and 5-nm Au nanocrystals both deposited on HOPG substrate are compared with the corresponding calculated value by using resonant ultrasound (RUS) method. This approach advantageously allows accounting for the elastic anisotropy of the material of which the simulated objects are composed [56, 57]. The longitudinal and transverse sound velocities are used to simulate such “virtual” directionally averaged isotropic materials. Table 2.2 shows the calculated frequency of fundamental radial breathing mode of 5.4-nm Au nanocrystals dispersed in chloroform and 7.3-nm Co nanocrystals self-assembled in 2D superlattices.

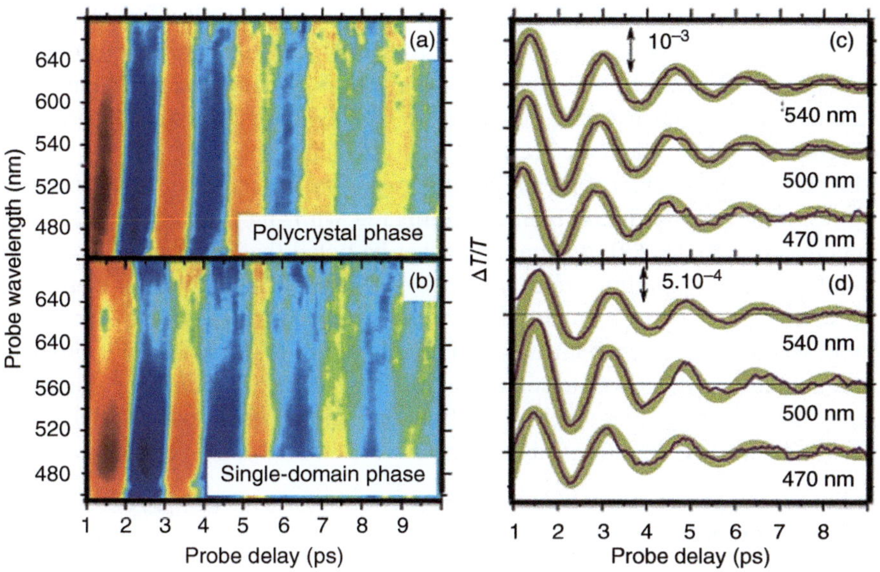

Figure 2.18 2D map showing the variation of the **Δ***T*/*T* signal amplitude as a function of both probe delay and wavelength for (a) polycrystalline and (b) single-crystalline Co nanocrystals after subtracting the slowly varying background signal. Oscillating component of the **Δ***T*/*T* signal extracted from the 2D map of (c) polycrystalline and (d) single-crystalline Co nanocrystals at three different selected probe wavelengths, i.e. 470, 500, and 540 nm. Green thick curves are the best fist to the dynamics curves using the function defined by equation given below: $f(t) = A \cdot \exp\left(-\frac{t}{\tau}\right) \cdot \sin(2\pi \nu t + \varphi)$ where A, ϕ, τ, and ν are the initial amplitude, the phase, the damping time constant, and the oscillation frequency, respectively. The traces plotted in (c) and (d) are vertically shifted for clarity. *Source:* Based on Goubet et al. [27]. Copyright 2013 American Chemical Society.

Table 2.2 Oscillation period of 5.4-nm Au and 7.3-nm Co nanocrystals, in (GHz), deduced by averaging the data measured at several probe wavelengths and different positions and the corresponding calculated value (*) by using the resonant ultrasound (RUS) method.

Material	Au nanocrystals 4 nm	Co nanocrystals 7.3 nm
$\upsilon_{\text{single domain}}$	506 ± 3	566 ± 10
$\upsilon_{\text{single domain}}{}^{*}$	599	692
$\upsilon_{\text{polycrystal}}$	517 ± 3	606 ± 8
$\upsilon_{\text{polycrystal}}{}^{*}$	554	682
$\delta\nu$	2.2%	2.3%
$\delta\nu^{*}$	1.5%	1.5%

Source: Based on Refs. [57, 45].

The mismatch between the measured and calculated frequencies are similar for both Au and Co nanocrystals differing by their nanocrystallinities and cannot be attributed to chemical reaction or the self-assemblies of nanocrystals. It could due to the use of bulk elastic constants in this model, as these constants are probably different from that of the material confined at the nanoscale. Note that the change in the physical parameters of metal (Young's modulus, dielectric constant, stiffness, etc.) resulting from such confinement compared with the bulk ones remains an open question [50, 58, 59]. It should however be noted that the slight decrease by around 1.5% of the breathing mode frequency predicted using RUS calculation from poly- to single-crystalline nanocrystals (Au and Co) well agrees with the experimentally observed decrease, which is found to be around 2.3%. This result provides therefore additional support to interpret the slight change in breathing mode frequency between poly- and single-crystalline nanocrystals due to the influence of crystallinity on their vibrational dynamics.

2.3.2 Quadrupolar Mode

Here, we use Au nanocrystals produced from inorganic chemical reaction [60, 61]. These nanocrystals differ by their average diameters (4.3, 5.1, and 4.3 nm). They are characterized by low size distribution (6%, 6%, and 7%, respectively) and a mixture of single-domain and polycrystalline nanoparticles [62, 63]. Furthermore, we used 5.8-nm Au polycrystalline Au nanocrystals with 13% as size distribution [64]. Nanocrystals deposited on HOPG, with low size distribution, self-assemble in long distance, whereas those with larger one (8 nm) is locally assembled.

The low-frequency Raman scattering (LFRS) spectrum of 5.1-nm Au nanocrystals (mixture of single domain and polycrystals) shown on Figure 2.19a exhibits a profile with two sharp bands of comparable intensities showing up at 156 and 229 GHz. The higher-frequency band appears to be asymmetric and broader than the former. For convenience, these two bands will be indicated in the following as "B1" and "B2," respectively. The observation of such a double band is surprising because it is in contrast with the results commonly obtained by LFRS from self-organized Au nanocrystals, whereas with 5.8-nm polycrystalline Au nanocrystals, the LFRS spectrum shows mainly one peak as usually observed (Figure 2.19b).

Figure 2.19c shows that both B1 and B2 bands inertly shift toward higher frequencies with nanocrystal diameter in the size domain from 5.3 to 4.3 nm. From other studies, we know that the Stokes/anti-Stokes low-frequency Raman spectra strongly depend on nanocrystal size. Consequently, such size dependence of the frequencies measured at the maximum of both B1 and B2 bands permits to assign these two bands to Au nanocrystal vibrations. To avoid any ambiguous identification of the vibrational modes revealed by double bands of the LFRS spectrum, several mechanisms can be considered [65–67]:

(i) *A size effect could involve a bimodal size distribution.* This can be excluded because of single Gaussian profile with a low size distribution of Au nanocrystals (≤7%). Besides, such double bands B1 and B2 are not related to well-organized assemblies of in 3D superlattices. This is supported by the fact that

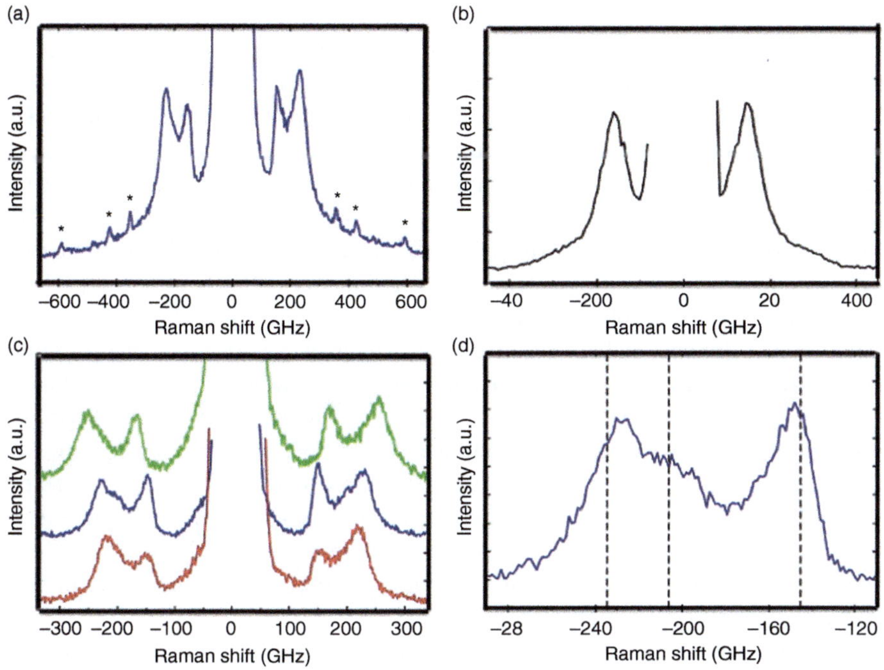

Figure 2.19 Stokes/anti-Stokes LFRS spectra of Au nanocrystals produced with 523 nm as the excitation wavelength. (a) LFRS of sample 5.1 nm Au nanocrystals where peaks at 353, 424, and 590 GHz (marked with asterisks) are likely to be due to light scattering by molecules from air (oration of oxygen and nitrogen) that are adsorbed on the sample. (b) LFRS spectrum of 5.8-nm multiply-twinned Au nanocrystals produced by Brust method [64]. (c) Comparison of Stokes/anti-Stokes LFRS spectra of samples 4.3 nm (green curve), 5.1 nm (blue curve), and 5.3 nm (red curve) Au nanocrystals illustrating the dependence of the LFRS signal on the size of the Au nanocrystals. (d) Anti-Stokes LFRS spectrum of sample 4.1-nm Au nanocrystals clearly showing the structured profile of the double band. Dashed lines are vertically plotted to indicate the vibrational frequencies of 5.1 nm Au nanocrystals, as calculated for the T_{2g}, the $l = 2$ quadrupolar, and the E_g modes (in decreasing order of Raman shifts). Spectra in (c) have been vertically shifted and amplitudes independently scaled for clarity. *Source:* Portalès et al. [60].

in both disorder and 3D superlattices of Au nanocrystals, the position and width of the Raman bands remain unchanged. Therefore, any possible effects of vibrational coherence as those being theoretically predicted [68] and already observed [69] in cubic face-centered 3D superlattices of Ag nanocrystals are ruled out.

(ii) *A shape anisotropy of nanocrystals* is excluded if we consider the nanocrystals are quite spherical without any shape anisotropy.

(iii) *The detection of the nonspherically symmetric ($l = 2$) quadrupolar mode of Au nanoparticles* [70].

Such effect is significant but should be observed for both single-domain and polycrystalline nanocrystals. Thus, we rule out such a phenomenon to be the cause of the double band.

At this point, the crystallinity of the Au nanocrystals appears likely to be at the origin of the double band in the LFRS spectra. The theoretical arguments and computational calculations of vibrational frequencies of Au spheres strongly argue in favor of this assumption. Lamb introduced the standard theory used to predict the frequencies of the vibration of nanoparticles more than one century ago [46]. This approach is based on the continuum approximation and assumes that the nanoparticles have a spherical shape with an elasticity isotropy. To go beyond this last limitation, RUS approach is used [56]. The results are double-checked both with the Lamb's model for isotropic system and with a finite element mesh sequence (FEMS) approach [71]. To account for the twinning of the nanocrystals, we use an isotropic approximation of Au with three-dimensional averaged sound velocities, i.e. $\langle v_L \rangle = 3330\,\text{m}\,\text{s}^{-1}$ and $\langle v_T \rangle = 1250\,\text{m}\,\text{s}^{-1}$. Individual polycrystal nanocrystals oscillate as isotropic spheres [69]. Therefore, the isotropic approximation provides a good description of this ensemble. In the following, the twinned nanocrystals are referred as fictive isotropic spheres.

It is expected on the basis of group theoretical considerations [72] and confirmed by RUS calculations that the fivefold degenerate spheroidal $l = 2$ mode of an elastically isotropic sphere is broken into two different levels by lowering the elastic symmetry to that of a cubic crystal. The corresponding representations are E_g (degeneracy two) and T_{2g} (degeneracy three).

Interestingly, using RUS, this lifting of degeneracy can be progressively followed by varying the elastic tensor C_{ij} of Au spheres expressed here as a linear combination of the elastic tensors of a "fictive" fully isotropic Au sphere, $C_{ij}^{[I]}$, and a perfectly anisotropic one, $C_{ij}^{[A]}$:

$$C_{ij} = (1 - x)C_{ij}^{[I]} + xC_{ij}^{[A]} \tag{2.3}$$

where x is real, ranging between 0 and 1. The splitting of the quadrupolar mode is illustrated in Figure 2.20, which shows the dependence of the vibrational frequencies of an Au sphere versus the variation of the Au elastic anisotropy.

It is worth noting that, even if the HRTEM images clearly reveal that most Au nanocrystals are faceted, the same splitting is calculated (with deviations of the ratio of the T_{2g} and E_g frequencies lower than 4%) for nanocrystals having spherical or polyhedral shape (cuboctahedra and truncated cuboctahedra), at variance with a previous report [72]. A schematic representation of the motions of a vibrating sphere involved in few selected modes is given in Figure 2.21. Note that for these modes, the displacements are almost identical for isotropic and anisotropic gold spheres, yet the frequencies are significantly different (Figure 2.19).

The vibrational frequencies of Au nanocrystals measured by LFRS from three different samples are listed in Table 2.3 and compared with those calculated using RUS for both isotropic and anisotropic Au spheres with a size corresponding for each sample to the mean TEM diameter (D_{TEM}). From this comparison, the frequencies measured at the maximum of the two bands agree with those calculated for the E_g and T_{2g} modes of a gold anisotropic sphere. A close inspection of the LFRS spectrum of 5.1-nm Au nanocrystals plotted in Figure 2.19d reveals the profile of the double band to be structured, with a shoulder around 200 GHz making the band B2 to

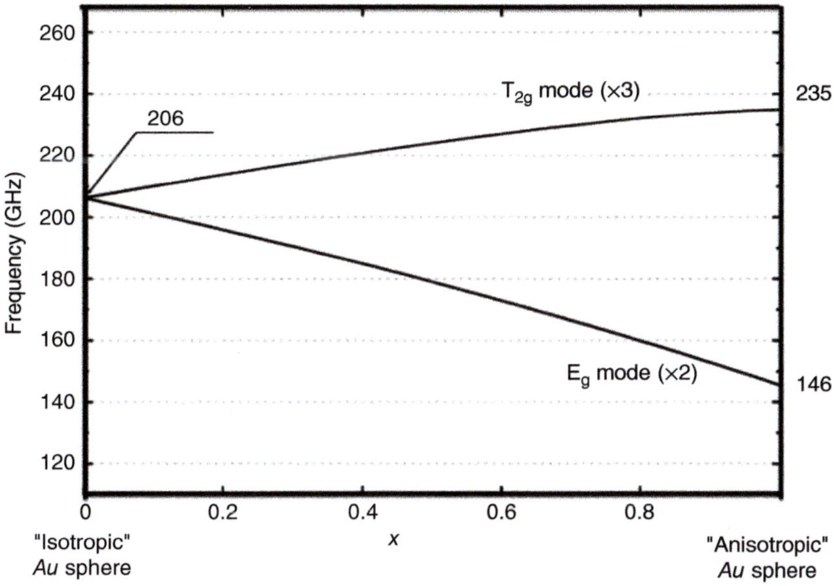

Figure 2.20 Splitting of the quadrupolar vibrational mode due to elastic anisotropy in a gold sphere of 5.1 nm in diameter. The plotted curves show the dependence of the $l = 2$ quadrupolar mode on the gold elastic anisotropy when the elastic tensor of gold is linearly varied according to Eq. (2.1). From the case of a directionally averaged "isotropic" metal ($x = 0$) to the one of a truly "anisotropic" medium ($x = 1$), the fivefold degenerate $l = 2$ mode of the gold sphere splits into two different levels corresponding to the twofold E_g and threefold T_{2g} degenerate modes. *Source:* Portalès et al. [60].

Figure 2.21 Schematic representation of the motions of a gold sphere for different vibrational modes. From top to bottom, the two E_g modes and one of the three T_{2g} modes are shown. Other T_{2g} modes are obtained by cyclic permutations of the three axes. In these pictures, arrows of different colors are used to represent either compression (red arrows) or expansion (blue arrows) motions of the sphere. Green symbols designate quasi-fixed points on the surface of the sphere, i.e. places where displacements are very limited in amplitude. *Source:* Portalès et al. [60].

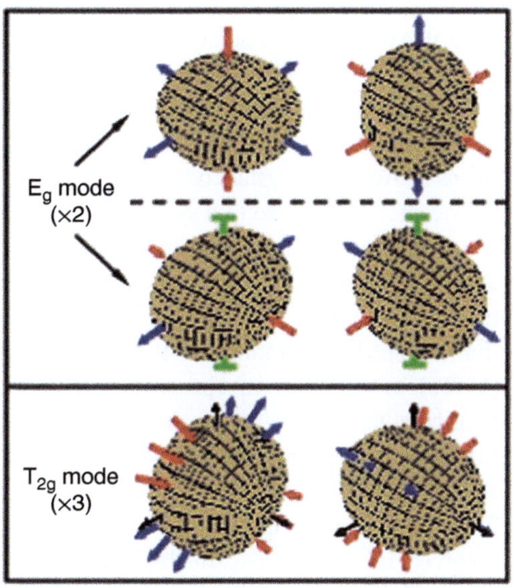

Table 2.3 Comparison of the vibrational frequencies of Au nanocrystals experimentally measured by LFRS with those calculated by using the RUS approach for the fivefold quadrupolar mode of a directionally averaged isotropic gold sphere and for the twofold E_g and threefold T_{2g} degenerate modes of an elastically anisotropic gold sphere.

	Isotropic gold sphere		Anisotropic gold sphere		Experiments
Size (mm)	Mode	Frequency GHz (degeneracy)	Mode	Frequency GHz (degeneracy)	Frequency GHz
5.3	$l=2$, $n=0$	197 (×5)	E_{3g}	139 (×2)	140
			T_{3g}	225 (×2)	211
5.1	$l=2$, $n=0$	206 (×5)	E_{3g}	146 (×2)	156
			T_{3g}	235 (×2)	229
4.3	$l=2$, $n=0$	243 (×5)	E_{3g}	172 (×2)	168
			T_{3g}	277 (×2)	253

D is the diameter determined by TEM. Iso-mode, Iso-Freq., Aniso-mode, and Aniso-Freq. are isotropic and anisotropic calculated modes and frequencies in GHZ, respectively. Exp-Freq. are the experimental frequencies. Samples are 5.5, 5.1, and 4.3 nm Au nanocrystals deposited on HOPG substrate.

appear asymmetric. This last frequency well matches that of the fivefold degenerate quadrupolar mode from a "fictive" isotropic Au sphere. Therefore, the several components of the LFRS (Figure 2.19d) from the nanocrystals assemblies can be interpreted through the following assignments: The band B1 corresponds to the scattering by the E_g mode of single-domain gold NCs. The band B2 and its shoulder on the low-frequency side are ascribed to the respective contributions of single- and polycrystalline nanocrystals to the LFRS signal. This assumption makes sense since, even if Au nanocrystals are mainly single domain in the samples from which the double band is observed, some other polycrystalline NCs also coexist.

To confirm the crucial role of the nanocrystal crystallinity, and equivalently, of the elastic anisotropy, let us consider 5.1-nm Ag nanocrystals produced by similar synthesis as Au nanocrystals used in this present study (Section 1.3.2.1). The major difference between these two syntheses is related to nanocrystallinity: Au nanocrystals are a mixture of single-domain and polycrystal nanoparticles, whereas Ag nanocrystals are MTP (see Figures 2.8 and 2.11 and Chapter 1). The corresponding LFRS spectrum shows only one high-intensity band (Figure 2.21), centered at 244 GHz. This unique band corresponds to the global contribution of multiply-twinned particles (MTPs) to the Raman spectrum, similar to what is observed from 5.8-nm Au nanocrystals (Figure 2.19b). In both cases, only one band is observed because of the inhomogeneous broadening that is inherent to the large variety of nanocrystalline structures in the population of the multiply-twinned nanocrystals. This clearly confirms that no splitting of the quadrupolar mode is observable when mainly MTPs are involved in the light-scattering mechanism. Moreover, when such a splitting is observed, it acts as a reliable indicator of the good nanocrystallinity of the nanocrystals from which the LFRS is coming, most of them being, therefore, single-domain nanocrystals.

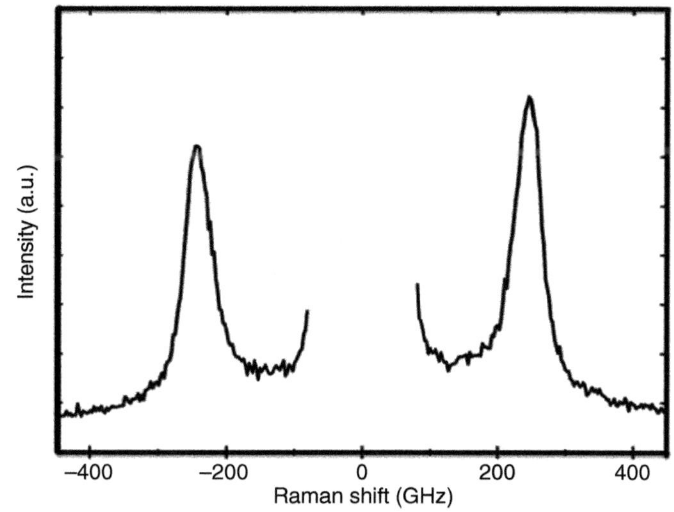

Figure 2.22 Stokes/anti-Stokes LFRS spectrum of 5.1-nm multiply-twinned Ag nanocrystals. The quality of the atomic ordering, through its associated elastic properties, sensitively affects. *Source:* Portalès et al. [60].

The vibrations of metal nanocrystals enable the clear observation by Raman scattering of a splitting of the intense quadrupolar mode when nanocrystals are mainly single domain, whereas no splitting is observed for their multiply-twinned counterparts.

Let us consider the Stokes/anti-Stokes LFRS spectra of the deposited 5.3-nm Au nanocrystals recorded at various excitation wavelengths (Figure 2.22). As in Figure 2.19, they exhibit two dominant bands. We know from experimental data (Figure 2.12) that the measured LSPR spectra of polycrystalline nanocrystals is characterized by a weaker intensity and larger linewidth than single-domain nanocrystals having the same average size. Taking into account the calculated absorption spectra (Figure 2.11), the slight changes between single domain and polycrystals could be attributed a slight proportion of decahedron nanocrystals produced experimentally.

This is consistent with the observation of similar features as observed above by using 532 nm as the excitation wavelength. Comparison between measured vibrational frequencies and those calculated (Table 2.4) makes it possible to assign the two contributions coming out from LFRS spectra (Figure 2.23): (i) the sharp band

Table 2.4 Comparison of the vibrational frequencies of 5.3-nm Au nanocrystals measured by LFRS with those calculated by using the RUS approach for the twofold E_g and threefold T_{2g} degenerate modes of an Au sphere with anisotropic elasticity, for the fundamental $l = 2$ quadrupolar mode of a "fictive" elastically isotropic one.

Elasticity	Shape	Mode	Calculated frequency (GHz)	Measured frequency (GHz)
Anisotropic	Sphere	E_{3g}	141 (×2)	137 ± 10
		T_{3g}	257 (×2)	210 ± 10
Isotropic	Sphere	$l = 2$	200	183 ± 10

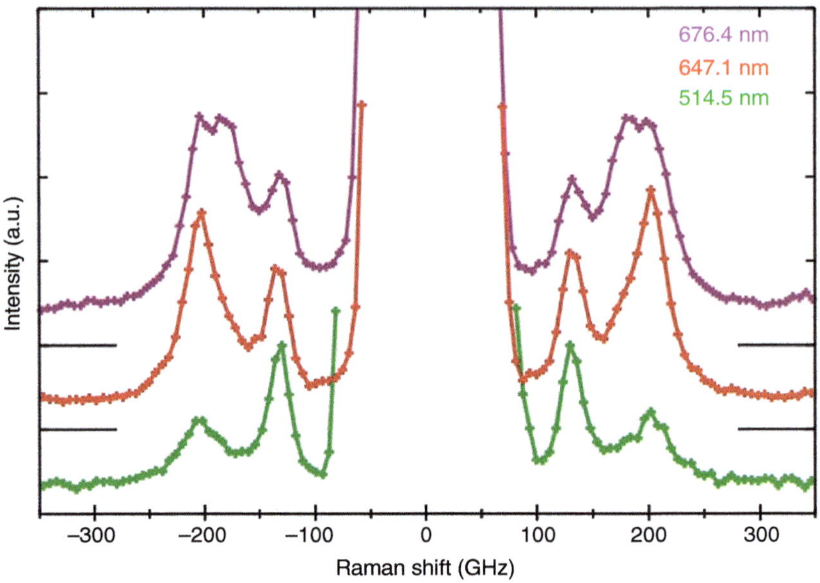

Figure 2.23 Stokes/anti-Stokes LFRS spectra of 5.3-nm Au nanocrystals. The wavelengths of the different laser lines used as the excitation source are successively 514.5, 647.1, and 676.4 nm, from the bottom to the top. Spectra have been vertically shifted and amplitudes normalized to the maximum of the lowest-frequency band for clarity. *Source:* Portalès et al. [61]. Copyright 2010 American Chemical Society.

peaked at 137 ± 10 GHz ascribed to the twofold degenerate E_g modes of monodomain Au nanocrystals and (ii) the second band, at higher frequency, which is broader than the former and also exhibits a structured profile. This is more pronounced in the spectrum recorded for an excitation wavelength of 676.4 nm. Actually, this band is the sum of two different contributions to the Raman signal: (α) the Raman scattering from the threefold degenerate T_{2g} modes of monodomain nanocrystals, corresponding to the peak centered at 210 ± 10 GHz, and (β) the light scattering due to the "$l = 2$" vibrations of MTPs and polycrystalline nanocrystals. From the calculated frequencies listed in Table 2.4, one can predict this last contribution to correspond to the shoulder that appears at around 180 GHz, on the low-frequency side of the T_{2g} band.

The LFRS spectrum produced with 532 nm as the excitation wavelength (Figure 2.24) is characterized by three-component profile that fits by ascribing to each band a Lorentzian profile and by considering the additional contribution of a constant background. This latter contribution is limited to a few percents of the whole signal attributed to some impurities and substrate effect. The $l = 2$ band is significantly broader than the E_g and T_{2g} bands (Figure 2.20). This feature is due to the various origins for the broadening of the Raman bands such as nanocrystal shape [57] and size [73] and the crystalline defects [66]. The narrow bandwidth of E_g and T_{2g} bands well agrees with their assignment to monodomain nanocrystals of quasi-spherical shapes, like cuboctahedra and truncated octahedra.

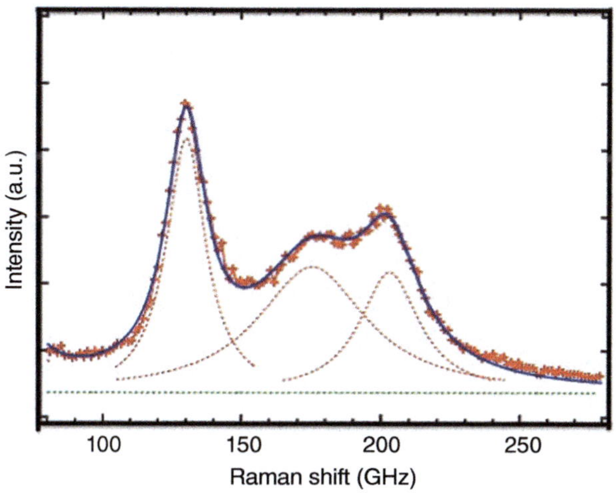

Figure 2.24 High-resolution Stokes LFRS spectrum of 5.3-nm Au nanocrystals (red curve and crosses online) recorded for an excitation at 532 nm. The entire profile of the spectrum is satisfactorily reproduced (blue curve) by ascribing to each component a Lorentzian profile (red dotted curves) represented along with the constant background (green dotted line) that is also involved in the multi-peak fitting. *Source:* Portalès et al. [61]/ American Chemical Society.

To examine the evolution of the LFRS spectrum profile on the excitation wavelength, the ratio $R(\lambda)$ of the integrated intensity of the higher-frequency structured band to that of the E_g band has been calculated. The ratios presented in Table 2.5 correspond to averaged values as derived from the fit of various LFRS spectra recorded at each wavelength. As indicated in Table 2.5, the smallest value of the $R(\lambda)$ is obtained for an excitation at 514.5 nm. This should reflect the predominant contribution of monodomain Au nanocrystals in the scattering process at this wavelength [74]. Regarding the calculated extinction spectra plotted in Figure 2.11, it appears that decahedra mainly absorb light via interband electronic transitions at 514.5 nm, which have been actually shown to not couple with the nanocrystals vibrations, thus inducing no enhancement of the Raman scattering [75].

The more efficient contribution of monodomain nanocrystals to the LFRS signal at 514.5 nm can therefore result from the selective excitation via plasmon–vibration coupling of quasi-spherical nanocrystals, while most of decahedral MTPs do not contribute to the scattering. When turning the excitation to 532 nm, the contribution of decahedra to the Raman scattering process significantly increases while other polycrystalline nanoparticles, like icosahedra, still remain resonantly excited and efficiently scatter the light. The significant increase of the $R(\lambda)$ ratio observed when changing the excitation from 514.5 to 532 nm is consistent with the increasing contribution of

Table 2.5 Ratios at four different wavelengths of the integrated intensity of the Raman band corresponding to the additional contributions of both the threefold T_{2g} degenerate modes of monodomain Au NCs and the low-frequency Raman active modes of Au MTPs to the integrated intensity of the band ascribed to the twofold E_g modes of monodomain Au NCs.

λ (nm)	514.5	532	647.1	676.4
$R(\lambda) = I_{tg}/I_{Eg}$	1.4±0.4	3.2±1.1	3.0±0.9	3.4±0.6

decahedra to the Raman signal. Whatever the excitation wavelength in the red is, 647.1 or 676.4 nm, the $R(\lambda)$ ratio almost remains unchanged, keeping the same value as the one determined at 532 nm. The contribution of quasi-spherical nanocrystals to the LFRS signal is logically expected to progressively decrease when shifting the excitation to the red. As a matter of fact, the low extinction cross sections calculated for both cuboctahedra and truncated octahedra at 647.1 and 676.4 nm (Figure 2.11) make it possible to predict the vanishing of the LFRS signal from monodomain nanocrystals at these two last wavelengths. The invariance of the $R(\lambda)$ ratio however indicates that monodomain nanocrystals still contribute quite efficiently to the Raman spectra in the red, in spite of their low extinction cross section. Furthermore, let us mention again that the amount of monodomain nanocrystals is estimated by TEM analysis to not exceed 30% of the full nanocrystal population in the studied samples. The persistence of the signal from monodomain nanocrystals suggests therefore their strong capacity for light scattering probably originating from their highly efficient plasmon–vibration coupling.

2.4 Crystal Growth Process

The influence of the nanocrystallinity plays a key factor in the final 3D superlattice structure [76]. In particular, their surfaces have a great potential as nanoscale interaction platforms. In Chapter 5, it is shown that spontaneous formation of upper vicinal surfaces for single-domain Au 3D superlattices appeared on a sublayer of ordered single-domain Co nanocrystals. These planes exhibit well-defined terraces and steps with a periodicity of c. 20 nm in most cases. These surfaces can be either stepped or kinked vicinal to the {100}, {110}, and {111} planes observed to be extended on the micrometer range. The formation of such high-index planes is explained by a heteroepitaxial relationship between both Co and Au nanocrystal superlattice. Unlike bulk vicinal surfaces, the high-index 3D superlattice planes are spontaneously formed. This is explained by the confinement of the Au nanocrystal superlattices on top of an underlying Co 3D superlattice film. It is thus possible to template the growth of high-index planes that are normally thermodynamically unstable. Thanks to the stronger single-nanocrystal interactions and large superlattice misfit between Au and Co nanocrystals (due to the difference in nanocrystal diameters and coating alkyl chain length), it is suggested that the Au vicinal surfaces are produced by a "soft" heteroepitaxial growth on compact planes of Co nanocrystal superlattice.

2.5 Mechanical Properties

As shown in Chapter 12, the mechanical properties of self-assembled nanocrystals in three-dimensional (3D) superlattices called 3D superlattices markedly change with the crystalline structure of the metal nanocrystals [77–80]. Face-to-face orientational ordering-related morphologies of nanocrystals improve the strength of the elastic moduli. This is in most of the case obtained with single-domain nanocrystals but can also be obtained with other nanocrystals having surface large edges. With single-domain nanocrystals, both translational and orientational orderings in the superlattices exist, whereas their polycrystalline counterparts behave like spheres with no orientational ordering. This is well demonstrated with Au nanocrystals. With Co nanocrystals, it is demonstrated a hierarchy in ordering at various scales. The atomic lattice within the nanocrystals to the nanocrystal superlattices within 3D superlattices is correlated with marked changes in the Young's modulus of 3D superlattices: from 0.7 ± 0.4 to 1.7 ± 0.5 to 6.6 ± 1.5 GPa as the crystalline structure of Co nanoparticles changes from *amorphous-phase* Co to epsilon-phase Co to hexagonal compact packing (hcp) Co, respectively. Note that epsilon-phase and hcp Co nanocrystals are both single-domain nanocrystals. This is mainly due to the fact that well-defined facets are observed with hcp Co nanocrystals, whereas epsilon-phase nanocrystals are spherical.

2.6 Conclusion

In this chapter, we demonstrated that the crystalline structure of nanocrystals called nanocrystallinity is one of the key parameters in some of their chemical and physical properties. We know that with regard to the diffusion of atoms, this also plays a crucial role in the final structure. A slight change in the localized surface Plasmon resonance is observed. Nevertheless, it can also induce important changes in the vibrational properties; the frequency of the breathing modes of nanocrystals is not markedly perturbed by the type of nanocrystals used. However, the damping process is markedly dropped with polycrystalline nanoparticles. The quadripolar vibrational modes markedly change between single domain and polycrystals and can be used as a tool to detect the nanocrystallinity of noble metals. However, many more chemical and physical properties have to be discovered.

References

1. Lisiecki, I., Turner, S., Bals, S. et al. (2009). Enhanced stability against oxidation due to 2D self organization of hcp Co nanocrystals: a new intrinsic property. *Chem. Mater.* 21: 2335–2338.
2. Yang, Z., Cavalier, M., Walls, M. et al. (2012). A phase-solution annealing strategy to control the Co nanocrystal anisotropy: structural and magnetic investigations. *J. Phys. Chem. C* 116: 15723–15730.
3. Yang, Z., Lisiecki, I., Walls, M., and Pileni, M.P. (2013). Nanocrystallinity and the ordering of nanoparticles in two-dimensional superlattices: controlled formation of either core/shell (Co/CoO) or hollow CoO nanocrystals. *ACS Nano* 7: 1342–1350.
4. Yang, Z., Walls, M., Lisiecki, I., and Pileni, M.P. (2013). Unusual effect of an electron beam on the formation of core/shell (Co/CoO) nanoparticles differing by their crystalline structures. *Chem. Mater.* 25: 2372–2377.

5 Yang, Z., Yang, N., Bergstrom, J. et al. (2013). Crystal polymorphism: dependence of oxygen diffusion through 2D ordered Co nanocrystals. *Phys. Chem. Chem. Phys.* 16: 9791–9796.
6 Yang, Z., Yang, N., Bergstrom, J., and Pileni, M.P. (2015). Control of the oxygen and cobalt atoms diffusion through Co nanoparticles differing by their crystalline structure and size. *Adv. Funct. Mater.* 25: 891–897.
7 Yang, Z., Yang, N., and Pileni, M.P. (2015). Nano Kirkendall related to nanocrystallinity of Co nanocrystals: influence of the atomic outward and inward diffusion in the final nanoparticle structure. *J. Phys. Chem. C* 119: 22249–22260.
8 Dinega, D.P. and Bawendi, M.G. (1999). A solution-phase chemical approach to a new crystal structure of cobalt. *Angew. Chem.* 38: 1788–1791.
9 Yang, J., Khazen, K., and Pileni, M.P. (2014). How nanocrystallinity and order define the magnetic properties of ε-Co supracrystals. *J. Phys. Condens. Matter* 23–26: 295303.
10 Tracy, I.B., Weiss, D.N., Dinega, D.P., and Bawendi, M.G. (2005). Exchange biasing and magnetic properties of partially and fully oxidized colloidal cobalt nanoparticles. *Phys. Rev. B* 72: 064404.
11 Huntington, H.B. and Seitz, F. (1942). Mechanism for self-diffusion in metallic copper. *Phys. Rev.* 61: 315–325.
12 Kirkendall, E.O. (1942). Diffusion of zinc in alpha brass. *Trans. Am. Inst. Min. Metall. Pet. Eng. Soc.* 147: 104–109.
13 Smigelskas, A.D. and Kirkendall, E.O. (1947). Zinc diffusion in alpha-brass. *Trans. Am. Inst. Min. Metall. Pet. Eng. Soc.* 171: 130–142.
14 Kim, S.W., Kim, M., Lee, M.Y., and Hyeon, T. (2002). Fabrication of hollow palladium spheres and their successful application to the recyclable heterogeneous catalyst for Suzuki coupling reactions. *J. Am. Chem. Soc.* 124: 7642–7643.
15 Hu, M., Chen, J., Li, Z.Y. et al. (2006). Gold nanostructures: engineering their plasmonic properties for biomedical applications. *Chem. Soc. Rev.* 35: 1084–1094.
16 Lou, X.W., Archer, L.A., and Yang, Z. (2008). Hollow micro-/nanostructures: synthesis and applications. *Adv. Mater.* 20: 3987–4019.
17 Caruso, F., Caruso, R.A., and Mohwald, H. (1998). Nanoengineering of inorganic and hybrid hollow spheres by colloidal templating. *Science* 282: 1111–1114.
18 Yin, Y., Rioux, R.M., Erdonmez, C.K. et al. (2004). Formation of hollow nanocrystals through the nanoscale Kirkendall effect. *Science* 304: 711–714.
19 Teo, J.J., Chang, Y., and Zeng, H.C. (2006). Fabrications of hollow nanocubes of Cu_2O and Cu via reductive self-assembly of CuO nanocrystals. *Langmuir* 22: 7369–7377.
20 Ha, D.H., Moreau, L.M., Honrao, S. et al. (2013). The oxidation of cobalt nanoparticles into Kirkendall-hollowed CoO and Co_3O_4: the diffusion mechanisms and atomic structural transformations. *J. Phys. Chem. C* 117: 14303–14312.
21 Fan, H.J., Goesele, U., and Zacharias, M. (2007). Formation of nanotubes and hollow nanoparticles based on Kirkendall and diffusion processes: a review. *Small* 3: 1660–1671.
22 Zhang, Q., Wang, W., Goebl, J., and Yin, Y. (2009). Self-templated synthesis of hollow nanostructures. *Nano Today* 4: 494–507.

23 An, K. and Hyeon, T. (2009). Synthesis and biomedical applications of hollow nanostructures. *Nano Today* 4: 359–373.

24 Wang, W., Dahl, M., and Yin, Y. (2003). Hollow nanocrystals through the nanoscale Kirkendall effect. *Chem. Mater.* 25: 1179–1189.

25 Hung, L.I., Tsung, C.K., Huang, W.Y., and Yang, P.D. (2010). Room-temperature formation of hollow Cu_2O nanoparticles. *Adv. Mater.* 22: 1910–1914.

26 Yang, P., Portalès, H., and Pileni, M.P. (2011). Dependence of the localized surface plasmon resonance of noble metal quasispherical nanoparticles on their crystallinity-related morphologies. *J. Chem. Phys.* 134: 024507. https://doi.org/10.1063/1.3523645.

27 Goubet, N., Yan, C., Polli, D. et al. (2013). Modulating the physical properties of isolated and self-assembled nanocrystals by change in their nanocrystallinity. *Nano Lett.* 13: 504–508.

28 Goubet, N., Tempra, I., Yang, J. et al. (2015). Size and nanocrystallinity controlled gold nanocrystals: synthesis, electronic and mechanical properties. *Nanoscale* 7: 3237–3324.

29 Kreibig, U. and Vollmer, M. (1995). *Optical Properties of Metal Clusters*. Berlin: Springer.

30 Charle, K.P., Frank, F., and Schulze, W. (1984). The optical properties of silver microcrystallites in dependence on size and the influence of the matrix environment. *Ber. Bunsen. Ges. Phys. Chem.* 88: 350.

31 Jin, R.C., Cao, W.Y., Mirkin, C.A. et al. (2001). Photoinduced conversion of silver nanospheres to nanoprisms. *Science* 294: 1901.

32 Kelly, K.L., Coronado, E., Zhao, L.L., and Schatz, G.C. (2003). The optical properties of metal nanoparticles: the influence of size, shape, and dielectric environment. *J. Phys. Chem. B* 107: 668.

33 Murphy, C.I., San, T.K., Gole, A.M. et al. (2005). Anisotropic metal nanoparticles: synthesis, assembly, and optical applications. *J. Phys. Chem. B* 109: 13857.

34 Salzemann, C., Brioude, A., and Pileni, M.P. (2006). Collections of copper nanocrystals characterized by different sizes and shapes: optical response of these nanoobjects. *J. Phys. Chem. B* 110: 7208.

35 Ino, S. and Ogawa, S.I. (1967). Multiply twinned particles at earlier stages of gold film formation on alkalihalide crystals. *J. Phys. Soc. Jpn.* 22: 1365–1374.

36 Purcell, E.M. and Pennypacker, C.R. (1973). Scattering and absorption of light by nonspherical dielectric grains. *Astrophys. J.* 186: 705–714.

37 Draine, B.T. and Flatau, P.J. (1994). Discrete-dipole approximation for scattering calculations. *J. Opt. Soc. Am. A* 11: 1491–1499.

38 Rodriguez-Fernandez, I., Novo, C., Myroshnychenko, V. et al. (2009). Spectroscopy, imaging, and modeling of individual gold decahedra. *J. Phys. Chem. C* 113: 18623–18631.

39 Lermé, L., Baida, H., Bonnet, C. et al. (2010). Size dependence of the surface plasmon resonance damping in metal nanospheres. *J. Phys. Chem. Lett.* 1: 2922–2928.

40 Zhang, I.Z. and Noguez, C. (2008). Plasmonic optical properties and applications of metal nanostructures. *Plasmonics* 3: 127–150.

41 Voisin, C., Del Fatti, N., Christofilos, D., and Vallée, F. (2001). Ultrafast electron dynamics and optical nonlinearities in metal nanoparticles. *J. Phys. Chem. B* 105: 2264–2280.

42 Arbouet, A., Voisin, C., Christofilos, D. et al. (2003). Electron–phonon scattering in metal clusters. *Phys. Rev. Lett.* 90: 177401.

43 Chen, K.P., Drachev, V.P., Borneman, I.D. et al. (2010). Drude relaxation rate in grained gold nanoantennas. *Nano Lett.* 10: 916–922.

44 Baida, H., Christofilos, D., Maioli, P. et al. (2011). Surface plasmon resonance spectroscopy of single surfactant-stabilized gold nanoparticles. *Eur. Phys. J. D* 63: 293–299.

45 Polli, D., Lisiecki, I., Portalès, H. et al. (2011). Low sensitivity of acoustic breathing mode frequency in Co nanocrystals upon change in nanocrystallinity. *ACS Nano* 5: 5785–5791.

46 Lamb, H. (1982). On the vibrations of an elastic sphere. *Proc. London Math. Soc.* 13: 189–212.

47 Del Fatti, N., Voisin, C., Chevy, F. et al. (1999). Coherent acoustic mode oscillation and damping in silver nanoparticles. *J. Chem. Phys.* 110: 11484–11487.

48 Hartland, G.V. (2002). Coherent vibrational motion in metal particles: determination of the vibrational amplitude and excitation mechanism. *J. Chem. Phys.* 116: 8048.

49 Crut, A., Maioli, P., Fatti, N.D., and Vallée, F. (2009). Anisotropy effects on the time-resolved spectroscopy of the acoustic vibrations of nanoobjects. *Phys. Chem. Chem. Phys.* 11: 5882–5888.

50 Tang, Y. and Ouyang, M. (2007). Tailoring properties and functionalities of metal nanoparticles through crystallinity engineering. *Nat. Mater.* 6: 754–759.

51 Newhouse, R.J., Wang, H., Hensel, J.K. et al. (2011). Coherent vibrational oscillations of hollow gold nanospheres. *J. Phys. Chem. Lett.* 2: 228–235.

52 Hartland, G.V. (2011). Optical studies of dynamics in noble metal nanostructures. *Chem. Rev.* 111: 3858–3887.

53 Hartland, G.V. (2006). Coherent excitation of vibrational modes in metallic nanoparticles. *Annu. Rev. Phys. Chem.* 57: 403–430.

54 Hodak, J.H., Henglein, A., and Hartland, G.V. (2000). Photophysics of nanometer sized metal particles: electron–phonon coupling and coherent excitation of breathing vibrational modes. *J. Phys. Chem. B* 104: 9954–9965.

55 Groeneveld, R.H.M., Sprik, R., and Lagendijk, A. (1992). Effect of a nonthermal electron distribution on the electron-phonon energy relaxation process in noble metals. *Phys. Rev. B* 45: 5079–5082.

56 Visscher, W.M., Migliori, A., Bell, T.M., and Reinert, R.A. (1991). On the normal-modes of free-vibration of inhomogeneous and anisotropic elastic objects. *J. Acoust. Soc. Am.* 90: 2154–2162.

57 Saviot, L. and Murray, D.B. (2009). Acoustic vibrations of anisotropic nanoparticles. *Phys. Rev. B* 79: 214101.

58 Scaffardi, L.B. and Tocho, J.O. (2006). Size dependence of refractive index of gold nanoparticles. *Nanotechnology* 17: 1309–1315.

59 Gu, Q.F., Krauss, G., Steurer, W. et al. (2008). Unexpected high stiffness of Ag and Au nanoparticles. *Phys. Rev. Lett.* 100: 045502.

60 Portalès, H., Goubet, N., Saviot, L. et al. (2008). Probing atomic ordering and multiple twinning in metal nanocrystals through their vibrations. *Proc. Natl. Acad. Sci. U.S.A.* 105: 14784–14789.

61 Portalès, H., Goubet, N., Saviot, L. et al. (2010). Crystallinity dependence of the plasmon resonant Raman scattering by anisotropic gold nanocrystals. *ACS Nano* 4: 3489.

62 Zheng, N., Fan, J., and Stucky, G.D. (2006). One-step one-phase synthesis of monodisperse noble-metallic nanoparticles and their colloidal crystals. *J. Am. Chem. Soc.* 128: 6550–6551.

63 Goubet, N., Richardi, J., Albouy, P.A., and Pileni, M.P. (2011). Which forces do control the supracrystal nucleation in organic media? *Adv. Funct. Mater.* 21: 2693–2704.

64 Brust, M., Walker, M., Bethell, D. et al. (1994). Synthesis of thiol-derivatized gold nanoparticles in a two-phase liquid/liquid system. *J. Chem. Soc., Chem. Commun.* 801–802.

65 Palpant, B., Portalès, H., Saviot, L. et al. (1999). Quadrupolar vibrational mode of silver clusters from plasmon-assisted Raman scattering. *Phys. Rev. B* 60: 17107–17111.

66 Duval, E., Boukenter, A., and Champagnon, A. (1986). Vibration eigenmodes and size of microcrystallites in glass: observation by very-low-frequency Raman scattering. *Phys. Rev. Lett.* 56: 2052–2055.

67 Margueritat, J., Gonzalo, J., Afonso, C.N. et al. (2006). Surface plasmons and vibrations of self-assembled silver nanocolumns. *Nano Lett.* 6: 2037–2042.

68 Duval, E., Mermet, A., Courty, A. et al. (2005). Coherence effects on Raman scattering from self-organized Ag nanocrystals: theory. *Phys. Rev. B* 72: 085439.

69 Courty, A., Mermet, A., Albouy, P.A. et al. (2005). Self-organized Ag-nanocrystals in fcc "supra" crystals: vibrational coherence. *Nat. Mater.* 4: 395–398.

70 Van Dijk, M.A., Lippitz, M., and Orrit, M. (2005). Detection of acoustic oscillations of single gold nanospheres by time-resolved interferometry. *Phys. Rev. Lett.* 95: 267406.

71 Murray, D.B., Laarakker, A.S., and Saviot, L. (2006). High precision numerical estimates of acoustic phonon frequencies of nonspherical nanoparticles. *Phys. Status Solidi C* 3: 3935–3938.

72 Stephanidis, B., Adichtchev, S., Etienne, S. et al. (2007). Vibrations of nanoparticles: from nanospheres to fcc cuboctahedra. *Phys. Rev. B* 76: 121404.

73 Bachelier, G., Margueritat, J., Mlayah, A. et al. (2007). Size dispersion effects on the low-frequency Raman scattering of quasispherical silver nanoparticles: experiment and theory. *Phys. Rev. B* 76: 235419.

74 Duval, E., Portalès, H., Saviot, L. et al. (2001). Spatial coherence effect on the low-frequency Raman scattering from metallic nanoclusters. *Phys. Rev. B* 63: 075405.

75 Portalès, H., Saviot, L., Duval, E. et al. (2001). Resonance and composition effects on the Raman scattering from silver-gold alloy clusters. *Eur. Phys. J. D* 16: 197.

76 Goubet, N., Yang, J., Albouy, P.A., and Pileni, M.P. (2014). Spontaneous formation of high-index planes in Au single domain nanocrystal superlattices. *Nano Lett.* 14: 6632–6638.

77 Yan, C., Arfaoui, I., Goubet, N., and Pileni, M.P. (2013). Soft supracrystals of Au nanocrystals with tunable mechanical properties. *Adv. Funct. Mater.* 23: 2315–2232.

78 Yan, C., Portalès, H., Goubet, N. et al. (2013). Assessing the relevance of building block crystallinity for tuning the stiffness of gold nanocrystal superlattices. *Nanoscale* 5: 9523–9527.

79 Gauvin, M., Yang, N., Barthel, E. et al. (2015). Morphology, nanocrystallinity and elastic properties of single domain ε-Co supracrystals. *J. Phys. Chem. C* 119: 7483–7490.

80 Gauvin, M., Yang, N., Yang, Z. et al. (2015). Hierarchical mechanical behavior of cobalt supracrystals related to nanocrystallinity. *Nanoresearch* 8: 3480–3487.

3

Au 3D Superlattices Produced by Solvent Evaporation Process

3D superlattices are produced under various solvent vapor pressures as follows:

(i) *Glove box*: zero toluene vapor pressure is obtained by pouring the colloidal solution into a beaker with a silicon wafer at the bottom and depositing it in a glove box having a large volume compared with that of the beaker.
(ii) *Glass vessel*: to increase and control the solvent vapor pressure during the solvent evaporation process of the colloidal solution, the following setup is used. The beaker is placed in a glass vessel having a small volume (Scheme 3.1). The toluene vapor pressure is estimated to be 39%. By passing nitrogen through a flask containing toluene solution before reaching the colloidal solution, the evaporation time increases to 25 hours, and the solvent vapor pressure to 75%. The evaporation time evolves from 7–8 to 8–9 and 25 hours by changing the vapor pressure to 0%, 39%, and 75%. In the glass vessel, the temperature is controlled and kept to 25 °C, respectively. On increasing the temperature to 50 °C, the toluene vapor pressure becomes 25%.

Here, we concentrate on Au nanocrystals produced by revisiting Stucky method [1, 2] (Chapter 1). This method permits to produce Au nanocrystals differing by their sizes from 4 to 8 nm and coated with various thiol derivatives such as dodecanethiol ($C_{12}H_{25}$–SH), tetradecanethiol ($C_{14}H_{29}$–SH), and hexadecanethiol ($C_{16}H_{33}$–SH). By depositing a drop of the colloidal solution, Figure 3.1 shows, whatever the alkyl chain length of the coating agent from C_{12} to C_{16}, well-defined 2D superlattices.

3.1 Au 3D Superlattice Morphology of Au Nanocrystals Coated with Thiol Derivatives

The morphology of Au 3D superlattices markedly changes with the coating agents and the solvent vapor pressures [1–4].

Metal Nano 3D Superlattices: Synthesis, Properties, and Applications, First Edition.

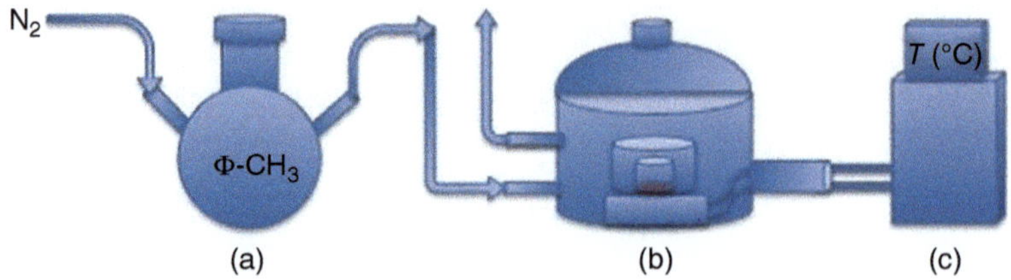

Scheme 3.1 The setup of a glass vessel made of (a) a flask with two thin exits for passing the N_2 flow through the toluene solution, (b) a glass vessel containing an evaporation beaker on the copper holder, and (c) a heater controlling temperature in the range of 15–60 °C. *Source:* Wan et al. [1, 2]/American Chemical Society.

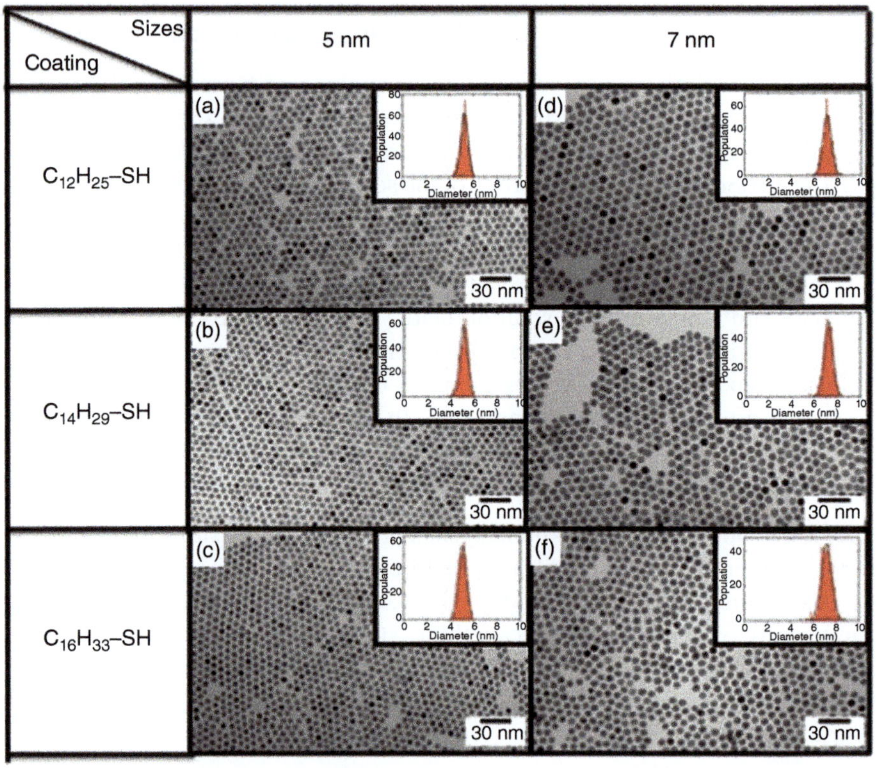

Figure 3.1 TEM images and size distribution (inset) of Au nanocrystals deposited in a TEM grid having an average diameter of 5-nm (a–c) and 7-nm (d–f) and differing by their coating agents as (a, d) dodecanethiol ($C_{12}H_{25}$–SH), (b, e) tetradecanethiol ($C_{14}H_{29}$–SH), and (c, f) hexadecanethiol ($C_{16}H_{33}$–SH). *Source:* Wan et al. [2]/American Chemical Society.

3.1.1 Au 3D Superlattice Morphology Produced at Zero Solvent Vapor Pressure

The colloidal solution of Au nanocrystals coated with dodecanethiol, differing by their sizes and dispersed in toluene, is evaporated under zero toluene vapor pressure. The 3D superlattice morphologies of various nanocrystal sizes used as building blocks are shown (Figures 3.2 and 3.3). The scanning electron microscopy (SEM) images of 4- and 5-nm nanocrystals as average diameters show uniform films covering most of the substrate with an average thickness of about a few μm (Figure 3.2a–d). The film surface reveals lines crossing at nearly 60° and 120° with some terraces with triangular structures. The corresponding linear steps having various heights from one nanocrystal thickness are observed. The small-angle X-ray diffraction (SAXRD) patterns (Figure 3.2e,f) are characteristic of face-centered cubic (fcc) structure.

On increasing nanocrystal diameter from 6 to 8 nm, the nanocrystals form aggregates characterized by very flat surface with well-defined edges, whereas others show fivefold symmetry and/or no characteristic symmetry (Figure 3.3a,b,d,e,g,h). These data agree with the previous ones [6–12]. The small-angle X-ray scattering (SAXS) patterns (Figure 3.3c,f,i) show continuous rings with discontinuous reinforcements and *d*-spacings with fcc structures as observed above with films of smaller nanocrystals. The rings are attributed to randomly oriented 3D superlattices and the reinforcements in the [111] axis to well-developed 3D superlattices sitting on their larger flat facets. From the experimental reflection coordinates, the

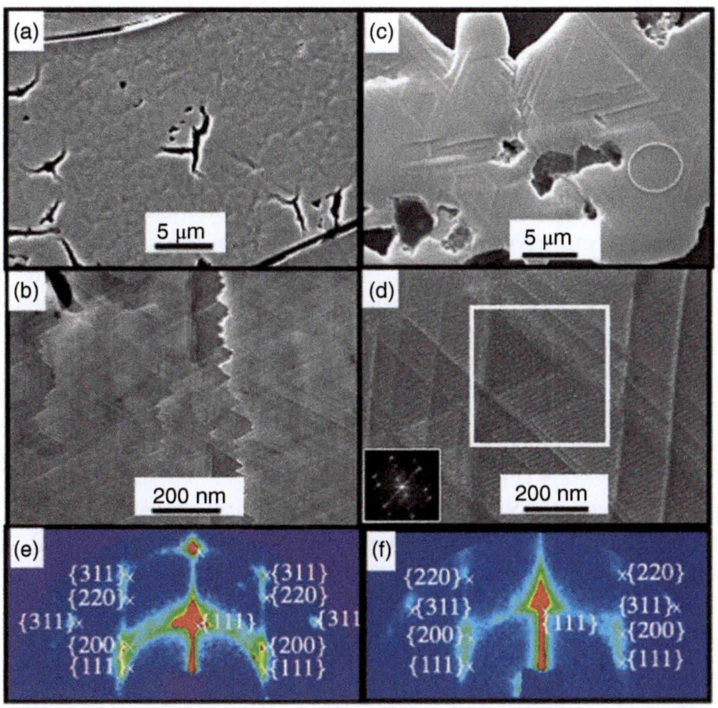

Figure 3.2 SEM images at various scales (a, b, d, e) and the corresponding SAXRD patterns (c, f) of fcc 3D superlattices built with Au nanocrystals characterized by an average diameter of 4 (a–c) and 5 nm (d–f), coated with dodecanethiol and dispersed in toluene and deposited on silicon substrate by solvent evaporation at zero toluene vapor pressure. *Source:* (a, b, e, f) Goubet et al. [5]/John Wiley & Sons, Inc. (c, d) Wan et al. [2]/American Chemical Society.

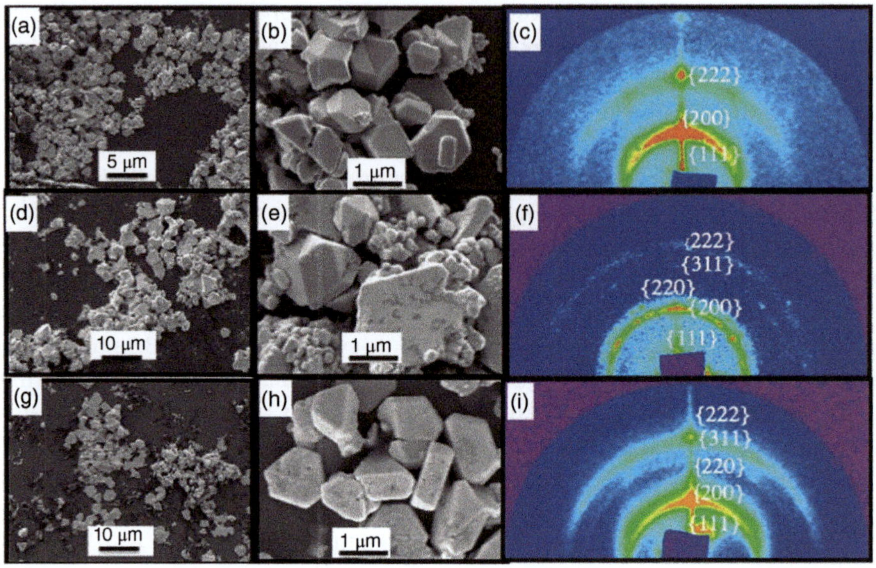

Figure 3.3 SEM images at various scales (a, b, d, e, g, h) and the corresponding SAXRD patterns (c, f, i) of fcc 3D superlattices built with Au nanocrystals characterized by an average diameter of 6 (a–c), 7 (d–f), and 8 nm (g–i) coated with dodecanethiol and dispersed in toluene and deposited on silicon substrate by solvent evaporation at zero toluene vapor pressure. *Source:* Goubet et al. [5]/John Wiley & Sons, Inc.

sample–detector distance obtained from SAXS patterns, the Bragg angle, and thus the modulus of the diffracted vector q from the Bragg relation are obtained.

At this stage, we can ask ourselves: What is the influence of the solvent on the formation 3D superlattices and their morphologies and properties? [5]. To answer this question, let us replace toluene by hexane, keeping Au nanocrystals described above coated with dodecanethiol and differing by their sizes. Note that the solubility of Au nanocrystals in hexane markedly drops, compared with toluene, to reach no dispersion of 8-nm Au nanocrystals preventing any 3D superlattice formation. The SEM patterns obtained for Au nanocrystals differing by their sizes show the formation of thin films on the substrate (Figure 3.4).

The insets in Figure 3.4 show the SAXRD patterns and confirm that similar fcc 3D superlattices are present, displaying the same type of texture as found in 4- and 5-nm Au films grown in toluene (Figure 3.4). No individually grown micrometric 3D superlattices are observed even at larger particle sizes (6- and 7-nm Au nanocrystals). As with 5-nm Au nanocrystals dispersed in toluene (Figure 3.2), the 3D superlattice surface of 6-nm Au nanocrystals dispersed in hexane (Figure 3.5a,b) is made of terraces with linear steps having various heights (from one nanocrystal to several nanocrystals' thickness).

The mono-nanocrystal step with an irregular edge covers the entire surface.

The similarity in the film morphologies, with a layer-by-layer growth, observed for 4- and 5-nm Au films produced by using toluene to disperse the nanocrystals (Figure 3.2b,d) with those obtained with larger nanocrystal sizes (6- and 7-nm Au

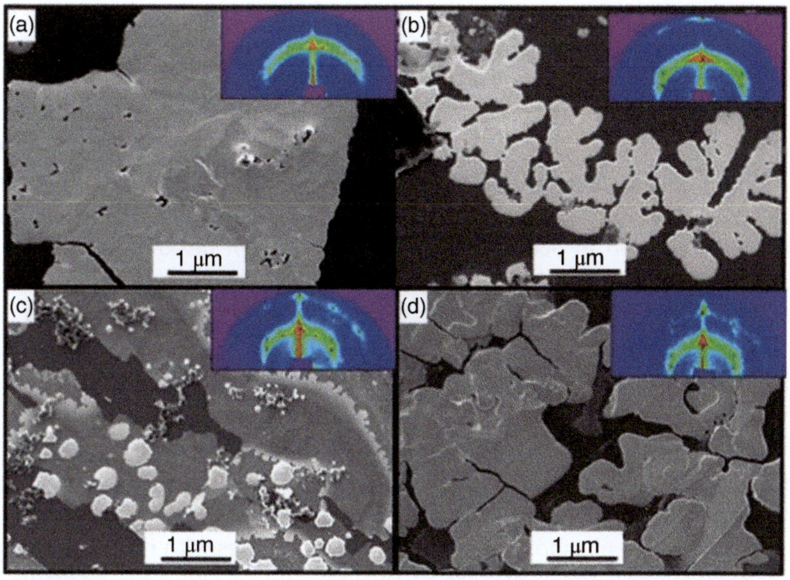

Figure 3.4 SEM pictures of 3D superlattices built with 4 nm (a), 5 nm (b), 6 nm, and 7 nm (d) of nanocrystals dispersed in hexane and deposited on silicon substrate. Inset: the corresponding SAXRD characterization. *Source:* Goubet et al. [5]/John Wiley & Sons, Inc.

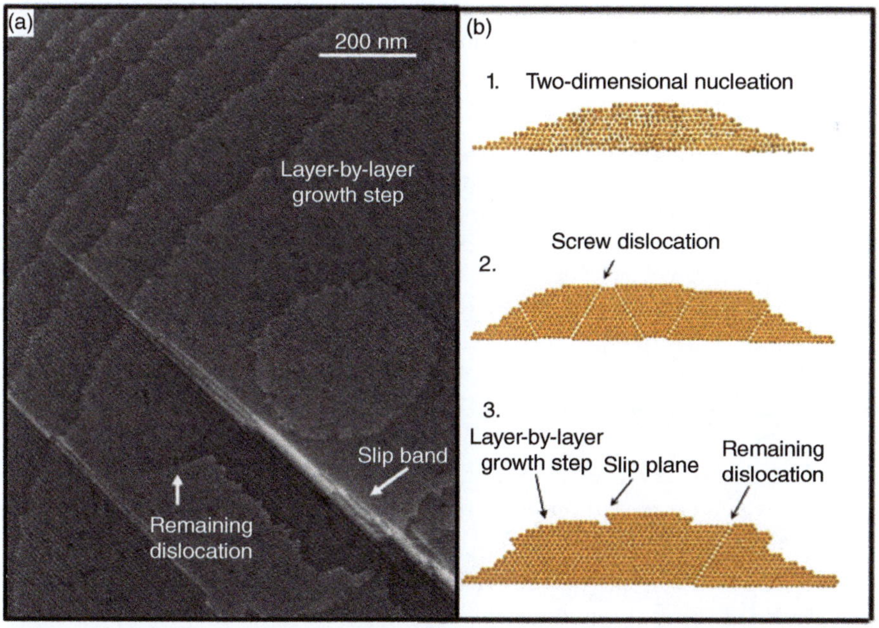

Figure 3.5 High resolution of 3D superlattice surface of 6-nm Au nanocrystals dispersed in hexane and deposited on TEM grid (a). Scheme illustrating the proposed mechanism of the 3D superlattice plastic deformation (b). *Source:* Goubet et al. [5]/John Wiley & Sons, Inc.

nanocrystals) in hexane (Figure 3.4a) makes it possible to conclude that the solvent in which the nanocrystals are dispersed is key parameter on the crystal growth mechanism.

Even in a quasi-saturated regime, the evaporation rate of hexane remains faster (five hours) than that of toluene (seven hours). To confirm that the change in the 3D superlattice growth mechanism is not related to the evaporation rate, let us consider another solvent with a boiling point closer to toluene ($b_p = 110\,°C$). The 6-nm Au nanocrystals are dispersed in octane ($b_p = 125\,°C$) instead of hexane ($b_p = 69\,°C$). The evaporation time is around eight hours for toluene. Insets in Figure 3.6a,b clearly show the formation of films with very bright SAXRD reflections normal to the substrate and well-defined Bragg peaks at the resolution limit in the [111] growth axis (Figure 3.6 insets), indicating a layer-by-layer growth in fcc 3D superlattices, as already observed for 6-nm Au nanocrystals dispersed in hexane and also 4- and 5-nm Au nanocrystals dispersed in toluene. However, it can be seen that the film area is markedly reduced compared with that obtained previously. For 6-nm Au nanocrystals in cumene ($b_p = 152\,°C$) where the evaporation time is close to 48 hours, the SEM images (Figure 3.6b) show the formation of aggregates with regular shapes. The SAXRD patterns and Bragg plot (inset Figure 3.6) indicate a fcc crystalline structure with no preferential orientation. Note that cumene and toluene have closely related structures. Nevertheless, from these data, it is concluded that the growth mechanism is controlled by the solvent nature and not by the evaporation rate. Also, note that the 3D superlattice growth mechanisms do not depend on the substrate effect. When highly oriented pyrolytic graphite replaces the silicon wafer, the crystal morphologies remain the same.

Figure 3.6 SEM picture of 3D superlattices builtwith 6-nm Au nanocrystals dispersed in octane (a,c) and cumene (b,d) and their corresponding SAXRD characterization inset. *Source:* Goubet et al. [5]/John Wiley & Sons, Inc.

These data show that two growth mechanisms of fcc Au 3D superlattice take place. The solvent in which Au nanocrystals are dispersed and their sizes are major parameters determining the final morphology of nanocrystal assemblies with either films by layer-by-layer growth (heterogeneous growth), characterized by their plastic deformation, or well-defined shapes grown in solution (homogeneous growth). These data are supported by Brownian dynamic and Monte Carlo methods [5] from which the 3D superlattice nucleation is mainly driven by solvent-mediated interactions and not solely by the van der Waals attraction between nanocrystal cores, as widely assumed in the literature.

Keeping the same experimental conditions and replacing dodecanethiol ($C_{12}H_{25}$–SH) by tetradecanethiol ($C_{14}H_{29}$–SH) as coating agent, i.e. corresponding to an increase of the alkyl chain length of the coating agent by two carbon atoms, the SEM patterns (Figure 3.7) show aggregates sitting on films. The HRSEM image corresponding to the zone limited by a circle (Figure 3.7B,D) shows that the aggregate is highly ordered in a hexagonal pattern as confirmed by the fast Fourier transform (FFT) (Figure 3.7B, FFT pattern of c, and Figure 3.7D, FTT pattern a), whereas the underlying film is disordered (Figure 3.7B, FFT pattern of a, and Figure 3.7D, FFT pattern of b). Note that at the edge of the aggregate, the nanocrystals are highly ordered near some disordered nanocrystals. The wide full width at half maximum (FWHM) (see below) of the main peak of the SAXRD pattern confirms the formation of fcc 3D superlattices simultaneously with some amorphous films.

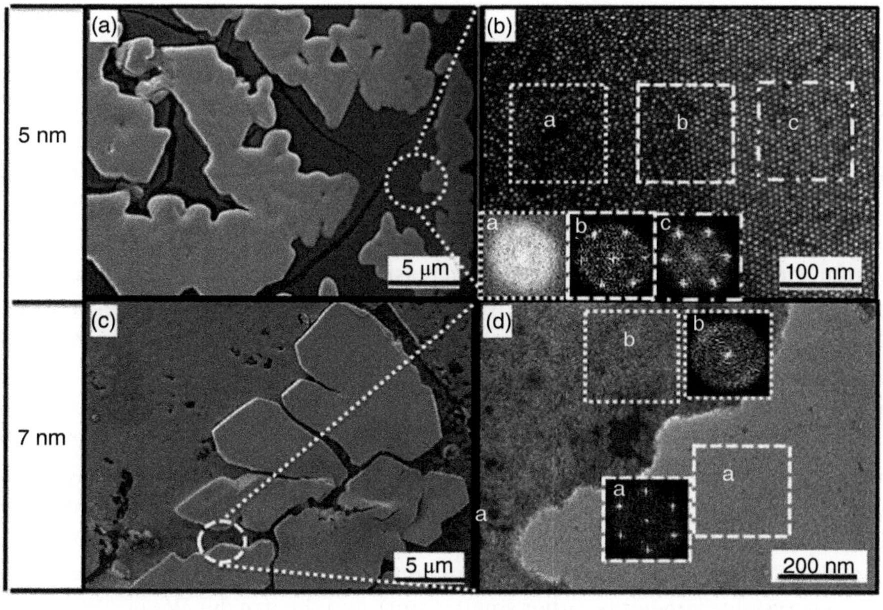

Figure 3.7 SEM and HRSEM images respectively obtained at the end of toluene evaporation on a silicon substrate at room temperature (T = 25 °C) of 5-nm (a, b) and 7-nm (c, d) Au colloidal solution of Au nanocrystals coated with tetradecanethiol, $C_{14}H_{29}$–SH. The vapor pressure P = 0 and evaporation time is about eight hours. *Source:* Wan et al. [2]/ American Chemical Society.

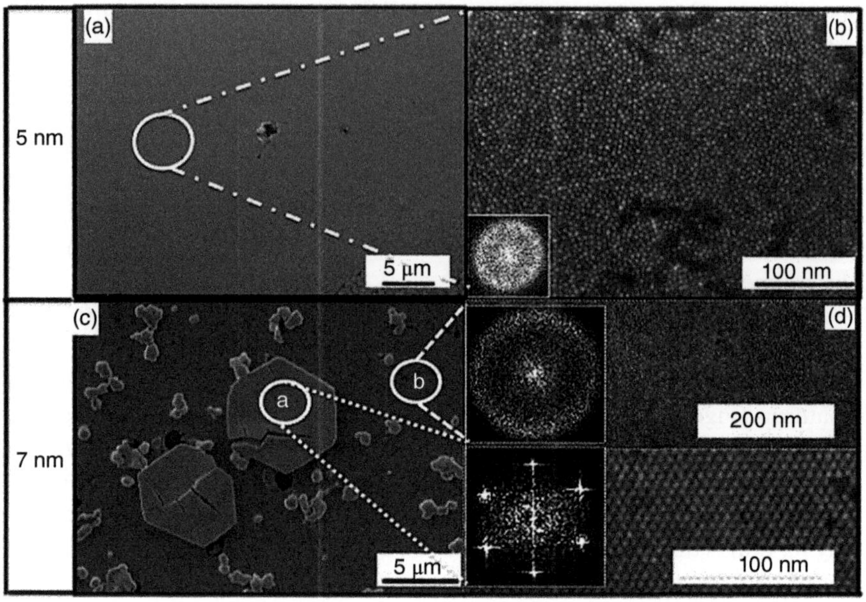

Figure 3.8 SEM and HRSEM images obtained at the end of toluene evaporation on a silicon substrate at room temperature ($T = 25\,°C$) of 5 nm (a, b) and 7 nm (c, d) of Au nanocrystals coated with hexadecanethiol, $C_{16}H_{33}$–SH. The vapor pressure $P = 0$ and evaporation time is about eight hours. *Source:* Wan et al. [2]/American Chemical Society.

A further increase, by two carbon atoms, of the coating agent chain length such as hexadecanethiol ($C_{16}H_{33}$–SH), a homogeneous film of 5-nm Au nanocrystals, is obtained (Figure 3.8a). The HRSEM image of the zone denoted by a circle in Figure 3.8b reveals that the film is amorphous with a random distribution of nanocrystals. The nanocrystal diameter is increased from 5- to 7-nm fcc aggregates sitting on amorphous film (Figure 3.8c,d).

From these, it is concluded that by increasing the chain length of the coating agent, the system evolves progressively from fcc 3D superlattices with dodecanethiol to disordered films with hexadecanethiol. Tetradecanethiol as coating agent is an intermediate between the two extremes. This could be explained as follow: from a theoretical study [13], the crystallization temperatures for different chain lengths are −7, 28, and above 42 °C for dodecanethiol, tetradecanethiol, hexadecanethiol, respectively. So it is easy to understand why nanoparticles could crystallize layer-by-layer films of 3D superlattices at room temperature for dodecanethiol, ordered film plus disordered film for tetradecanethiol, and only amorphous film for hexadecanethiol. Such behavior confirms the data obtained previously [14, 15]. Here, the Au nanocrystals diameter is rather small (5 nm), and the van der Waals attractions between nanocrystals are weak. Consequently, the ligand shells especially dominate the order–disorder transition.

On *increasing the nanocrystal sizes* from 5 to 7 nm, similar data as those derived with tetradecanethiol ($C_{14}H_{29}$–SH) are obtained: the SEM patterns (Figure 3.7A,C) show well-defined flat aggregates with an underneath film of Au nanocrystals coated with $C_{14}H_{29}$–SH and $C_{16}H_{33}$–SH, respectively. From HRSEM patterns (Figure 3.7B,D) and the corresponding FFT patterns recorded in various areas, the aggregates are fcc 3D superlattices sitting on disordered (amorphous) films. The van der Waals interactions are not strong enough to induce change in the morphology of the nanocrystals assemblies. Similarly with hexadecanethiol, there is a transition from amorphous film to fcc aggregate sitting on amorphous film (Figure 3.8).

3.1.1 Au 3D Superlattice Morphologies Produced Under Various Solvent Vapor Pressures

The experimental conditions remain the same as those described above. The only difference compared with data obtained at zero vapor pressure is related to the fact that the colloidal solution is subjected to either 39% or 75% of toluene vapor pressure, and consequently, the evaporation time evolves from 7–8 (0%) to 8–9 (39%) and 25 hours (75%), respectively.

Let us consider 5- and 7-nm Au nanocrystals as building blocks subjected to various toluene vapor pressure during the evaporation process, while the other experiment conditions remain the same.

Whatever the experimental conditions, the SAXS experiments show formation of fcc 3D superlattices. As already mentioned, at zero vapor pressure, 5-nm Au nanocrystals dispersed in toluene produce well-defined layer-by-layer fcc 3D superlattices with terraces of nanocrystal layers having linear steps, various heights, and dislocations (Figure 3.9a). Similar behavior is observed on increasing the solvent

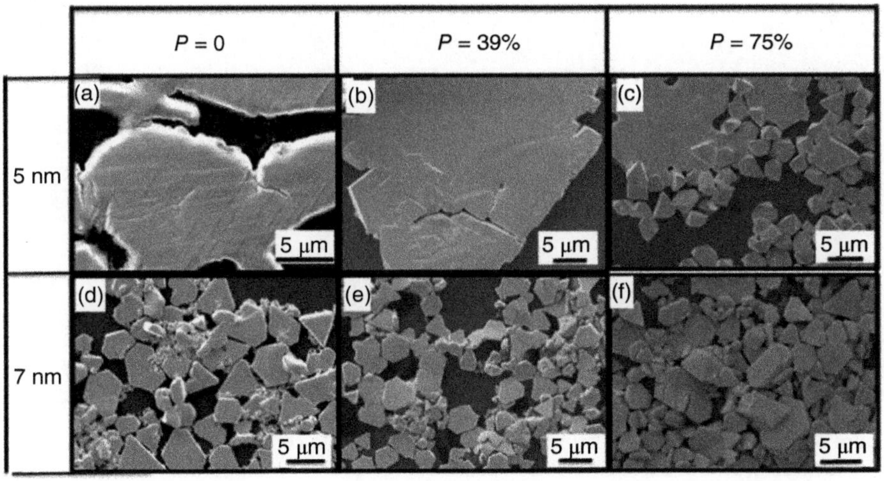

Figure 3.9 SEM images of fcc 3D superlattices built with Au nanocrystals coated with dodecanethiol and dispersed in toluene characterized by an average diameter of 5 nm (a–c) and 7 nm (d–f). The colloidal solutions are subjected to various solvent vapor pressures (*P*). At *P* = 0 (a, d), *P* = 39% (b, e) and *P* = 75% (c, e, f). Temperature is 25 °C. *Source:* Wan et al. [2]/American Chemical Society.

vapor pressure to 39% (Figure 3.9b). However, a further increase in the solvent vapor pressure to 75% induces the formation of shape in coexistence with films (10%) fcc 3D superlattices (Figure 3.9c). On increasing the nanocrystal size to 7 nm, whatever the solvent vapor pressure, well-defined shaped fcc 3D superlattices are produced (Figure 3.9d–f). Hence, with 5-nm Au nanocrystals, layer-by-layer film with a few shaped 3D superlattices is observed, whereas with 7-nm nanocrystals, shaped fcc 3D superlattices are produced.

From these data, it is concluded that toluene vapor pressure plays an important role in the morphology of fcc 3D superlattices produced with 5- and 7-nm *Au nanocrystals coated with dodecanethiol.* However, shaped 3D superlattices are observed with 5-nm Au nanocrystals subjected to 75% toluene vapor pressure, whereas at lower one films are produced.

In contrary to what is observed with dodecanethiol, the 3D superlattice morphology markedly changes with tetradecanethiol ($C_{14}H_{29}$–SH). However, the SAXRD patterns indicate that the crystalline structure of 3D superlattices remains fcc assemblies. Hence, the SEM images produced from a colloidal solution subjected to 75% solvent vapor pressure during the evaporation exhibit coexistence of a well-defined truncated-tetrahedral shapes coexisting and film 3D superlattices (Figure 3.10a). The shaped (Figure 3.10b) and film (Figure 3.10c) 3D superlattice are well observed and the corresponding high-resolution SEM images (Figure 3.10d,e) show that these two different assemblies are composed of well-ordered superlattices

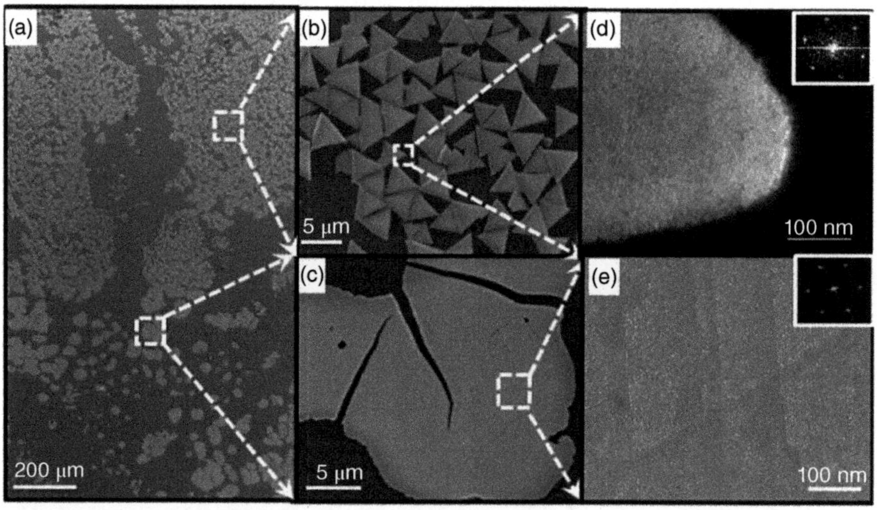

Figure 3.10 (a) Low-magnification SEM image of a sample composed of Au 3D superlattices grown at 25 °C on a silicon substrate for a toluene evaporation time of 25 hours. (b, c) SEM images of two areas selected from the first SEM image showing, at higher magnification, (b) truncated-tetrahedral 3D superlattices, and (c) supracrystalline films. (d, e) Further magnification applied using high-resolution SEM on areas selected from the images shown in (b) and (c) panels, respectively. For each HRSEM image, a power spectrum obtained by FFT analysis is presented as an inset. *Source:* Wan et al. [3]/ Springer Nature.

of Au nanocrystals deposited. At this stage, we can ask: What is the driving forces to control the 3D superlattice shapes? We first have to remember that the synthesis of Au nanocrystals used through inorganic reaction (see Chapter 1, Section 3.2) with 5-nm Au nanocrystals 30% of nanocrystals is single-domain nanocrystals, whereas the others are 65% polycrystals. However, 5% is poorly twinned nanocrystals.

From Chapter 2, we know that the fivefold degenerate quadrupolar vibrational mode of single-domain Au nanocrystals splits in two different levels, i.e. the two- and threefold degenerate E_g and T_{2g} modes, respectively, whereas one peak is observed for multiple-twinned and polycrystalline nanoparticles [16]. To explain such changes in the 3D superlattice morphology, let us assume that a segregation process between single-domain and polycrystalline nanocrystals takes place. To support such hypothesis, let us measure the micro-low-frequency Raman scattering (M-LFRS) spectrum selectively on the triangular (Figure 3.11a) and film (Figure 3.11d) fcc 3D superlattices.

Figure 3.11b,e shows the Stokes/anti-Stokes M-LFRS spectra of 5.1-nm Au nanocrystals issued from truncated-tetrahedral 3D superlattices (Figure 3.11a) and supracrystalline films (Figure 3.11b). This is viewed in Figure 3.11c showing where the laser spot is localized on the sample. Looking first at the spectrum plotted in Figure 3.11b, two peaks centered at the frequency shifts of 140 and 221 GHz are observed. These measured frequencies agree with data described in Chapter 2 for the E_g and T_{2g} modes of Au nanospheres that are 146 and 236 GHz, respectively.

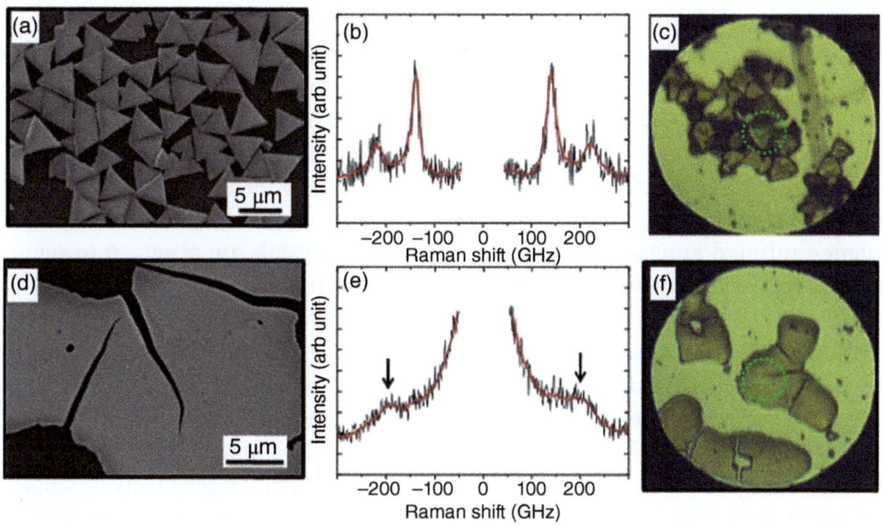

Figure 3.11 Truncated-tetrahedral 3D superlattices and films (a–c) and (d–f), respectively: STEM images (a, d), Stokes/anti-Stokes low-frequency Raman scattering spectra of Au nanocrystals as recorded by focusing the laser on (b, e). For clarity, each spectrum is plotted along with a guide for the eyes (red curve) obtained by smoothing the original spectrum. A camera was used to capture image in (c) and (f) where the laser spot was localized on the sample (delimited by the green circle) for the acquisition of the spectra (b) and (b, e), respectively. *Source:* (a, c, d, f) Wan et al. [3]/Springer Nature.

Furthermore, the intensity of the E_g peak is significantly higher than the T_{2g} one, as observed in the low-frequency Raman scattering (LFRS) spectra of single-domain nanocrystals recorded by using the same excitation source at 532 nm. From such comparison, these two peaks are thus accurately assigned to the signal from single-domain Au nanocrystals. As the M-LFRS spectrum plotted in Figure 3.11b is recorded by focusing the laser beam onto truncated-tetrahedral 3D superlattices, it is therefore concluded that the latter are almost exclusively composed of single-domain building blocks. At variance with this result, the M-LFRS spectrum that is recorded by adjusting the laser spot onto supracrystalline films (Figure 3.11d) only exhibits one Raman band centered around 200 GHz (Figure 3.11e). This value is close to the vibrational frequency calculated for the fivefold degenerated quadrupolar mode of 5.1-nm Au nanospheres. An additional low-intensity band seems to emerge from the signal background at a frequency shift of around 145 GHz. In spite of the long acquisition duration used for recording each spectrum, the M-LFRS signal detectable from the films exhibits a low signal-to-noise ratio, making it difficult to ensure for this second band. Let us mention that this hypothetical signal could correspond to the E_g mode contribution of single-domain nanocrystals and reveals their residual presence in the supracrystalline films. For that reason, we do not actually exclude that the supracrystalline films could be composed of a mixture of single-domain and polycrystalline nanoparticles. The higher contribution of the polycrystalline nanoparticles to the LFRS signal would nevertheless indicate that they are present in a larger proportion in the films. Self-assembly of nanocrystals is sensitive to the variation in size of the building blocks, and potential size segregation of nanocrystals should not be definitely excluded to explain the segregation of single-domain nanocrystals in truncated-tetrahedral 3D superlattices. Due to the narrow size dispersion (<7%) of the whole population of nanocrystals, any size segregation seems nevertheless hardly achievable from a unique batch of nanocrystals. The separation of single from polycrystalline nanocrystals based on crystallinity-related shape segregation appears therefore as the most probable mechanism. With similar 5-nm Au nanocrystals, as shown in Chapter 4, at the equilibrium (under toluene-saturated atmosphere), single-domain nanocrystals are observed to segregate in 3D superlattices, whereas polycrystals of same size did not self-assemble [17]. Note that, in the present case, the solvent slowly evaporates and fcc truncated-tetrahedral 3D superlattices are produced. At variance, no orientational ordering is observed in 3D superlattices composed of polycrystalline nanoparticles. These data confirm those obtained above from M-LFRS from which the truncated-tetrahedral superlattice appears as being mainly composed of single-domain nanocrystals with fcc structures presenting both {111} and {100} facets. The larger facets of single-domain nanocrystals with fcc structures than of multiply-twinned particles (MTPs) favor face-to-face interactions via the tetradecanethiol molecules. Hence, only translational order is observed with 3D superlattice film of MTP nanoparticles. From these results, one expects that crystallinity segregation of nanocrystals takes place during their self-assembling in 3D superlattices on the substrate.

The data described above are obtained when the colloidal solution is subjected to high solvent vapor pressure during the evaporation process. Let us decrease the solvent vapor pressure from 75% to 39%, keeping the same setup under nitrogen and free of toluene. Consequently, the evaporation time decreases from 25 to 9 hours. Figure 3.12a,b shows thin well-ordered fcc layer-by-layer films while no more truncated-tetrahedral 3D superlattices are observed. This highlights the key role of the evaporation rate and/or solvent vapor pressure in producing simultaneously the truncated-tetrahedral and film 3D superlattices. Hence, segregation of nanocrystallinity through 3D superlattice formation is only observed in systems under slow evaporation.

On increasing the Au nanocrystals from 5 to 7 nm, keeping the same solvent vapor pressure (39%), Figure 3.12c clearly shows the coexistence of ordered and disordered films. Hence, colloidal Au nanocrystals coated with tetradecanethiol dispersed in toluene and subjected to a slow evaporation process under high solvent vapor pressure favor ordered mesostructures either in truncated-tetrahedral 3D superlattices or in supracrystalline films.

With hexadecanethiol used as coating agent, instead of evaporation under at 39% and 75% solvent pressure, of 5- and 7-nm Au nanocrystals, amorphous films are obtained (Figure 3.13).

Hence, the morphologies of the fcc 3D superlattices markedly change with the solvent vapor pressure. On increasing it from 0% to 75%, with 5-nm Au nanocrystals coated with dodecanethiol, the morphology evolves from film to shaped 3D superlattices. The major changes are observed when nanocrystals are coated with tetradecanethiol. The morphology evolves from an fcc 3D superlattice film sitting on a disordered film to highly ordered films and finally to equilateral fcc 3D superlattices with underneath fcc 3D superlattice films.

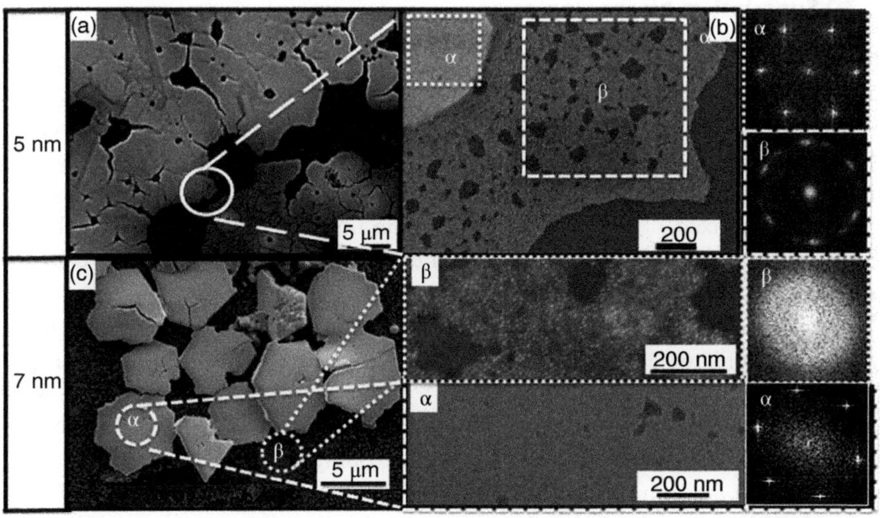

Figure 3.12 SEM images of 5-nm (a, b) and 7-nm (c) Au nanocrystals coated with various tetradecanethiol **α** and **β** zones of the SEM images and the corresponding Fourier transform. The solvent pressure during the evaporation process is 39% corresponding to the evaporation time of nine hours. *Source:* Wan et al. [2]/American Chemical Society.

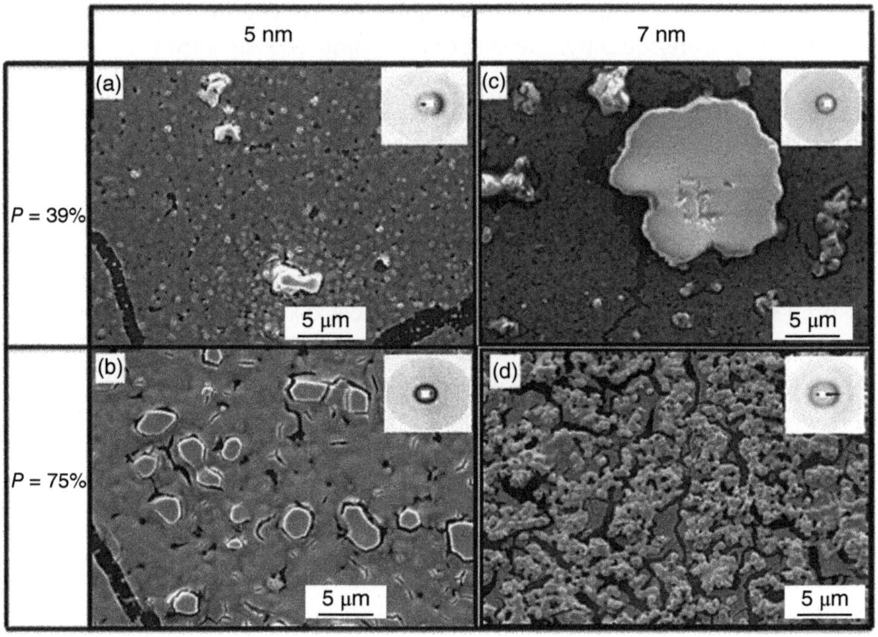

Figure 3.13 SEM images of assemblies obtained by toluene evaporation process, under 39% (a, c) and 75% (b, d) solvent vapor pressure at 25 °C of 5-nm (a, b) and 7-nm (c, d) Au nanocrystals coated with hexadecanethiol. *Source:* Wan et al. [2]/American Chemical Society.

3.1.2 Influence of Temperature During the Evaporation Process on Au 3D Superlattice Morphologies

The experiments above were performed at room temperature. We could ask ourselves: What is the temperature effect on the evolution of both morphology and interparticle distances? We use similar procedure as that described for 75% of toluene vapor pressure at 25 °C. (Nitrogen flow passes through a flask containing toluene solution before reaching the colloidal solution (Scheme 3.1a), and the substrate temperature is fixed.) At 50 °C, the evaporation time is around four to five hours. The 3D superlattice morphologies of 5- and 7-nm Au nanocrystals coated by the various coating agents used above is given in Figures 3.14 and 3.15.

With 5-nm Au nanocrystals coated with dodecanethiol or tetradecanethiol, layer-by-layer films (Figure 3.14) with a few percent of shaped 3D superlattices are present (Figure 3.14a–f). Furthermore, in contrary to what was obtained at 25 °C, fcc layer-by-layer 3D superlattices are produced for 5-nm Au nanocrystals coated with hexadecanethiol. Hence, on increasing the substrate temperature from 25 to 50 °C, a transition from amorphous (disordered) to 3D superlattices (highly ordered) films is observed. This is explained as follows: (i) the crystallization temperature increases with the chain length of coating ligands, especially above 42 °C for $C_{16}H_{33}$–SH [18], so only on increasing the substrate temperature up to 50 °C, the 5-nm Au nanocrystals coated with hexadecanethiol can be self-organized into a superlattice. (ii) The increase in temperature induces a dynamic process with transmission of the gauche defect

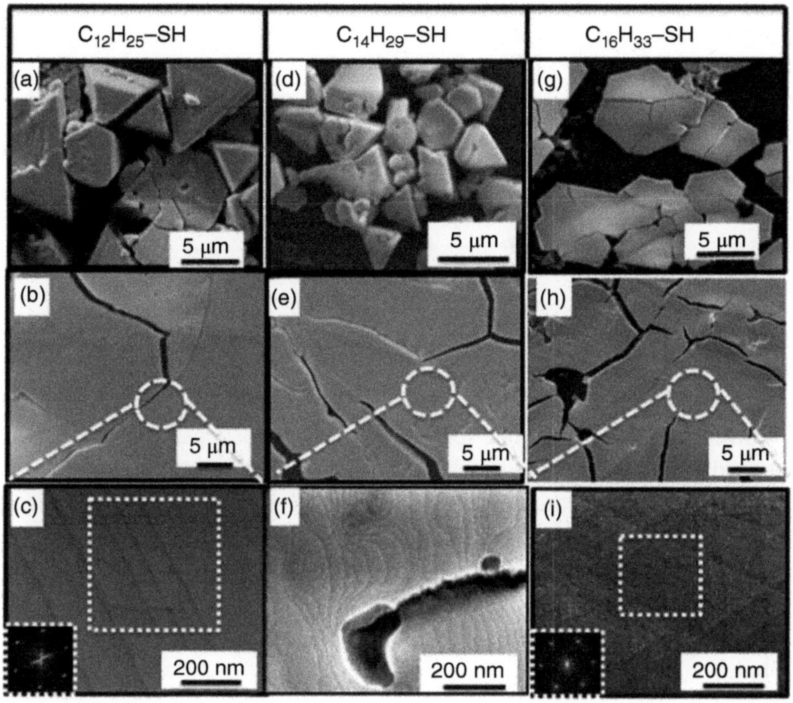

Figure 3.14 SEM images obtained for 5-nm Au nanocrystals coated with dodecanethiol (a–c), tetradecanethiol (d–f), and hexadecanethiol (g–i) produced by controlling the substrate temperature at 50 °C, evaporation time of five hours. *Source:* Wan et al. [2]/ American Chemical Society.

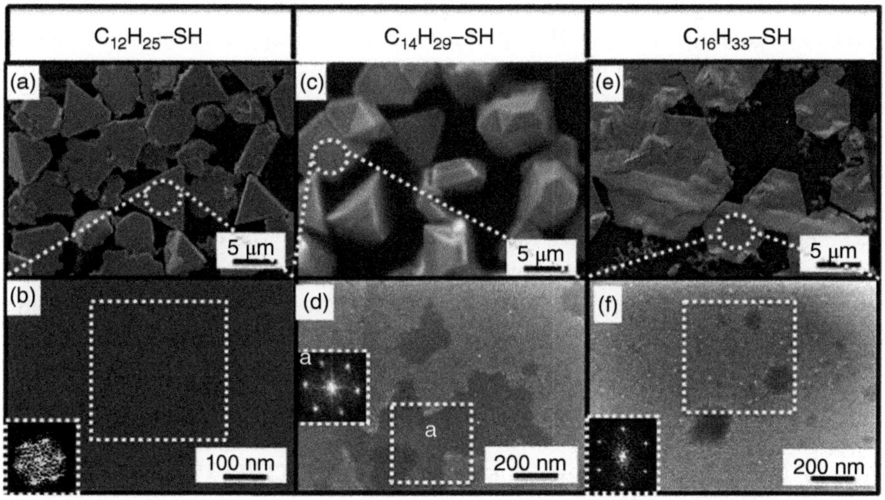

Figure 3.15 SEM images obtained for 7-nm Au nanocrystals coated with dodecanethiol (a, b), tetradecanethiol (c, d), and hexadecanethiol (e, f) produced by controlling the substrate temperature at 50 °C, evaporation time of five hours. *Source:* Wan et al. [2]/ American Chemical Society.

Table 3.1 Morphologies corresponding to the various experimental conditions with pressure (P) and temperature (T), film of self-ordered nanocrystals (F), amorphous film of disordered nanoparticles (A), shapes (S), and faceted film (FF).

Coating agent	$C_{12}H_{25}$–SH		$C_{14}H_{29}$–SH		$C_{16}H_{33}$–SH	
***d* (nm)**	**5.3**	**7.0**	**4.8**	**7.0**	**5.1**	**6.7**
$P = 0$, $T = 25\,°C$	F	S	F+A	F+A	A	F+A
$P = 39$, $T = 25\,°C$	S+F	S+F	A	S	A	A
$P = 75$, $T = 25\,°C$	S+F	S	S+F	S	A	A
$P = 25$, $T = 75\,°C$	S+F	S	S+F	S+F	F	F

from the end of the long-chain cone to the head direction [19]. Hence, these gauche defects preventing the 3D superlattice formation at 25 °C disappear, and dynamic offers the needed driving force for the chains to interdigitate and consequently to form an ordered film. The increase in the dynamic process indicates that the configuration of capping ligand chains is sensitive to temperature as already observed [20]. Crystalline agglomerates of 3D superlattices occur at high temperature [13].

On increasing the nanocrystal size from 5 to 7 nm, for any coating agent, more shaped 3D superlattices are obtained (Figure 3.15a,c,e) with well-ordered surfaces (Figure 3.15b,d,f).

3.1.3 Hierarchy in the 3D Superlattice Crystallinity

An overview of the data presented here shows a hierarchy in the mesoscopic ordering of nanocrystals. Sequential transitions disordered/disordered with ordered film/ordered films/shaped assemblies are observed (Table 3.1).

This sequential hierarchy in 3D superlattice growth is supported by the half width of the Bragg peak deduced from the diffraction patterns. Figure 3.16 shows that the half width of the Bragg peak is larger when nanocrystals form fcc films than shaped fcc assemblies.

From these data, it is claimed that a hierarchy in the 3D superlattice growth process takes place from disordered film to layer-by-layer to shaped fcc 3D superlattices. The kinetic process governs such hierarchy in the nanocrystal ordering with a transition from heterogeneous (layer-by-layer) to homogeneous (shaped 3D superlattices) growths. Furthermore, it is demonstrated that the increase in the chain length of coating agents induces a decrease in the ability of nanocrystals to self-assemble in fcc 3D superlattices.

3.2 Interparticle Distance Between Nanocrystals Coated with Thiol Derivatives

Let us first consider the interparticle distance of 3D superlattices produced under zero *solvent vapor pressure*. The SAXS measurements show that crystalline structures are fcc 3D superlattices. From such packing, the center-to-center distance D is

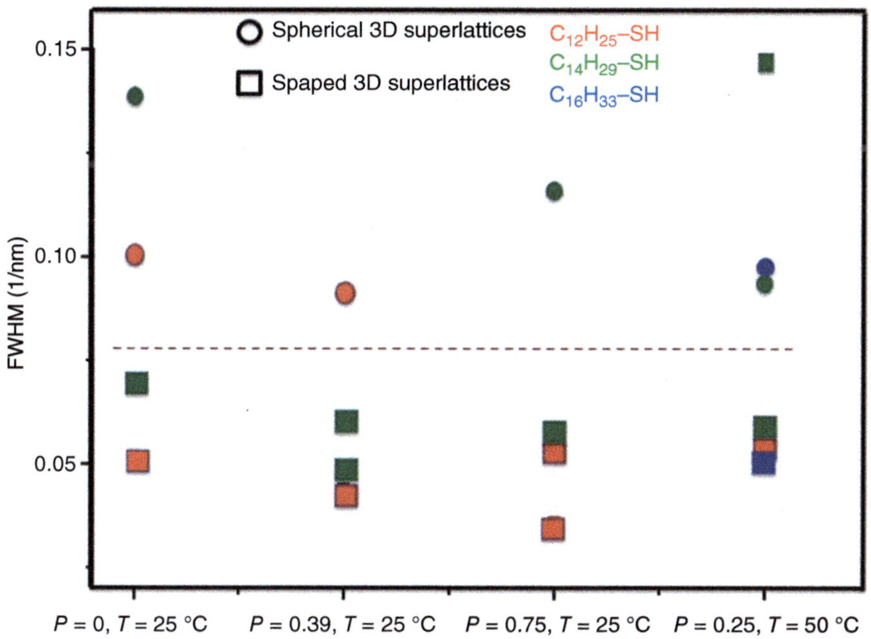

Figure 3.16 The full width at half maximum (FWHM) of all SAXS diffraction patterns given in this figure, expressed in various experimental conditions (*P*, *T*). Spheres represent films and square shapes represent 3D superlattices. Au nanocrystals coated with C_{12}, C_{14}, and C_{16} are red, green, and blue, respectively. *Source:* Wan et al. [2]. Copyright 2013 American Chemical Society.

deduced, and the interparticle spacing δ is given by $\delta = D - d$, with d being the average diameter of the Au nanocrystals determined from transmission electron microscopy (TEM) images. The width of the reflections is resolution-limited indicating average 3D superlattice dimensions of at least a few tenths of a micrometer. Table 3.2 shows that, within the experimental error, the average interparticle distance δ calculated from the d_{111} spacing is closed to the length of the all trans configuration thiol derivations (1.78, 2.03, and 2.28 nm for $C_{12}H_{25}$–SH, $C_{14}H_{29}$–SH, and $C_{16}H_{33}$–SH, respectively) [22], indicating a "quite" perfect interdigitation of the coating agents. From these data, a transition from film to shaped fcc 3D superlattices takes place on increasing the nanocrystal size, keeping similar crystalline structure (fcc) and average diameter between nanocrystals ($\delta = 2.0 \pm 0.1$ nm).

By plotting the interparticle distance with the number of carbon atoms, a linear dependence is observed with a standard deviation of 1.7% (Figure 3.17 [■]). These data confirm that there is no significant effect of the nanocrystals' sizes on the final interparticle distance after the formation of the 3D superlattices. It is also noteworthy that such linear dependence, when extrapolated to a coating agent's chain of zero carbon atom, reaches $\delta_{pp} = 0.51$ nm. This value is in full agreement with data obtained previously [23].

Schapotschnikow et al. [21] develop a geometrical model supported by Monte Carlo simulations and based on the assumption that the motion of the ligands

Table 3.2 Lattice parameters obtained for 3D superlattices dried under zero toluene vapor pressure for Au nanocrystals differing by the length of their coating agents and their sizes: d (nm) is the average diameter, D (nm) is the center–center inter-NC distance, L (nm) is the length of alkyl chain (L_{C12} = 1.77, L_{C14} = 2.03, L_{C16} = 2.28), the scaled distance is $\tau = D/d$, and the ratio is $\lambda = L/R = 2\ L/d$. $\tau_{(theo)} = 1 + \lambda$ and $\lambda_{(theo)}$ are the estimated values from the computations. $\tau_{(exp)}$, $\lambda_{(exp)}$ are the values deduced from experiments. L and S indicate layer-by-layer and shape 3D superlattices, respectively, whereas A is related to amorphous film.

Coating agent	$C_{12}H_{25}$–SH		$C_{14}H_{29}$–SH		$C_{16}H_{33}$–SH	
d (nm)	5.3	7.0	4.8	7.0	5.1	6.7
D (nm)	7.1	8.9	6.9	9.0	A	9.0
δ (nm)	1.8	1.9	2.1	2.0	A	2.3
σ (%)	7.9	5.1	7.0	5.1	5.1	4.8
δ/L	1.02	1.07	1.03	0.99	A	1.01
$\tau_{4b(theo)}$	1.32	1.27	1.40	1.28	A	1.34
τ_{exp}	1.34	1.27	1.44	1.29	A	1.34
$\lambda_{4b(theo)}$	0.68	0.51	0.85	0.58	A	0.68
λ_{exp}	0.67	0.51	0.85	0.58	A	0.68
Morphology	L	S	L+A	S	A	L+A

Based on Schapotschnikow et al. [21].

adsorbed onto the particle's surface is limited in space and assumes a conical shape, overlapping when two or more nanoparticles come in close proximity. The alkyl chains used to coat the nanocrystals are assumed to be in a semicrystalline state in their trans configuration and to control the cone's volume and length and consequently the average distance between the nanocrystals. Considering the two variables developed in this model, i.e. the scaled equilibrium distance $\tau = D/d$ and the scaled ligand length $\lambda = 2L/d$, where D (nm) is the center–center inter-nanocrystal distance, d (nm) is the average diameter, and L (nm) is the length of alkyl chain, the best agreement with the experimental data corresponds to a four-body interaction process (Table 3.2). This indicates that, with 3D superlattices produced by solvent evaporation *under zero vapor pressure*, the length of the cone's portion that overlaps one another is directly related to the length of the alkyl chains used to passivate the nanocrystals. This clearly shows that the interparticle distances are well predicted by Monte Carlo simulations when the solvent vapor pressure is kept to zero during 3D superlattice growth with rather efficient interdigitation process.

At this point we could ask ourselves: Does the *solvent vapor pressure* play a role on the interparticle distance? To answer this question, let us start with a dried fcc 3D superlattice of 5-nm Au nanocrystals coated with dodecanethiol and characterized by 1.8 nm as the δ_{pp} value. The well-defined Bragg reflection width (Figure 3.18a) confirms long-range ordering of the Au nanocrystals in the plane parallel to the surface of the substrate. To increase the toluene vapor pressure

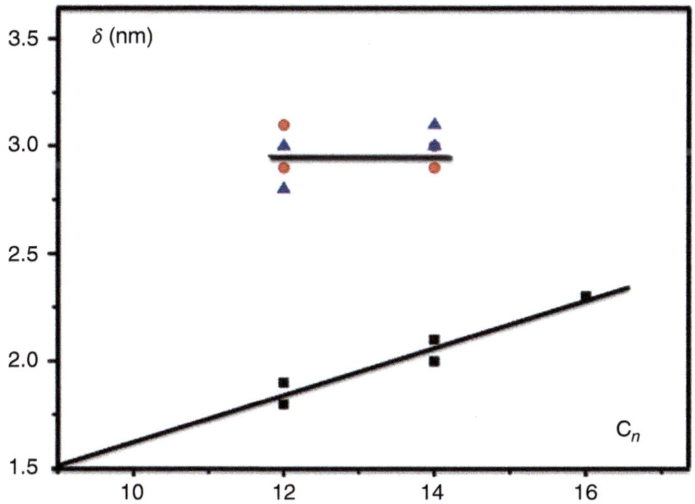

Figure 3.17 Variations of the interparticle edge-to-edge distance **δ** (nm) with the number of carbon atoms of the alkyl chain of the coating agent obtained through various procedures characterized at $P_t = 0$ (■), $P_t = 39\%$ (•), and $P_t = 75\%$ (▲). Black indicates 3D superlattices grown from 5-nm nanocrystals. Red indicates 3D superlattices of 7-nm nanocrystals. *Source:* Wan et al. [1]. Copyright 2013 American Chemical Society.

from 0% to 100%, cotton swabs are placed within the chamber containing the dried 3D superlattices. In a two- to three-second timeframe, the {111} Bragg reflection shifted toward lower angles and no longer varied with time (Figure 3.18b) indicating a rapid increase in the interparticle distance between Au nanocrystals from 1.8 to 2.5 nm (Table 3.3). If the cotton swabs are taken off the chamber and the sample is allowed to dry in air for 15 hours, the Bragg peak is shifted back toward large angles (Figure 3.18c). The interparticle distance reaches 1.9 nm, which is closed to the value obtained before saturating the atmosphere with toluene (1.8 nm). This clearly shows that the average distance between nanocrystals decreases when P_t is reduced from 100% to 0%. By repeating this experience several times, it is deduced the reversibility of the interparticle distances. Those results are depicted in Figure 3.18d,e. The 3D superlattices expand or retract reversibly through elastic deformations by tuning the solvent vapor pressure. This observation is consistent with a model developed previously [24] describing the elastic contribution of the ligand chains that can interpenetrate and be compressed when the value of the scaled distance $\tau_{exp} = D/d$ is below 1.36. Removing the toluene molecules from the nanocrystal arrays takes longer time than including them within the superlattices. This shows the strong ability of the 3D superlattice to trap the solvent molecules even though it is initially dried and leads to a rapid increase in δ_{pp} values. These variations in the internal structure of the 3D superlattices are in good agreement with those obtained with PbSe and PbS nanocrystals [24] for which δ values increase with solvent vapor pressure. This clearly indicates that dried 3D superlattices do not entirely keep their internal integrity even though they remain ordered.

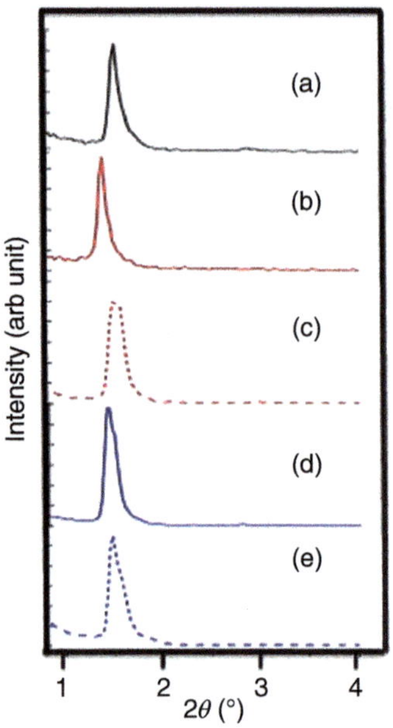

Figure 3.18 *In situ* SAXRD patterns. (a) Original pattern with interparticle distance, $\boldsymbol{\delta}_{pp}$ = 1.8 nm; (b) first increase of solvent vapor pressure, $\boldsymbol{\delta}_{pp}$ = 2.5 nm; (c) first decrease of solvent vapor pressure, $\boldsymbol{\delta}_{pp}$ = 1.9 nm; (d) second increase of solvent vapor pressure, $\boldsymbol{\delta}_{pp}$ = 2.3 nm; and (e) second decrease of solvent vapor pressure, $\boldsymbol{\delta}_{pp}$ = 1.9 nm. *Source:* Wan et al. [1]. Copyright 2013 American Chemical Society.

Table 3.3 Interparticle distance ($\boldsymbol{\delta}_{pp}$) varying reversibly while increasing/decreasing the toluene vapor pressure within the chamber by adsorption/desorption of a cotton swab impregnated with toluene.

Time	0	2–3 s	15 h	2–3 s	15 h
δ (nm)	1.8	2.5	1.9	2.3	1.9

Note: s, seconds; h, hours; δ_{pp}, the edge-to-edge distance between nanocrystals.

At such point, one may wonder if a disordered dried nanocrystal film with a low size distribution could form 3D superlattices under solvent exposure. To answer this question, let us consider 5-nm Au nanocrystals coated with hexadecanethiol forming an amorphous film (Figure 3.8). The SAXRD pattern of the dry film shows, as expected, a diffuse ring characteristic of an amorphous structure (Figure 3.19a). By submitting the sample of 100% solvent vapor pressure as described above, during 15 hours, the SAXRD pattern shows a retraction of the ring and the appearance of shadow peaks, indicating an increase in the average distance between nanocrystals and the emergence of local nanocrystal ordering (Figure 3.19b). The sample is then kept in the absence of toluene for 24 hours and allowed to dry. The corresponding SAXRD pattern shows similar features of that depicted in Figure 3.19a, characteristic of a disordered film (Figure 3.19c). Such switch between disordered and locally ordered films is successfully repeated several times, and similar structural

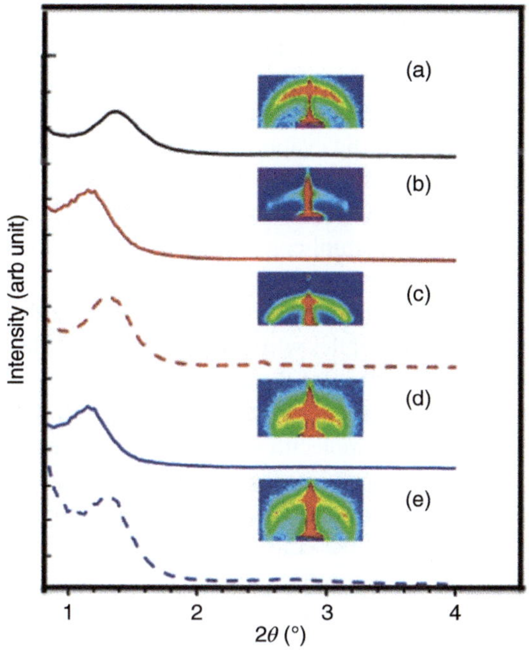

Figure 3.19 *In situ* SAXRD patterns. (a) Original pattern 3D superlattice grown at $P_t = 0$ of 5-nm Au nanocrystals coated with hexadecanethiol, (b) first increase of solvent vapor pressure for 15 hours, (c) first decrease of solvent vapor pressure for 24 hours, (d) second increase of solvent vapor pressure for 15 hours, and (e) second decrease of solvent vapor pressure for 24 hours. *Source:* Wan et al. [1]. Copyright 2013 American Chemical Society.

variations are observed (Figure 3.19d,e), indicating a reversible process. However, the SAXRD patterns presented in Figure 3.19b,d only show the presence of a local order when the film is submitted to toluene saturated vapor pressure, and the emergence of a long-range order within the 3D superlattices is not observed. Nevertheless, the overall ordering within the structure can be greatly improved by swelling.

These data clearly show that the solvent vapor pressure plays an important role on the interparticle distance between nanocrystals used as building blocks to produce 3D superlattices.

Solvent molecules stick to the alkyl chain and blow up the interdigitation of the alkyl chain and consequently the interparticle distance. Such behavior is observed when the nanocrystals were deposited under zero vapor pressure.

On increasing toluene vapor pressure during the solvent evaporation process and using a glass vessel (see above), the interparticle distance markedly increases compared with what is observed at zero vapor pressure and remains independent to the length of the alkyl chains used as coating agents and consequently to the 3D superlattice growth mechanism. Built on the results depicted above, it is obvious that the solvent vapor pressure during the 3D superlattice growth process and/or after its growth has a major impact on its internal structure and on the interparticle distances within it in particular. This inefficiency of the alkyl chain length and the large value of the interparticle distance compared with data obtained at zero vapor pressure cannot be attributed to an adsorption of solvent molecules on the alkyl chains. Another effect is importance in the structure of the superlattices: in spite of the purification process after the nanocrystal synthesis, the presence of free thiols within the colloidal solutions, and hence during the evaporation process, has to be considered. The chemical analysis of the samples shows a large excess of free alkanethiol molecules. Hence, during the evaporation process and under zero toluene vapor pressure, a part of the residual "unreacted" thiol-containing molecules physiosorbed on the alkyl chains used as coating agents and could be likely expelled with the solvent during evaporation. The depletion forces created by these thiol-containing molecules favor the nanocrystal assemblies while the length of the alkyl chains controls the interparticle distances. This process markedly drops on increasing the solvent vapor pressure to reach a total drop at 75%. Then, the excess thiol derivatives cannot be excluded with the solvent molecules by depletion forces. In these conditions, the solvent and thiol-containing molecules remain in the superlattices and render the contribution of the coating agent's alkyl chain length negligible. Hence, both residual thiols and solvent molecules control the length of the cone and consequently the interparticle distances. A tentative explanation of such confinement effect could be as follows: during the evaporation process, a liquid–gas-phase transition takes place, and the 3D superlattices are formed before the end of the evaporation process [25]. The increase in the vapor pressure (39% and 75%) could induce as previously observed with various microemulsion changes in the chemical potential of the system and consequently in the composition of the microphases. Hence, the chemical potential at zero vapor pressure markedly drops compared with higher values. Here, one can assume that the change in the chemical potential induces a variation in the phase compositions and consequently an evolution of the system from a liquid–gas-phase transition to a triple-point (liquid–gas–solid) transition. By analogy to what was observed with 7.8-nm silica nanoparticles, at high-volume fractions [26], the size of the geometrical cones could be regulated to induce a vitrification process, hence keeping the cone's length "quasi" constant for the various alkyl chain lengths used. The large standard deviation of the interparticle distance is explained by the fact that the amount of thiol-containing molecules could slightly evolve from one sample to the other.

The interparticle distance of 3D superlattices having 5-nm Au nanocrystals, produced by increasing the substrate temperature during the evaporation process, drops on increasing the chain length of the alkyl chain (Table 3.4, Figure 3.20). This is related to the fact that even though the chain lengths are predominantly in the all trans zigzag conformation, the end-gauche defects markedly increase with increasing the alkyl chain length, whereas no internal (surface) gauche defects are detected [22, 27–29]. The long alkyl chains rotate more freely, leading to an increase in the residence time on increasing the temperature above the melting temperature of the organic chains [13]. Hence, large amplitude motions about the long axes of the chains and molecular mobility increase toward the unbound ends. The increase in the alkyl chain length induces an increase in the population of gauche conformer and consequently a higher mobility of the chains. Upon increasing the substrate

Table 3.4 Lattice parameters obtained for 3D superlattices grown at 50 °C of Au nanocrystals differing by the length of their coating agents and their sizes: *d* (nm) is the average diameter, **σ** (%) is the size distribution, *D* (nm) is the center–center inter-NC distance, and $\boldsymbol{\delta}_{pp}$ (nm) is the edge-to-edge distance.

Coating agent	$C_{12}H_{25}$–SH		$C_{14}H_{29}$–SH		$C_{16}H_{33}$–SH	
d (nm)	5.3	7.0	4.8	7.0	5.1	6.7
σ (%)	7.9	51	7.0	51	7.0	4.8
δ (nm)	2.9	2.5	21	2.8	2.3	2.8

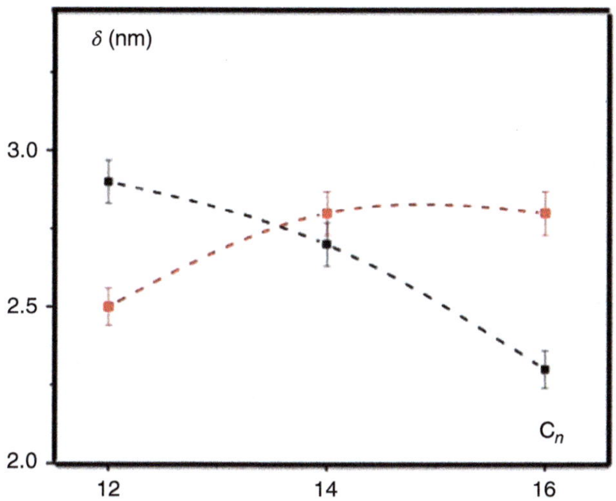

Figure 3.20 Plot of the $\boldsymbol{\delta}_{pp}$ values at 50 °C (P_t = 25%) versus number of carbons in the alkyl chain of the various thiol-containing coating agents. *Source:* Wan et al. [4]. Copyright 2014 American Chemical Society.

temperature from room temperature to 50 °C, the alkyl chains bound to the Au surfaces form bundles during the evaporation time and consequently when the dry 3D superlattices are formed. The formation of bundles along the alkyl chains is likely to be favored by the end-gauche defects. On increasing the alkyl chain length, the defects move along the chain and the bundles formed. Hence, the number of carbon atoms involved in the bundles increases with the length of the alkyl chains. Consequently, the distance between nanocrystals decreases with increasing length of the alkyl chains.

On increasing the nanocrystal's size to 7 nm, the interparticle distance remains unchanged at room temperature, whereas at 50 °C, it increases with increasing number of carbon atom of the alkyl chains. Such behavior is the opposite of what was observed with 5-nm Au nanocrystals at the same growth temperature. Figure 3.20 shows such opposite behavior of the interparticle distance dependence with the length of the alkyl chain used as coating agent of 3D superlattices having building blocks of either 5 and 7 nm as diameters. The only difference between these two processes is related to the core size of nanocrystals. To explain such difference, one could assume that the increase in the van der Waals attractive forces due to an increase in the nanocrystal metallic core size leads to a reduction of the chain dynamics and consequently a decrease in the formation of bundles. In fact, one could assume that both the van der Waals interactions and the dynamic processes of the chains control the interparticle distance. Hence, the interparticle distance within 5- and 7-nm Au nanocrystals is governed by both the dynamics of the coating agent chains and the van der Waals attractive forces occurring between the nanocrystals.

From this set of data, it seems reasonable to claim that, even though some general rules can be drawn, the interparticle distance remains very challenging to predict since it is dependent on many experimental parameters.

3.3 Au 3D Superlattices Coated with *N*-Heterocyclic Carbene

In the last decades, *N*-heterocyclic carbenes (NHCs) have emerged as an essential class of ligands in organometallic chemistry [30–32]. Metal–NHCs have been widely investigated for catalytic applications [33], and their potential as biologically active complexes has been highlighted [34–36]. In both areas, the strength of the Au–$C_{carbene}$ bond has been exploited to get stable Au–NHC complexes. For the same reason, and because of their high synthetic flexibility, NHCs are expected to be highly promising ligands for the development of functionalized materials [35]. A few reports described the formation of NHC-based self-assembled monolayers (SAMs) on Au surfaces [37–38]. It was demonstrated that benzimidazole-derived NHCs provided SAMs with high stability in various conditions. In addition, the use of appropriate NHCs has enabled to modify the properties of NHC-coated Au surfaces by chemical post-functionalizations. Besides, NHCs have been involved as free NHCs or metal–NHC complexes for the preparation/stabilization of metal nanoparticles (metal = Pd, Ru, Ag, Au) [39–46]. However, the access to NHC-coated NCs with a narrow size distribution, a prerequisite for self-assembly into organized superstructures, remains a critical issue. Only one report described the emergence of 3D organized Au nanocrystal superstructures with three nanocrystal layers, obtained by slow evaporation of a concentrated solution of relatively monodisperse Au nanocrystals on a TEM grid [42].

Various benzimidazole-based NHC ligands as coating agents are designed to study the influence of the NHC structure on the average size, size distribution, and self-assembling properties of NHC-coated Au nanocrystals (Figure 3.21). We introduce long alkyl chains at different positions (R_1 and R_2) of the NHC ligands. This points in various directions of the space. Consequently, this shows the influence of long alkyl chains with various spatial arrangement.

Let us consider nanocrystals coated by a ligand with long chains ($C_{14}H_{29}$) both on the nitrogen atoms and at position 5 of the aromatic ring, called for simplicity Au–NHC (L_3) as shown in Figure 3.22a.

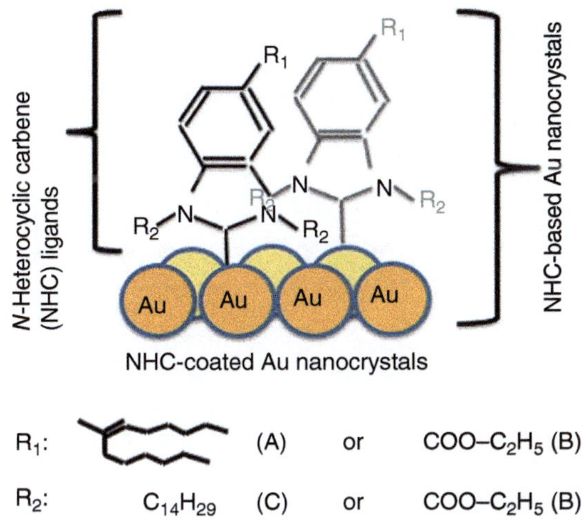

Figure 3.21 *N*-Heterocyclic carbene (NHC) ligands used in this study to investigate the preparation of NHC-coated Au nanocrystals and their 3D assembling into ordered superstructures.

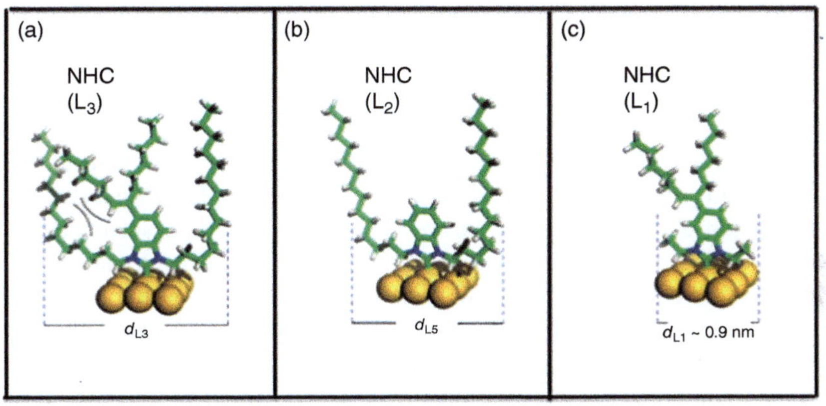

Figure 3.22 3D representations of Au–NHC (L_3) (a), Au–NHC (L_2) (b), and Au–NHC (L_1) (c).

Note that the nanocrystal size produced by this procedure is 4 nm. This corresponds to a smaller size than the synthesis of nanocrystals coated with dodecanethiol. This is attributed to greater steric hindrance. Therefore, the maximum distance between the alkyl chains and its volume is rather large, suggesting a greater covering of Au surface and greater steric overlap with neighboring ligands. At the end of solvent evaporation, amorphous colloidal film with a homogeneous distribution of very small scale 3D superlattice is obtained (Figure 3.23a). HRSEM images of a zone (bright zone) show very small organized domains (Figure 3.23b, FFT pattern of zones a^1 and a^2), whereas the dark areas (zone b in Figure 3.23a) are disordered (Figure 3.23c, FFT pattern of zones b^1).

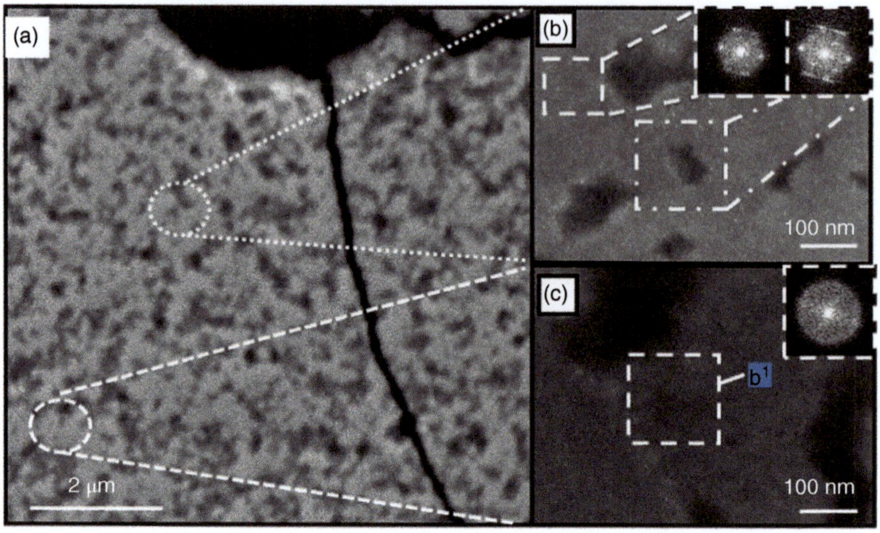

Figure 3.23 TEM images at various scales of Au–NHC (L3) amorphous films with small zone of 3D superlattices. (a) overall TEM image (b,c) increase of the resolution in two specific zones of the TEM image shown in (a), inset (b,c) enhancement of zones shown in b and c and HRSEM images of Au–NHC (L_3) amorphous films with small zone of 3D superlattices. *Source:* Ling et al. [32]/American Chemical Society.

If we remove the long chain on the nitrogen atoms and at position 5 of the aromatic ring keeping the two long chains ($C_{14}H_{29}$) called for simplicity Au-NHC (L_2) as shown is Figure 3.22b, the nanocrystal size is 5.4 nm. At the end of solvent evaporation, thick homogeneous films (c. 0.5–1 μm) are characterized by domains up to 8–10 μm with small surface defects and separated by cracks (Figure 3.24). These domains are highly ordered in a hexagonal pattern as clearly demonstrated by the HRSEM images and confirmed by FFT (insets). Similar behavior is obtained with Au–NHC (L_2) nanocrystals (Figure 3.22c).

The distance of benzimidazole-based NHC is 0.84 nm [38]. This distance is directly related to the volume occupied by the ligand at the surface of the nanocrystals. In Au–NHC (L_2), long alkyl chains are placed on the nitrogen atoms and point on both sides of the NHC ligand. Due to the geometry, these long chains are more distant to each other than in Au–NHC (L_2). Moreover, gauche defects are likely to occur at a distance from the nitrogen atom exceeding two carbons. Therefore, the maximum distance between the alkyl chains in Au–NHC (L_2) and its volume might be higher than in Au–NHC (L_1) ($d_{L2} > d_{L1}$) (Figure 3.23b,c). This suggests that long alkyl chains on the nitrogen atoms induce greater covering of Au surface and greater steric overlap with neighboring ligands, resulting in smaller nanocrystal. In the case of Au–NHC (L_3) (Figure 3.23a), the steric repulsion between the alkyl chains on the benzimidazole backbone and the C_{14} chains on the nitrogen atoms may push the *N*-alkyl chains outwardly leading to an increase of the NHC volume by comparison with L_5 and an increase of steric overlap ($d_{L3} > d_{L2} > d_{L1}$). From that, it is concluded that a limit in the ligand distance is needed to produce nanocrystals.

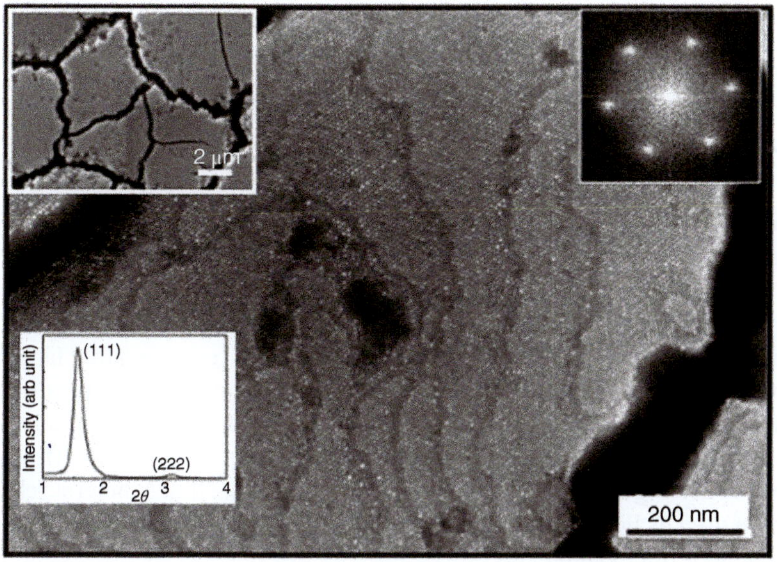

Figure 3.24 HRSEM image of Au 3D superlattices coated by Au–NHC (L_2). Insets: SEM image, SAXRD pattern, and FFT of image. *Source:* Roland et al. [30]/American Chemical Society.

The edge-to-edge distances in the 3D superlattices are determined from the main peak of the SAXRD patterns by using the Bragg's law for Au–NHC (L) and Au–NHC (L_2) nanocrystals, which are 1.8 and 2.0 nm respectively. An estimation of the distance between the Au atom and the extremity of the aromatic part of the NHC gave a value of about 0.75 nm. This implies a remaining distance of 0.5 nm between the aromatic cores of two NHC ligands belonging to two neighboring Au nanocrystals. To stabilize 3D superlattices, the long C_{14} alkyl chains on the nitrogen atoms must be involved in stabilizing interdigitations. For the interdigitation to take place, the long alkyl chains cannot adopt all-trans zigzag configuration but likely contain near-surface gauche defects so that the chains point outward with the appropriate geometry. This possible mode of action is presented in Figure 3.25. The estimated

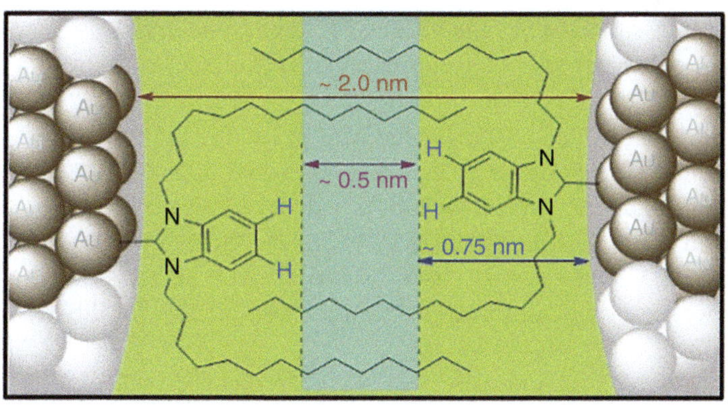

Figure 3.25 Proposed model for the self-assembly of 5-nm Au nanocrystals capped with Au–NHC (L_2) nanocrystals. *Source:* Ling et al. [32]/American Chemical Society.

distance of 0.5 nm between the aromatic NHC cores (blue area) corresponds in the 3D superlattice to the interdigitation of at least five carbons of the alkyl chains.

3.4 Conclusions

In this chapter, we show that the solvent used to disperse nanocrystals and its vapor pressure during the evaporation processes are key parameters on the 3D superlattice growth mechanism inducing marked changes in the final 3D superlattice morphology. The interparticle distance between nanocrystals increases linearly with the length of the alkyl chain at zero vapor pressure, whereas it remains rather larger and unchanged when the evaporation process takes place at solvent vapor pressure larger than zero. By slow evaporation, the crystallinity of nanocrystals called nanocrystallinity induces a hierarchy in the 3D superlattice growth processes (from homogeneous to heterogeneous). Furthermore, we demonstrate that *N*-heterocyclic carbene can be used as ligand to produce 3D superlattices when the volume occupied by the ligand is not too large.

References

1 Wan, Y.F., Goubet, N., Albouy, P.A., and Pileni, M.P. (2013). Hierarchy in Au nanocrystal ordering in supracrystal: II. The control of inter-nanocrystal distances. *Langmuir* 29: 13576–13581.

2 Wan, Y.F., Goubet, N., Albouy, P.A., and Pileni, M.P. (2013). Hierarchy in Au nanocrystal ordering in supracrystals: a potential approach to detect new physical properties. *Langmuir* 29: 7456–7463.

3 Wan, Y.F., Portales, H., Goubet, N. et al. (2013). Impact of nanocrystallinity segregation on the growth and morphology of nanocrystal superlattices. *Nano Res.* 6: 611–618.

4 Schaeffer, N., Wan, Y., and Pileni, M.P. (2014). Hierarchy in Au nanocrystal ordering in supracrystal: III. Competition between van der Waals and dynamic processes. *Langmuir* 30: 7177.

5 Goubet, N., Richardi, J., Albouy, P.A., and Pileni, M.P. (2011). Which forces do control the supracrystal nucleation in organic media? *Adv. Funct. Mater.* 21: 2693–2704.

6 Compton, O.C. and Osterloh, F.E. (2007). Evolution of size and shape in the colloidal crystallization of gold nanoparticles. *J. Am. Chem. Soc.* 129: 7793–7798.

7 Rupich, S.M., Shevchenko, E.V., Bodnarchuk, M.I. et al. (2010). Size-dependent multiple twinning in nanocrystal superlattices. *J. Am. Chem. Soc.* 132: 289–296.

8 Stoeva, S.I., Prasad, B.L.V., Uma, S. et al. (2013). Face centered cubic and hexagonal closed-packed nanocrystal superlattices of gold nanoparticles prepared by different methods. *J. Phys. Chem. B* 107: 7441–7448.

9 Sigman, M.B., Saunders, A.E., and Korgel, B.A. (2003). Metal nanocrystal superlattice nucleation and growth. *Langmuir* 20: 978–983.

10 Abecassis, B., Testard, F., and Spalla, O. (2008). Gold nanoparticle superlattice crystallization probed *in situ*. *Phys. Rev. Lett.* 100: 115504.

11 Lalatonne, Y., Richardi, J., and Pileni, M.P. (2004). Van der Waals versus dipolar forces controlling mesoscopic organizations of magnetic nanocrystals. *Nat. Mater.* 3: 121–125.

12 Yan, H., Cingarapu, S., Klabunde, K.J. et al. (2009). Nucleation of gold nanoparticle superclusters from solution. *Phys. Rev. Lett.* 102: 095501.

13 Geyer, T., Born, P., and Kraus, T. (2012). Switching between crystallization and amorphous agglomeration of alkyl thiol-coated gold nanoparticles. *Phys. Rev. Lett.* 109: 128302.

14 Prasad, B.L.V., Sorensen, C.M., and Klabunde, K.J. (2008). Gold nanoparticle superlattices. *Chem. Soc. Rev.* 37: 1871–1883.

15 Motte, L. and Pileni, M.P. (1998). Influence on the 2D and 3D self-organization of the length of the alkyl chains used to passivate silver sulfide nanoparticles. *J. Phys. Chem.* 102: 4104.

16 Portalès, H., Goubet, N., Saviot, L. et al. (2008). Probing atomic ordering and multiple twinning in metal nanocrystals through their vibrations. *Proc. Natl. Acad. Sci. U.S.A.* 105: 14784.

17 Goubet, N., Portalès, H., Yan, C. et al. (2012). Simultaneous growths of gold colloidal crystals. *J. Am. Chem. Soc.* 134: 3714–3719.

18 Luedtke, W.D. and Landman, U. (1996). Structure, dynamics, and thermodynamics of passivated gold nanocrystallites and their assemblies. *J. Phys. Chem.* 100: 13323–13329.

19 Jiménez, A., Sarsa, A., Blazquez, M., and Pineda, T.A. (2010). Molecular dynamics study of the surfactant surface density of alkanethiol self-assembled monolayers on gold nanoparticles as a function of the radius. *J. Phys. Chem. C* 114: 21309–21314.

20 Wang, Y., Dai, Q., Wang, L. et al. (2010). Mutual transformation between random nanoparticles and their superlattices: the configuration of capping ligand chains. *J. Phys. Chem. C* 114: 11425–11429.

21 Schapotschnikow, P., Pool, R., and Vlugt, T.J.H. (2008). Molecular simulation of interacting nanocrystals. *Nano Lett.* 8: 2930–2934.

22 Badia, A., Gao, W., Singh, S. et al. (1996). Structure and chain dynamics of alkanethiol-capped gold colloids. *Langmuir* 12: 1262–1269.

23 Martin, J.E., Wilcoxon, J.P., Odinek, J., and Provencio, P. (2000). Control of the interparticle spacing in gold nanoparticle superlattices. *J. Phys. Chem. B* 104: 9475–9486.

24 Hanrath, T., Choi, J.J., and Smilgies, D.M. (2009). Structure/processing relationships of highly ordered lead salt nanocrystal superlattices. *ACS Nano* 3: 2975–2988.

25 Germain, V. and Pileni, M.P. (2005). Size distribution of cobalt nanocrystals: a key parameter in formation of columns and labyrinths in mesoscopic structure. *Adv. Mater.* 17: 1424–1429.

26 Robbes, A.S. (2009). Osmotic stress on concentrated colloidal suspensions: a path towards equilibrium? *Braz. J. Phys.* 39: https://doi.org/10.1590/S0103-97332009000200006.

27 Hostetler, M.J., Templeton, A.C., and Murray, R.W. (1999). Dynamics of place-exchange reactions on monolayer-protected gold cluster molecules. *Langmuir* 15: 3782–3789.

28 Mitra, S., Binoj Nair, S., Pradeep, T. et al. (2002). Alkyl chain dynamics in monolayer-protected clusters (MPCs): a quasielastic neutron-scattering investigation. *J. Phys. Chem. B* 106: 3960–3967.

29 Pradeep, T., Mitra, S., Sreekumaran Nair, A., and Mukhopadhyay, R. (2004). Dynamics of alkyl chains in monolayer-protected Au and Ag clusters and silver thiolates: a comprehensive quasielastic neutron scattering investigation. *J. Phys. Chem. B* 108: 7012–7020.

30 Roland, S., Ling, X., and Pileni, M.P. (2016). *N*-Heterocyclic carbene ligands for Au nanocrystal stabilization and three-dimensional self-assembly. *Langmuir* 32: 7683–7696.

31 Ling, X., Schaeffer, N., Roland, S., and Pileni, M.P. (2015). Superior oxygen stability of *N*-heterocyclic carbene-coated Au nanocrystals: comparison with dodecanethiol. *Langmuir* 31: 12873–12882.

32 Ling, X., Roland, S., and Pileni, M.P. (2015). Supracrystals of *N*-heterocyclic carbene-coated Au nanocrystals. *Chem. Mater.* 27: 414–423.

33 Bhattacharyya, A., Hwang, W.S., and Lin, I.J.B. (2009). Beyond conventional *N*-heterocyclic carbenes: abnormal, remote, and other classes of NHC ligands with reduced heteroatom stabilization. *Chem. Rev.* 109: 3445–3478.

34 Bourissou, D., Guerret, O., Gabbai, F.P., and Bertrand, G. (2000). Stable carbenes. *Chem. Rev.* 100: 39–92.

35 Herrmann, W.A. (2002). Heterocyclic carbenes: synthesis and coordination chemistry. *Angew. Chem. Int. Ed.* 41: 1290.

36 Diez-Gonzalez, S., Marion, N., and Nolan, S.P. (2009). *N*-Heterocyclic carbenes in late transition metal catalysis. *Chem. Rev.* 109: 3612.

37 Hindi, K.M., Panzner, M.J., Tessier, C.A. et al. (2009). The medicinal applications of imidazolium carbene–metal complexes. *Chem. Rev.* 109: 3859–3884.

38 Mercs, L. and Albrecht, M. (2010). Beyond catalysis: *N*-heterocyclic carbene complexes as components for medicinal, luminescent, and functional materials applications. *Chem. Soc. Rev.* 39: 1903.

39 Oehninger, L., Rubbiani, R., and Ott, I. (2013). *N*-Heterocyclic carbene metal complexes in medicinal chemistry. *Dalton Trans.* 42: 3269.

40 Zhukhovitskiy, A.V., Mavros, M.G., Van Voorhis, T., and Johnson, J.A. (2013). Addressable carbene anchors for gold surfaces. *J. Am. Chem. Soc.* 135: 7418.

41 Crudden, C.M., Horton, J.H., Ebralidze, I.I. et al. (2014). Ultra stable self-assembled monolayers of N-heterocyclic carbenes on gold. *Nat. Chem.* 6: 409.

42 Melaiye, A., Sun, Z., Hindi, K. et al. (2005). Silver(I)–imidazole cyclophane gem-diol complexes encapsulated by electrospun tecophilic nanofibers: formation of nanosilver particles and antimicrobial activity. *J. Am. Chem. Soc.* 127: 2285.

43 Huang, R.T.W., Wang, W.C., Yang, R.Y. et al. (2009). Liquid crystals of gold(I) *N*-heterocyclic carbene complexes. *Dalton Trans.* 36: 7121.

44 Vignolle, J. and Tilley, T.D. (2009). *N*-Heterocyclic carbene-stabilized gold nanoparticles and their assembly into 3D superlattices. *Chem. Commun.* 46: 7230.
45 Hurst, E.C., Wilson, K., Fairlamb, I.J.S., and Chechik, I.V. (2009). *N*-Heterocyclic carbene coated metal nanoparticles. *New J. Chem.* 33: 1837.
46 Serpell, C.J., Cookson, J., Thompson, A.L. et al. (2013). Haloaurate and halopalladate imidazolium salts: structures, properties, and use as precursors for catalytic metal nanoparticles. *Dalton Trans.* 42: 1385.

4

3D Superlattice Growth in a Thermodynamic Equilibrium

4.1 Simultaneous 3D Superlattices Heterogeneous and Homogeneous Growth Processes

From a colloidal solution of nanocrystals, two growth processes of 3D superlattices have been identified [1, 2]. The heterogeneous growth is layer by layer process whereas the homogeneous one takes place in the colloidal solution.

4.1.1 General Behavior of the Two Simultaneous Supracrystal Growth Processes of Au Nanocrystals

Dodecanethiol-coated Au nanocrystals, dispersed in toluene, with sizes varying from 5 to 8 nm and 6% size distribution are kept in a beaker under saturated toluene atmosphere. After seven days, *simultaneously appearing*, flocculation of Au nanocrystals occurred at the air–toluene interface, resulting in a bright thin interfacial film and precipitates of nanocrystals. Furthermore, the colloidal solution remains slightly colored, indicating that Au nanocrystals are still dispersed in solution (Figure 4.1). As the nanocrystal aggregation in the form of film and precipitate continued, the Au nanocrystal concentration in suspension decreases, as revealed by the color of solution. Such decrease is faster for larger nanocrystals; this has to be related to the fact that the depth and range of the attraction interaction between nanocrystals increase with the particle diameter. It indicates their ability to aggregate with the faster rate (Table 4.1). Note that with 4-nm nanocrystal size, the colloidal solution remains stable without the formation of both film and shaped 3D superlattices.

The interfacial film, withdrawn using a tungsten ring, is deposited on a substrate. Whatever the nanocrystal diameters, the scanning electron microscopy (SEM) images of interfacial films remain similar. Figure 4.2a–c shows the formation of cracks. Domains larger than several 10 μm are observed on low magnification of the interfacial film covering large surfaces of the substrate.

Metal Nano 3D Superlattices: Synthesis, Properties, and Applications, First Edition.

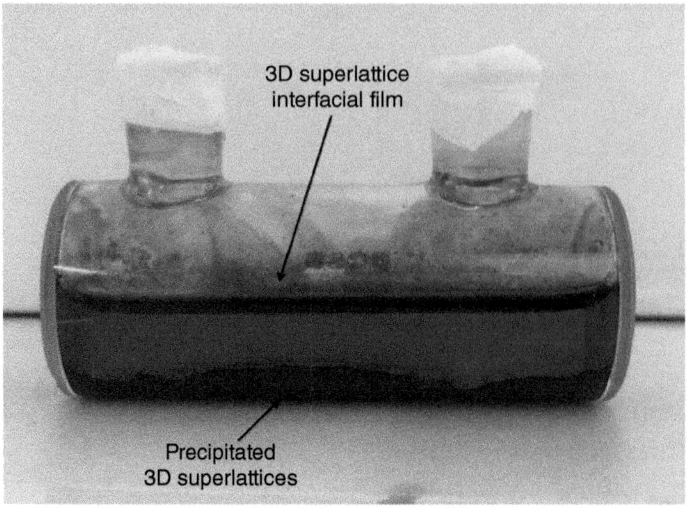

Figure 4.1 Photograph showing the formation of both the air–toluene interfacial and precipitated 3D superlattices in the colloidal solution placed under toluene-saturated atmosphere in a cylindrical vial. *Source:* M. P. Pileni (2019)/The Chemical Society of Japan.

Table 41 Lattice parameters obtained for film and shaped 3D superlattices grown at room temperature by keeping colloidal solution, under toluene saturation during seven days, composed of Au nanocrystals differing by their diameters and coated with dodecanethiol: *d* (nm) is the average diameter, and $\boldsymbol{\delta}_{\text{films}}$ (nm) and $\boldsymbol{\delta}_{\text{shape}}$ (nm) are the edge–edge distances of interfacial and precipitate 3D superlattices. *P* is the percentage of Au nanocrystals remaining in the colloidal solution (%).

Building block size	4.3	5.0	6.0	6.9	7.5
Film 3D superlattices	—	bcc + fcc	fcc	fcc	fcc
Shaped 3D superlattices	—	fcc	fcc	fcc	fcc
$\delta_{\text{film 3D superlattices}}$ (nm)	—	1.7	1.8	1.8	1.8
$\delta_{\text{shaped 3D superlattices}}$ (nm)		1.6	1.0	2.1	2.1
P (%)	100	51	13	5	4

Figure 4.3a–c shows the formation of terraces with linear steps having various heights, from one nanocrystal to several nanocrystals thickness, induced by screw dislocations. By tilting the samples (Figure 4.3d–f), the thickness of interfacial films having building blocks up to 6 nm remains unchanged (from 200 to 300 nm). For any building block size, the high-resolution SEM pictures, as shown in Figure 4.3g, shows the formation of a well-ordered Au nanocrystal film with very large uniform orientation areas indicating a long coherence length (several μm) in the nanocrystal assembly. This demonstrates that the van der Waals interactions between nanocrystals are not key parameter in the interfacial film formation.

The small-angle X-ray diffraction (SAXRD) patterns corresponding to the interfacial films (Figure 4.2d–f) show various diffraction spots that are indexed by using a

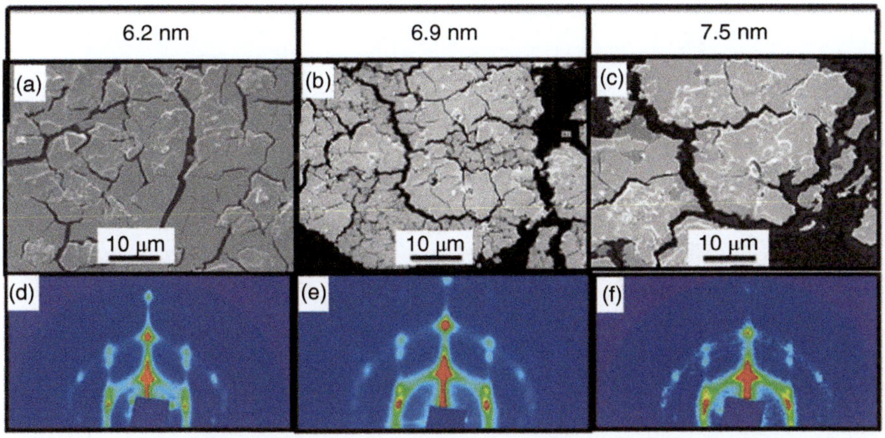

Figure 4.2 SEM images (a–c) and small-angle X-ray scattering (SAXS) patterns (d–f) of 3D superlattice films of 6.2 nm (a, d), 6.9 nm (b, e), and 7.5 nm (c, f) obtained at room temperature under toluene saturation regime.

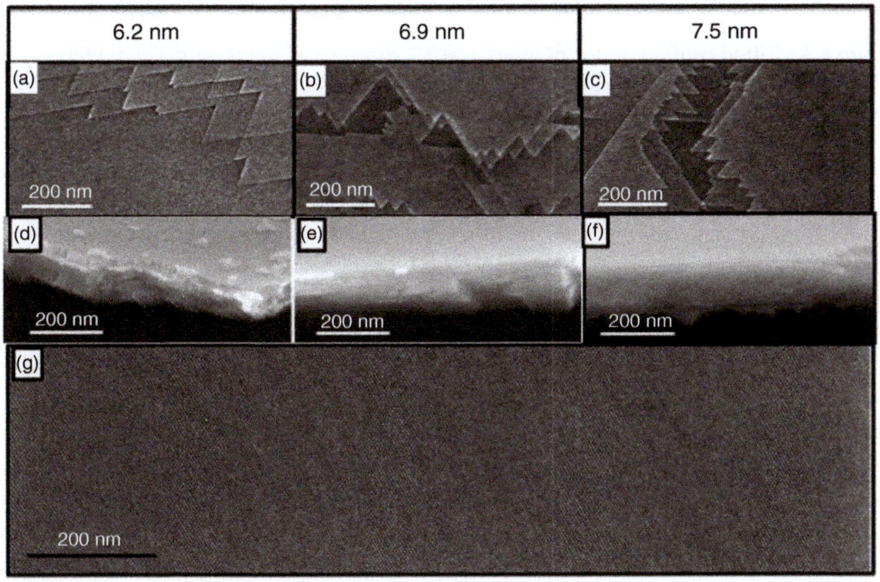

Figure 4.3 High-magnification SEM images for interfacial films of 6-nm (a), 7-nm (b), and 8-nm (c) Au nanocrystals and the corresponding tilted configurations in (d)–(f), respectively, and (g) large coherent length of the films. *Source:* Goubet et al. [1]. Copyright 2012 American Chemical Society.

face-centered cubic (fcc) lattice. This indexation reveals that the 3D superlattices have their [111] axis perpendicular to the substrate. The width of the spots is resolution-limited indicating very few defects in the nanocrystal ordering in fcc 3D superlattices. This is confirmed by the absence of defects over a large surface area (Figure 4.3g).

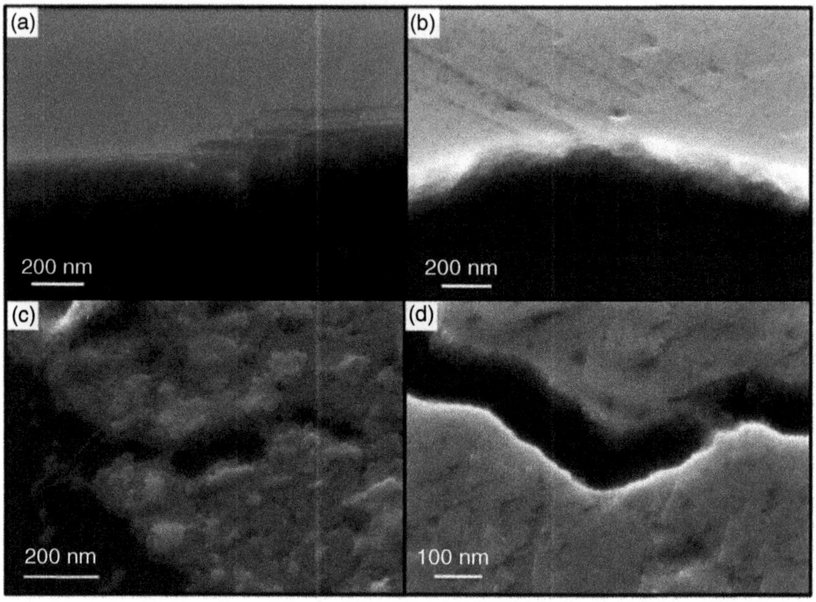

Figure 4.4 Tilted configurations of interfacial 3D superlattice made with a colloidal solution at 10^2 M (a), 5.10^{-3} M (b), and 10^{-3} M (c, d). Goubet et al. [1]/American Chemical Society.

The average interparticle distance, δ, of interfacial films calculated from the d_{111} spacing (Table 4.1) is similar to the length of the dodecanethiol all-trans configuration (1.8 nm). From the high-resolution SEM images, evaluation of the correlation length is deduced by recording the various Fourier transform patterns on various regions of the film. Similar orientation of the nanocrystals along a distance of 10 μm is observed (Figure 4.3g) permitting to deduce a long coherence length (around 10 μm).

On decreasing the initial concentration of the colloidal solution, the thickness of the interfacial films remains unchanged (Figure 4.4). This is explained as follows.

Due to the lower surface energy of NC coating ($\sigma_l = 25.4 \times 10^{-3}$ Jm^{-2}) than of toluene ($\sigma_s = 28.4 \times 10^{-3}$ Jm^{-2}), the overall surface energy decreases, leading to the attraction between the nanocrystals and air–toluene interface. Thus, the energy loss due to the larger surface is compensated by the smaller surface tension of the coating. Once the surface is covered by a Au nanocrystal monolayer, further adsorption of nanocrystals is governed by the balance between the interaction energy and entropy loss resulting from the nanocrystallization. This explains why the film thickness decrease with decreasing initial concentration of nanocrystals.

These data clearly show that with Au nanocrystals, up to 5-nm 3D superlattices are formed by a layer-by-layer (heterogeneous) growing process at the toluene–air interface. The interfacial films are composed by highly ordered nanocrystals with fcc structures.

Whatever the Au nanocrystal size from 5 to 8 nm, precipitates are collected at the bottom of the beaker and collected on a substrate. Figure 4.5a–c shows individual

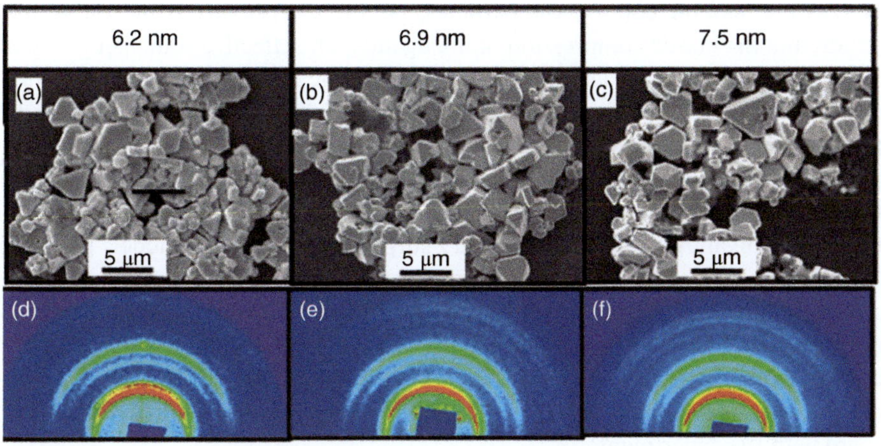

Figure 4.5 SEM images (a–c) and SAXS patterns (d–f) of shaped 3D superlattices of 6.2 nm (a, d), 6.9 nm (b, e), and 7.5 nm (c, f) obtained at room temperature under toluene saturation regime. *Source:* Goubet et al. [1]. Copyright 2012 American Chemical Society.

aggregates with well-defined shaped aggregates of nanocrystals having a size up to 6 nm. The shaped aggregate sizes are from 1 to 10 μm. The SAXRD patterns remain similar (Figure 4.5d–f) with continuous rings that show the absence of preferential orientation. Each line of the different profiles can be indexed using a fcc structure. The width of the reflections is also resolution-limited that implies long range extending upon at least 200 nm. This randomly oriented crystalline texture agrees perfectly with SEM images. A similar homogeneous nucleation for large Au nanocrystals has been widely observed in various solvents [3, 4].

The average interparticle distances, δ, obtained for the shaped 3D superlattices of nanocrystals with various sizes remain similar (1.9 ± 0.1 nm). Note the interparticle distances remain similar to those deduced with film 3D superlattices (Table 4.1).

Contrary to the film 3D superlattices, the decrease of the mother colloidal solution induces a decrease in the amount of precipitated 3D superlattices, whereas their sizes remain similar. This clearly indicates that the growth mechanism of 3D superlattices remains dependent on van der Waals interactions but not on the amount of initial material used.

From that it is deduced that the growth mechanisms of interfacial and precipitated 3D superlattices markedly differ.

Note that with hexane as solvent, neither interfacial nor shaped 3D superlattices are obtained. This is attributed to the strong affinity of hexane with the alkyl chains and consequently decreases the interactions between nanocrystals. Furthermore, the hexane surface tension ($\sigma_{hexane} = 18.4\,mN\,m^{-1}$) markedly drops compared with the toluene ($\sigma_{toluene} = 26.5\,mN\,m^{-1}$), preventing the migration of nanocrystals to the interface.

The assembly of Au nanocrystals dispersed in solution is modeled at a nanocrystal level by Brownian dynamics [5, 6]. The interaction potential between the nanocrystals is defined as a sum of three terms [1, 7, 8]: the van der Waals attraction between the metallic cores of the particles, the free energy of mixing of the thiol ligands, and the elastic compression of the ligands. The effective Hamaker constant for the van der Waals attraction between the metallic nanocrystals does not influence the simulation and is chosen equal to 1.95 eV. The interaction between the nanocrystals and the air–toluene interface is modeled by using two approaches. The first one is based on the change in surface energy due to the nanocrystal penetration of the air–toluene interface, whereas the second one takes into account the presence of some free ligands at the interface. Very surprisingly these two models [9] reach similar behavior. Both approaches predict two growth mechanisms, one within the solution and the other at the air–toluene interface only for nanocrystal diameters equal or larger than 5 nm in toluene as observed in the experiments. Please note that additional calculations have shown that in hexane the nanocrystal–interface potentials obtained are purely repulsive and no 3D superlattice growth is observed in accord with the experiment.

From the good agreement between experiments and simulations, the formation of the nanocrystal film at the air–toluene interface is explained as follows: due to the attractive potential between nanocrystals and the interface, nanocrystals are trapped at the solution surface and form monolayers. The attraction between nanocrystals in toluene leads to a further growth on this first layer. Once the surface is covered by an Au nanocrystal monolayer, further adsorption of nanocrystals is governed by the diffusion of the nanocrystals or aggregates to the air–toluene interface. This explains why the film thickness becomes thinner with decreasing initial concentration of nanocrystals. Some solvent stays within the nanocrystal film. Only when this film is deposited on a substrate, the rest of the solvents evaporate, leading to the formation of cracks and dislocation. In fact, during such a drying process, small defects within the film lead to dislocations that reduce the final mechanical stability of the films. In contrast to the precipitated 3D superlattices, no dislocations appear.

4.1.2 Analogy with Nature: Air–Solvent Interface Acts as Perfect Substrate

Figure 4.3 shows that the film 3D superlattices are characterized by a long coherence length with several hundred nanometers. We have to remain that these fcc film 3D superlattices of Au nanocrystals having up to 5.5 nm, with very few defects, are produced when the colloidal solution is subjected to solvent saturation (Figure 4.3g). This has to be associated to Mg-calcite crystals observed in cave pools in Central Texas (USA) produced at the air–water interface (Figure 4.6a) [10].

The arrows show that the surface of calcite grown at the air–water interface is highly flat (Figure 4.6a). Similar flat and quite perfect surface is observed when Au nanocrystals self-assembled at the toluene–air interface (Figure 4.6b). In the Texas pools, the calcite grows under saturation of water. Similarly, 3D superlattice films grow under toluene saturation. In both cases, the surface are "quite" perfect. This is attributed to the fact that air–solvent is highly flat and behaves as a perfect substrate.

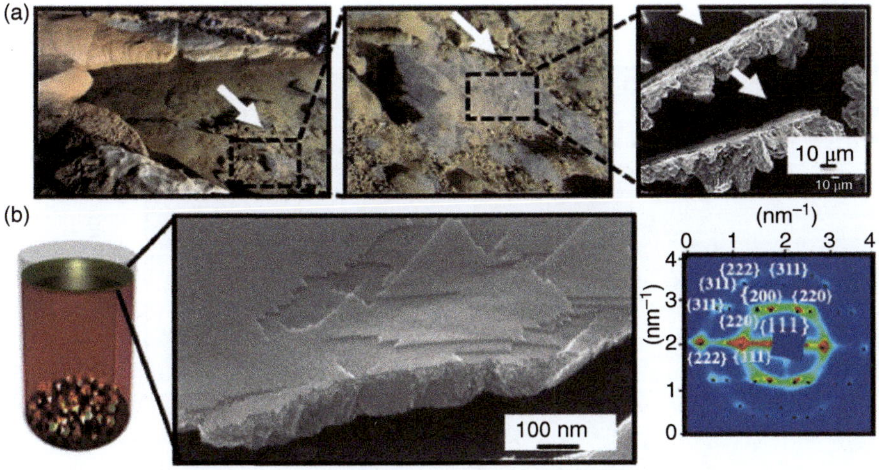

Figure 4.6 (a) Central Texas pools with calcite formation at the interface and (b) 3D superlattice film of Au nanocrystals having size from 5.5 to 8 nm.

4.1.3 Case of 5-nm Au Nanocrystals in Toluene Saturation

With 5-nm Au nanocrystals similar behavior (interfacial film [Figure 4.7a], precipitate aggregates [Figure 4.7c]) are produced. The 3D superlattice film shows cracks, and the diffraction pattern (Figure 4.7a,b) is characterized by a mixture of fcc and body-centered cubic (bcc) structure. Note that some holes are observed on the surface of the 3D superlattice film. This is developed in Chapter 2.

Other marked changes compared with what was described above with nanocrystals having a diameter equal or up to 6 are observed. Even though the diffraction pattern of the precipitates (Figure 4.7d) remains characteristic of a fcc structure, the size of shaped 3D superlattices with 5.1-nm Au nanocrystals as building blocks (Figure 4.7c) is markedly larger than that observed with larger Au nanocrystals, i.e. up or equal to 6 nm (Figure 4.5). Furthermore, the various shapes are characterized by threefold and fivefold symmetries, forming single-domain 3D superlattices and truncated-tetrahedral 3D superlattices on one of the {111} surfaces. Hence, octahedron, icosahedron, and truncated tetrahedron isolated or associated are shown in Figure 4.8.

The surface of such polyhedron 3D superlattices markedly changes compared with the heterogeneous growth process, and 2D nucleation occurs on the surface of each facet (Figure 4.9).

To understand such difference in 3D superlattice shape and size, let us consider 5-nm Au nanocrystals. In Chapter 1, it is mentioned that 5-nm Au nanocrystals produced by inorganic chemistry developed by Stucky and coworkers [11] are composed by homogeneous and heterogeneous contrasts in dark-field transmission electron microscopy (TEM) images corresponding to single-domain and multiply-twined particle (MTP) nanocrystals, respectively. To identify the composition of the two 3D superlattices (Figure 4.10a) and the remaining solution in equilibrium

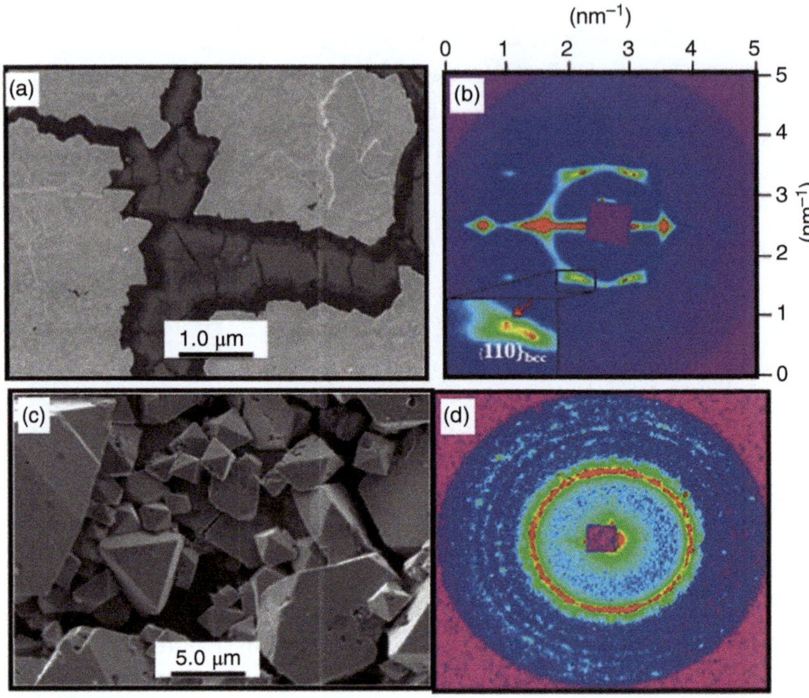

Figure 4.7 SEM images (a, c) and SAXS patterns (b, d) of film (a, b) and shaped (c, d) 3D superlattices of 5.1-nm nanocrystals obtained at room temperature under toluene saturation regime.

Figure 4.8 Various shaped 3D superlattices of 5-nm Au nanocrystals: (a) monocrystalline shaped as octahedron, (b) fivefold symmetry shapes as icosahedral terminated by (111) planes also formed corresponding to multiply twinned crystals, (c) multitwinned 3D superlattices, (d) association of two truncated-tetrahedral single 3D superlattices on a spinel twin, (e) twin penetration, and (f) multitwinned nanocrystals. *Source:* (a, c, d, f) N. Goubet (2011), p. 1024–1031/American Chemical Society. (b) Goubet et al. [1]/American Chemical Society.

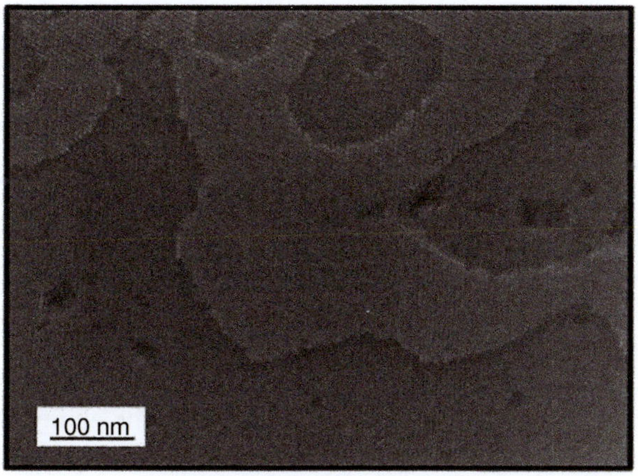

Figure 4.9 High-resolution SEM image of surface of shaped 3D superlattices. *Source:* M. P. Pileni, 2019/The Chemical Society of Japan.

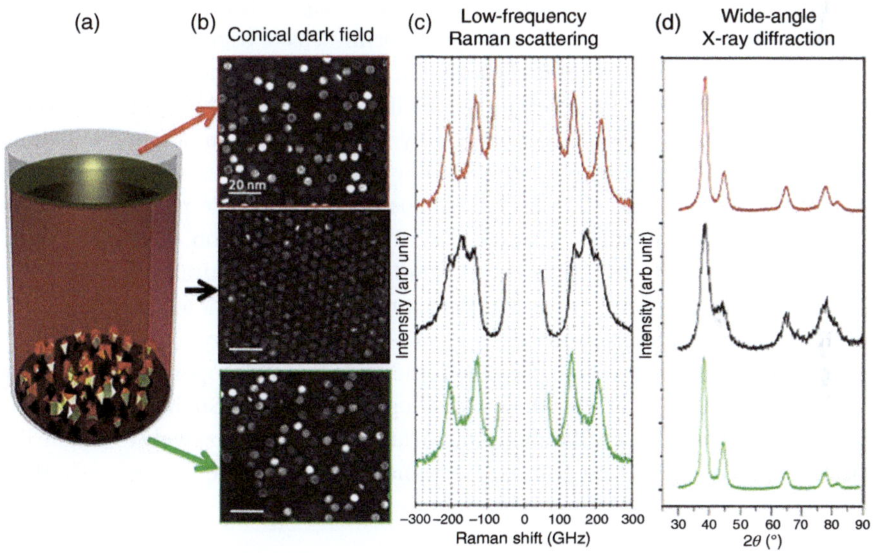

Figure 4.10 (a) Scheme of Figure 4.1 with two film and precipitated 3D superlattices and the remaining solution, (b) TEM bright field images of the corresponding nanocrystals, (c) LFRS corresponding, and (d) wide-angle X ray diffraction. *Source:* (b) Goubet et al. [1]/ American Chemical Society.

between them, the 3D superlattices are dissolved in hexane, and the remaining solution is evaporated and then redispersed in the same solvent. From the TEM dark-field images shown in Figure 4.10b, homogeneous contacts of the nanocrystals are obtained from 3D superlattices, whereas heterogenous one is obtained with nanocrystals coming from the solution. We know that low-frequency Raman scattering (LFRS) permits to identify single-domain nanocrystals to their MTP counterparts. As mentioned in Chapter 2, the quadrupolar modes of single-domain Au nanocrystals are split in two distinct (E_g and T_g) modes, whereas the MTP nanocrystals shows only one peak. Figure 4.10c shows the corresponding LFRS spectra with a splitting in the bands for film and precipitate 3D superlattices, whereas the remaining solution is composed by MTP Au nanocrystals. The size deduced from wide X-ray diffraction is similar to those determined from TEM images for 3D superlattices (around 5 nm), whereas from the remaining solution, the size markedly decreases (around 2.5 nm), as shown in Figure 4.10d. These overall data confirm that the film and precipitated 3D superlattices are composed of single-domain nanocrystals, whereas those from the remaining solution are MTP nanoparticles.

One question arises: What is the cause of segregation of single-domain nanocrystals in both interfacial and precipitate 3D superlattices? We assume that the nanocrystal morphology, which depends on crystallinity, influences the ability of nanocrystals to interact. It is either due to their environment (air–toluene interface) or to the neighboring nanocrystals to self-assemble in close-packed 3D superlattices.

Let us come back to the crystalline structure of nanocrystals: MTPs are well known to exhibit either decahedral or icosahedral shapes [12]. The decahedral and icosahedral MTPs are fully enclosed by uniformly sized triangular facets corresponding to {111} crystallographic surfaces. On the other side, the possible shapes of single nanocrystals were described by the Wulff construction [13], which is a reliable tool for predicting the most energetically stable clusters in single-component systems beyond a few hundred atoms. Among the variety of existing shapes, two typical examples of commonly observed shapes for metal single nanocrystals with fcc structure are the cuboctahedron and the truncated octahedron, where both {111} and {100} facets are present [14]. That is in contrast with decahedral and icosahedral MTPs. Depending on their crystallographic surface, the facets of a single nanocrystal differ not only in shape and size but also in surface energy and atom density (Figure 4.11). As a consequence, the electronic and bonding properties of the facets should naturally also depend on their crystallographic surface. Such variation in the facet properties from one crystallographic surface to the others is likely to dictate the respective ability of MTPs and single nanocrystals in interacting with their

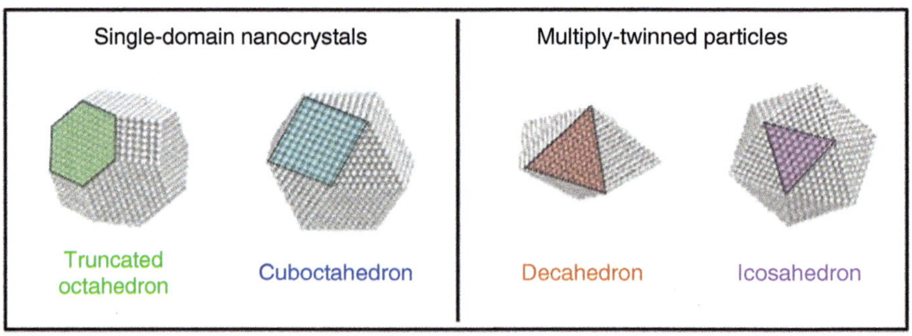

Figure 4.11 Scheme of single-domain and MTP surfaces of 5-nm Au nanocrystals.

environment via the coating molecules. Additionally, the presence of facets in typical morphologies of single nanocrystals with relatively larger area than the facets of MTPs could also play a key role in the crystallinity segregation by strengthening the interactions between single nanocrystals, thus favoring their self-assembly with, in addition to van der Waals interaction, the interaction between nanocrystal and interface.

The low-angle diffraction patterns, recorded from a thin film of single and polycrystal 5-nm Au nanocrystals, show hexagonal symmetry confirming the compact structure of the superlattice. The high-angle diffraction patterns confirm the compact structure of the nanocrystal superlattice. However, with the film of single-domain nanocrystals, Figure 4.12a exhibits several rings corresponding to electron diffraction by specific planes of atomic lattices with fcc structure. In this pattern, small arcs are distinctly observed for most diffraction rings. Similar arcs in the high-angle diffraction pattern were previously observed in superlattices of self-assembled nonspherical building blocks [16] and especially consisting of Ag [17] and Au [18, 19] nanocrystals.

Such a feature provides evidence for an average coherent alignment of the atomic lattice planes of nanocrystals to one another within the superlattice. This orientational ordering of the atomic lattices of Au nanocrystals is indicative of their faceted morphology and is presumably favored by large facets in single-domain nanocrystals. These latter facets are indeed expected to exhibit large enough area to induce their face-to-face orientation between neighboring nanocrystals despite the presence of dodecanethiol molecules tethered to their surface. In contrast, when looking at the high-angle diffraction pattern obtained from polycrystalline nanocrystals (Figure 4.12b), diffraction rings exhibit no similar arcs as those previously mentioned although these nanocrystals are also close-packed in compact superlattice. This result means that polycrystalline nanocrystals self-assemble into arrays with long-range translational but no orientational ordering and confirms previous observations from Au nanocrystal superlattices whose structures were found to depend

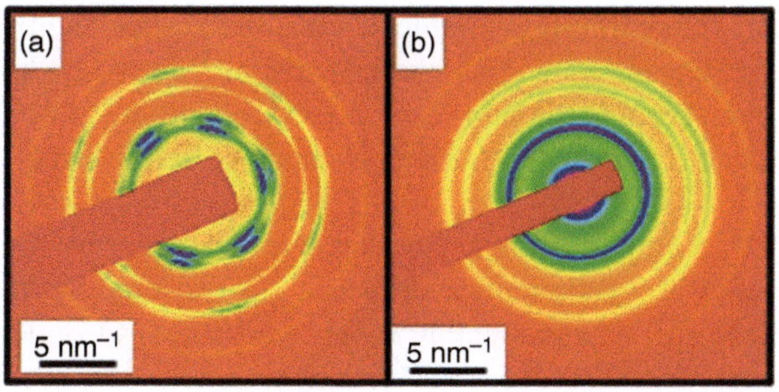

Figure 4.12 High-angle electron diffraction patterns recorded from a thin film of close-packed (a), single-crystalline, and (b) polycrystalline 5-nm Au nanocrystals. In (a) and (b), the patterns are colored to help in viewing the arcs, and diffractions by {111}, {200}, {220}, and {311} planes are indexed. *Source:* Portales et al. [15]. Copyright 2012 American Chemical.

upon the crystallinity of the nanocrystals via the method used for their preparation [19]. Actually, in addition to the crystallinity of nanocrystals and their inherent shape anisotropy, the van der Waals interactions between the surfactant chains and their possible directional distribution should also play an important role in the orientational ordering of nanocrystals, as suggested by previous studies on Ag nanocrystal superlattices [17, 20]. The crystallinity segregation of colloidal faceted nanocrystals can be therefore ascribed to a shape-directed crystallization process mediated by the interactions of surfactant molecules tethered on the facets of neighboring nanocrystals.

From these data, different packing behaviors between single and polycrystalline nanocrystals are revealed by electron diffraction. After being deposited onto a substrate, single nanocrystals are found to self-assemble into close-packed superlattices with long-range translational and orientational orderings, while polycrystalline nanocrystals behave like spheres with no preferential orientation in the nanocrystal array. This difference in packing behavior clearly indicates the major role that crystallinity-related faceting and shape anisotropy of nanocrystals play in the structure and subsequent properties of their superlattices. The crystallinity segregation reported here definitely reveals the direct influence of the morphology of nanocrystals on their ability to interact and self-assemble. The separation of single nanocrystal from polycrystalline nanocrystal by such crystallinity segregation interestingly appears like an efficient alternative to the development of complex synthetic route for controlling nanocrystallinity.

4.1.4 Nanocrystals of 4-nm Dispersed in Toluene

Instead of using nanocrystals having a size equal or up to 5 nm, the same experiment as described above is performed with 4-nm Au nanocrystals dispersed in toluene. Whatever the time used to keep the colloidal solution under toluene saturation, the 3D superlattices do not grow. This is attributed to the fact that the nanocrystal/nanocrystal interactions and nanocrystal interface are too weak to induce precipitation or film formation. Note that the synthesis to produce 4-nm nanocrystals slightly changes and a deficit on free surfactant could take place. By the addition of dodecanethiol to the colloidal solution, keeping the same experimental conditions, a film and cubic 3D superlattices appear. However, it is difficult to assign crystalline structures to such assemblies.

Instead of adding dodecanethiol to the colloidal solution, various substrates are deposited on the toluene/air interface: (i) raw mica surface, a mica covered by Au layer either with (ii) or without (iii) a self-assembly monolayer (SAM) of dodecanethiol, and (iv) Teflon. Except with Teflon, very surprisingly well-defined assemblies are produced, and no differences in shapes are observed (Figure 4.13). Whatever the substrate above, it has not been possible to identify what type of crystalline structure is produced, as well as the mechanism of 3D superlattice growth process at the interface. In each case, it seems that it could be a mixture of bcc and fcc.

Figure 4.13 Various views of 3D superlattices deposited on various substrates such as mica on which has been deposited or not a Au layer containing or not a SAM. *Source:* Goubet et al. [21]. Copyright Royal Chemical Society.

4.2 Submillimeter-Size Single 3D Superlattices of 5-nm Au Nanocrystals

To produce submillimeter-size single-crystals of Au nanocrystals, the concept to mix good and bad solvents to select either polymers [22], nanocrystals size [23], or nanocrystallinity [1, 24] is used. This target involves two successive crystallization processes (Figure 4.14):

(i) *First crystallization:* the addition of small amount of ethanol to the colloidal solution of a mixture of single-domain and polycrystalline nanocrystals is deposited on a tube containing a capillary glass and maintained under saturated toluene vapor during a week. The colloidal solution remains red colored indicating that some Au nanocrystals remain in solution. However, shiny reflections are observed on the capillary glass and on the vessel walls indicating the growth of large well-defined nanocrystal aggregates (Figure 4.15).

The corresponding SEM image shows the formation of shaped aggregates around 50 μm in the longer dimension (Figure 4.16). These aggregates are then dissolved into a good solvent (chloroform) before deposition on a grid. The TEM images confirm that the aggregates are composed of Au nanocrystals

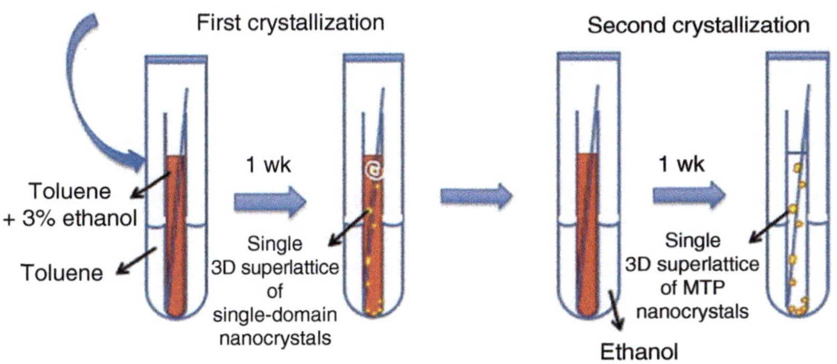

Figure 4.14 Scheme of the crystallization process.

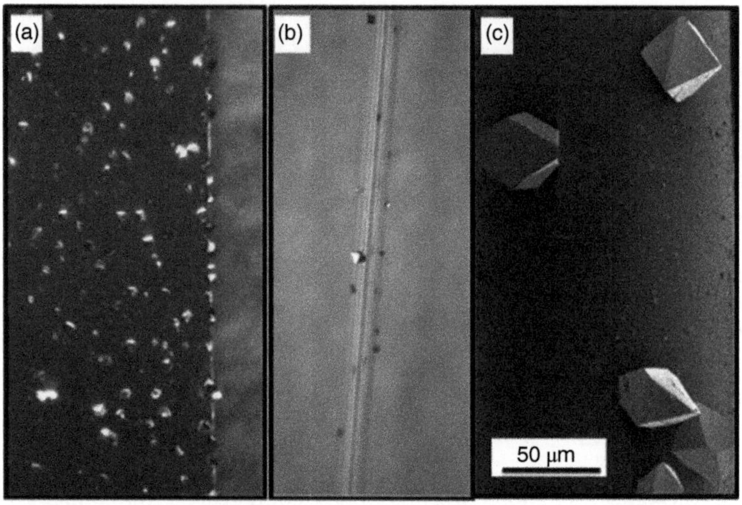

Figure 4.15 (a) Photo, (b) optical micrograph, and (c) SEM images of the capillary glass deposited in the colloidal solution of mixed single-domain and polycrystal nanocrystals. *Source:* M. P. Pileni, 2019/The Chemical Society of Japan.

Figure 4.16 Submillimeter Au 3D superlattices. *Source:* M. P. Pileni, 2019/The Chemical Society of Japan.

having similar diameter (5 nm) as those initially used. From High Resolution Transmission Electron Microscope (HRTEM) images, it is demonstrated that truncated-octahedral single-domain nanocrystals make up the large aggregates. X-ray diffraction patterns demonstrate a bcc crystalline structure of such submillimeter assemblies. A remarkable feature is the preferential orientation of the symmetry axes of the single-domain nanocrystals along those of the colloidal 3D superlattice with orientation correlation within the {110} colloidal crystal planes. From the first crystallization, bcc submillimeter single 3D superlattices of single-domain nanocrystals are produced.

(ii) *Second crystallization:* the remaining colloidal solution is poured into another tube and placed in an enclosure saturated with ethanol vapor (bad solvent). The slow diffusion of ethanol into the colloidal solution induces the formation of shiny selections indicating nanocrystal aggregation with a progressive discoloration of the solution (Figure 4.14).

The SEM images (Figure 4.16) show similar well-defined shaped aggregates as shown above with single 3D superlattices of single-domain nanocrystals (Figure 4.8). By subjecting them to the same treatment as above, it is demonstrated that the submillimeter assemblies, produced by the second crystallization, are composed by MTP nanocrystals mainly in the form of icosahedron and decahedron. From X-ray diffraction patterns, such assemblies are characterized by a fcc crystalline structure. Hence, from second crystallization process, single 3D superlattices of polycrystalline decahedral and icosahedral nanocrystals as building blocks are produced, displaying random orientation. The study of submillimeter 3D superlattices has been performed with other nanocrystals [25–27].

4.3 Conclusions

Au nanocrystals, in the size range size from 5 to 8 nm, with low size distribution (<6%) coated with dodecanethiol and dispersed in toluene and maintained in a saturated solvent atmosphere during seven days, self-assemble simultaneously both at the interfacial toluene/solvent and in solution to produce a film and shaped 3D superlattices, respectively. Furthermore, a colloidal solution of Au nanocrystals remains in solution. The thickness of the film is independent on the mother concentration of the colloidal solution, whereas the amount of shaped 3D superlattices increases. This clearly demonstrates that the growth mechanism of these two 3D superlattices markedly differs. For nanocrystals having a diameter of up to 5.5 nm, both film and shaped 3D superlattice are fcc crystalline phase and made of a mixture of single domain and polycrystals. For 5 nm, both film and shaped 3D superlattices are of single-domain nanocrystals, whereas the nanocrystals remaining in solution are MTP nanoparticles. To explain the mechanism of these two simultaneous 3D superlattice growths, we have to take into account in the one hand the attractive interactions between nanocrystals and in other hand the surface tension at the toluene interface. For large nanocrystals sizes, due to the solvent-mediated attraction between the nanocrystals by the interparticle potential, 3D superlattices are formed within the solution by homogeneous nucleation. This clearly explains the growth of the precipitated 3D superlattices. Concerning the interfacial 3D superlattices, the trapping of nanocrystals at the interface is governed by the attractive interaction potential between the nanocrystals and the interface. This favors the formation of the first close-packed monolayer. With increasing time, additional nanocrystals from suspension are attracted by the first nanocrystal monolayer inducing a progressive stacking of nanocrystal layer to reach the formation of a well-defined 3D superlattices. Such interfacial 3D superlattice growth mechanism is consistent with the fact that (i) the thickness of film decreases with diluting the Au mother suspension and (ii) the film thickness does not show any correlation with the nanocrystal size. It is found that for any nanocrystal sizes and growth mechanisms, the 3D superlattices are usually characterized by fcc structures, except with 5-nm Au nanocrystal interfacial film 3D superlattices that comprise a mixed fcc/bcc structure. The mixed structure is explained by a recently developed model based on competitive effect between dense packing and minimization of contact surface between nanocrystals [28, 29].

Finally, the crystalline structures of submicrometer size of single-domain 3D superlattices depend on the nanocrystallinity of the nanocrystals used as building blocks: bcc crystalline structure is obtained with single-domain nanocrystals, whereas hexagonal close packed (hcp) is obtained with the polycrystalline counterpart.

References

1 Goubet, N., Portalès, H., Yan, C. et al. (2012). Simultaneous growths of gold colloidal crystals. *J. Am. Chem. Soc.* 134: 3714–3719.

2 Goubet, N., Yang, P., Albouy, P.A., and Pileni, M.P. (2014). Spontaneous formation of high-index planes in Au single domain nanocrystal superlattices. *Nano Lett.* 14: 6632–6638.

3 Compton, O.C. and Osterloh, F. (2007). Evolution of size and shape in the colloidal crystallization of gold nanoparticles. *J. Am. Chem. Soc.* 129: 7793–7798.

4 Abecassis, B., Testard, F., and Spalla, O. (2008). Gold nanoparticle superlattice crystallization probed *in situ*. *Phys. Rev. Lett.* 100: 115504.

5 Lalatonne, Y., Richardi, J., and Pileni, M.P. (2004). Van der Waals versus dipolar forces controlling mesoscopic organizations of magnetic nanocrystals. *Nat. Mater.* 3: 121–125.

6 Allen, M. and Tildesley, D. (1993). *Computer Simulation of Liquids*. Oxford: Clarendon Press.

7 Goubet, N., Richardi, J., Albouy, P.A., and Pileni, M.P. (2011). Which forces do control the supracrystal nucleation in organic media? *Adv. Funct. Mater.* 21: 2693–2704.

8 Khan, S.J., Pierce, F., Sorensen, C.M., and Chakrabarti, A. (2009). Self-assembly of ligated gold nanoparticles: phenomenological modeling and computer simulations. *Langmuir* 25: 13861–13868.

9 Goubet, N., Richardi, J., Albouy, P.A., and Pileni, M.P. (2013). Simultaneous interfacial and precipitated supracrystals of Au nanocrystals: experiments and simulation. *J. Phys. Chem. B* 117: 4510–4516.

10 Taylor, P. and Chafetz, H.S. (2004). Floating rafts of calcite crystals in cave pools, central Texas, U.S.A.: crystal habit vs. saturation state. *J. Sediment. Res.* 74: 328–341.

11 Zheng, N., Fan, J., and Stucky, G.D. (2006). One-step one-phase synthesis of monodisperse noble-metallic nanoparticles and their colloidal crystals. *J. Am. Chem. Soc.* 128: 6550–6551.

12 Ino, S. and Ogawa, S.J. (1967). Multiply twinned particles at earlier stages of gold film formation on alkalihalide crystals. *J. Phys. Soc. Jpn.* 22: 1365–1374.

13 Wulff, G. (1901). Thermodynamic interpretation of the morphology individuality of natural and synthesized apatite single crystals. *Z. Kristallogr. Krystallogr. Mineral* 34 (5/6): 449–530.

14 Wang, Z.L. (2000). Transmission electron microscopy of shape- controlled nanocrystals and their assemblies. *J. Phys. Chem. B* 104: 1153–1175.

15 Portales, H., Goubet, N., Sirotki, S. et al. (2012). Crystallinity segregation upon selective self-assembling of gold colloidal single nanocrystals. *Nano Lett.* 12: 5292–5298.

16 Quan, Z. and Fang, J. (2010). Superlattices with non-spherical building blocks. *Nano Today* 5: 390–411.

17 Harfenist, S.A., Wang, Z.L., Alvarez, M.M. et al. (1996). *J. Phys. Chem.* 100: 13904–13910.

18 Brown, L.O. and Hutchison, J.E. (2001). Formation and electron diffraction studies of ordered 2-D and 3-D superlattices of amine-stabilized gold nanocrystals. *J. Phys. Chem. B* 105: 8911–8916.

19 Stoeva, S.I., Prasad, B.L.V., Uma, S. et al. (2003). Face centered cubic and hexagonal closed-packed nanocrystal superlattices of gold nanoparticles prepared by different methods. *J. Phys. Chem. B* 107: 7441–7448.

20 Korgel, B.A., Fullam, B.A., Connolly, S., and Fitzmaurice, D.J. (1998). Assembly and self-organization of silver nanocrystal superlattices: ordered “soft spheres”. *J. Phys. Chem. B* 102: 8379–8388.

21 Goubet, N., Albouy, P.A., Thompson, A., and Pileni, M.P. (2016). Polymorphism in nanoparticle-based crystals depending upon their single or polycrystalline character. *CrystEngComm* 18: 6166–6155.

22 Paine, A.J., Luymes, W., and McNulty, J. (1990). Dispersion polymerization of styrene in polar solvents. 6. Influence of reaction parameters on particle size and molecular weight in poly(*N*-vinylpyrrolidone)-stabilized reactions. *Macromolecules* 23: 3104–3109.

23 Taleb, A., Petit, C., and Pileni, M.P. (1997). Synthesis of highly monodisperse silver nanoparticles from AOT reverse micelles: a way to 2D and 3D self-organization. *Chem. Mater.* 9: 950.

24 Bang, J. and Lodge, T.P. (2003). Mechanisms and epitaxial relationships between close-packed and BCC lattices in block copolymer solutions. *J. Phys. Chem. B* 107: 12071–12081.

25 Li, R., Bian, K., Wang, Y. et al. (2015). An obtuse rhombohedral superlattice assembled by Pt nanocubes. *Nano Lett.* 15: 6254–6260.

26 Simon, P., Rosseeva, E., Baburin, I.A. et al. (2012). PbS–organic mesocrystals: the relationship between nanocrystal orientation and superlattice array. *Angew. Chem. Int. Ed.* 51: 10776–10781.

27 Li, R., Bian, K., Hanrath, T. et al. (2014). Decoding the superlattice and interface structure of truncate PbS nanocrystal-assembled supercrystal and associated interaction forces. *J. Am. Chem. Soc.* 136: 12047–12055.

28 Courty, A., Richardi, J., Albouy, P.A., and Pileni, M.P. (2011). How to control the crystalline structure of supracrystals of 5-nm Ag nanocrystals. *Chem. Mater.* 23: 4186–4192.

29 Rupich, S.M., Shevchenko, E.V., Bodnarchuk, M.I. et al. (2011). Size-dependent multiple twinning in nanocrystal superlattices. *J. Am. Chem. Soc.* 132: 289–296.

5

Ag 3D Superlattices

5.1 Control of the Crystalline Structure of Ag 3D Superlattices

Here, we will use 5-nm Ag nanocrystals produced from reverse micelles [1, 2]. Ag nanocrystals are coated with either dodecanethiol (C_{12}) or decanethiol (C_{10}) and dispersed in various solvents like hexane, toluene, octane, and dodecane characterized by various evaporation rates and boiling points (Table 5.1).

The Ag 3D superlattices are produced by solvent evaporation under a slow flow of nitrogen of the colloidal suspension. The substrate temperature is controlled from 15 and 50 °C. Depending on the boiling point of the solvent and on the substrate temperature, the solvent evaporation time can vary from around one hour (case of hexane heated at 50 °C) to three weeks (case of dodecane cooled at 15 °C). The 3D superlattice crystalline structure is deduced by comparison between the observed and calculated diffraction spot coordinates in the small-angle X-ray diffraction (SAXRD) pattern. The center-to-center interparticle distance, d, is deduced from the position of the intense and well-defined Bragg reflections.

Except for hexane, the film morphologies and the SAXRD patterns are highly reproducible and remain stable over more than four weeks. The substrate temperature is controlled during the solvent evaporation process. Below, it is shown that the crystalline structures depend on the solvent used to disperse nanocrystals before evaporation and on the evaporation rate.

Let us first consider 5-nm Ag nanocrystals coated with dodecanethiol and dispersed in various solvents:

* At 15 °C, for any solvent used to disperse 5-nm Ag nanocrystals coating with dodecanethiol, disordered 3D assemblies are observed.
* At 25 °C, with either dodecane, octane (Figure 5.1), or decane (Figure 5.2), i.e. long evaporation time, hexagonal close-packed (hcp) structure is produced. Figure 5.1 shows the ones with the lowest boiling point such as hexane and toluene, i.e. with shorter evaporation time, and well-defined spots indexed using face-centered cubic (fcc) lattice with the (111) plane of each 3D superlattice parallel to the substrate are obtained. The width of the first-order Bragg

Metal Nano 3D Superlattices: Synthesis, Properties, and Applications, First Edition.
Marie-Paule Pileni.

Table 5.1 Boiling points of solvents.

Solvent	Hexane	Toluene	Octane	Decane	Dodecane
Boiling point (°C)	68	111	125	174	216

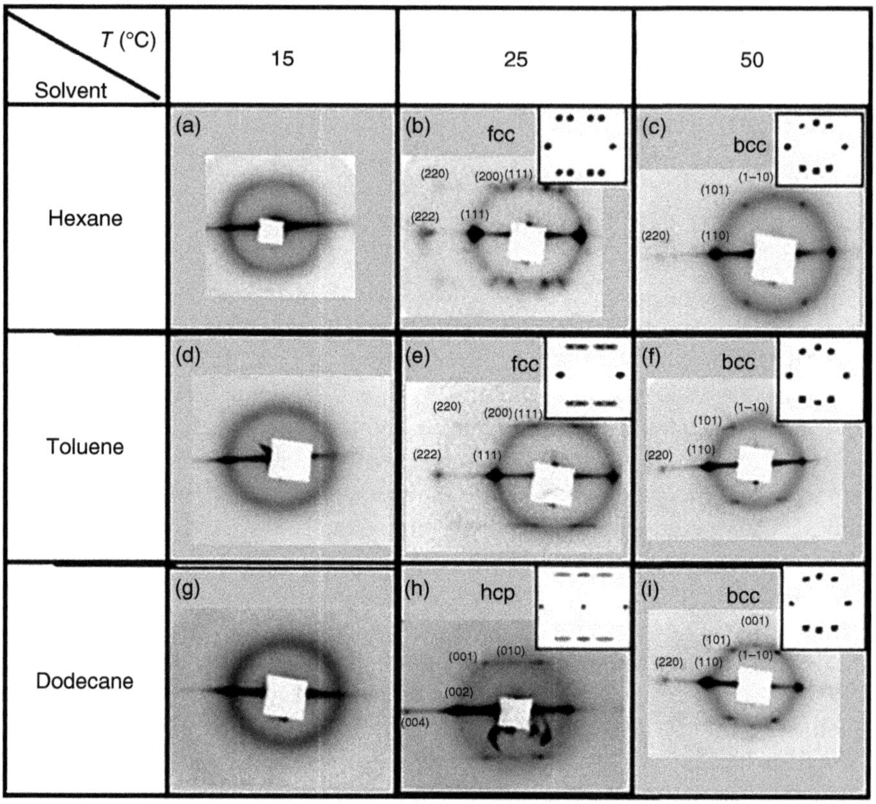

Figure 5.1 Crystalline phase diagram of 3D self-assemblies made of silver nanocrystals coated by dodecanethiol dispersed in various solvents such as Hexane (a–c), toluene (d–f) and dodecane (g–i) and at various temperatures 15 °C (a, d, g), 25 °C (b, e, h) and 50 °C (c, f, i). *Source:* Courty et al. [1]/American Chemical Society.

reflection is nearly resolution-limited. Note that, with hexane, the results are not highly reproducible and amorphous assemblies can be produced. Time-to-time fcc feature is observed. Hence, less volatile solvents like dodecane, octane, and decane (Table 5.1) appear to be optimal to produce hcp arrangements instead of fcc with toluene and hexane.

* At 50 °C, for any solvent used to disperse nanocrystals, well-defined SAXRD spots indexed with the body-centered-cubic (bcc) lattice with the (110) plane of each 3D superlattices parallel to the substrate are produced (Figures 5.1 and 5.2). Except for hexane, the formation of bcc 3D superlattices is highly in time reproducible, and they remain stable, keeping the SAXRD patterns weeks unchanged over several. Note that with hexane as solvent, it is still possible to

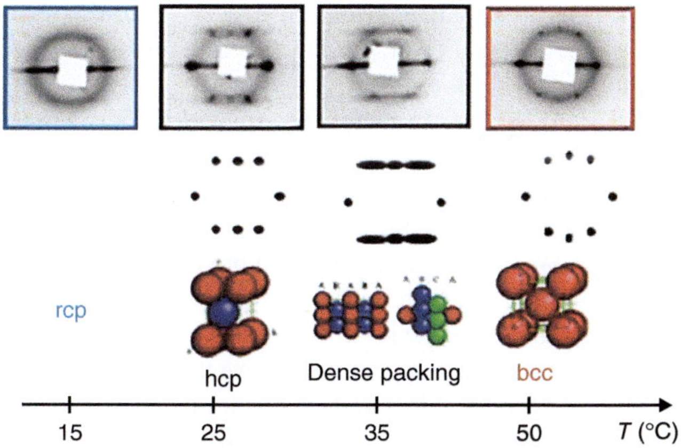

Figure 5.2 Crystalline phase diagram of 3D self-assemblies made of silver nanocrystals coated by dodecanethiol dispersed in decane (r.c.p = random close packing structure). *Source:* Henry et al. [2, 3]/American Chemical Society.

obtain disordered aggregates keeping all conditions similar. This is associated to the fast evaporation time (at 50 °C, a few hours compared with 1, 1.5, and 5 days for toluene, octane, and dodecane, respectively). Nevertheless, when ordered arrangements are obtained with hexane as solvent, they remain stable.

Hence, whatever the solvent, the transitions from hcp to bcc and from fcc to bcc are observed by increasing the substrate temperature from 25 to 50 °C. Table 5.2 gives the center-to-center distances, *D*, between nanocrystals (5-nm diameter) in

Table 5.2 The center-to-center distance *D* in 3D superlattices of nanocrystals coated with dodecanethiol calculated from d_{111}, d_{002}, and d_{110} spacings corresponding to fcc, hcp, and bcc 3D superlattices at 25 and 50°C substrate temperature. For hcp, fcc, and bcc arrangements, *D* are calculated from the $d_{(002)}$, $d_{(111)}$, and $d_{(110)}$ spacings to give $D = d_{(002)}\ [3/2]^{0.5}$, $D = d_{(111)}[3/2]^{0.5}$, and $D = d_{(110)}[3/2]^{0.5}$.

	Temperature	
Solvent	**25 °C**	**50 °C**
Dodecane	hcp $D = 7.4$ nm	bcc $D = 7.6$ nm
Decane	hcp $D = 7.4$ nm	bcc $D = 7.6$ nm
Octane	hcp $D = 7.4$ nm	bcc $D = 7.6$ nm
Toluene	fcc $D = 7.4$ nm	bcc $D = 7.6$ nm
Hexane	fcc $D = 7.4$ nm	bcc $D = 7.6$ nm

fcc, hcp, or bcc arrangements. They are calculated from the d_{111}, d_{002}, and d_{110} spacings, respectively. Within the experimental error, the D value remains unchanged with the 3D superlattice structure.

In contrast to former model [4–6], let us assume that the transitions between fcc and bcc are due to van der Waals interactions between nanocrystals and elastic compression of the ligands [7]. The details of the interaction between the ligand chains and the solvent may play a role. However, it is difficult to account for this phenomenon with a non-atomistic model. We, therefore, assume in a preliminary approach that this energy term is similar in both structures. The entropy difference between fcc and bcc is estimated by an expansion of the partition function assuming a harmonic system with $3N$ degrees of freedom at temperatures above the highest normal mode frequency [5]. The ratio between the Einstein frequencies in both lattices is calculated using a nearest-neighbor central force model [8]. Note that within this model, the entropy always favors the presence of a bcc structure. In the calculation of the free energy difference between fcc and bcc, we assume, for both structures, the same distance between the nearest neighbors due to the van der Waals attraction between the nanoparticles. In this case, configurational entropy terms due to the different volume packings of fcc and bcc structures is neglected [9]. In Figure 5.3, the free energy difference is given as a function of the edge–edge distance, δ, for the three temperatures used here. The model predicts the fcc phase for $\delta < 2.4$ nm at 15 and 25 °C. At 50 °C, the bcc structure is expected for $\delta > 2.3$ nm. Actually, the fcc structures are experimentally observed for an average edge–edge distance of 2.4 nm (C_{12}) while the bcc structures are found for $\delta = 2.6$ nm. Obviously, the model correctly predicts the onset of the bcc structure as a function of the edge–edge distance. It shows that the van der Waals interaction between the nanocrystal cores markedly influences the fcc- to bcc-phase transition in 3D superlattices. Note that these results

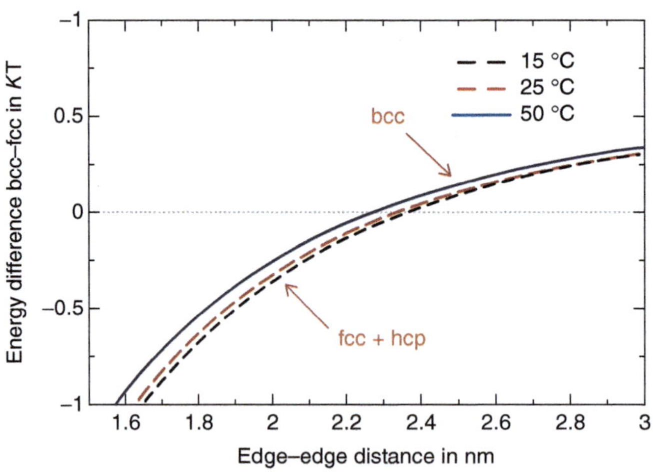

Figure 5.3 The calculated free energy difference between bcc and fcc structures for silver nanocrystals (5 nm in diameter). *Source:* Courty et al. [1]. Copyright 2011 American Chemical Society.

agree with the phenomenological approach proposed in the mid-1990s from which bcc packing is favored for a large ratio $2L/d$, while compact packing of the fcc or hcp type occurs at smaller values for $2L/d$ where L and d are the alkyl chain of the coating agent [10] and the nanocrystal diameter, respectively. The $2L/d$ thresholds evolve from 0.65 to 0.75. Here, with 5-nm nanocrystals, $2L/d$ is 0.71 indicating that bcc structures are expected. Within this model, the free energy differences of the transitions fcc → bcc and hcp → bcc is identical. This is due to the fact that both the van der Waals energy and the entropy change depend only on the number and the absolute distances of the neighbors that are identical for the fcc and hcp structures. Therefore, the results obtained for the fcc → bcc transition are also valid for hcp → bcc.

Let us replace dodecanethiol (C_{12}–SH) by decanethiol (C_{10}–SH) as coating agent. At 15 °C, with toluene and hexane as the solvent, Ag nanocrystals self-assembled in fcc 3D superlattices. Hence, with hexane, by decreasing the length of the alkyl chain used as coating agent, a transition from amorphous aggregates to fcc 3D superlattices is observed. With decane (Figure 5.4), a transition from disordered aggregates to hcp 3D superlattices takes place. These results confirm that the highest-volatile solvents like toluene and hexane (Table 5.1) appear to be optimum to produce fcc arrangements instead of hcp. At 25 and 50 °C, whatever the solvent used, an fcc structure with the (111) plane of each 3D superlattice parallel to the substrate is observed. On the contrary to nanocrystals coated with dodecanethiol, with decanethiol, nanocrystals self-assemble in fcc instead of bcc structure. Table 5.3 gives the center-to-center distance, D, between nanocrystals (5-nm diameter) in fcc 3D superlattices calculated from the d_{111} spacing. Within the experimental error, the center-to-center distance between nanocrystals is slightly shorter

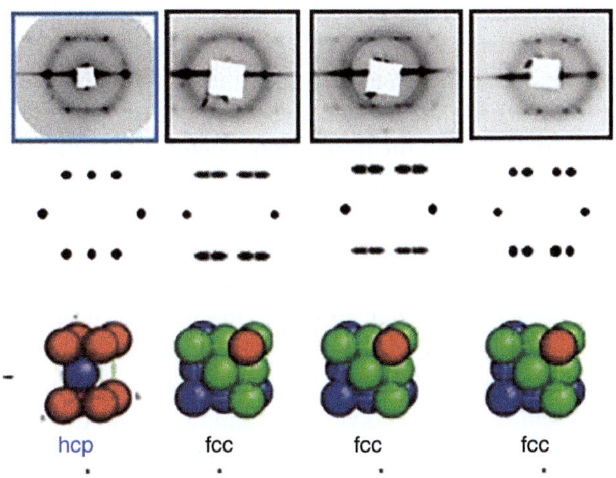

Figure 5.4 Crystalline phase diagram of 3D self-assemblies made of silver nanocrystals coated by decanethiol dispersed in decane. *Source:* Henry et al. [2, 3]/American Chemical Society.

Table 5.3 Center-to-center distance between nanocrystals coated with decanethiol and self-assembled in 3D superlattices. For hcp, fcc, and bcc arrangements, d are calculated from the $d_{(002)}$, $d_{(111)}$, and $d_{(110)}$ spacings: $D = d_{(002)}[3/2]^{0.5}$, $D = d_{(111)}[3/2]^{0.5}$, and $D = d_{(110)}[3/2]^{0.5}$.

	Temperature	
Solvent	**15 °C**	**50 °C**
Decane	hcp $D = 7.1$ nm	fcc $D = 7.1$ nm
Toluene	fcc $D = 7.2$ nm	fcc $D = 7.3$ nm
Hexane	fcc $D = 7.3$ nm	fcc $D = 7.3$ nm

with C_{10}–SH than C_{12}–SH and remains unchanged whatever the solvent used to disperse the nanocrystals and the substrate temperature (Table 5.2). This can be explained by the decrease in chain length ($\Delta L = 0.25$ nm) [10]. The fcc structure obtained at 50 °C is consistent with the model, from which van der Waals interactions play a major role in the transition between compact structures (fcc/hcp) to bcc 3D superlattices. The theoretical model (Figure 5.3) predicts fcc structures for edge–edge distances smaller than 2.4 nm, while the experimental distance is about 2.3 nm. Furthermore, a clear transition from hcp to fcc between 15 and 50 °C is observed for a high-boiling-point solvent (decane), whereas for solvents having rather low boiling points (toluene, hexane), only fcc 3D superlattices are obtained.

Hence, a difference in the alkyl chain length as small as two carbon atoms has a strong effect on the supracrystalline structures.

Temperature is also a major parameter. Let us consider the concept of "hard cores" and "soft corona" to describe the interactions between nanocrystals [4]. The balance between the hard-core repulsion and the thiol coronas determines whether nanocrystals self-assemble in dense or loose lattices. The interdigitation between the alkyl chains is known to reduce their orientational entropy within the corona. Nanocrystal assembly is then controlled primarily by hard-core repulsion. In this case, in a first approximation, dense lattices such as fcc or hcp are favored. Conversely, the soft corona favors structure that minimize the interfacial energy with the formation of the loose structure (bcc).

The interdigitation markedly depends on the dynamic of the alkyl chains. In the temperature range between 15 and 50 °C, C_{10} alkyl chains are not dynamics and can interdigitate [11]. Below the melting point (52 °C), C_{12} alkyl chains interdigitate, whereas above it, they cannot. From these data, it is expected to observe close-packed structures with C_{10} alkyl chains in the range of 15–50 °C and with C_{12} counterparts below the melting point. This explains clearly the data shown

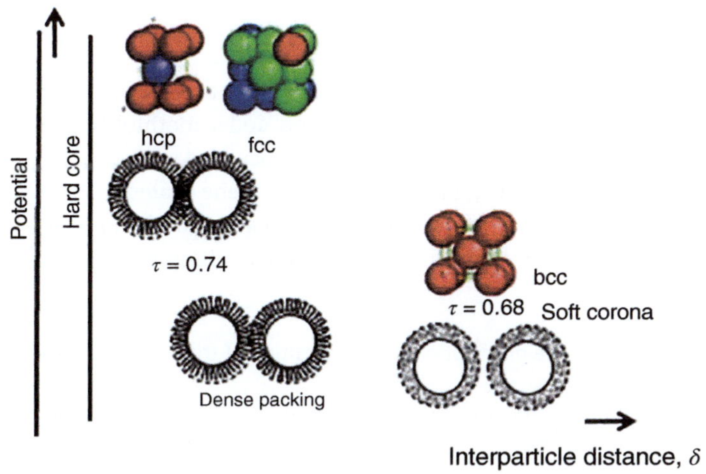

Figure 5.5 Interaction potential between silver nanocrystal and 3D superlattice formation. *Source:* Courty et al. [1]. Copyright 2011 American Chemical Society.

above. Dense superlattices (hcp, fcc) are obtained at up to 50 °C for nanocrystals coated with C_{10} alkyl chains (Figure 5.4), whereas with C_{12} counterparts, bcc lattices are produced at 50 °C while dense fcc or hcp lattices are below 50 °C. These data confirm the calculations [12] from which the stable solid phase is fcc lattice (below the melting temperature) and the liquid phase is bcc lattice (above the melting point). Thus, the interparticle attraction forces dominated by the alkylthiol interactions are clearly shown to control the 3D superlattice formation (Figure 5.5).

At this point, a question arises: What factor favors hcp or fcc structure? They are both compact arrangements of spheres with the same compacity (0.74) and differing by the sequence of the stacking planes: sequence ABC for fcc and AB for hcp (Figure 5.6). According to theoretical studies [13], the difference between these two structures is related to small difference in the entropy ($+10^{-3}\ k_B$) favoring the fcc phase against hcp one. No easy explanation can be given concerning the hcp/fcc transition.

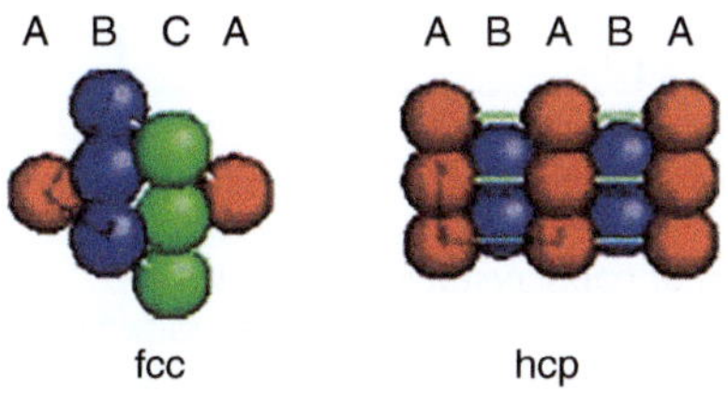

Figure 5.6 ABC and AB sequences in fcc and hcp structures, respectively, and polyhedron of the first 12 neighboring of these structures.

5.2 Optical Properties

Two type of ligands – amine (NH) or thiol (SH) ligands – are used to stabilize Ag nanocrystals of various sizes: oleylamine (C_{18}–NH_2) and dodecanethiol (C_{12}–SH). Their size and size distribution are given in Table 5.4. The localized surface plasmon resonance (LSPR) spectra of Ag nanocrystals dispersed in toluene coated with either oleylamine (Figure 5.7a) are shifted toward smaller wavelength and higher energies on increasing the nanocrystal size [15]. Similar behavior is observed with dodecanethiol (see Chapter 15).

Table 5.4 Nanocrystal diameter, size distribution (σ), and position of the maximum peak of LSPR with two types of coating agent.

Coating agent	C_{12}–NH_2				C_{18}–NH_2				
d (nm)	2.9	4.1	5.1	6.3	2.2	4.0	5.9	8.7	12.9
σ (%)	9	9	8	8	9	9	8	9	9
LSPR (eV)	2.91	2.98	3.01	3.02	2.9	2.98	3.02	3.03	3.03

Source: Based on Wei et al. [14].

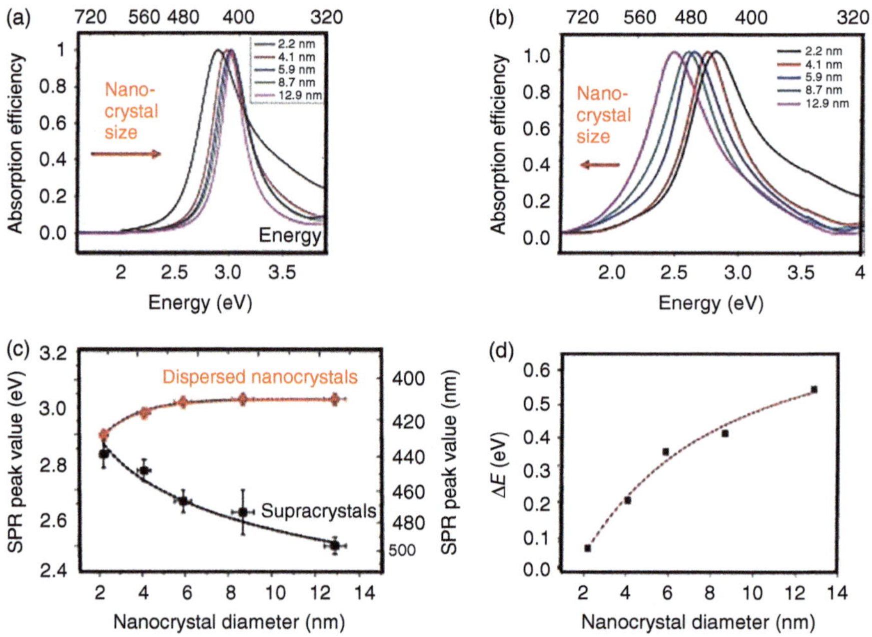

Figure 5.7 (a and b) Absorption spectra (normalized) of nanocrystals coated with oleylamine differing by their average sizes either dispersed in toluene (a) or deposited on a glass wafer and forming a thin film (b). (c) LSPR peak position values of dispersed nanocrystals (red sphere) and corresponding supracrystalline films (black square) as a function of the nanocrystal diameter (d) plot of the ΔE (the energy difference of the LSPR peak position between the dispersed samples and the supracrystalline films) against nanocrystal diameter. *Source:* (a, d) Based on Wei et al. [14].

5.2.1 Thin Film

Let us deposit dilute colloidal solutions of Ag nanocrystals coated with C_{18}–NH_2 on either transmission electron microscopy (TEM) grid or a glass wafer [14]. Figure 5.8 shows long-range order (1–3 μm) confirmed by the corresponding fast Fourier transform (FFT) pattern as shown in insets. Note that insets in Figure 5.8b–e show hexagonal patterns, whereas inset in panel Figure 5.8a shows only a rectangular one. The thickness of the films was evaluated to be two to seven layers from those images.

Figure 5.9 shows the rather inhomogeneous color films obtained by depositing the colloidal solutions on the glass wafer instead of TEM grid. The corresponding grazing-incidence small-angle X-ray scattering (SAXS) patterns confirm that the nanocrystals forming the films are well ordered in a supracrystalline structure. The interparticle distance, δ_{pp}, between nanocrystals remains in the same order of magnitude (Table 5.5).

The LSPR spectra of the 3D superlattice *thin* films deposited on a glass wafer are recorded using a horizontally polarized light source and keeping the incident light parallel to the substrate. The LSPR spectra of nanocrystals are redshifted (low energy shift) on increasing the nanocrystal diameters, as depicted in Figure 5.7b. This is in complete opposition with what has been observed with the same nanocrystals dispersed in toluene (Figure 5.7a) with a blueshift observed on increasing the nanocrystal sizes in solution [15–17]. This is explained by the lower conductivity of the outer layer in smaller nanocrystals. There are marked changes on the LSPR maximum absorption wavelength with nanocrystal either dispersed in solution or self-assembled in fcc thin 3D superlattices (Figure 5.7c). Similar data are obtained with C_{18}–NH_2 and C_{12}–NH_2 as coating agents. Table 5.5 and Figure 5.7d show the energy differences in the LSPR band peak position between the dispersed nanocrystals and the corresponding thin supracrystalline films, ΔE.

These nonnegligible energy differences are attributed to the interactions between the Ag nanocrystals embedded in the 3D superlattices. In fact, due to an increase in the dipole moments led by an increase in the nanocrystal diameters, the electromagnetic coupling between nanocrystals within the 3D superlattices increases and consequently induces a change in the optical properties of these assemblies.

Absorption spectra of 3D superlattices are recorded by using polarized light parallel to the plane of incidence at various incident angles θ (Figure 5.10). On increasing θ angle, a secondary peak, at higher energy than LSPR, progressively appears.

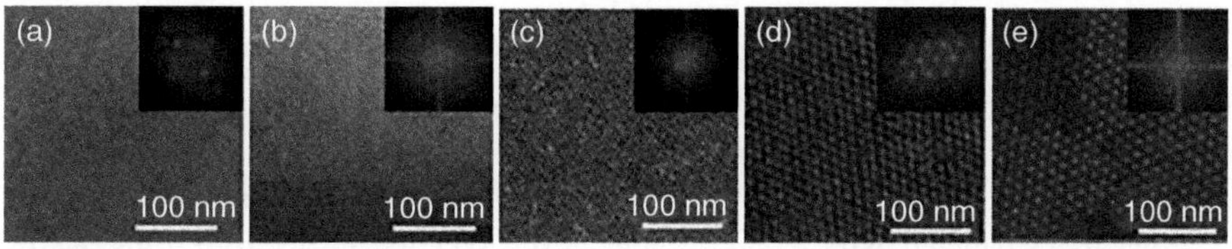

Figure 5.8 TEM images of thin supracrystalline films of Ag nanocrystals coated with oleylamine (C_{18}–NH_2) differing by their diameters: (a) 2.2 nm, (b) 4.1 nm, (c) 5.9 nm, (d) 8.7 nm, and (e) 12.9 nm. The inserted scale bars are 100 nm. *Source:* Wei et al. [14]/American Chemical Society.

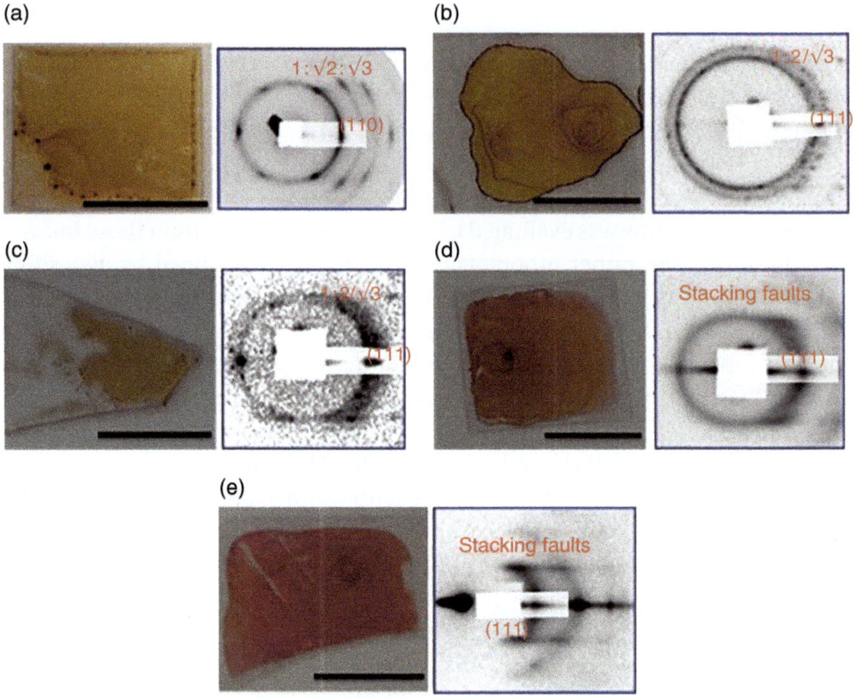

Figure 5.9 Optical images of the supracrystalline films on glass substrate and corresponding X-ray diffraction patterns for thin 3D superlattices of Ag nanocrystals coated with oleylamine ($C_{18}-NH_2$) and differed by their diameters: (a) 2.2 nm, (b) 4.1 nm, (c) 5.9 nm, (d) 8.7 nm, and (e) 12.9 nm. The scale bars in the optical images all are 1 cm. $1 : \sqrt{2} : \sqrt{3}$ (or $1 : 2/\sqrt{3}$) positions for first diffraction lines are presently typical for bcc (or fcc) structures. Stacking faults in fcc result in streaks. *Source:* Wei et al. [14]/American Chemical Society.

Figure 5.11 shows such splitting of the LSPR band more pronounced with large nanocrystal sizes than with smaller ones. At $\theta = 60°$, this splitting is also observed for small nanocrystal diameter. The presence of such splitting reveals that the LSPR spectra in the visible range are markedly dependent on the incidence of the light source. These data confirm those predicted by discrete dipole approximation (DDA) simulation [17] and observed experimentally by reflectivity [18] for 5-nm Ag nanocrystals self-assembled in *2D superlattices*.

The presence of a higher energy band is interpreted as follows: for incident angle $\theta \neq 0°$, the electric field of the incident radiation projects in both parallel and perpendicular directions to the nanocrystal arrays. The amplitudes of these two bands evolve in the opposite trends when varying the incident angle. The perpendicular component increases with θ, and the resonant excitation of both the longitudinal and transverse surface plasmon eigenmodes of the 3D superlattice *thin* films results in the presence of two bands in the absorption spectra. This feature points out the optical anisotropy of the 3D array of the nanocrystals. Such variations in the amplitude of both longitudinal and transverse LSPR are well observed by plotting their

Table 5.5 Interparticle distance (δ_{pp}) between Ag nanocrystals coated with $C_{18}-NH_2$ within the 3D superlattices, deduced from small-angle X-ray scattering measurements. Maximum energy of the LSPR ($E_{3D\ superlattice\ peak}$) spectrum and the corresponding maximum wavelength (λ_{max}) of Ag nanocrystals embedded within the 3D superlattices and measured at 0° incident angle. ΔE is the difference of E values between Ag nanocrystals dispersed in toluene (Table 5.4) and deposited on a glass substrate forming 3D superlattices. LSPR peak position of the longitudinal ($E_{longitudinal\ peak}$) and transverse ($E_{transverse\ peak}$) peaks and the corresponding wavelengths ($\lambda_{longitudinal}$) and transverse ($\lambda_{transverse}$) with various Ag nanocrystal sizes coated with oleylamine ($C_{18}-NH_2$) for an incident angle of 60°.

Nanocrystal size (nm)	2.2	41	5.9	8.7	12.9
δ_{pp}	3.2	3.2	3.1	2.8	2.5
$E_{3D\ superlattice}$ (eV)	2.83	2.77	2.67	2.62	2.50
$\lambda_{3D\ superlattice}$ (nm)	438	448	4.64	473	496
ΔE	0.07	0.21	0.35	0.41	0.34
$E_{longitudinal\ peak}$ (eV)	2.84	2.78	2.67	2.62	2.50
$\lambda_{longitudinal\ peak}$ (nm)	436	446	464	473	496
$E_{transverse\ peak}$ (eV)	3.13	3.37	3.38	3.45	3.48
$\lambda_{transverse\ peak}$ (nm)	396	370	367	359	356

Source: Based on Wei et al. [14].

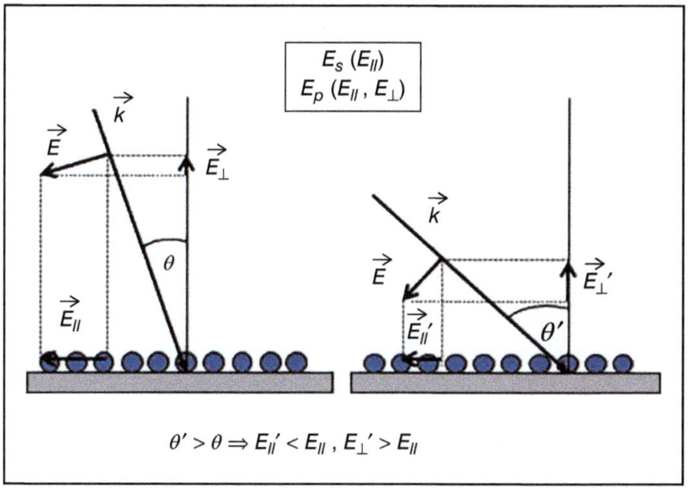

Figure 5.10 Schematic representation of the modeled system. *Source:* Based on Wei et al. [14].

correspondent energies with γ ($\gamma = \delta_{pp}/d$), where d and δ_{pp} are the nanocrystal diameter and the interparticle distance between nanocrystals within the 3D superlattices (Figure 5.12 and Table 5.5). The two LSPR peaks tend, on increasing the γ value, to reach values close to their counterparts in 3D superlattices of 2.2-nm Ag nanocrystals.

Such a feature, observed experimentally with nanocrystals differing by their sizes self-assembled in 3D superlattices with δ_{pp} rather constant (Table 5.5), was recently

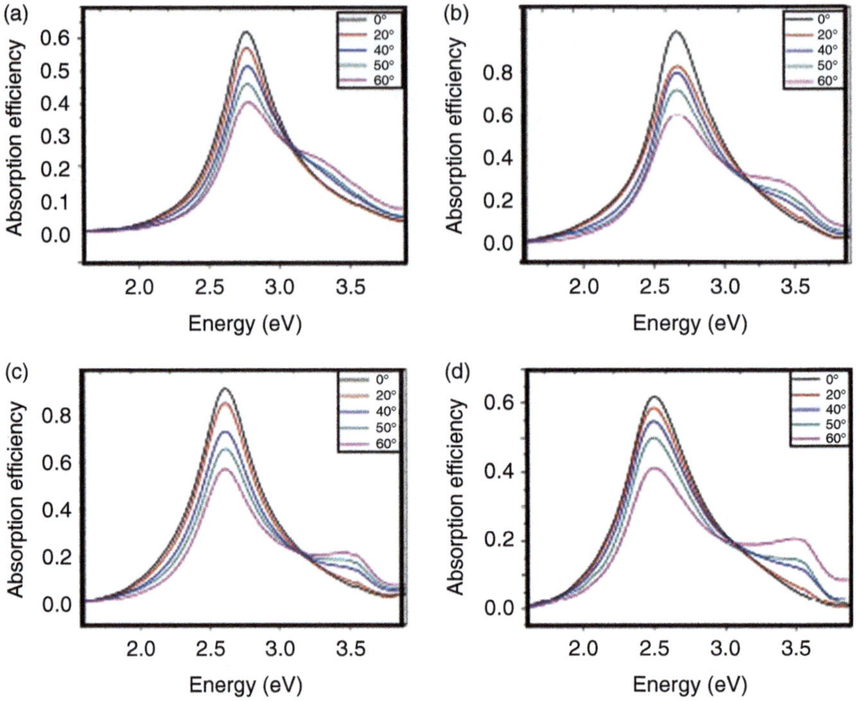

Figure 5.11 LSPR spectra of the 3D superlattice films of nanocrystals differing by their diameters and coated with C_{18}–NH_2 at various incidence angles: (a) 4.1 nm, (b) 5.9 nm, (c) 8.7 nm, and (d) 12.9 nm. *Source:* Wei et al. [14]. Copyright 2015 American Chemical Society.

predicted by DDA simulation of 2D superlattices and is attributed to an increase in the dipolar interactions [17]. Comparison between the data presented hereby and those deduced from simulations shows a good agreement for energy values of the transverse LSPR energy band for C_{18}–NH_2-coated Ag nanocrystals differing by their sizes, shown in Figure 5.12, whereas discrepancies are observed between experimental and theoretical energy values for the longitudinal LSPR bands. This can be attributed to the fact that the longitudinal peak is more sensitive to the coating agent than the transverse one. In fact, DDA simulations mentioned above did not take the surface coating agent into account, only assuming spheres surrounded by a continuous medium. These data also show that the transversal LSPR peak does not markedly depend on a weak coating agent such as C_{18}–NH_2, whereas the longitudinal modes could be highly sensitive to this factor.

At this stage, we raise a question on what the influence of the coating agent on the LSPR spectra of "thin 3D superlattices" is. To answer this, let us replace oleylamine (C_{18}–NH_2) by dodecylamine (C_{12}–NH_2) or by dodecanethiol (C_{12}–SH), keeping a similar average diameter (c. 6-nm). Unfortunately, it has been impossible to produce 3D superlattices with various sizes of Ag nanocrystals coated with C_{12}–NH_2 or C_{12}–SH. In a similar manner to what was described above with C_{18}–NH_2-coated nanocrystals, these Ag nanocrystals are deposited on a glass wafer and form thin films. Figure 5.13a,b shows that the films on glass are relatively inhomogeneous.

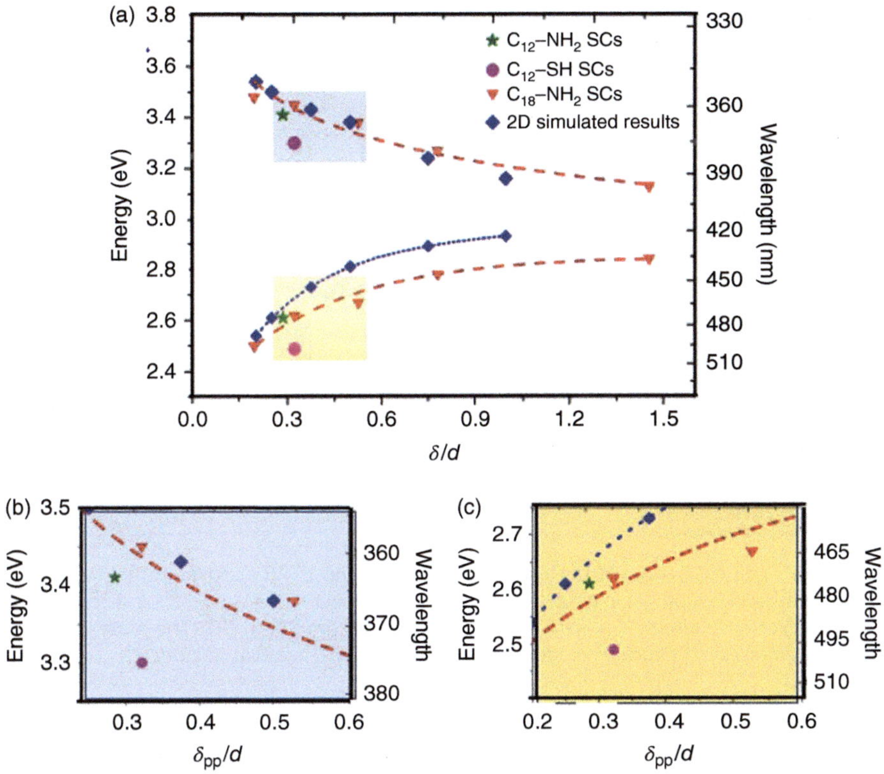

Figure 5.12 (a) Comparison of the experimental and simulated longitudinal and transversal LSPR band position [15] values against $\gamma = \delta_{pp}/D$ of nanocrystals differing by their coating agents and sizes. (b and c) Local magnifications of the (a). *Source:* Wei et al. [14] Copyright 2015 American Chemical Society.

Nevertheless, the corresponding SAXRS patterns (Figure 5.13c,d) show that the ~6-nm nanocrystals coated with C_{12}–NH_2 and C_{12}–SH are ordered in fcc crystalline structure as observed with C_{18}–NH_2.

Table 5.6 shows that the interparticle distance, δ_{pp}, remains similar (1.8–1.9 nm) for those specimens, values being smaller than those obtained with C_{18}–NH_2 (3.1 nm). The optical properties of these C_{12}–NH_2 and C_{12}–SH 3D superlattices (Figure 5.14) differing by the nature of their coating agent show similar behavior as described above with C_{18}–NH_2 (Figure 5.7b,c). Figure 5.14 compares the LSPR spectrum of nanocrystals having similar diameter (~6 nm) coating with similar alkyl chains (C_{12}) and differing by their head polar group (NH_2 or SH), i.e. by their interactions with the nanocrystal metallic surface. Table 5.6 shows that the average distance between nanocrystals remains similar (~1.8-nm). As shown with C_{18}–NH_2 (Figure 5.7b,c), for a given coating agent, the redshift of the LSPR spectrum of 3D superlattices compared with the same nanocrystals dispersed in solution is attributed to dipolar interactions. Furthermore, the LSPR redshift of both dispersed nanocrystals and 3D superlattices of C_{12}–SH compared with C_{12}–NH_2-coated nanocrystals cannot be attributed to the difference in the nanocrystal diameter

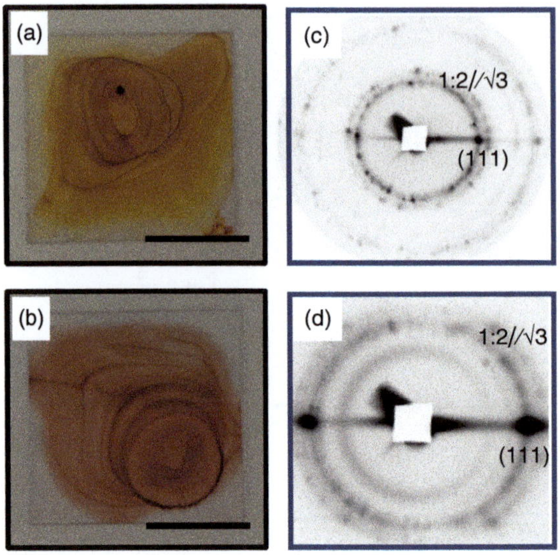

Figure 5.13 Optical images, X-ray diffraction patterns, and LSPR spectra of the thin supracrystalline films on glass made from 5.9-nm Ag nanocrystals with different coating agents: (a, c) dodecylamine (C_{12}–NH_2) and (b, d) dodecanethiol (C_{12}–SH). The scale bar in the optical images is 1 cm. *Source:* Wei et al. [14]/American Chemical Society.

Table 5.6 The interparticle distance (δ_{pp}) between Ag nanocrystals in the 3D superlattices differing by the coating agents 6.0-nm C_{12}–NH_2 and 6.3-nm C_{12}–SH. The maximum LSPR wavelength and the corresponding energy of the LSPR peak of Ag nanocrystals dispersed either in toluene or within the 3D superlattices recorded when the incident angle is 0°. LSPR peak position of the longitudinal and transverse peaks with ~6-nm Ag-nanocrystal sizes coated with either dodecylamine (C_{12}–NH_2) or dodecanethiol (C_{12}–SH) for an incident angle of 60°.

Coating agent	δ_{pp}	$E_{3D\ superlattice}$ (eV)	ΔE	$E_{longitudinal\ peak}$ (eV)	$\lambda_{longitudinal\ peak}$ (nm)	$E_{transverse\ peak}$ (eV)	$\lambda_{transverse\ peak}$ (nm)
C_{12}–SH	1.9	2.49	0.40	2.5	496	3.3	376
C_{12}–NH_2	1.8	2.61	0.41	2.61	475	3.41	364

Source: Based on Wei et al. [14].

(6 and 6.3 nm, respectively) but to a decrease in the conductivity of the outer layer of the Ag nanocrystals within the 3D superlattices compared with the inner core. That confirms the influence of the coating agent on the LSPR spectra.

For an incident angle $\theta = 60°$ and by replacing C_{18}–NH_2 by C_{12}–NH_2 as coating agent, a rather good agreement is obtained between the data, whereas a marked discrepancy is observed when replacing C_{12}–NH_2 by C_{12}–SH (Figure 5.12). This is consistent with the fact that in the case of NH_2 when the coating agent keeps the same head polar group, the change in the conductivity of the outer layer of the nanocrystals is negligible. In the case of SH, when the strength of the interactions between Ag atoms at the surface and the head polar group is increased, the LSPR

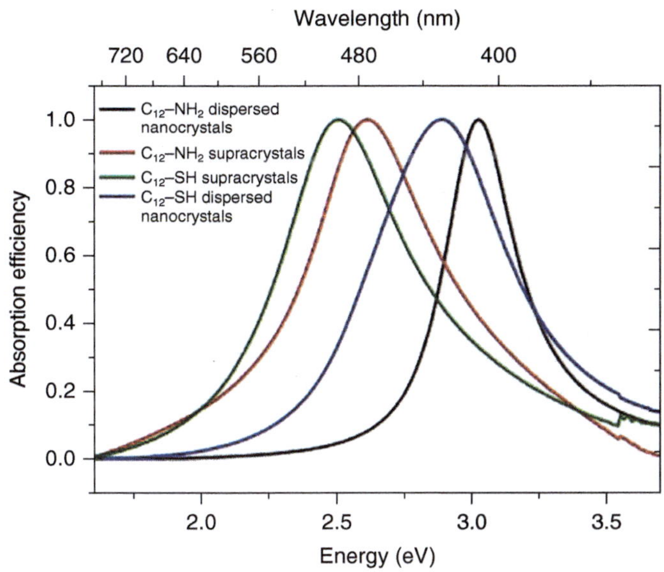

Figure 5.14 LSPR spectra of the supracrystalline films (red and green) deposited on a substrate and their building blocks dispersed in toluene (black and blue). Nanocrystals coated with C_{12}–NH_2 (red and black) or C_{12}–SH (green and blue). *Source:* Wei et al. [14] Copyright 2015 American Chemical Society.

resonance shifts to the lower energy. From these data, it is concluded that dipolar interactions play the major role in the plasmonic properties of 3D thin 3D superlattices. However, the contribution of coating agents of nanocrystals embedded in 3D superlattices cannot be neglected. There is high similitude between the DDA simulation of 2D superlattices [15] and C_{12}–NH_2 and C_{12}–SH compared with the same nanocrystals dispersed in toluene. Table 5.6 shows that the differences in the energies determined from the maximum of the LSPR spectra between deposited and dispersed remain the same ($\Delta E \approx 0.40\,\text{eV}$), when nanocrystals within the 3D superlattice keep similar interparticle distance (~1.8-nm), i.e. nanocrystals are coated with C_{12} derivatives. This difference is larger with 3D superlattices of 5.9-nm Ag nanocrystals coated with longer chain length of C_{18}–NH_2 ($\Delta E = 0.35\,\text{eV}$). The experimental data obtained with thin film 3D superlattices indicates that the substrate effect is negligible.

Note that the mode of producing 3D thin film could play a role in the optical properties of the assemblies. A recent study shows that, by deposition of a colloidal solution at the diethylene glycol (DEG)/air interface, the nanocrystal ordering is markedly improved [19]. Hence, for comparison, the same amount of Ag nanocrystals coated with C_{18}–NH_2 colloidal solution as used above was deposited on DEG liquid in a Teflon well. The supracrystalline film is formed at the interface of DEG/air and is transferred either on a TEM grid or on a glass substrate. The TEM images confirm a larger coherence length of Ag nanocrystal films as observed previously [19]. By deposition on glass, a better homogeneity than drop deposition process is observed. However, some inhomogeneities can still be observed. The X-ray diffraction patterns is typical for a restricted number of well-ordered stacking monolayers. The interparticle distances are almost the same for the supracrystalline films obtained by the two different methods. Optical measurements similar as those described above are carried out on those supracrystalline films. Similar trends are observed with a normal incident light to the substrate; that is, a redshift (low energy shift) occurs when increasing the nanocrystal sizes. However, Figure 5.15a shows a nonnegligible decrease of the maximum energy of the LSPR band that does not follow the same trend as that observed in the absence of DEG. Similarly, by increasing θ from 0° to 60°, a splitting in the LSPR spectrum takes place as observed in the absence of DEG. However, a careful analysis of these spectra shows that the evolution of the longitudinal and transverse LSPR maximum energies with γ markedly change compared with what is shown in Figure 5.12 (Figure 5.15b). These differences cannot be attributed to the changes in the

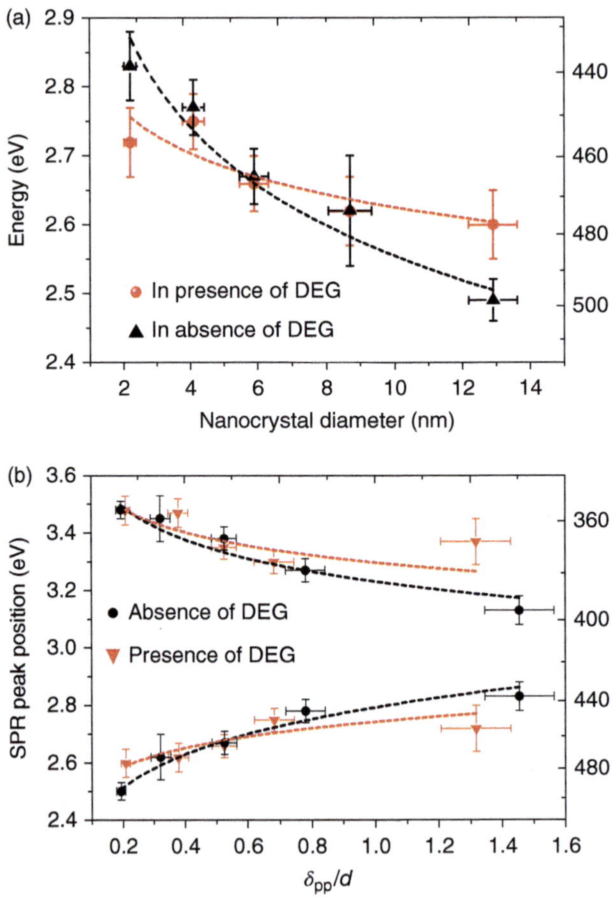

Figure 5.15 (a) Plots of the LSPR peak positions with various Ag nanocrystal sizes: (red sphere) 3D superlattices produced in the presence of DEG and (black triangle) 3D superlattices produced in the absence of DEG. (b) Plot of the longitudinal and transverse LSPR peak position value against δ_{pp}/d in the absence (black) and presence of DEG (red). *Source:* Based on Wei et al. [14].

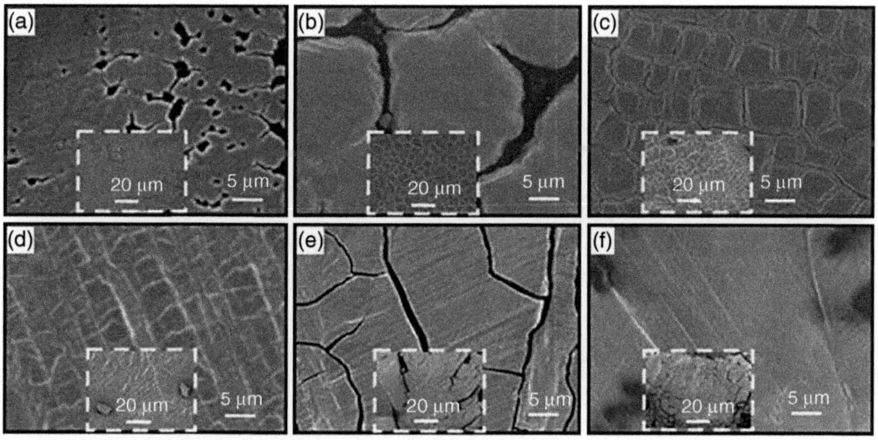

Figure 5.16 Scanning electron microscopy (SEM) images of 3D superlattice films of 8.2-nm Ag nanocrystals coated with oleamine with various thicknesses: (a) 0.2–0.6, (b) 0.4–1.0, (c) 0.6–1.4, (d) 0.9–1.6, (e) 1–1.4, and (f) 1.3–1.9 μm. *Source:* Marie Paule Pileni, 2020/Royal Society of Chemistry.

interparticle distances that remain similar to those obtained by drop deposition. Hence, such differences are attributed to diffusion of DEG molecules inducing a change in the nanocrystal environments and consequently in the optical properties of nanocrystals within the 3D superlattices. This is also supported by the fact that it is impossible to image by scanning tunneling microscopy (STM) the 3D superlattice film produced using the DEG method and deposited on highly oriented pyrolytic graphite (HOPG), whereas by simple solvent evaporation of same amount of colloidal solutions on HOPG, the film is very easily imaged with the presence of well-defined Ag nanocrystals.

5.2.2 Thick Films

With a large nanocrystal concentration of the colloidal solution used before solvent evaporation process, thick films of fcc 3D superlattices are produced [20]. At the end of the evaporation, the drying process induces film shrinking with the restriction of the film adhesion on the surface and consequently crack formation. Figure 5.16 shows crack patterns observed at various film thicknesses. The geometries of different crack patterns are observed in a very large variety of materials, over a wide range of scales with specific scaling laws [21–29]. These cracks are observed by optical microscopy in transmission modes (Figure 5.17). They form a network of channels separating islands (of a few tens of μm) of thick film. Light coming out of channels is represented in blue while light from islands is reddish (no transmission is observed). At the opposite, with ferrite nanocrystals, the bottom of the cracks did not show any waiting layers of nanocrystals [28, 29]. At the opposite, here by transmission microscopy, we observe various color contrasts in the channels. This indicates the presence of nanocrystal layers at the bottom of the cracks. This could be due to the change in nanocrystal size, materials, and coating agents. Consequently,

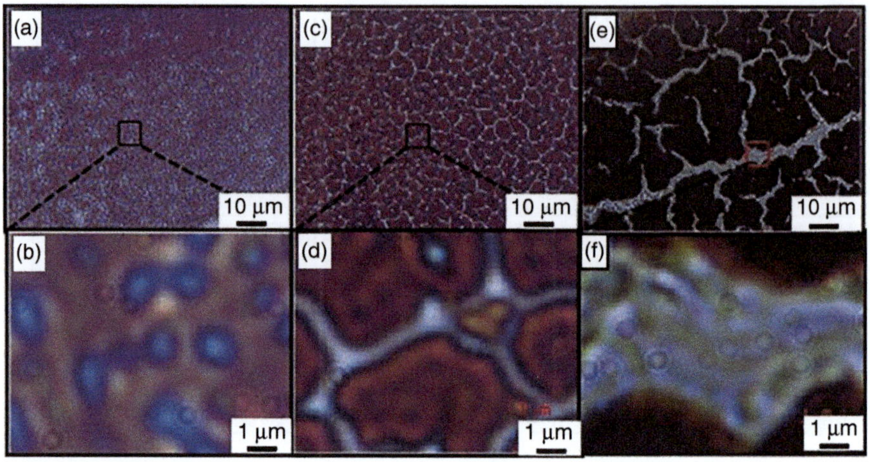

Figure 5.17 Optical micrographs in transmission mode of the 0.2- to 1.4-μm 3D superlattice film: (a, b) 0.2–0.6, (c, d) 0.4–1.0, and (e, f) 1.0–1.4 μm. *Source:* Marie Paule Pileni, 2020/Royal Society of Chemistry.

the blue and white colors in the crack channels demonstrate the presence of wetting layers at the bottom of each crack. Similar patterns are observed in bright-field reflection mode; cracks appear dark due to the light diffraction at the film edges (not shown here). The loss of uniformity in the transmission mode clearly indicates the presence of several wetting layers at the bottoms of the cracks.

On increasing the number of nanocrystal layers, the LSPR spectra should reach saturation, which explains why most of the optical properties of 3D superlattices were investigated for less than 10 layers [30–35]. One attempt, in limit of Beer–Lambert law, was performed on 3D superlattices having 30–35 layers [35]. By using effective medium theory (EMT) without *features on their surface* and by taking into account the crystalline structure of the film (fcc) with 8.2 nm, 2.8 nm, and 0.3 as average nanocrystal diameter, nanocrystal interparticle distance, and filling factor, respectively, the calculated LSPR spectra of homogeneous 3D superlattice film (flat surface) differing by their thicknesses, from 28 to 180 nanocrystal layers, reach saturation in the wavelength range from 300 to 550 nm. At larger wavelength, the transmission intensity through the 3D superlattice films decreases with increasing the film thickness, and a wider linewidth is observed. Figure 5.18 shows transmission spectra for various film thicknesses. For fcc 3D superlattice films formed by 27–78 layers, a direct measurement of absorption, reflection, and transmission is still possible (Figure 5.18a). By increasing the number of layers from 27 to 180, the transmission spectra of Ag 3D superlattice film (Figure 5.18a–d, dashed line) reveal a decrease in the transmission with an increase in the notch full width at half maximum (FWHM) as the film thickness increases. Hence, the calculated spectra totally disagree with the experimental data.

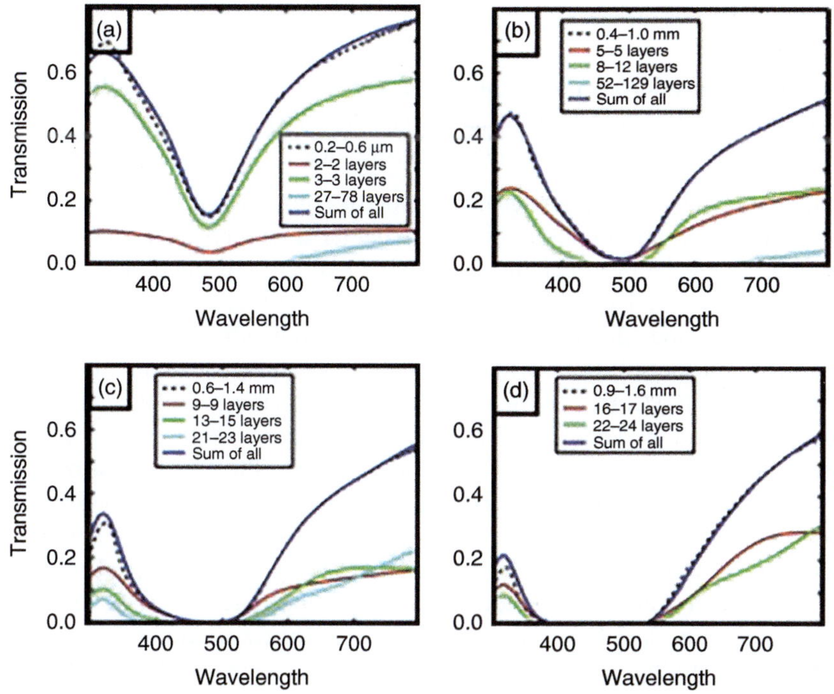

Figure 5.18 Calculated (blue solid traces) and experimental (dashed traces) transmission spectra of 3D superlattice films with various thicknesses of 8.2-nm Ag nanocrystals: (a) 0.2–0.6, (b) 0.4–1, (c) 0.6–1.4, and (d) 0.9–1.6 μm. The calculated spectra account for wetting layers with different thicknesses. Colors correspond to different number of monolayers. *Source:* Wei et al. [20]. Copyright 2019 American Chemical Society.

According to the optical micrographs in transmission mode (Figure 5.17), we assumed, at the bottom of each crack, presence wetting layers of nanocrystals which are characterized by steps having various thicknesses. In the total transmission calculation, the thickness and the weighted sum of the transmissions of each zone are the two free parameters taken into account. Figure 5.18 reveals the excellent agreement between calculations accounting for the cracks and experiments for films of various thicknesses (0.2–0.6, 0.4–1, 0.6–1.4, and 0.9–1.6 μm). The solid red, green, and cyan lines represent the transmission spectra of different wetting layers that are introduced in the cracks for each film. The blue trace is the sum of all colored lines.

From these data, we conclude that cracks dominate the optical response of the films of different thicknesses. Wetting layers with different thicknesses, sitting at the bottom of the cracks, explain the optical properties observed with thick fcc films of Ag nanocrystals with a large number of layers. From these data, it seems impossible, with dried fcc 3D superlattices, to observe any collective optical modes.

5.3 Stability

Nanocrystal organizations represent a new generation of materials with specific properties compared with those of isolated nanocrystals and of the bulk material [3, 36, 37]. Chapter 11 presents two intrinsic properties:

(i) The first intrinsic crystalline growth properties of highly ordered mono- and multilayers of 5-nm Ag nanocrystals. Triangular single crystals with fcc structures are obtained by annealing the ordered nanocrystals under atmospheric pressure at 50 °C. The triangles are mixed with well-crystallized, coalesced particles of various shapes. Their size depends on the initial nanocrystal ordering range on the substrate, which is local on amorphous carbon and highly extended on HOPG. Hence, the single-crystal size is larger on HOPG than on amorphous carbon. These observations show that the crystalline growth properties of silver nanocrystals can be tailored by controlling their organization. Furthermore, on HOPG, an epitaxial orientation of the triangles is observed.

(ii) Solvent–ligand interactions in colloidal nanocrystals used to modulate the way they pack into superlattices. The crystal structures of the nanocrystal superlattices made of 2.2-nm Ag nanocrystals can be controlled by using different carrier solvents. Specifically, the superlattice structures are tuned from bcc to fcc when varying solvents from hexane to tetrachloroethylene (TCE). Furthermore, by simultaneously annealing these two samples at different temperatures, bcc structures originating from hexane solutions are dominated by simple coalescence mechanism, while fcc structure stemming from TCE solutions undergoes Ostwald ripening process that can produce a variety of binary nanocrystal superlattices such as NaCl, AlB_2, $NaZn_{13}$, and $MgZn_2$, the formation of those structures being well explained by a pure entropy-driven process. This is believed to be due to the variations in the ligand coverage ratio of the nanocrystals in different solvents that are changing the stability of superlattice structures. Those findings provide insights into the solvent-mediated nanocrystal superlattices and the Ostwald ripening process in nanocrystal superlattices.

5.4 Conclusions

The control of the crystalline structures of nanocrystals self-assembled in 3D superlattices depends on the solvent used to dispersed nanocrystals in solution, the nanocrystal size, the chain length of the coating agents, and the substrate temperature.

The optical properties of Ag nanocrystals, dispersed in solution, depend on their average size and coating agent. Due to dipolar interactions, this dependence markedly differs when the nanocrystals are deposited on a substrate. With thick 3D superlattice films, the optical properties are dominated by the presence of cracks and wetting layers at the bottom of the cracks.

References

1 Courty, A., Richardi, J., Albouy, P.A., and Pileni, M.P. (2011). How to control the crystalline structure of supracrystals of 5-nm silver nanocrystals. *Chem. Mater.* 23: 4186–4192.

2 Henry, A.I., Courty, A., Pileni, M.P. et al. (2008). Tuning of solid phase in supracrystals made of silver nanocrystals. *Nano Lett.* 8: 2000–2005.

3 Henry, A.I., Courty, A., Goubet, N., and Pileni, M.P. (2008). How do self-ordered silver nanocrystals influence their growth into triangular single crystals? *J. Phys. Chem. C* 112: 48–52.

4 Whetten, R.L., Shafigullin, M.N., Khoury, J.T. et al. (1999). Crystal structures of molecular gold nanocrystal arrays. *Acc. Chem. Res.* 32: 397–406.

5 Korgel, B.A. and Fitzmaurice, D. (1999). Small-angle X-ray-scattering study of silver-nanocrystal disorder-order phase transitions. *Phys. Rev. B* 59: 14191–14201.

6 Landman, U. and Luedtke, W.D. (2004). Small is different: energetic, structural, thermal, and mechanical properties of passivated nanocluster assemblies. *Faraday Discuss.* 125: 1–22.

7 Evans, R., Smitham, J.B., and Napper, D.H. (1977). Theoretical prediction of the elastic contribution to steric stabilization. *Colloid. Polym. Sci.* 255: 161–167.

8 Friedel, J. (1974). Stability of body centered cubic phase in metals at high temperatures. *J. Phys. France Lett.* 35: L 59.

9 Ziherl, P. and Kamien, R.D. (2001). Maximizing entropy by minimizing area: towards a new principle of self-organization. *J. Phys. Chem. B* 105: 10147.

10 Bain, C.D., Troughton, E.B., Tao, Y.T. et al. (1989). Formation of monolayer films by the spontaneous assembly of organic thiols from solution onto gold. *J. Am. Chem. Soc.* 111: 321–335.

11 Pradeep, T., Mitra, S., Sreekumaran Nair, A., and R. Mukhopadhyay (2004). Dynamics of alkyl chains in monolayer-protected Au and Ag clusters and silver thiolates: a comprehensive quasielastic neutron scattering investigation. *J. Phys. Chem. B* 108: 7012–7020.

12 Robbins, M.G., Kremer, K., and Grest, G.S. (1998). Molecular dynamics simulation for polymers in the presence of a heat bath. *J. Chem. Phys.* 88: 3286.

13 Koch, H., Radin, C., and Sadun, L. (2005). Most stable structure for hard spheres. *Phys. Rev. E* 72: 061708.

14 Wei, J., Schaeffer, N., Albouy, P.A., and Pileni, M.P. (2015). Surface plasmon resonance properties of silver nanocrystals differing in size and coating agent ordered in 3D supracrystals. *Chem. Mater.* 27: 5614–5621.

15 Wei, J., Schaeffer, N., and Pileni, M.P. (2014). Ag nanocrystals: 1. Effect of ligands on plasmonic properties. *J. Phys. Chem. B* 118: 14070–14075.

16 Romann, J., Wei, J., and Pileni, M.P. (2015). Computational matching of surface plasmon resonance: interactions between silver nanoparticles and ligands. *J. Phys. Chem. C* 119: 11094–11099.

17 Yang, P., Portales, H., and Pileni, M.P. (2011). Dependence of the localized surface plasmon resonance of noble metal quasispherical nanoparticles on their crystallinity-related morphologies. *J. Chem. Phys.* 134: 024507.

18 Taleb, A., Russier, V., Courty, A., and Pileni, M.P. (1999). Collective optical properties of silver nanoparticles organized in two-dimensional superlattices. *Phys. Rev. B* 59: 13350–13358.

19 Dong, A.G., Chen, J., Vora, P.M. et al. (2010). Binary nanocrystal superlattice membranes self-assembled at the liquid–air interface. *Nature* 466: 474–477.

20 Wei, J., Deeb, C., Pelouard, J.L., and Pileni, M.P. (2019). Influence of cracks on the optical properties of silver nanocrystals supracrystal films. *ACS Nano* 13: 573–581.

21 Allain, C. and Limat, L. (1995). Regular patterns of cracks formed by directional drying of a colloidal suspension. *Phys. Rev. Lett.* 74: 2981–2984.

22 Shorlin, K.A., de Bruyn, J.R., Graham, M., and Morris, S.W. (2000). Development and geometry of isotropic and directional shrinkage-crack patterns. *Phys. Rev. E* 61: 6950–6957.

23 Dufresne, E.R., Corwin, E.I., and Greenblatt, N.A. (2003). Flow and fracture in drying nanoparticle suspensions. *Phys. Rev. Lett.* 91: 224501.

24 Heinrich, M., Gruber, P., Orso, S. et al. (2006). Dimensional control of brittle nanoplatelets. A statistical analysis of a thin film cracking approach. *Nano Lett.* 6: 2026–2030.

25 Groisman, A. and Kaplan, E. (1994). An experimental study of cracking induced by desiccation. *Europhys. Lett.* 25: 415–420.

26 Sasa, T.S.K. (1997). Pattern selection of cracks in directionally drying fracture. *J. Appl. Phys.* 36: 391–395.

27 Leung, K.T. and Néda, Z. (2000). Pattern formation and selection in quasistatic fracture. *Phys. Rev. Lett.* 85: 662–665.

28 Ngo, A.T., Richardi, J., and Pileni, M.P. (2008). Cracks in magnetic nanocrystal films: do directional and isotropic crack patterns follow the same scaling law? *Nano Lett.* 8: 2485–2489.

29 Ngo, A.T., Richardi, J., and Pileni, M.P. (2013). Crack patterns in superlattices made of maghemite nanocrystals. *Phys. Chem. Chem. Phys.* 15: 10666–10672.

30 Taleb, A., Petit, C., and Pileni, M.P. (1997). Synthesis of highly monodisperse silver nanoparticles from AOT reverse micelles: a way to 2D and 3D self-organization. *Chem. Mater.* 9: 9500.

31 Tao, A.R., Ceperley, D.P., Sinsermsuksakul, P. et al. (2008). Self-organized silver nanoparticles for three-dimensional plasmonic crystals. *Nano Lett.* 8: 4033–4038.

32 Ross, M.B., Ku, J.C., Vaccarezza, V.M. et al. (2015). Nanoscale form dictates mesoscale function in plasmonic DNA-nanoparticle superlattices. *Nat. Nanotechnol.* 10: 453–458.

33 O'Brien, M.N., Lin, H.X., Girard, M. et al. (2016). Programming colloidal crystal habit with anisotropic nanoparticle building blocks and DNA bonds. *J. Am. Chem. Soc.* 138: 14562–14565.

34 Mason, J.A., Laramy, C.R., Lai, C.T. et al. (2016). Contraction and expansion of stimuli-responsive DNA bonds in flexible colloidal crystals. *J. Am. Chem. Soc.* 138: 8722–8725.

35 Aubertin, P., Aissa, M.A.B., Raouafi, N. et al. (2015). Optical response and SERS properties of individual large scale supracrystals made of small silver nanocrystals. *Nano Res.* 8: 1615–1626.

36 Courty, A., Henry, I., Goubet, N., and Pileni, M.P. (2007). Large triangular single crystals formed by mild annealing of self-organized silver nanocrystals. *Nat. Mater.* 6: 900–907.

37 Wei, J., Schaeffer, N., and Pileni, M.P. (2016). Solvent-mediated crystallization of nanocrystal 3D assemblies of silver nanocrystals: unexpected superlattice ripening. *Chem. Mater.* 28: 293–302.

6

Mesostructures of Magnetic Nanocrystals Subjected to Applied Magnetic Field

In solution, magnetic nanocrystals are subjected to van der Waals, dipolar, and also nanocrystal/substrate attractive forces.

With weak dipolar interactions, large isotropic van der Waals interactions induce the formation of spherical aggregates. For weak van der Waals interactions, the strength of the dipolar moments controls the formation of chain-like structures. With no applied magnetic field, an abrupt transition from separate nanoparticles to randomly oriented chain-like aggregates takes place on increasing the nanoparticle size [1]. This is well demonstrated with CdTe nanocrystals: such semiconductor self-organization in pearl-necklace aggregates. This is explained by strong electrical dipole interactions [2].

When dipolar interactions are predominant, the formation of columns or labyrinths with submicron dimensions is observed in a variety of experimental systems like ferrofluids, films of magnetic garnets, di-block copolymers, phospholipid monolayers, superconductors, semiconductors, and Rayleigh Bernard instabilities [3–8]. The structure, thermodynamics, and dynamics in many physical and chemical systems are determined by an interplay of short-range isotropic and long-range anisotropic forces. For the study of this interplay, magnetic nanoparticles dispersed in solution are ideal model systems, since they are subjected to isotropic van der Waals and anisotropic dipolar forces. The properties of these ferrofluids attract the researcher interested in studying dipolar liquids, colloidal systems, and the self-organization at nanometric and mesoscopic scales. When a ferrofluid is confined with an immiscible nonmagnetic fluid between closely spaced parallel plates and subjected to a uniform magnetic field applied normal to the plates, the so-called labyrinthine instability occurs [3]. Static patterns appear that exhibit labyrinthine structures having walls of opaque ferrofluids separated by analogous bands of clear immiscible fluids. The phenomenon was observed for the first time in the late 1970s. In 1983, Rosensweig et al. [3] proposed a simple theory for spacing between the walls of the labyrinthine pattern, which is discussed in detail in Rosensweig's

Metal Nano 3D Superlattices: Synthesis, Properties, and Applications, First Edition.

classical textbook [4]. They assumed that the demagnetization field within the ferrofluid is uniform and equal to that in the center of the labyrinthine strips.

Here by using either g-Fe_2O_3 or Co nanocrystals, we try to reach a better understanding of the mechanism inducing the formation of various structures by applying, during the evaporation process, a magnetic field either parallel or perpendicular to the substrate. We point out the influence of the coating agent on the final structure that could tune the dipole–dipole and van der Waals interactions and consequently the influence of the final structure.

6.1 Maghemite Nanocrystals

6.1.1 Stripes Formed by Applying a Magnetic Field: van der Waals Versus Dipolar Forces Controlling Mesoscopic Organizations of Magnetic Nanocrystals

Dipolar and van der Walls interactions play a major role on the mesoscopic structures [9–15]. Whatever the maghemite coating agents, γ-Fe_2O_3 nanocrystals are characterized by the same average size (10 nm), polydispersity (20%), and magnetic properties with a reduced remanence, saturation magnetization, and coercive field of 0.30, 77 emu g^{-1}, and 360 Oe, respectively. This clearly indicates that the nanocrystals differing by their coating are characterized by the same magnetic dipolar moments.

The ferrofluids are γ-Fe_2O_3 nanocrystals coated with either octanoic (C_7H_{15}–COOH), decanoic (C_9H_{19}–COOH), or dodecanoic acid ($C_{11}H_{23}$–COOH) and dispersed in cyclohexane. For simplicity, the nanocrystals are named C_8, C_{10}, and C_{12}, respectively. In diluted dispersion, whatever the nanocrystals coatings, no aggregation is observed.

A transmission electron microscopy (TEM) grid is deposited in the bottom of the beaker containing the ferrofluids. During the solvent evaporation process, the system is subjected or not to a magnetic field (0.59 T) applied parallel to the substrate. Figure 6.1 shows drastic changes in the TEM patterns with the various coatings:

(i) With no applied magnetic field, spherical aggregates differing by various sizes appear with C_8 coating (Figure 6.1a). In contrast, the C_{10}- and C_{12}-coated nanocrystals are randomly deposited on the TEM grid (Figure 6.1b,c, respectively).
(ii) *Under a 0.59 T magnetic field*, nanocrystals coated with C_8 form thick striped structures in the direction of the applied field (Figure 6.1d,e). The average length of chain-like structures is several tens of micrometers, whereas the average diameter is around 300 nm. Note the presence of some nanocrystals between the stripes. Similar behavior is observed with C_{10} (Figure 6.1f,g) Conversely, with C_{12}, the TEM patterns do not show any nanocrystal organization (Figure 6.1h,i). A very small number of nanocrystal aggregates appear to be organized along the direction of the applied field.

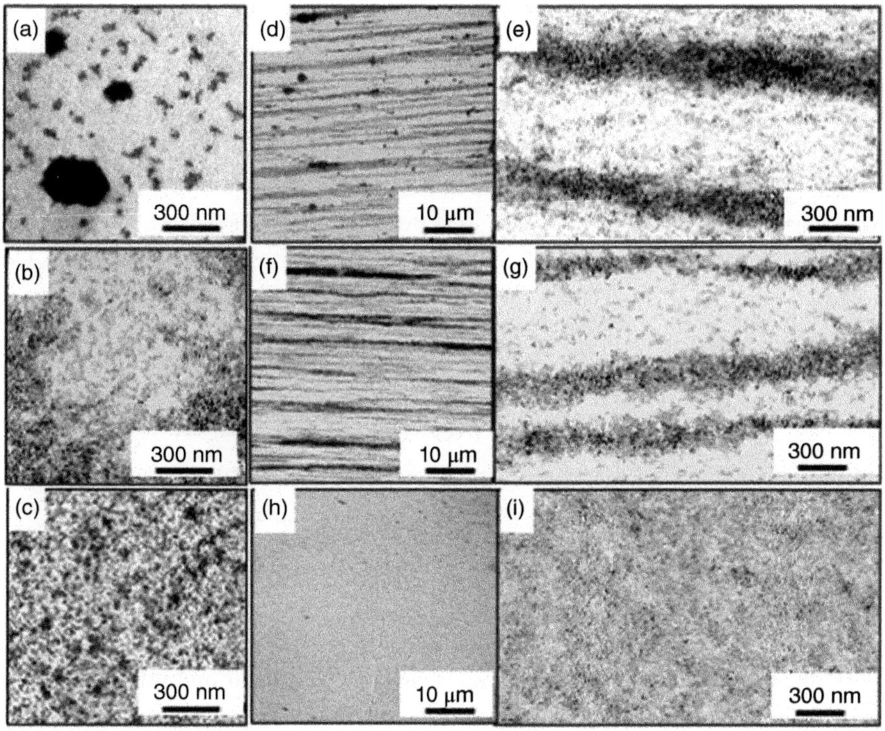

Figure 6.1 Transmission electron microscopy images of γ-Fe_2O_3 nanocrystals deposited with (0.59 T) and without magnetic field. (a–c) Nanocrystals coated with octanoic acid (C_8), decanoic acid (C_{10}), and dodecanoic (C_{12}) acid and deposited *without a magnetic field* form isotropic clusters during the evaporation process. (d–i) The same nanocrystals deposited *under a magnetic field* organize in either chain-like structures (d–g) or forming film (h, i). Note the presence of isolated nanocrystals between the stripes for the same deposition at a higher resolution. Nanocrystals coated with C_8 (a, d, e), C_{10} (b, f, g), and C_{12} (c, h, i). *Source:* Y. Lalatonne (2005), p. 18/American Physical Society.

Because the nanocrystals coated with C_8, C_{10}, and C_{12} are characterized by the same magnetic dipolar moments, the changes in the nanocrystal assemblies with the coating (Figure 6.1) cannot be directly attributed to the magnetic properties of the γ-Fe_2O_3 nanocrystals. Usually, high dipolar parameter (λ) defined as the ratio of the magnetic dipolar to thermal energies is needed for a chain formation of the nanoparticles. Assuming that the edge–edge core spacing between the nanoparticles is about 1 nm due to their coating, the calculated dipolar parameter is 0.69 for the average diameter of 10 nm. Due to the low size dispersion of the nanoparticles, even the largest particles of about 15 nm (2% of all nanoparticles) have a λ value of about 2.5, which is smaller than the threshold λ equal to 3 for chain formation [16–19]. Moreover, a simulation study shows that chains of larger dipolar particles do not induce the aggregation of smaller nanoparticles in polydisperse systems [20].

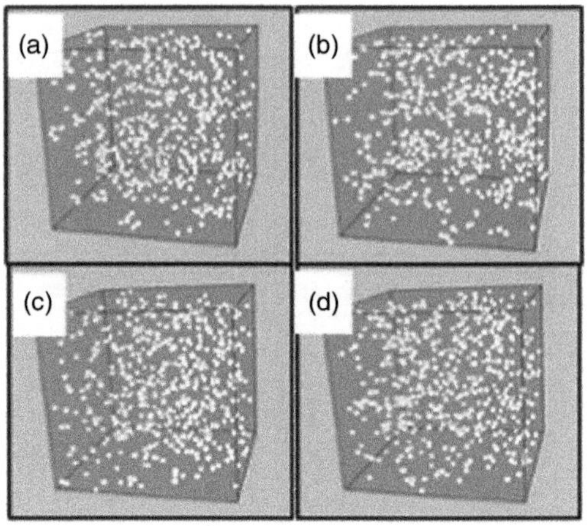

Figure 6.2 Snapshots of the configurations by the Brownian dynamics simulations. For the coating layer thickness $d = 1.2$ and 1.6 nm (a, c) before evaporation without (b, d) or with an applied field (0.59 T), no significant aggregation of particles is observed. *Source:* Lalatonne et al. [9]/Springer Nature.

Therefore, the chain formation cannot be directly explained by the presence of larger nanoparticles. To explain the formation of the chain-like structures in spite of the weak dipolar interaction, Brownian dynamics simulations at a particle level investigate the evaporation of magnetic nanoparticle solutions. The translation and rotation of the magnetic nanoparticles are described by Langevin equations [21] that make it possible to consider the influence of the solvent upon the solute by a combination of random forces and frictional terms. The Langevin equations are integrated using an algorithm proposed by Allen [22]. The forces and torques between the nanoparticles are calculated from the interparticle interaction, which is defined as a sum of the steric repulsion due to the coating molecules, the dipole–dipole potential, and the van der Waals attraction. We use the interaction model proposed elsewhere [17, 23], where the explicit equations of all interaction terms are given. The interaction parameters are adapted to the system experimentally used. The double thickness of the layer of coating molecules, denoted by d in the following, appears as a parameter in the equation for the steric repulsion. By changing d, the influence of the coating molecule size on the simulation results can be studied. Simulations are carried out varying d from 1.0 to 3.0 nm. Before evaporation, no significant aggregation of nanoparticles is observed when they are (Figure 6.2a,b) or not (Figure 6.2c,d) subjected to a magnetic field (0.59 T). For $d = 1.1$ and 1.6 nm, similar configurations are obtained.

The evaporation process is modeled as follows: a wall at the bottom of the simulation box represents the substrate. The wall–particle interaction is defined by a steric repulsion due to the coating molecules and a van der Waals interaction between a sphere and a half-space [24]. At the top of the box, there is a liquid–gas interface that goes slowly down during evaporation. The interface is made of the solvent and the coating of the nanoparticles crossing the interface. When the nanoparticle crosses the interface, the interfacial area and, therefore, its energy increase. This energy increase confines the nanoparticles in the solution layer. The evaporation speed used in the simulations is much faster than the experimental one, which is determined by the boiling point of the solvent. A systematic variation of the evaporation speed in the simulations has shown that the speed used here is sufficiently large to predict qualitatively the morphology of structures. However, the assembly size increases at slower evaporation that may explain the differences in the size of aggregates observed by simulations and experiments. Furthermore, the use of a very low evaporation speed in simulations leads to qualitatively the same mesostructures, but the nanoparticles are well organized in a 3D network. The size distribution of the γ-Fe_2O_3 nanocrystals is too large to induce internal organization. All simulations started from a configuration obtained for $d = 3.0$ nm where no aggregation of nanoparticles occurs. Figure 6.3 shows the configuration after an equilibration of 10 million times steps for a coating layer from 1.1 to 1.6 nm. At the end of solvent evaporation, major changes are observed.

For $d = 1.1$ and 1.2 nm with no applied magnetic field, spherical clusters of nanoparticles are observed (Figures 6.3a,b). This aggregation is explained by a larger interparticle attraction. Under an applied field during evaporation process, nanoparticles tend to align (Figure 6.3e). A progressive increase of d from 1.2 to 1.6 shows disappearance of clusters in absence of an applied field (Figures 6.3c,d) and strips presence (Figures 6.3g,h). At 1.6 only films are observed (Figure 6.3d,h).

The energy of the most stable head–tail configuration decreases from −3.0 to $-4.5kT$ on reducing d from 1.6 to 1.2 nm (Figure 6.4). This large variation is mainly due to the marked sensitivity of the van der Waals potential to the interparticle distance. The stronger attraction leads to a condensation of the nanoparticles. To explain the formation of striped structures with γ-Fe_2O_3 nanocrystals in a magnetic field in spite of a low dipole interaction with respect to the thermal energy, both van der Waals attraction and magnetic dipoles have to be taken into account. When the coating molecule is small enough, "clusters" are formed due to the van der Waals forces. The cluster formation considerably enhances the long-range dipolar forces compared with isolated particles with the creation of a "macro-dipolar moment" leading to an anisotropic organization of the nanoparticles. On increasing

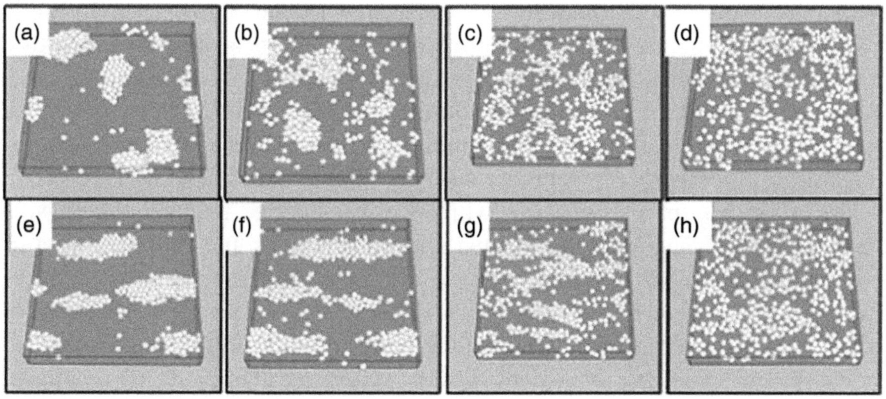

Figure 6.3 Snapshots of the configurations by the Brownian dynamics simulations after evaporation, for the nanocrystal coating layer thickness $d = 1.1$ nm (a, e), 1.2 nm (b, f), 1.3 nm (c, g), and 1.6 nm (d, h) without (a–d) and with (e–h) applied field (0.59 T). *Source:* Y. Lalatonne (2005), p. 18/American Physical Society.

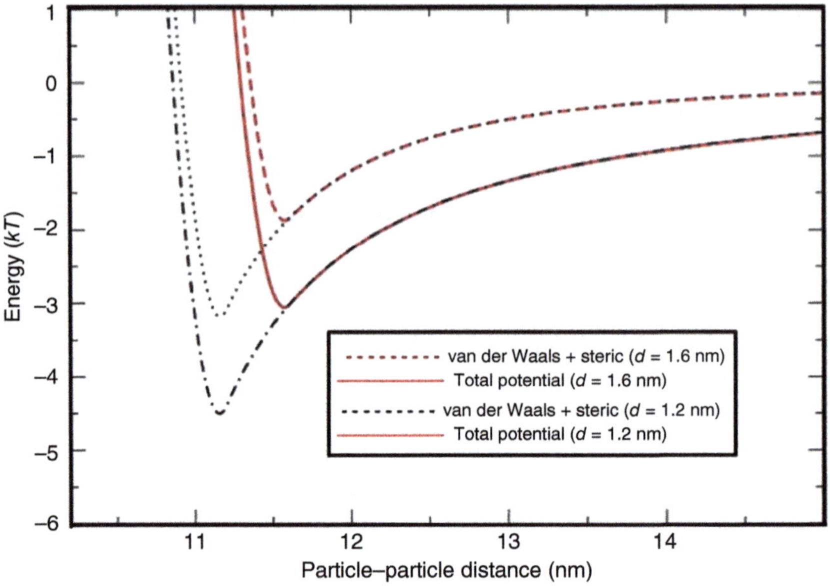

Figure 6.4 Dependence of the interaction energy on the interparticle distance. The total energy and the sum of the van der Waals attraction and the steric repulsion are shown for two values of the coating layer thickness *d*. For the calculation of the dipolar interaction, the most stable head–tail configuration is chosen. The minimum of the total energy decreases from −3.0 to −4.5*kT* on reducing *d* from 1.6 (red) to 1.2 nm (black) due to the sensitivity of the van der Waals potential to the interparticle distance. *Source:* Lalatonne et al. [9]. Copyright Nature.

progressively d to reach $d = 1.6$ nm, strips disappear, and no aggregation of the nanoparticles is observed even during the evaporation when particles are subjected (Figure 6.3h) or not (Figure 6.3d) to an applied magnetic field (0.59 T). The minima of the potentials in Figure 6.4 indicate that the distance of closest contact increases from 1.1 to 1.6 nm. For the contact distance of 1.6 nm, the van der Waals interactions are not strong enough to induce the formation of nanoparticle clusters, preventing any enhancement of the dipolar forces for such an assembly of nanoparticles.

Actually, in simulations, the aggregation disappears when the edge–edge core spacing between the particles is varied from 1.2 to 1.6 nm. A similar disappearance of organization is experimentally observed when C_{12} instead of C_8 is used as coating. An idea of the interparticle spacing, which can be expected for these coatings, is obtained from Ag_2S nanocrystals coated with alkanethiols and self-assembled in hexagonal networks [25]. Thus, an increase in the core spacing from 1.4 to 1.7 nm is observed in this system, when octanethiol (C_8H_{17}–SH) is replaced by dodecanethiol ($C_{12}H_{25}$–SH) as coating agent. On the one hand, these experimentally observed spacings are in reasonable agreement with the values used here in the simulations. On the other hand, the increase of 0.3 nm in spacing experimentally expected between C_8 and C_{12} is similar to the small variation of 0.4 nm necessary for the disappearance of organization in simulations.

Rather good agreement between the structures obtained from experiments (Figure 6.1) and those obtained from simulations (Figure 6.3) can be observed: (i) with γ-Fe_2O_3 nanocrystals coated with C_8, large clusters are observed from the TEM pattern (Figure 6.1a). This structure is in qualitative agreement with the particle aggregation found by simulations for $d = 1.1$ and 1.2 nm (Figure 6.3e,f). Similarly, the TEM images show a striped structure when the nanocrystals are subjected to a magnetic field during the deposition process (Figure 6.1d,e), whereas aligned particles are seen by simulations (Figure 6.3f). (ii) With γ-Fe_2O_3 nanocrystals coated with C_{12}, random distributions of nanoparticles are observed from the TEM pattern (Figure 6.1c) as obtained by simulation for $d = 1.6$ nm (Figure 6.3h). When the system is subjected to a magnetic field during the deposition process, no changes in the nanocrystals dispersion are observed, either experimentally (Figure 6.1f,h) or by simulation (Figure 6.3d,h).

Let us *increase* the maghemite nanocrystal concentration:

(i) With C_8 as coating agent, Figure 6.5a shows, in the absence of any applied magnetic field, a very rough thick film of highly compact spherical clusters with an average diameter of 1.5 nm. The number and size of clusters markedly increase compared with dilute solution. Under a magnetic field (0.59 T), the

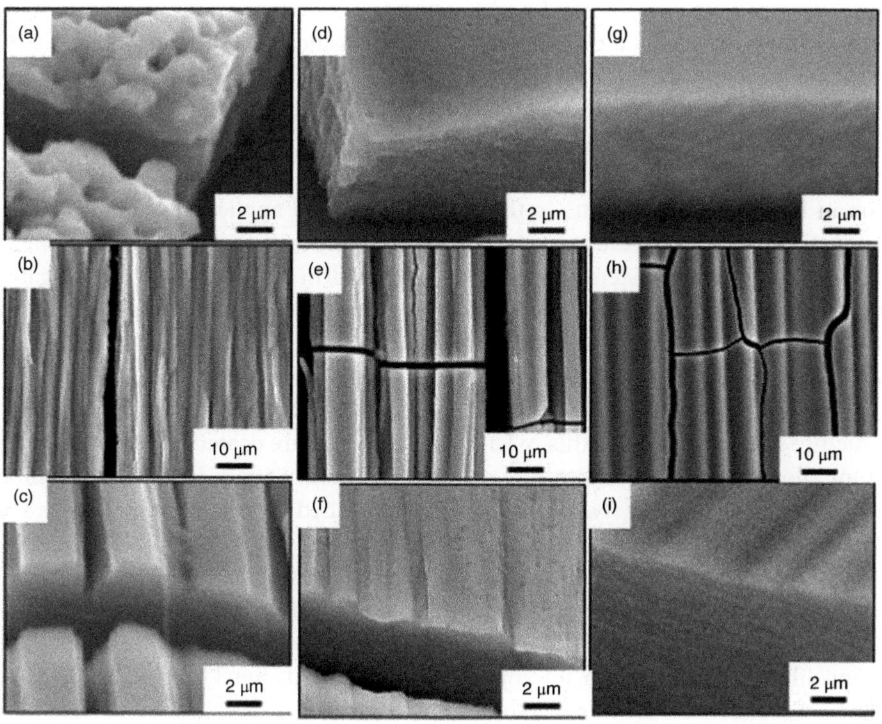

Figure 6.5 SEM images of γ-Fe_2O_3 nanocrystals coated with C_8 (a–c), with C_{10} (d–f), and with C_{12} (g–i) deposited in the absence (a, d, g) and presence of an applied field (b, c, e, f, h, i). B = 059 T. *Source:* Y. Lalatonne (2005), p. 18/American Physical Society.

large clusters induce large dipolar forces allowing the nanoparticles to form cylinders (Figure 6.5b,c). Because of the high nanocrystal concentration, the cylinders tend to superimpose with a diameter 10 times larger than that observed in dilute solution.

(ii) With C_{12} as coating agent, as expected from above, the film behavior markedly differs from that of the C_8 derivative. In the absence of applied magnetic field during the evaporation process, thick film with a very flat surface and high compacity is observed (Figure 6.5d,g), whereas under 0.59 T field, slight undulations on the surface of the thick film are observed (Figure 6.5e,f,h,i). On a macroscopic scale, the film does not change (Figure 6.5i) compared with that obtained with no applied magnetic field (Figure 6.5g). This is in good agreement with results obtained in dilute solution either experimentally or by simulation.

From these data, it is concluded that, because the γ-Fe_2O_3 nanocrystals with different surface coatings keep the same average size, crystalline structure, and magnetic properties, the coating is the major cause of the change in the mesoscopic structures observed in Figure 6.1. Moreover, the magnetic dipolar interactions are too small to induce a nanocrystal self-ordering as observed [1]. With no magnetic field, the aggregations observed with C_8 and C_{10} nanocrystals are caused by vdW interactions and not by a dipolar force, for which more chain-like aggregates are expected. The simulations, shown in Figure 6.3, provide evidence that when the distance between nanoparticles is short enough, the van der Waals attraction induces their aggregation. In this case, the application of a magnetic field leads to chain-like structures, which is observed by experiment in Figure 6.1 and by simulation in Figure 6.3. This is due to the fact that the field application induces a total magnetic dipole in each cluster, which is much larger than that of a single nanoparticle. Therefore, the clusters attract each other to form larger elongated aggregates. The marked dependence of the van der Waals interaction on the interparticle distance explains why a small increase in the alkyl chain length leads to the disappearance of nanoparticle aggregation observed in the experiments.

A question arises: Do other types of interactions play similar role in the mesostructure shape? To answer this, γ-Fe_2O_3 nanocrystals are coated with a passivating agent allowing them to disperse in aqueous solution. Citrate ions and propanoic acid are used. For simplicity, the nanocrystals are called Cit and C_3, respectively. The deposition process differs from that used previously. As above, the solvent (water) is evaporated by applying or not a magnetic field parallel (0.59 T) to the substrate. Furthermore, diluted and concentrated solutions of γ-Fe_2O_3 nanocrystals are deposited.

For dilute solutions, with no applied magnetic field, the TEM image obtained after evaporation of Cit nanocrystals shows spherical aggregates of nanocrystals (Figure 6.6a), as observed above in cyclohexane for C_8 nanocrystals (Figure 6.1a). Conversely, C_3 nanocrystals are randomly deposited on the TEM grid (Figure 6.6b). This is observed whatever the magnification. In a magnetic field, the TEM patterns markedly differ. Long chain-like structures are formed with Cit nanocrystals

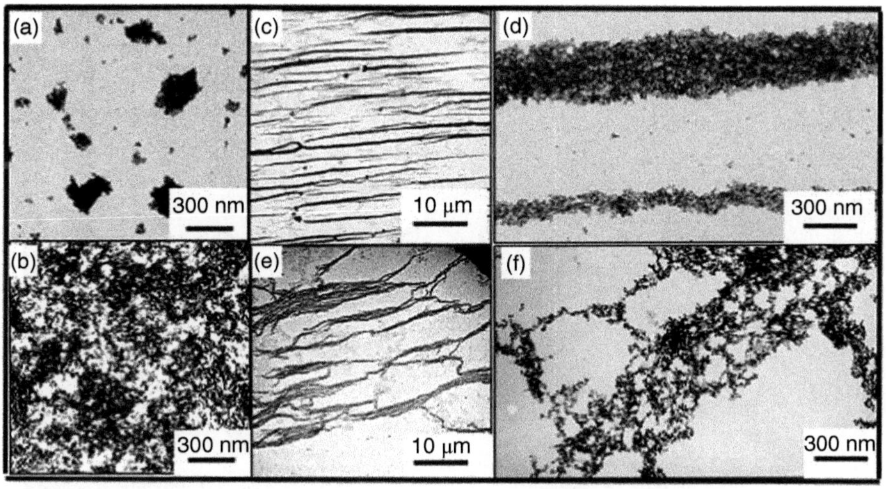

Figure 6.6 TEM images of γ-Fe_2O_3 nanocrystals coated with citrate ions (a, c, d) and propanoic (b, e, f) acid (CiT and C_3) in the absence (a, b) and presence of an applied magnetic field (0.59T) during the evaporation process (c–f). *Source:* Y. Lalatonne 2005, p. 18/American Physical Society.

(Figure 6.6c,d). The average length (40 mm), diameter (300 nm), and average distance between the chains (3 mm) are similar to those observed for C_8 nanocrystals (Figure 6.1d,e). Note that very few particles are between these ribbons. By replacing Cit with C_3 nanocrystals, a drastic change in the self-organization is observed: on a large scale, Figure 6.6e,f shows interconnected chain-like structures up to 100 mm in length.

For concentrated solutions, with Cit nanocrystals, the SEM image shows a very rough surface (Figure 6.7a) with spherical, highly compact agglomerates, having an average diameter of 1.5 ± 1 mm. Conversely with C_3, the scanning electron microscopy (SEM) pattern shows a very flat surface and high compacity of the film (Figure 6.7b). In a magnetic field, Cit nanocrystals self-organize in long cylinders with a very regular structure (Figure 6.7c,d). By tilting the sample (Figure 6.7b), it is seen that the structure corresponds to superimposed tubes with an average diameter of 3 ± 1 mm. Conversely, with C_3 nanocrystals, stripes are observed on a large scale with highly dense and undulated structures (Figure 6.7f), and the surface is very flat (Figure 6.7e). Hence, the SEM images obtained with Cit and C_3 are similar to those observed with C_8 and C_{12}. Of course, because these nanocrystals are dispersed in aqueous solution, new kinds of interactions such as ionic double-layer repulsion and hydrophobic interactions have to be taken into account, and the theoretical model developed above cannot be applied directly. Nevertheless, these simulations show that nanoparticle aggregation can be expected for these nanoparticles having a layer thickness smaller than those of C_8 and C_{10} nanocrystals. For Cit nanocrystals solubilized in water, the distance between the nanoparticles is determined by the overlap of the ionic double layers surrounding the particles [24]. During the evaporation, the distance between nanoparticles should be reduced, since the Debye length markedly

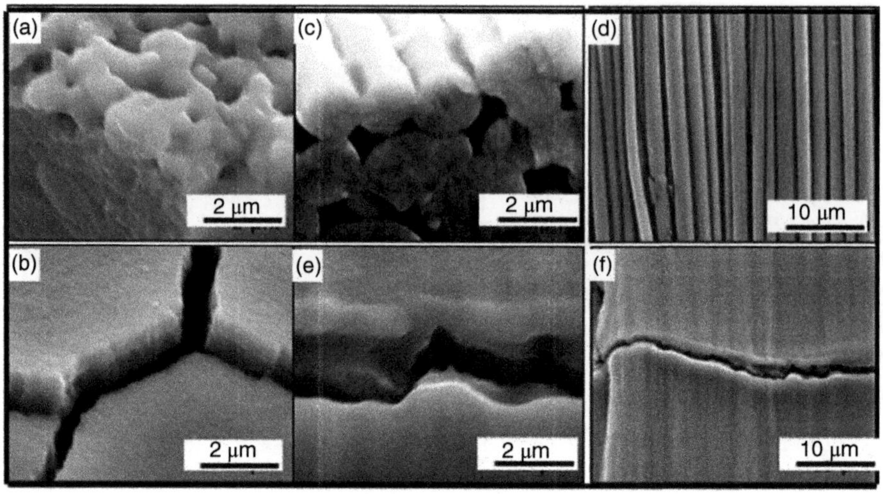

Figure 6.7 SEM images of γ-Fe_2O_3 nanocrystals coated with citrate ions (a, c, d) and propanoic (b, e, f) acid (CiT and C_3) in the absence (a, b) and presence of an applied magnetic field (0.59T) during the evaporation process (c–f). *Source:* Y. Lalatonne (2005), p. 18/American Physical Society.

decreases with increasing concentration. Therefore, at the end of the evaporation, aggregation can be expected like that observed with C_8 nanocrystals. This is in good agreement with the experiment. The case of C_3 nanocrystals seems to be more complicated. No aggregation is observed without a field, while chain-like structures appear when the field is applied during evaporation as observed with C_{10}. This might be a side effect of the hydrophobic interactions expected for this surfactant. For the citrate ions, hydrophobic attraction does not exist, since this coating is highly hydrophilic.

Hence, by nanocrystal deposition on a substrate, drastic changes in their organization are obtained on varying the coating. When the nanocrystal distance is rather small, as observed with citrate ions and octanoic and decanoic acids as passivating agents, the nanocrystals tend to aggregate without any applied magnetic field and tend to form ribbons in the direction of an applied field. Nanoparticle aggregation is attributed to strong van der Waals attraction. The chain-like mesostructures form in spite of a small magnetic dipole. This can be explained by the combination of van der Waals and dipolar interactions as demonstrated by the simulations in the case of carboxylic acid-coated nanocrystals. The van der Waals interaction is very sensitive to the interparticle distance. This explains the disappearance of any mesostructures, when the surfactant size is increased, as in the case of dodecanoic acid as coating agent. The details of the mesostructures are not yet well understood, in particular, for propanoic acid as coating agent. It will be necessary to develop more complex interaction models taking, for example, the hydrophobic interaction into account. Moreover, the use of other coating agents, such as hydrophilic nonionic surfactants, could yield valuable additional information.

At this stage, ask ourselves with the following question: What is the influence of the strength of the applied magnetic field on the mesostructures? To answer this, let

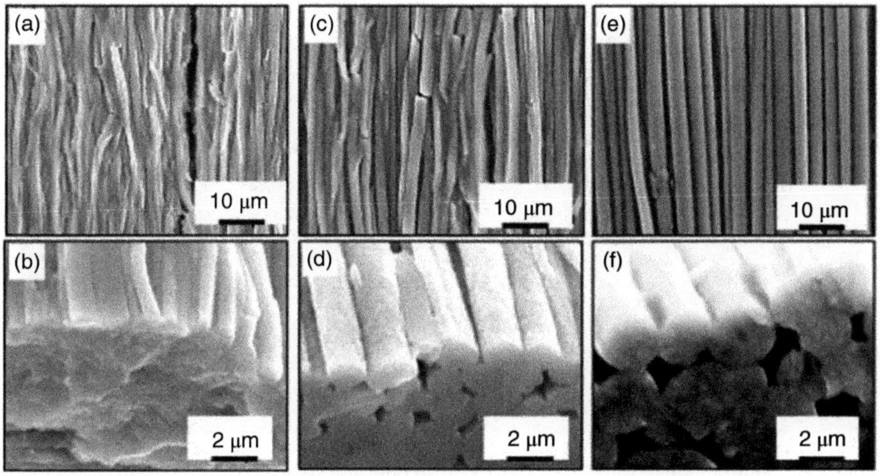

Figure 6.8 SEM images of γ-Fe_2O_3 nanocrystals coated with citrate ions, at fixed concentration, subjected to various magnetic applied fields during the evaporation process. Applied 0.1 T (a, b), 0.05 T (c, d), and 0.59 T (e, f). *Source:* Lalatonne et al. [9]/American Chemical Society.

us consider γ-Fe_2O_3 nanocrystals coated with citrate ion colloidal solution subjected to applied magnetic fields having various strengths, keeping the nanocrystal concentration constant. Consequently, the number of clusters of nanocrystals and the dipolar interactions remain unchanged. Figure 6.8 clearly shows the diameter of the cylinder increases with increasing the applied magnetic field. Hence, the cylinders are controlled by the applied field. Furthermore, they are more defined at large applied field.

6.1.2 Applied Magnetic Field Perpendicular to the Substrate: Liquid–Gas Phase Transition

By applying a magnetic field perpendicular to the substrate marked changes of the mesoscopic structures are observed [26–30]. Here, we consider the same maghemite nanocrystals as used above. The coating agent is dodecanoic acid [$C_{11}H_{23}$–COOH], and the ferrofluid is obtained by dispersion of dispersed in hexane. Their average diameter and a polydispersity are 11 nm and of 22%, respectively. Highly ordered pyrolytic graphite (HOPG) substrate is placed at the bottom of a glass beaker. During the solvent evaporation time, the colloidal solution of γ-Fe_2O_3 nanocrystals is subjected or not to an applied field perpendicular to the substrate with various strengths (0.17, 0.33, 0.46, and 0.59 T).

In the absence of applied field, the SEM image (Figure 6.9a) shows inhomogeneous structures forming islands without a well-defined shape.

By applying a magnetic field during the deposition process, the SEM images markedly change. In the range of the investigated field, its strength markedly modifies the structural shape with respect to the structure with no field.

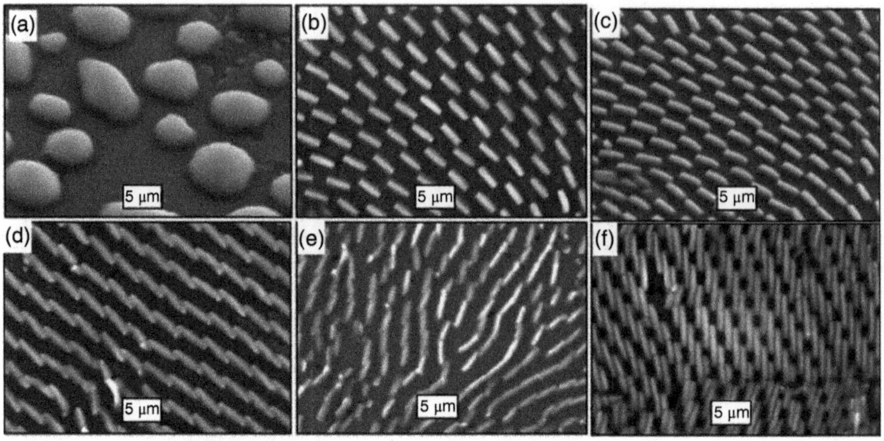

Figure 6.9 SEM images of mesostructures of maghemite nanocrystals deposited on HOPG (a) without an applied field at 45° tilt and (b) with an applied field of 0.59 T perpendicular to the substrate, at zero angle (b, d–f) and at 45° tilt (c). *Source:* Ngo et al. [15]/American Chemical Society.

Whatever the applied field, similar features are observed:

(i) Local array of hexagonal structures made of elongated aggregates are observed (Figure 6.9b). By tilting the samples, it is shown that cylinders are formed (Figure 6.9c).
(ii) The cylinders are interconnected in head-to-tail lines with and without preferential orientations (Figure 6.9d,e).
(iii) The cylinders are organized in a dense array of hexagonal structures (Figure 6.9f).
(iv) Two cylinders are associated to give stripes.

For any applied field strength, the observed cylinders are uniform in size in a given domain.

The free energy approach predicts the hexagonal structures of magnetic nanocrystals, which are observed after evaporation in a perpendicular field [31]. Within the framework of this theory, the hexagonal structures are idealized as a hexagonal array of cylinders made of magnetic nanocrystals.

To minimize the free energy per surface area F_s [3, 26, 32], several parameters obtained by SEM patterns have to be taken into account (Figure 6.9):

1) The variation of the cylinder radius, r, with the strength of the external field.
2) The ratio of the magnetic phase to the total volume, F, is calculated from the cylinder radius and the distance between two cylinders.
3) The variation of the cylinder height, L, with the strength of the external field.
4) The average surface area per cylinder, s_h, is measured by counting the cylinders in a given large area on the substrate.
5) The magnetic energy, F_m.

The free energy per surface area is given by

$$F_b = 2 \times F_s L / r + F_m / s_h$$

The cylinder average length varies from 2 to 5 mm from one domain to another on the substrate. No preferential length or structure is observed with the strength of the applied magnetic field during the evaporation process. We assume that the cylinders are upright during the formation of the structures and collapse due to capillary forces at the end of evaporation as shown by video with cobalt nanocrystals [33]. From these values, the best fits between experiments and calculation are obtained for an interfacial tension, s, evolving from 1.4×10^{-4} to 3.5×10^{-4} Nm^{-1}. These values are close to those expected for a liquid–gas phase transition [33] and agree with other prediction [26–28]. This confirms data described below with cobalt nanocrystal fluids and demonstrates that the mechanism proposed by the Rosensweig instability [34] cannot be retained. In the latter case, the surface tension of the solution is expected to be close to that of the solution in contact with air (which is close to that of the pure solvent: 1.84×10^{-2} Nm^{-1} for hexane) [24].

From these data, it is concluded that mesoscopic structures of collapsed cylinders organized in hexagonal, stripe, and wavy line-like structures are fabricated by slow evaporation of maghemite nanocrystals dispersed in hexane and subjected to an applied field perpendicular to the substrate. The sizes of the experimental structures are well described by a theory based on the minimization of the total free energy. Comparison between experiment and theory shows that the structures are explained by a labyrinthine instability enabled by a colloidal liquid–gas phase transition during the evaporation process.

6.2 Cobalt Nanocrystals

As described in Chapter 1, the control of Co nanocrystals and their size distribution are obtained by various methods. In all the experiments described here, amorphous Co nanoparticles, produced from reverse micelles, are used.

6.2.1 Alignments Induced by Dipolar Interactions Between Co Nanoparticles Subjected to a Magnetic Field Parallel to the Substrate

Cobalt nanoparticles (8 nm 14% size distribution) coated with dodecanoic acid ($C_{12}H_{25}$–COOH) are dispersed in hexane. The nanocrystal size distribution is too large [35, 36] to produce 3D superlattices. Consequently, the distance between nanoparticles is up to 2 nm, and van der Waals interactions between nanoparticles remain negligible. Co nanoparticles dispersed in solution are superparametic at room temperature, and at 3 K the thermal excitation vanishes. The magnetic curve shows that hysteresis with a saturation magnetization (120 emu g^{-1}) is reached at 2 T.

The Co nanoparticles dispersed in concentrated solution is evaporated under either air or "pre-"saturated hexane vapor. During the evaporation, a magnetic field parallel to the substrate can be applied. The behavior of the films obtained at the end of the evaporation process, observed on a very large scale (up to 0.25 mm^2), changes drastically with the experimental conditions. In all cases, a nanostructured film made of non-coalesced nanocrystals is formed (the film is dissolved in hexane, the TEM image of dispersed nanoparticles shows individual nanoparticles, and the magnetic properties remain unchanged). Whatever the experimental conditions, the X-ray reflectivity measurements show that a large width of the peak indicates a large size distribution as mentioned above. Furthermore, the lack of peaks at high q values indicates the absence of long-distance order in the film.

Let us first consider the *evaporation in air*: in the absence of applied field, the SEM pattern shows a homogeneous film (Figure 6.10a). Magnification of this image shows that it is highly porous (Figure 6.10b) and made of spherical aggregates with an average diameter of 70 nm (Figure 6.10c). These aggregates are mainly due to fast evaporation process and not to van der Waals interactions as observed above with γ-Fe_2O_3 nanocrystals. When a 0.78 T magnetic field is applied during evaporation process, linear structures of cobalt nanocrystals are observed (Figure 6.10d). The nanocrystals are aligned along the applied magnetic field direction. The increase of the magnetic field strength shows that the film is still highly porous (Figure 6.10e) and is made of around 70 nm spherical aggregates as observed in the absence of a magnetic field (Figure 6.10f). Hence, at high magnification, no differences in the SEM patterns obtained in the absence (Figure 6.10c) and in the presence

Figure 6.10 Film SEM patterns of 10 nm Co nanoparticles coated with dodecanoic acid obtained by solvent evaporation under air in the absence (a–c) and presence (d–f) of 0.78 T magnetic field. *Source:* Petit et al. [37]/AIP Publishing.

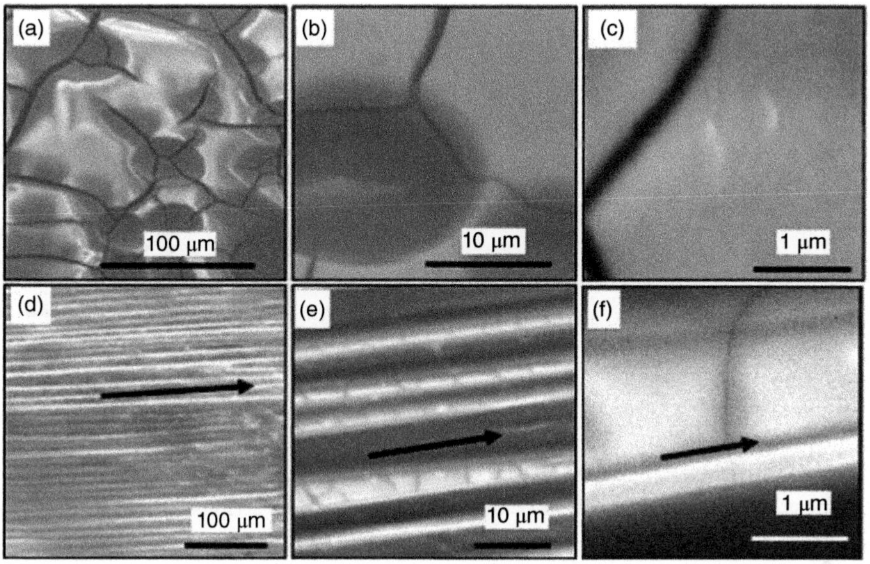

Figure 6.11 Film SEM patterns of 10 nm Co nanoparticles coated with dodecanoic acid obtained by solvent evaporation under pre-saturated hexane in the absence (a–c) and presence (d–f) of 0.78T magnetic field. *Source:* Petit et al. (2002) [37]/AIP Publishing.

(Figure 6.10f) of a magnetic field are observed, indicating that the organization takes place at the macroscopic level. The alignment is due to dipolar interactions between nanoparticles with an interparticle distance of up to 2 nm corresponding to the formation of 3D superlattices [35, 36].

When the evaporation occurs under hexane vapor, marked change in the film behavior is observed. It is more homogeneous with less roughness than that observed under air (Figure 6.11a). However, numerous discontinuities are present. There is a large domain formed by the stacking of several layers of nanocrystals, and holes are observed on the film. The height of the film is determined by tilting the sample (45°) and is around 2 mm (Figure 6.11b). The internal structure of the film appears homogeneous (Figure 6.11b,c) without any spherical aggregates as observed after evaporation under air (Figure 6.10b,c). As already mentioned, no long-distance order is observed. These changes in morphology with evaporation under either air or hexane vapor are related to the evaporation rate. Indeed, the evaporation time is 45 minutes or 12 hours when performed under air or a hexane vapor atmosphere, respectively. In the latter case, the nanocrystals freely diffuse in solution, thus leading to the formation of homogeneous structures. When a 0.78 T, the structure of the film changes drastically (Figure 6.11d). Long stripes in the direction of the applied field without any holes and cracks are formed. Magnification of the stripes shows a highly compact film (Figure 6.11e,f). The average film height remains around 2 mm. From these data, it can be concluded that evaporation in a high applied magnetic field leads to the formation of mesoscopic scale structures of nanocrystals aligned along the field direction. However, to obtain regular and homogeneous stripes, the

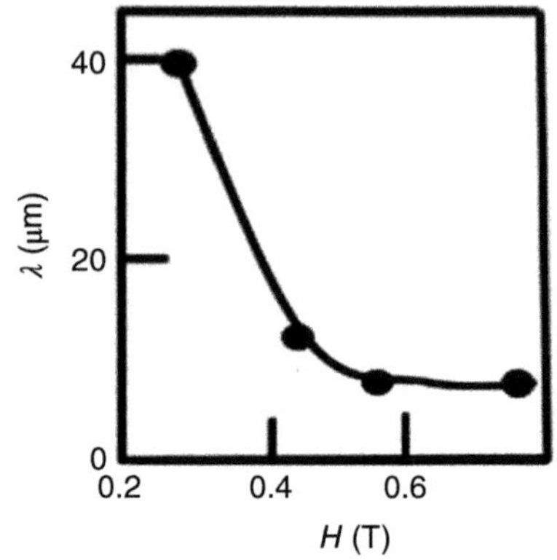

Figure 6.12 Variation of the critical wavelength, l_c, distance center to center between two adjacent stripes with the intensity of the field applied during the evaporation process.

evaporation process must be done under hexane vapor. At various applied fields, Figure 6.12 shows that stripes are formed with a periodic structure. However, the roughness and the average distance between two adjacent stripes vary. A characteristic wavelength l_c is defined as the average distance between two adjacent stripes. With the increasing strength of the applied field, the characteristic distance decreases to reach a plateau around 0.56 T (Figure 6.12). Simultaneously, the stripes are sharper. Again, such behavior can be related to that observed with magnetic fluids. The compacity of stripes increases with increasing applied field [38, 39]. The major difference can be noticed in the nature of the patterns: conversely to the ferrofluids, the array of cobalt nanocrystals persists after the field is switched off.

6.2.2 Columns and Labyrinths of Co Nanoparticles: Nanocrystal Size Distribution as a Key Parameter on the Mesostructures

Let us consider Co nanocrystals coated with lauric acid (lauric acid = dodecanoic acid = $C_{12}H_{25}$–COOH), having various sizes (from 5 to 8 nm) and size distribution (from 10% to 20%), dispersed in hexane and deposited in a beaker containing a substrate. The beaker is subjected (or not) to an external magnetic field (H) perpendicular to the substrate. The solvent slowly evaporates, and the magnetic field is applied until evaporation is complete. At the end of the evaporation, the TEM grid is removed, and the mesoscopic structures were observed [40, 41].

We know that when the size distribution is low enough, nanocrystals are able to self-assemble in 3D superlattices [35, 42]. Let us first consider Co nanocrystals of various average sizes with low size distributions. In the absence of an applied magnetic field, nanocrystals with an average diameter d = 5.7 nm and size distribution s = 13% produced large aggregates with sizes between 2.5 and 9 mm (Figure 6.13a). On magnifying the image of these aggregates, well-defined superlattices of nanocrystals are observed. The fast Fourier transforms (FFTs) of various area of the superlattices exhibit sixfold symmetry, indicating ordering of the nanocrystals in an fcc structure. This agrees with previous results, in which 3D superlattices in fcc structures are produced. On increasing the average size of the nanocrystals to 7.1 nm and keeping a low size distribution (s = 11.4%), similar behavior was obtained with the

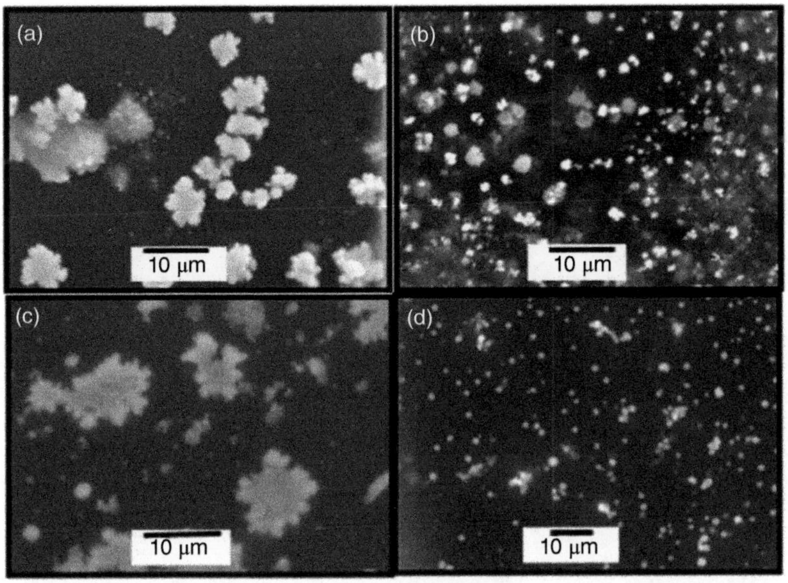

Figure 6.13 SEM (images of aggregates made of cobalt nanocrystals with different sizes of 5.7 nm (a), 7.1 nm (b), 8.1 nm (c), and 6.9 nm (d) and different size distributions of 13% (a), 11.4% (b), 10.3% (c), and 18% (d). No magnetic field was applied. *Source:* V. Germain (2005)/ John Wiley & Sons, Inc.

formation of large aggregates (Figure 6.13b) and the observation of well-defined nanocrystal ordering at higher magnification. Figure 6.13c confirms the same behavior for nanocrystals for which $d = 8.1\,\text{nm}$ and $s = 10.3\%$. These data clearly indicate the formation of large aggregates ordered in an fcc structure when cobalt nanocrystals with a low size distribution (≤13%) are deposited in the absence of a magnetic field (Table 6.1) and do not depend on the average nanocrystal size ranging from 5 to 8 nm. Note that, with nanocrystals having a larger size distribution, aggregates are still formed, whereas their size markedly decreases (Figure 6.13d).

In the presence of an applied magnetic field (0.25 T) during the evaporation process, the mesostructures show well-defined cylinders of nanocrystals differing by their average diameters (Figure 6.14a,d,g). Enhancements of the cylinders shows they are composed by nanocrystals (Figure 6.14b,c,e,f,h,i

Whatever the size distribution, the same type of video is recorded during the solvent evaporation process (around 12 hours). During the first hours of the evaporation, as seen in Figure 6.15a,b, there are defects due to dust (no particular care was taken to ensure clean conditions). After seven hours, a number of dots appear (see the arrows in Figure 6.15c). The number of columns increases progressively (Figure 6.15d), and these then migrate in the solution (Figure 6.15e,f) to form hexagonal arrays. This confirms that the columns form during the evaporation process.

At the end of the evaporation, a progressive collapse of the columns is induced by waves (Figure 6.16). From these data, it is concluded that, during the evaporation process, the applied magnetic field *perpendicular* to the substrate induces the

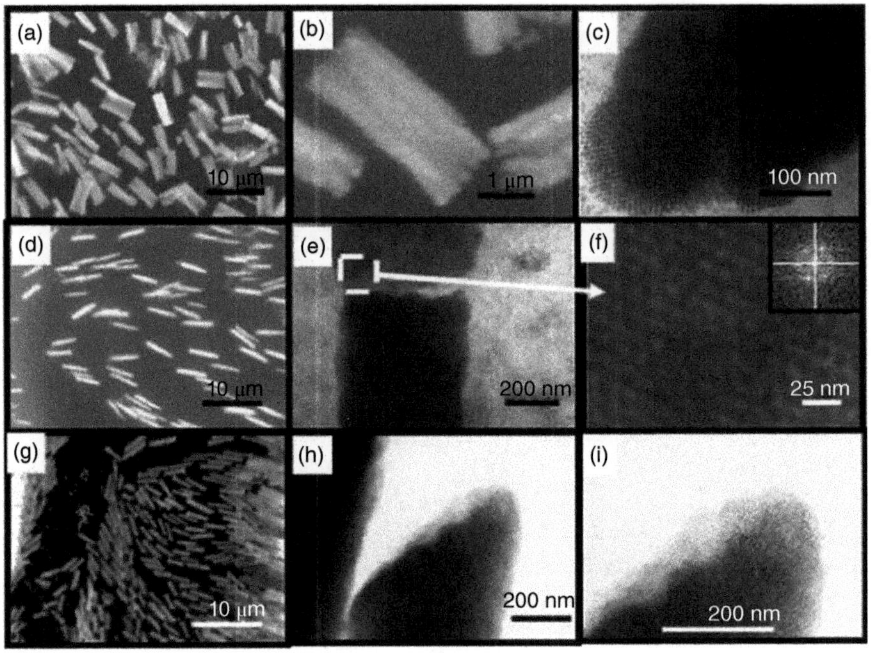

Figure 6.14 SEM (a, c, e, g) and TEM (b, d, f, h) images of mesostructures made of cobalt nanocrystals with different sizes of 5.7 nm (a, b), 7.1 nm (c, d), 8.1 nm and (e, f), 6.9 nm (g–i) 13% (a, b), 11.4% (c, d), 10.3% (e, f), 18% (g–i) subjected to a 0.25 T magnetic field applied perpendicularly to the substrate. Inset in (a): magnified view of a column. *Source:* V. Germain (2005)/John Wiley & Sons, Inc.

formation of columns. Some of them fall because of the waves induced by capillary forces.

Figure 6.14c,f,i shows that their extremities were made of well-ordered nanocrystals with fourfold symmetry FFT patterns confirming assemblies of nanocrystals in fcc crystalline structure. Due to the thickness of the patterns, it is impossible to observe the ordering inside the columns by TEM, and the lack of material prevented any small-angle X-ray scattering (SAXS) measurements. However, it is reasonable to conclude that the columns are made of nanocrystals ordered in an fcc structure. In some TEM images, "foot" of the columns remained stacked on the substrate. This remaining "foot" and the end of the fallen column (Figure 6.14e) are found to consist largely of well-ordered nanocrystals (Figure 6.14f).

Except for the sample shown in Figure 6.14a,b, the average aspect ratio of the column remained unchanged. This apparent discrepancy, observed in Figure 6.14a,b, is related to second-level hexagonal patterns. These patterns depend on the real-time phase transition in the structure formation as observed previously [38] with a magnetic fluid in a Hele–Shaw cell. In fact, each observed entity consists of two columns (Figure 6.14a,b). By changing the evaporation rate and/or lowering the diffusion of the nanocrystals, well-defined columns are produced. It can then be concluded that each column observed in Figure 6.14a is, in fact, made of two entities with an average

Table 6.1 Various parameters related to mesostructures produced at the end of solvent evaporation of a cobalt nanocrystal's colloidal solution. The mesostructures are obtained by subjecting or not a magnetic field (0.25 T) perpendicular to the substrate deposited at the bottom of the beaker containing the colloidal solution. In the absence of applied magnetic field, large aggregates are produced, whereas in the presence, columns are obtained. At the end of the solvent evaporation, either aggregates or columns are produced.

Diameter (nm)	5.7	7.1	8.1
Size distribution (%)	13	11.4	10.3
Size of the aggregates (μm) Assuming they are spherical	2.5–9	1.3–6	1.8–6
Height of the columns (μm)	5	5.2	3.5
Size distribution of the column height (%)	12	11.5	−17.1
Width of the columns (μm)	1.7 (0.85)	0.84	66
Size distribution of the column width (%)	17.9	24.2	25.7
Aspect ratio	6.2	6.15	7.5

Table 6.2 Influence of the applied magnetic field strength on the patterns obtained with cobalt nanocrystals having average diameter of 7.1 nm and size distribution of 11.4%.

Applied magnetic field (T)	0.25	0.46	0.59
Height of the columns (μm)	5.2	4	5.3
Size distribution of the column height (%)	11.5	16	21
Width of the columns (μm)	0.84	0.62	0.7
Size distribution of the column width (%)	18	24	26
Aspect ratio	6.2	6.15	7.5

height and width of 5 and 0.85 mm, respectively. Taking into account the formation of these two columns close to each other, the aspect ratios of columns made of nanocrystals of different diameters (from 5.5 to 8 nm) with low size distributions are of the same order of magnitude (Table 6.1). Furthermore, mesostructure patterns of cobalt nanocrystals of different sizes with low size distributions are columns of ordered nanocrystals in fcc structure. Table 6.1 shows that the columns produced keep similar aspect ratio. Furthermore, in the magnetic field ranging from 0.25 to 0.59 T, there are no changes in the average length, width, and aspect ratio (Table 6.2). This is probably due to the fact that the strength of the applied field is rather large and corresponds to the zone where the magnetic response of the nanocrystals is close to saturation (i.e. nonlinear regime). Note that in this size distribution range, no transition from columns to labyrinths took place on increasing the strength of the applied magnetic field. Instead of using TEM grid covered by an amorphous layer of carbon with a rough surface, let us use HOPG as the substrate that favors the diffusion freely on the surface. "Fallen" and upright columns containing nanocrystals are formed (Figure 6.17a,b).

Let us consider Co nanocrystals having similar sizes of above 6.9 nm and larger size distribution of 18%.

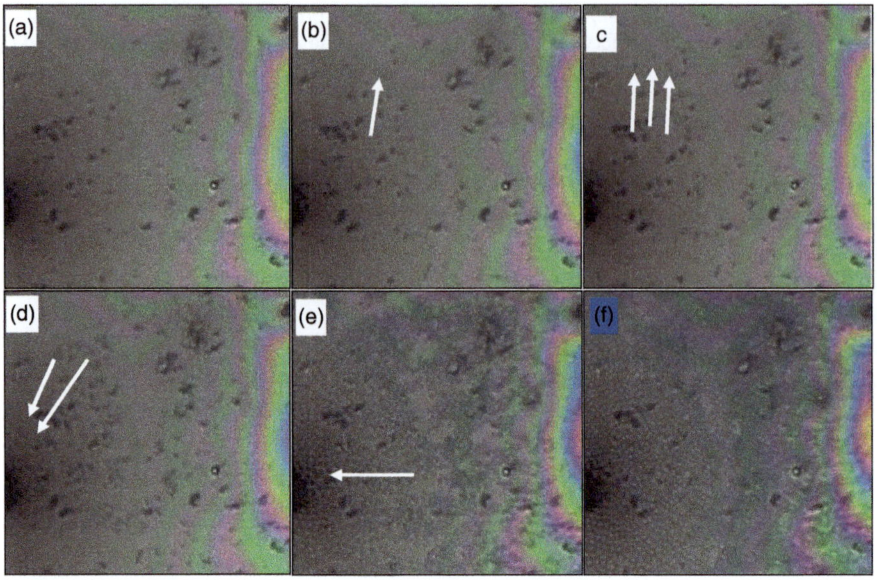

Figure 6.15 Video microscopy images obtained during the evaporation of a cobalt nanocrystal solution subjected to 0.25 T magnetic field at the moment when the columns appear. The substrate is silicon. The arrows indicate column formation. *Source:* V. Germain (2002), p. 5548–5553/American Chemical Society.

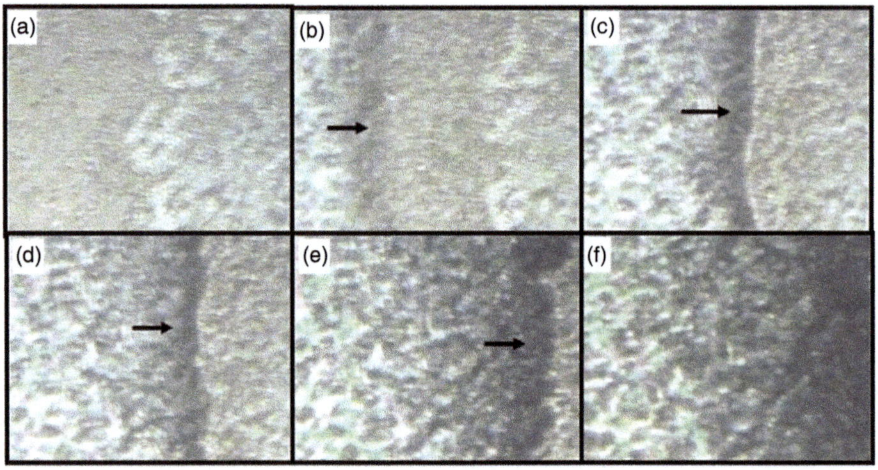

Figure 6.16 Video microscopy images with time (from a to f) obtained at the end of the evaporation of a Co nanocrystal solution in a 0.25 T magnetic field. The arrows indicate the direction of the wave created by the capillary forces. *Source:* V. Germain (2002), p. 5548–5553/American Chemical Society.

In the absence of the applied magnetic field, aggregates are produced as above. However, Figure 6.13d shows that their sizes drop (between 0.4 and 1.6 mm) compared with those of aggregates produced with a lower size distribution (between 2.5 and 9 mm). Furthermore, a magnified view of these aggregates shows that, as expected, nanocrystals are randomly deposited and not ordered. When the nanocrystals are subjected to a 0.25 T applied magnetic field at the end of the evaporation, columns are no longer well defined (Figure 6.18a) with a rather large disorder. In addition, at the ends of the columns and along the columns, "fingerlike" patterns appeared (Figure 6.18a,b).

Figure 6.19 also shows the formation of labyrinths on a large scale. Some are very long and others are wormlike. A similar pattern is observed by TEM (Figure 6.19b), and the labyrinths are formed with disordered nanocrystals (Figure 6.19c). This is related to the fact that small aggregates without any ordered structure are observed when these nanocrystals are not subjected to an applied magnetic field. The formation of columns with many defects is explained as follows: when the liquid–gas transition [26] takes place, columns start to grow on rough and deformed aggregates. This disorder induces defects during column growth and results in the patterns observed in Figure 6.14b,c.

Let us compare the patterns obtained, in the same experimental conditions, with nanocrystals having similar average diameter of 5.8 ± 1 nm and differing by their size distributions from 13% to 16.8%. With nanocrystals having low size distribution (13%), well-defined patterns with columns or dots and very few labyrinths (Figure 6.17a). Magnification of the patterns shows that some columns are upright (Figure 6.17b), whereas others have fallen (Figure 6.17c). With nanocrystals having a larger size distribution (16.8%), cylinders (Figure 6.17d) either isolated or associated with each other are obtained. The cylinder borders are not well-defined and include numerous defects (Figure 6.17e). In few part of the substrate, coalesced

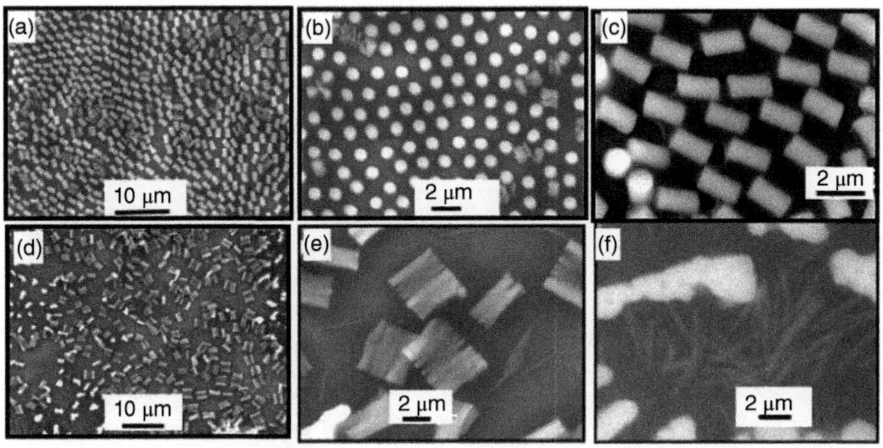

Figure 6.17 SEM images of mesostructures obtained on applying a magnetic field 0.25 T perpendicular to substrate of nanocrystals characterized by (a–c) (d = 5.7 nm, s = 13%) and (d–f) (d = 5.9 nm, s = 18%). *Source:* V. Germain (2005)/John Wiley & Sons, Inc.

cylinders are produced (Figure 6.17f). Furthermore, the length of the columns is longer with nanocrystals having a large size distribution than low ones.

Figure 6.20a shows, on a larger scale, the appearance of wormlike and labyrinth structures. A highly magnified view of these labyrinths shows that they are of coalesced cylinders (Figure 6.20c). In fact, when the size distribution is large, columns of various lengths and widths are formed. Because the columns are formed in solution, the entities of similar sizes tend to self-assemble via van der Waals interactions. The large dispersion in nanocrystal size induces defects at the border of the columns and consequently favors their coalescence. Furthermore, the large dispersion in the height of the columns induces the formation of labyrinths of various heights (Figure 6.20b). Because the nanocrystals diffuse freely during evaporation, some size segregation takes place, inducing the formation of well-defined columns. This explains why some columns are produced and their number decreases on increasing the size distribution. From this, it is concluded that the *size distribution* of nanocrystals is the key parameter that determines the formation of columns, wormlike structures, or labyrinths. The latter are produced by agglomeration of cylinders that have, more or less, the same size.

To understand the growth of the columns and the influence of the nanocrystal size distribution on the patterns, let us first consider the mesoscopic structure obtained in the absence of an applied magnetic field with nanocrystals having large size distribution ($s = 18\%$). The size of the aggregates markedly drops (between 0.4 and 1.6 mm) compared with those produced from nanocrystals that have a similar average diameter and a lower size distribution (between 2.5 and 9 mm). Furthermore, a magnified view of these aggregates shows that the nanocrystals are randomly deposited, whereas they are arranged in fcc structures for a low distribution. Note that the aggregates shown in Figure 6.18c are not compact. When the nanocrystals were subjected to a 0.25 T applied magnetic field at the end of the evaporation process, the columns were no longer well defined (Figure 6.18a) and had a rather large disorder. In addition, at the ends of the columns and along the columns, "fingerlike" patterns appeared (Figure 6.18). This is related to the fact that small aggregates without any ordered structure are observed when these nanocrystals are not subjected to an applied magnetic field. The formation of columns with many defects is

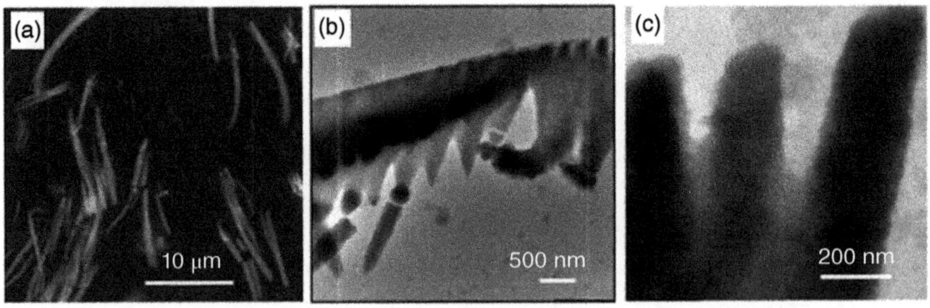

Figure 6.18 SEM (a) and TEM (b, c) images of mesostructures made of cobalt nanocrystals with average diameter and size distribution of 6.9 nm and 18% respectively, subjected to a 0.25 T magnetic field applied perpendicularly to the substrate. *Source:* V. Germain (2005)/John Wiley & Sons, Inc.

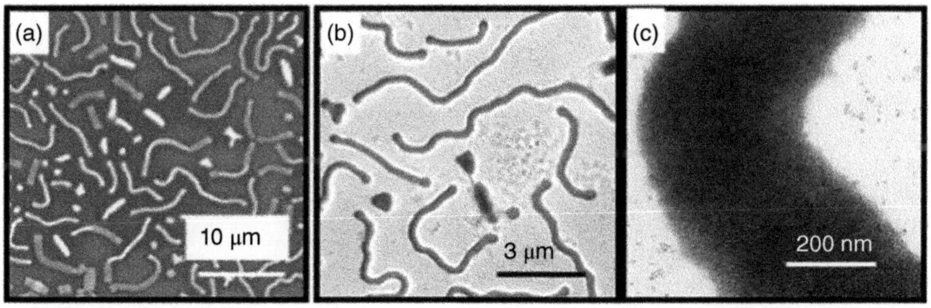

Figure 6.19 (a) SEM mesostructure obtained with cobalt nanocrystals (d = 8.7 nm, s = 20%). (b) TEM image of the same sample. (c) Enhancement of the TEM image. *Source:* V. Germain (2005)/John Wiley & Sons, Inc.

explained as follows: when the liquid–gas phase transition takes place [26], the columns start to grow on rough and deformed aggregates. This disorder induces defects during column growth and results in the patterns observed in Figure 6.18b,c. Hence, a large change in the mesoscopic patterns with the size distribution of the nanocrystals used to make the mesoscopic structures is observed. To confirm this claim, the mesoscopic pattern produced in a 0.25 T applied magnetic field is made of cobalt nanocrystals (8.7 nm) with a rather large size distribution (20%). Figure 6.19a shows, by SEM, the formation of labyrinths on a large scale. Some are very long and others are wormlike; the arrows point to columns and the fusion of two columns. A similar

Figure 6.20 SEM images of mesostructures of Co nanocrystals (d = 8.7 nm, s = 20%). (a) Overall pattern (b and c) local enhancements. *Source:* V. Germain (2005)/ John Wiley & Sons, Inc.

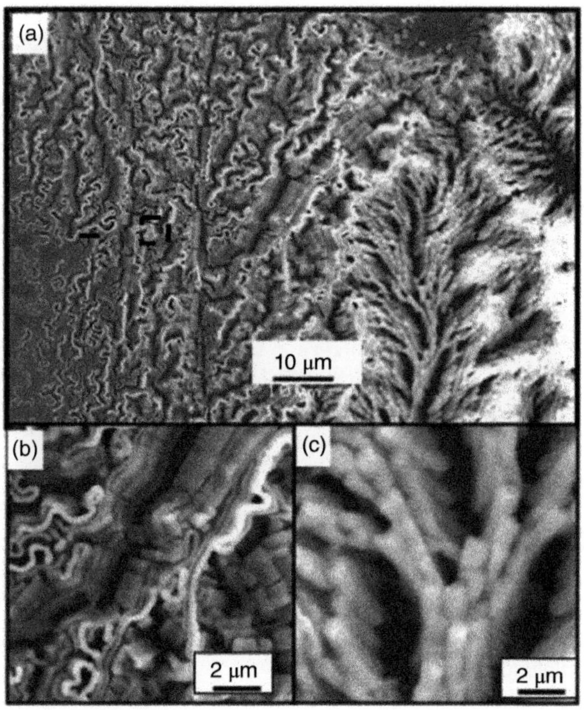

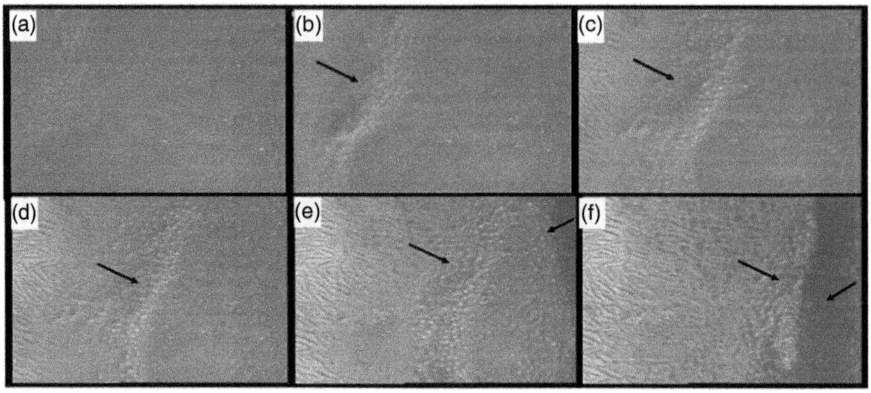

Figure 6.21 Video microscopy images with time (a–f) obtained at the end of the evaporation process of a cobalt nanocrystal solution in a 0.25 T magnetic field. A HOPG substrate is used. The arrows indicate the direction of the wave created by the capillary forces. *Source:* V. Germain (2005), p. 5548–5553/American Chemical Society.

pattern is observed by TEM (Figure 6.19b). Magnification of the TEM image shows that the labyrinths were formed with disordered nanocrystals (Figure 6.19c).

As shown above, whatever the size distribution, during the evaporation process, columns appear and migrate in the solution. At the end of the evaporation, a progressive collapse of the columns through waves is induced by capillary forces. The results of these two opposite waves are also observed at the end of the evaporation (Figure 6.21), giving rise to fernlike structures made of columns as shown on Figure 6.22.

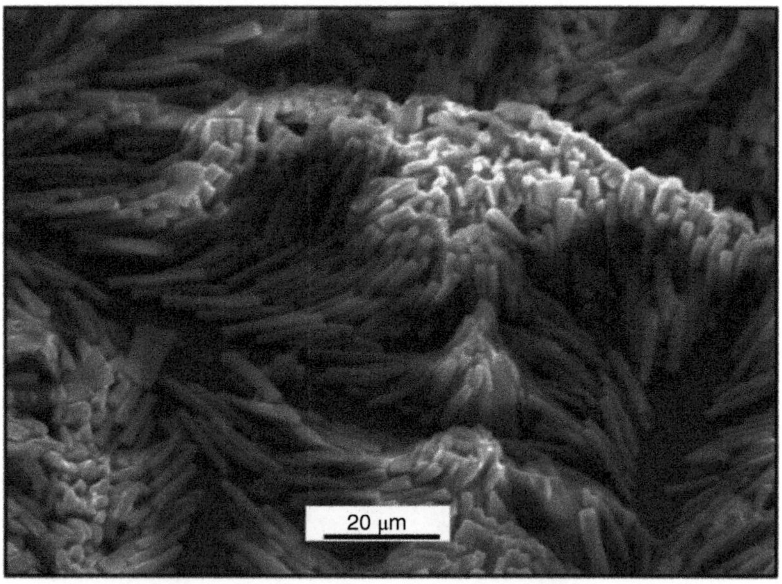

Figure 6.22 SEM image pattern in a fernlike structure obtained by applying a magnetic field (0.25 T).

For experimental reasons, it is impossible to record the video at the end of the evaporation. Figure 6.22 shows the formation of a fernlike structure made of columns, and some of them tend to fuse to form wormlike structures or labyrinths with a disorder in the organization obtained by applying 0.25 T magnetic field.

From these results and considerations, it is claimed that applying a magnetic field perpendicular to the substrate during evaporation produces well-defined columns of cobalt nanocrystals with a low size distribution. The size of the columns depends neither on the nanocrystal diameters in a 5.5–8 nm range nor on the applied field strength. The key parameter in column formation is the nanocrystal size distribution. At low size distributions, the nanocrystals self-assemble into fcc structures; they are closely packed. This induces strong cohesion inside columns with the same height, width, and aspect ratio for any applied field strength. The increase in the nanocrystal distribution destroys the cohesion because the system is unable to form a compact structure, and disordered aggregates are produced. This creates defects at the ends of the columns and at their edges, favoring their fusion in labyrinths. From this, it is concluded that labyrinths are a product of the fusion of columns. This agrees with calculations [27, 28, 43] showing that, energy being the same, columns and labyrinths are produced. To our knowledge, it is the first example from which the stiffness of self-assembled nanocrystals in crystalline structure is larger than the same assemblies in disordered form.

6.3 Conclusion

The mesoscopic structures of magnetic nanoparticles produced by applying a magnetic field either parallel or perpendicular to the substrate during the solvent evaporation process markedly differ. Chain-like and random structures, columns, and labyrinths are produced. The nanoparticle size distribution plays a role in the stability of the mesoscopic structures. They are formed in the solution when the system reaches liquid–gas phase transition.

References

1 Butter, K., Bomans, P.H.H., Frederik, P.M. et al. (2003). Direct observation of dipolar chains in iron ferrofluids by cryogenic electron microscopy. *Nat. Mater.* 2: 88–91.

2 Tang, Z., Kotov, N.A., and Giersig, M. (2002). Spontaneous organization of single CdTe nanoparticles into luminescent nanowires. *Science* 297: 237–240.

3 Rosensweig, R.E., Zahn, M., and Shumovich, R. (1983). Labyrinthine instability in magnetic and dielectric fluids. *J. Magn. Magn. Mater.* 39: 127.

4 Rosensweig, R.E. (1985). *Ferrohydrodynamics*. Cambridge: Cambridge University Press.

5 Seul, M. and Andelman, D. (1995). Domain shapes and patterns: the phenomenology of modulated phases. *Science* 267: 476.

6 Weis, R.M. and McConnell, H.M. (1984). Two-dimensional chiral crystals of phospholipid. *Nature* 310: 47.

7 Thomas, E.N., Anderson, D.M., Henke, C.S., and Hoffman, D. (1988). Periodic area-minimizing surfaces in block copolymers. *Nature* 334: 598.

8 Hasegawa, H. and Hashimoto, T. (1992). Morphology of block copolymers and mixtures of block copolymers at free surfaces. *Polymer* 33: 475.

9 Lalatonne, Y., Richardi, J., and Pileni, M.P. (2004). Van der Waals versus dipolar forces controlling mesoscopic organizations of magnetic nanocrystals. *Nat. Mater.* 3: 121–125.

10 Lalatonne, Y., Richardi, J., Motte, L., and Pileni, M.P. (2005). Influence of short-range interactions on the mesoscopic organization of magnetic nanocrystals. *Phys. Rev. E* 71: 011404.

11 Legrand, J., Petit, C., and Pileni, M.P. (2001). Domain shapes and superlattices made of 8 nm cobalt nanocrystals: fabrication and magnetic properties. *J. Phys. Chem. B* 105: 5643–5646.

12 Legrand, J., Ngo, A.T., Petit, C., and Pileni, M.P. (2001). Domain shapes and superlattices made of cobalt nanocrystals. *Adv. Mater.* 13: 57–59.

13 Ngo, A.T. and Pileni, M.P. (2000). Nanoparticles of cobalt ferrite: influence of the applied field on the organization of the nanocrystals on a substrate and on their magnetic properties. *Adv. Mater.* 12: 276–279.

14 Ngo, A.T. and Pileni, M.P. (2001). Assemblies of ferrite nanocrystals: partial orientation of the easy magnetic axes. *J. Phys. Chem. B* 105: 53–58.

15 Ngo, A.T., Richardi, J., and Pileni, M.P. (2005). Mesoscopic solid structures of 11-nm maghemite, γ-Fe_2O_3, nanocrystals: experiment and theory. *Langmuir* 21: 10234–10239.

16 Chantrell, R.W., Bradbury, A., Popplewell, J., and Charles, S.W. (1982). Agglomerate formation in a magnetic fluid. *J. Appl. Phys.* 53: 2742–2744.

17 Morimoto, H. and Maekawa, T. (1999). Dynamic analysis of a ferromagnetic colloidal system. *Int. J. Mod. Phys. B* 13: 2085–2092.

18 Tavares, J.M., Weiss, J.J., and Telo da Gama, M.M. (1999). Strongly dipolar fluids at low densities compared to living polymers. *Phys. Rev. E* 59: 4388–4395.

19 Tlusty, T. and Safran, S.A. (2000). Defect-induced phase separation in dipolar fluids. *Science* 290: 1328–1331.

20 Wang, Z. and Holm, C. (2003). Structure and magnetic properties of polydisperse ferrofluids: a molecular dynamics study. *Phys. Rev. E* 68: 041401.

21 Allen, M.P. and Tildesley, D.J. (1987). *Computer Simulation of Liquids*. Oxford: Oxford Science Publications.

22 Allen, M.P. (1980). Brownian dynamics simulation of a chemical reaction in solution. *Mol. Phys.* 40: 1073–1079.

23 Morimoto, M. and Maekawa, T. (2000). Cluster structures and cluster–cluster aggregations in a two-dimensional ferromagnetic colloidal system. *J. Phys. A: Math. Gen.* 33: 247–258.

24 Hunter, R.J. (2001). *Foundations of Colloid Science*. Oxford: Oxford University Press.

25 Motte, L. and Pileni, M.P. (1998). Influence on the 2D and 3D self-organization of the length of the alkyl chains used to passivate silver sulfide nanoparticles. *J. Phys. Chem.* 102: 4104.

26 Richardi, J., Ingert, D., and Pileni, M.P. (2002). Labyrinthine instability in magnetic fluids revisited. *J. Phys. Chem. B* 106 (7): 1521–1523.

27 Richardi, J. and Pileni, M.P. (2004). Towards efficient methods for the study of pattern formation in ferrofluid films. *Eur. Phys. Rev. E* 69 (1): 016304.

28 Richardi, J. and Pileni, M.P. (2004). Nonlinear theory of pattern formation in ferrofluid films at high field strengths. *Phys. Rev. E* 69: 016304.

29 Richardi, J., Pileni, M.P., and Weis, J.J. (2008). Self-organization of magnetic nanoparticles: a Monte Carlo study. *Phys. Rev. E* 77: 061510.

30 Richardi, J., Pileni, M.P., and Weis, J.J. (2009). Self-organization of confined dipolar particles in a parallel field. *J. Chem. Phys.* 130: 124515–124519.

31 Germain, V., Richardi, J., Ingert, D., and Pileni, M.P. (2005). Mesostructures of cobalt nanocrystals. 1. Experiment and theory. *J. Phys. Chem. B* 109: 5541–5547.

32 Cebers, A. and Maiorov, M.M. (1980). *Magnetohydrodynamics*, vol. 16, 21. University of Tatvia.

33 de Hoog, E.H.A. and Lekkerkerker, H.N.W. (1999). Measurement of the interfacial tension of a phase-separated colloid–polymer suspension. *J. Phys. Chem. B* 103: 5274.

34 Zahn, M. (2001). Magnetic nanoparticles: advantages of using, methods for preparation, characterization, application in pharmacy. *J. Nanopart. Res.* 3: 73.

35 Lisiecki, I., Albouy, P.A., and Pileni, M.P. (2004). "Supra" crystal: control of the ordering of self-organization of cobalt nanocrystals at the mesoscopic scale. *J. Phys. Chem. B.* 108: 20050–20055.

36 Lisiecki, I. and Pileni, M.P. (2003). Synthesis of well-defined and low size distribution cobalt nanocrystals: the limited influence of reverse micelles. *Langmuir* 19: 9486–9489.

37 Petit, C., Legrand, J., Russier, V., and Pileni, M.P. (2002). Three dimensional arrays of cobalt nanocrystals: fabrication and magnetic properties. *J. Appl. Phys.* 91: 1502–1508.

38 Hong, C.Y., Jang, I.J., Horng, H.E. et al. (1997). Ordered structures in Fe_3O_4 kerosene-based ferrofluids. *J. Appl. Phys.* 81: 4275.

39 Ivey, M., Liu, J., Zhu, Y., and Cutillas, S. (2000). Magnetic-field-induced structural transitions in a ferrofluid emulsion. *Phys. Rev. E* 63: 011403.

40 Germain, V. and Pileni, M.P. (2005). Mesostructures of cobalt nanocrystals. 2. Mechanism. *J. Phys. Chem. B* 109: 5548–5553.

41 Germain, V. and Pileni, M.P. (2005). Size distribution of cobalt nanocrystals: a key parameter in formation of columns and labyrinths in mesoscopic structures. *Adv. Mater.* 17: 1424–1429.

42 Motte, L., Lacaze, E., Maillard, M., and Pileni, M.P. (2000). Self-assemblies of silver sulfide nanocrystals on various substrates. *Langmuir* 16: 3803–3812.

43 Dickstein, A.L., Erramilli, S., Goldstein, R.E. et al. (1993). Labyrinthine pattern formation in magnetic fluids. *Science* 261: 1012–1015.

7

Binary 3D Superlattices

Self-assembly of micrometer-sized colloids has been intensively studied and is determined by minimizing the system's free energy $F = U - TS$. Since those colloids are forbidden to interpenetrate and the thermal energy U is constant, entropy S remains the key parameter [1, 2]. For hard spheres, binary crystal structures can be determined from free energy calculations using the Frenkel–Ladd–Einstein thermodynamic integration in computer simulations [1]. Based on the available theoretical calculation, NaCl, AlB_2, $NaZn_{13}$, and Laves phase ($MgZn_2$, $MgCu_2$, $MgNi_2$) structures can be produced by entropic force. They are called hard-sphere model structures. At the nanoscale, self-assemblies are controlled by nanoparticle interactions, entropic forces, and organic molecules attached at the nanocrystal surface to keep their integrities [1–3]. Hence, the emergence of these binary nanoparticle superlattices called binary 3D superlattices can partially be predicted. With uniform spherical nanoparticles, a close-packed pattern usually prefers to adopt a compact structure through either hexagonal close-packed (hcp) or face-centered cubic (fcc) packing mode to minimize the system's potential energy and produce the highest space filling of ~74%. While for binary structures consisting of two sizes, the packing efficiency is often higher than that of a single component because the vacancy generated by the closest packing still accommodates smaller spheres. The packing efficiency is governed by the effective size ratio of small (A) and large (B) nanoparticles, γ, and their stoichiometry. The surface coating layer, which can significantly contribute to the real nanocrystal size, cannot be directly analyzed from transmission electron microscopy (TEM) images. Instead, the effective nanocrystal size is measured from the center-to-center distance in the corresponding single-component hcp monolayer. The measured effective nanoparticle size is compared with the theoretical calculation $d_{eff} = d_{core} + 2L$, where d_{core} and L are the diameter of nanoparticles and the length of alkyl chain used to coat nanoparticles, respectively. Figure 7.1 shows the phase diagrams determined from free energy calculations in computer simulations for binary mixtures of hard spheres.

With nanoparticles having two different sizes, a large structural diversity in binary nanoparticle superlattices was discovered, which broadens the point of view in crystal engineering. Previous endeavors have constructed a number of binary

Metal Nano 3D Superlattices: Synthesis, Properties, and Applications, First Edition.

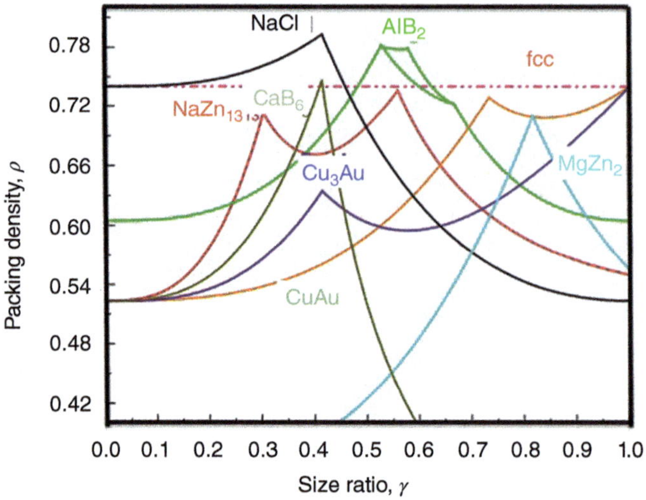

Figure 7.1 Theoretical predictions of various binary structures deduced from the hard sphere model. *Source:* Based on [4–23].

nanoparticle superlattices in analogy to binary atomic and ionic lattices, such as AuCu, NaCl, CsCl, AlB_2, $MgZn_2$, $MgNi_2$, $AuCu_3$, Fe_4C, $CaCu_5$, CaB_6, $NaZn_{13}$, cuboctahedra AB_{13}, A_6B_{19}, and the quasicrystalline phase (Figures 7.1 and 7.2) [4–23]. However, some discrepancies are observed, and the model requires to be further studied.

As already mentioned, to keep the integrity of nanoparticles, it is necessary to coat them with organic molecules. Below, it will be demonstrated that the role of the ligand remains an open question. To produce 3D superlattice binary structures, the interactions have to be taken into account: (i) between alkyl chains and (ii) between terminated ligand groups and nanoparticle surface atoms. In brief, we know that

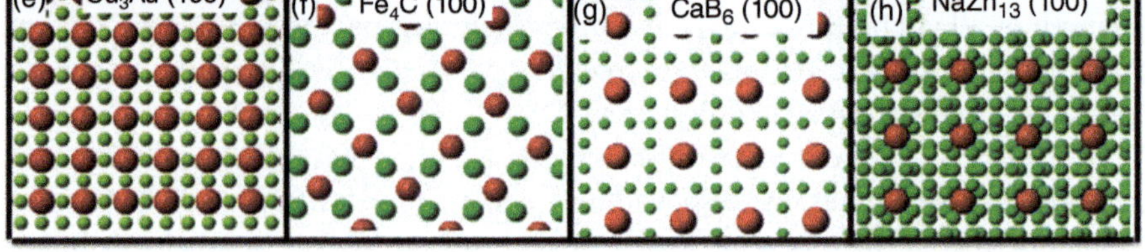

Figure 7.2 Various crystal models of binary nanoparticles superlattices. *Source:* Based on [4–23].

the interactions between alkylthiol ($C_{12}H_{25}$–SH) and noble metal (Au, Ag) are very strong [24–26]. Carboxylic group is covalently bound to Co and Fe atoms [27]. With amine-terminated ligands (NH_2) such as oleylamine (OAM), their bindings on the metallic surfaces are very weak compared with SH and COOH groups [28].

Here, we will mainly focus on metal binary 3D superlattices such as Ag/Ag, Co/Ag, and Co/Co. The Ag nanoparticles are polycrystalline phase, whereas Co nanoparticles are either amorphous-phase or single-domain nanoparticles such as the ε-phase or hcp structures. For simplicity, the latter will be denoted as Co_A, Coε, and Co_{HCP} nanoparticles, respectively. Note that all these Co nanoparticles in the studied size range are superparamagnetic at room temperature except Co_{HCP} nanoparticles of 9 nm.

Two colloidal solutions are prepared separately. The first one is composed of large (A) nanoparticles dispersed in a solvent, whereas the second one (B) contains the smaller nanoparticles. They are both characterized by a low size distribution. These colloidal solutions are then mixed usually at room temperature except when it will be mentioned. The nanoparticle concentration ratio $[A]/[B]$ varies from 1/12 to 1, keeping the total nanoparticle concentration ($[A] + [B]$) constant. During the evaporation time, the system temperature remains at room temperature except when it is mentioned in the text.

7.1 Structure of 3D Superlattices Predicted by the Hard-Sphere Model

A hard sphere model has been proposed to predict the structure of 3D superlattices [17, 20]. According to the hard-sphere model, the system is expected to adopt the crystal structure corresponding to the most efficient space filling, to reach the maximization of the packing density ρ for a given ratio of the sphere radii γ [29].

7.1.1 Co/Ag Binary 3D Superlattices of Co and Ag Nanoparticles

Amorphous Co_A nanoparticles of 7.2 nm are coated with oleic acid, whereas polycrystalline Ag nanoparticles of 4 nm are coated with dodecanethiol, i.e. the coating agents are strongly bounded to the nanoparticle surface. By changing the γ parameter various structures have been obtained [19–21]. Hence with $\gamma = 0.59$ the corresponding structure is expected to be AlB_2 type. Co_A nanoparticles are self-assembled into hexagonal pattern, and the Ag nanoparticles fill into the interstices between the Co_A nanoparticle layers that agrees well with the literature with AlB_2 structure (Figure 7.3a). For $[Co_A]/[Ag] = 1$, the domains are rather small, ranging from hundreds of nanometers to microns (Figure 7.3b), whereas large domains up to tens of microns (Figure 7.3d) on increasing the Ag nanoparticles by a factor 2 ($[Co_A]/[Ag] = 1/2$). The preferential orientation of the AlB_2 structure is with the (001) plane parallel to the substrate as it is observed through the TEM image (Figure 7.3c,e) and fast Fourier transform (FFT) patterns (Figure 7.3f,g).

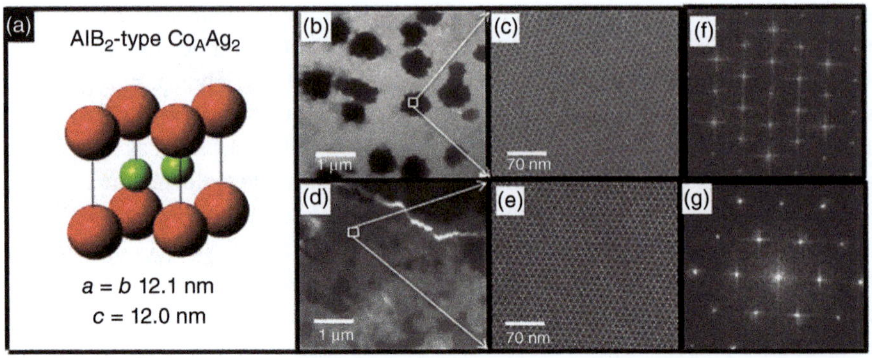

Figure 7.3 (a) Crystal model of AlB_2 structure. (b–e) TEM images of the domains size of binary 3D superlattices formed with different $[Co_A]/[Ag]$ keeping other conditions constant ($\gamma = 0.59$, $T = 25\,°C$). (b, c) $[Co_A]/[Ag] = 1$. (d, e) $[Co_A]/[Ag] = 1/2$. (f, g) The corresponding FFT patterns of (c, e). *Source:* (b–g) Yang et al. [20]/American Chemical Society.

Three sets of typical crystal superlattice planes are observed: (001) (Figure 7.4a), (110) (Figure 7.4c), and (100) (Figure 7.4e) superlattice planes. Moreover, the corresponding projections of the model structures are shown in Figure 7.4b,d,f, respectively. The superlattice spacings in the (100) and (001) planes measured from the corresponding TEM images are 10.4 and 12.0 nm, corresponding to lattice dimensions $a = b = 12.1$ nm and $c = 12.0$ nm. The center-to-center distance between Co_A nanoparticles in Co_AAg_2 superlattices is calculated for these various orientations: the average distance between Co_A nanoparticles is 12.0 ± 0.2 nm, and the edge-to-edge value between Co_A nanoparticles is estimated to be 4.8 ± 0.2 nm. This

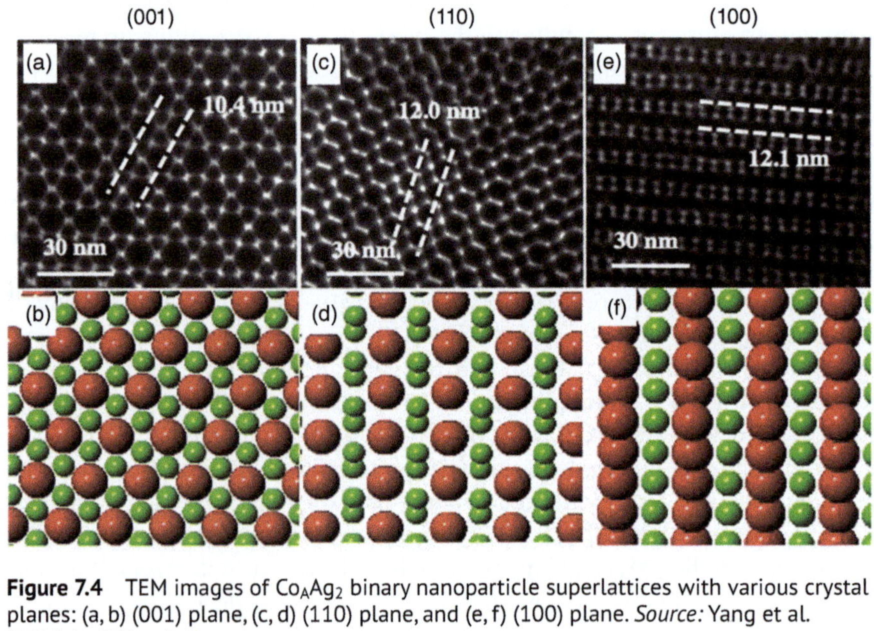

Figure 7.4 TEM images of Co_AAg_2 binary nanoparticle superlattices with various crystal planes: (a, b) (001) plane, (c, d) (110) plane, and (e, f) (100) plane. *Source:* Yang et al. [21]/John Wiley & Sons, Inc.

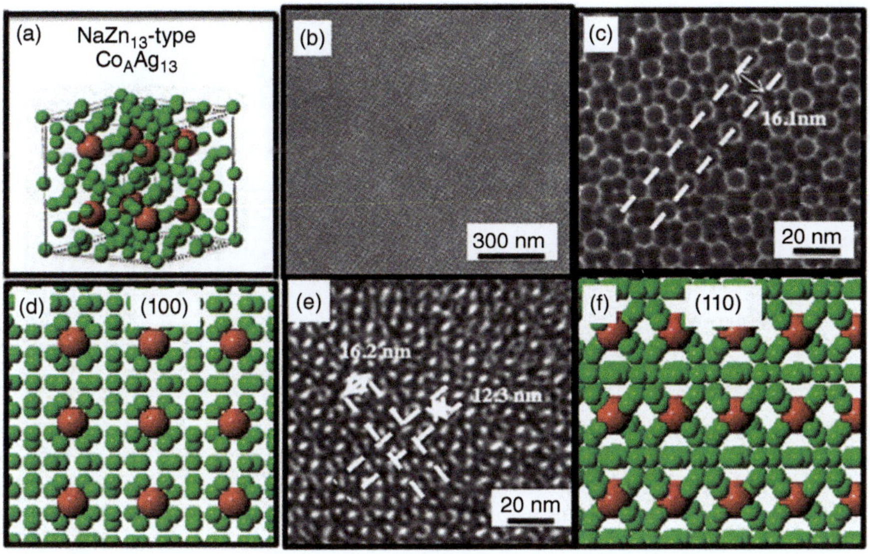

Figure 7.5 (a) Crystal model of $NaZn_{13}$ structure. (b) TEM images of Co_AAg_{13} binary nanoparticle superlattices with various crystal planes: (c, d) crystal (100) plane and (e, f) (110) plane. *Source:* (b, c, e) Yang et al. [21]/John Wiley & Sons, Inc.

value markedly differs from that in single-component fcc-phase Co superlattices (3.4 nm). The domain size of the $CoAg_2$ binary 3D superlattices is up to tens of microns as observed previously [30].

By increasing the relative amount of colloidal Ag nanoparticles with respect to the Co_A one ($[Co_A]/[Ag] = 1/10$), keeping the total nanoparticle concentration constant, a long range (up to 10 μm) of ordered binary system (Figure 7.5b) attributed to $NaZn_{13}$ structure (Figure 7.5a) is produced. Figure 7.5c,d shows a preferential crystal orientation with the (100) plane parallel to the substrate for the $CoAg_{13}$ 3D superlattices. In addition, the (110) plane parallel to the substrate shown in Figure 7.5e,f further confirms the $NaZn_{13}$-type $CoAg_{13}$ structures [30]. Hence, 8 Co_A nanoparticles are well isolated by 104 Ag nanoparticles in a unit cell (Figure 7.5a). The spacings between Co nanoparticles within the binary structure in the (001) and (110) planes are 16.1 ± 0.2 and 12.3 ± 0.2 nm, respectively.

7.1.2 Co/Co Binary 3D Superlattices of Amorphous Co Nanoparticles

Let us now consider Co_ACo_{A2} binary system produced from two colloidal solutions: one containing 4.6-nm Co_A nanoparticles and the other 7.2-nm Co_A nanoparticles [19]. The concentration of each colloidal solution is the same as above. According to the relative nanoparticle size ratio, $\gamma_{eff} = 0.76$, $MgZn_2$-type structure is most favorable based on the recent theoretical work showing that the $MgZn_2$ phase is stable in the size ratio range of $0.76 \leq \gamma \leq 0.84$ [8]. Here, by careful analysis of the TEM data, we only observe an $MgZn_2$ 3D superlattices (Figure 7.6). The structure of $MgZn_2$ contains four molecular units per unit cell with the smaller Zn atoms ordered in

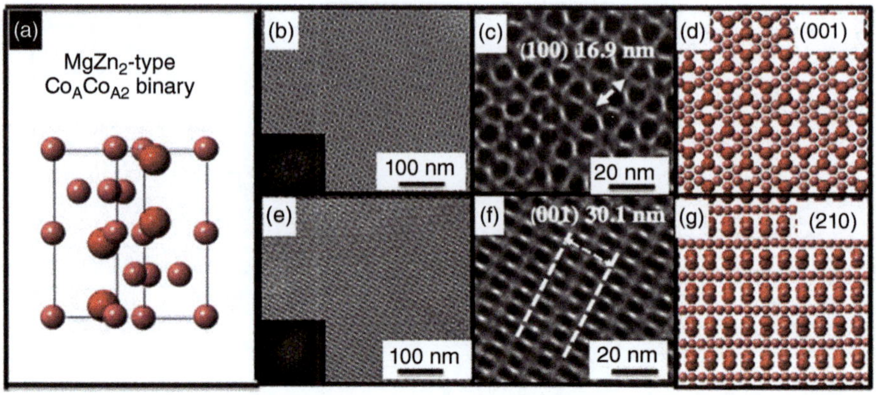

Figure 7.6 (a) Structural model of $MgZn_2$-type binary structure. (b, c, e, f) TEM image of binary 3D superlattices from the evaporation of 7.2- and 4.6-nm Co_A binary nanoparticle mixtures. (b–d) (00l) crystal planes. (e–g) (210) crystal planes. Insets in panels (b) and (e) are the corresponding FFT patterns. *Source:* (b, c, e, f) Yang et al. [21]/John Wiley & Sons, Inc.

tetrahedra, whereas the larger Mg atoms fall into the vacancies provided by the Zn tetrahedral (Figure 7.6a). Typical TEM images of these structures with the (001) plane parallel to the substrate are shown in Figure 7.6b,c, and the corresponding structure model projected along the same zone axis is shown in Figure 7.6d. The FFT pattern in the inset of Figure 7.6b shows the hexagonal symmetry. From this set of crystal planes, it is possible to evaluate the unit cell length $a = b = 16.9 \pm 0.2$ nm. Another crystal plane with indices (210), with a minor fraction parallel to the substrate, is shown in Figure 7.6e–g, along with the projection of the model structure in Figure 7.6a. The unit cell edge length measured from this image is $c = 30.1 \pm 0.2$ nm. Thus, a unit cell with cell parameters of $a = b = 16.9$ nm and $c = 30.1$ nm was built to evaluate the various Co_A–Co_A interparticle distances (Figure 7.6a): the nearest gap between 7.2-nm Co_A nanoparticles is 4.2 ± 0.4 nm, resulting in a Co_A–Co_A interparticle distance of 11.4 nm. The domain size of the Co_ACo_{A2} binary 3D superlattices is up to ~5 μm.

7.1.3 Ag/Ag Binary 3D Superlattices of Polycrystalline Ag Nanoparticles

With Co_A/Ag and Co_A/Co_A binary systems because of the limitation in the control of nanocrystal size, the overall hard-sphere models could not be discovered. Here, with Ag nanoparticles coated with single ligand, OAM = $C_{18}NH_2$, instead of dodecanethiol, a rather good agreement between experimental data and hard-sphere model is obtained [17]. The concentration ratio between large (*A*) and small (*B*) nanoparticles is 1/4.

As predicted by the hard-sphere model in which nanoparticles are densely packed as driven by entropy, various structures are obtained. Hence, NaCl (Figure 7.7), AlB_2 (Figure 7.8), $NaZn_{13}$ (Figure 7.9), and $MgZn_2$ (Figure 7.10) types are observed.

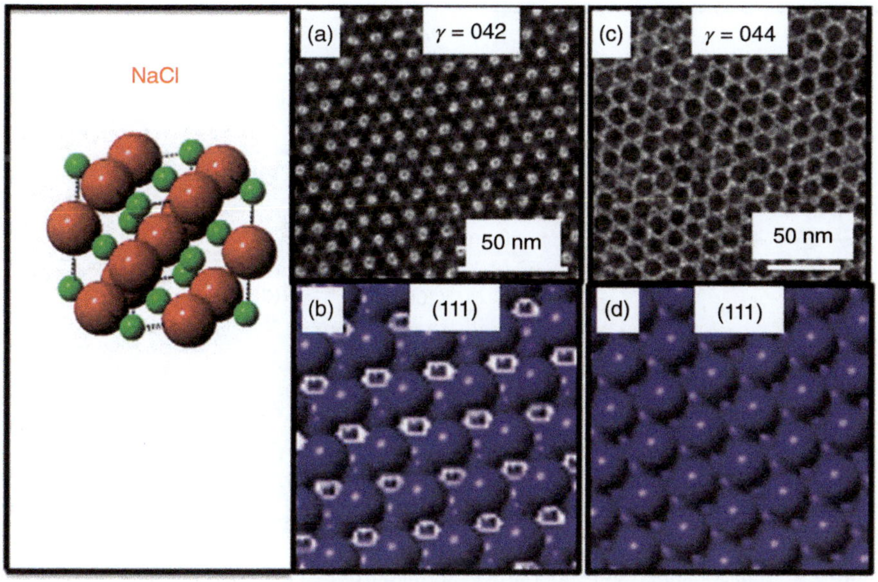

Figure 7.7 Binary nanocrystal superlattices formed from Ag nanoparticles coated with oleylamine as single ligand. (a, c) TEM images and (b, d) structural model. (a, b) $A = 11.9$ nm, $B = 2.9$ nm, $\gamma_{eff} = 0.42$. (c, d) $A = 11.9$ nm, $B = 3.7$ nm, $\gamma_{eff} = 0.4$. *Source:* Jingjing Wei 2015, p. 137/ American Chemical Society.

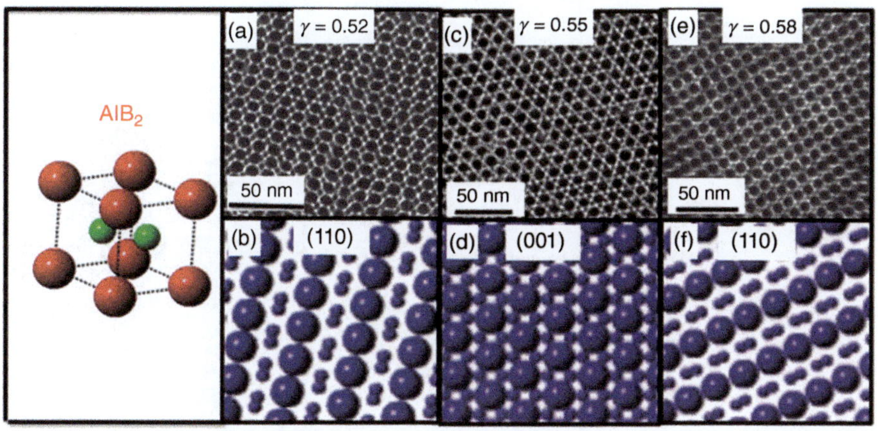

Figure 7.8 Binary nanocrystal superlattices formed from Ag nanoparticles coated with single ligand, OAM. (a, c) TEM images and (b, d) structural model. (a, b) $A = 9.6$ nm, $B = 3.7$ nm, $\gamma_{eff} = 0.52$. (c, d) $A = 8.2$ nm, $B = 2.9$ nm, $\gamma_{eff} = 0.55$. (e, f) $A = 8.2$ nm, $B = 3.7$ nm, $\gamma_{eff} = 0.58$. *Source:* Jingjing Wei 2015, p. 137/American Chemical Society.

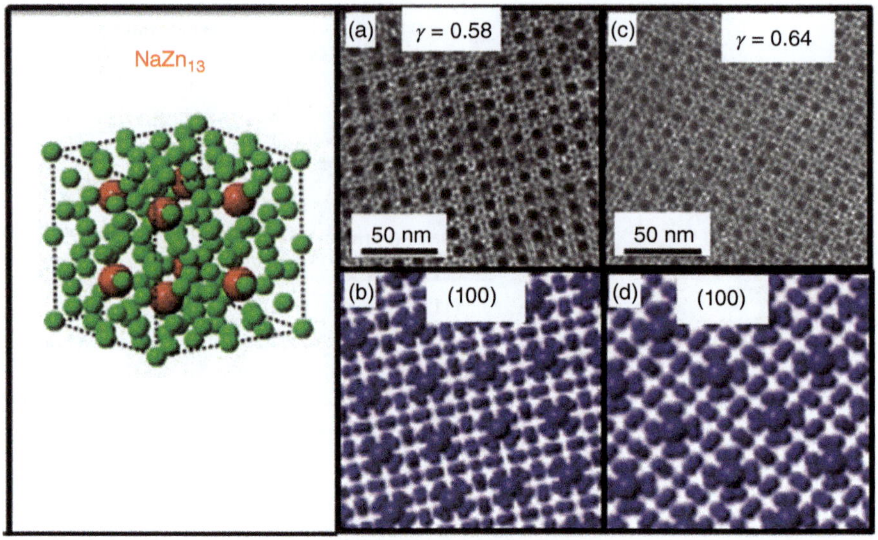

Figure 7.9 Binary nanocrystal superlattices formed from Ag nanoparticles coated with single ligand, OAM. (a, c) TEM images and (b, d) structural model. (a, b) $A = 8.2$ nm, $B = 3.7$ nm, $\gamma_{eff} = 0.58$. (c, d) $A = 7.5$ nm, $B = 3.7$ nm, $\gamma_{eff} = 0.64$. *Source:* Jingjing Wei 2015, p. 137/American Chemical Society.

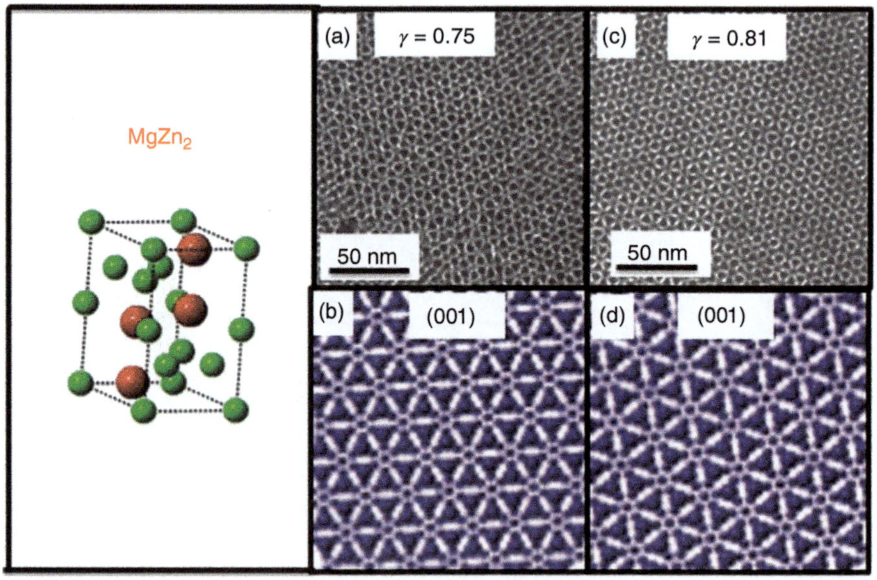

Figure 7.10 Binary nanocrystal superlattices formed from Ag nanoparticles coated with single ligand, OAM. (a, c) TEM images and (b, d) structural model. (a, b) $A = 5.5$ nm, $B = 3.7$ nm, $\gamma_{eff} = 0.79$. (c, d) $A = 7.5$ nm, $B = 5.5$ nm, $\gamma_{eff} = 0.81$. *Source:* Jingjing Wei 2015, p. 137/American Chemical Society.

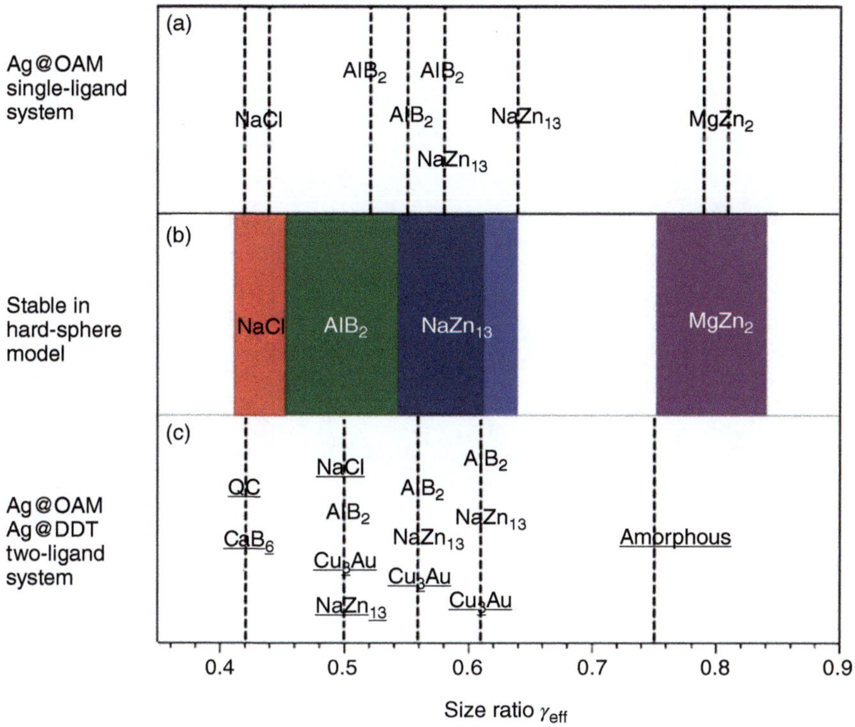

Figure 7.11 Various binary structures formed: (a) *A* and *B* coated with oleylamine as a single ligand, (b) hard-sphere theoretical predictions, and (c) *A* coated with oleylamine and *B* with dodecanethiol. *Source:* Yang et al. [19].

Figure 7.11a confirms a good agreement between the experimental data and the hard-sphere model (Figure 7.11b) with a sequential structural evolution of binary nanocrystal superlattices from NaCl to AlB_2 to $NaZn_{13}$ to $MgZn_2$ with the increase of size ratio γ.

7.2 Limitation of the Hard-Sphere Model

To the best of our knowledge, a systematic study of the assembly of binary 3D superlattices based on all metallic nanoparticles is still lacking and needs to be demonstrated. Here, we give some examples in which the hard-sphere model cannot be retained.

7.2.1 Ligand Exchange

In Figures 7.8–7.10 and 7.11a, we described the small and large Ag nanoparticles coated with the *same ligand*: OAM. Below, the large Ag nanoparticles remain coated with OAM, whereas the smaller ones are coated with dodecanethiol [17].

With 11.9- and 4-nm Ag nanoparticles, γ is 0.42. With the same ligand, Figure 7.7 shows the formation of NaCl-type structure. By replacing OAM by dodecanethiol

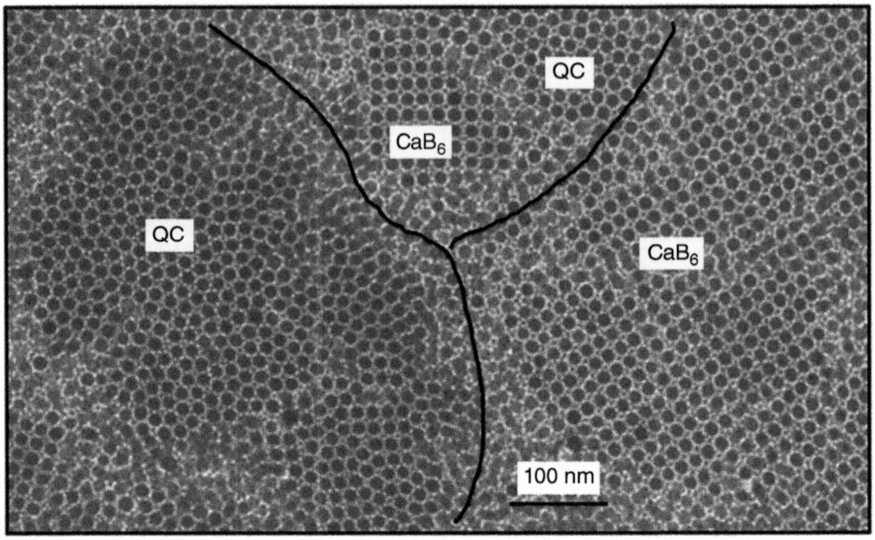

Figure 7.12 TEM image of binary superlattices formed with $A = 11.9$ nm coated with oleylamine and $B = 4.0$ nm coated with dodecanethiol: mixture of the quasicrystal (QC) order in binary nanocrystal superlattices and CaB_6-type binary nanocrystal superlattices. *Source:* Jingjing Wei 2015, p. 137/American Chemical Society.

for small nanoparticles, Figure 7.12 shows the formation of both CaB_6 and quasicrystal (QC) phases.

The crystal domain boundaries CaB_6 and quasicrystal phase agree with the reported growth mechanism based on entropic force [19]. Figure 7.13a, confirmed by FTT (Figure 7.13b), shows magnified image of ordered crystal domains of CaB_6-type 3D superlattices, in which six small Ag nanoparticles form the octahedral

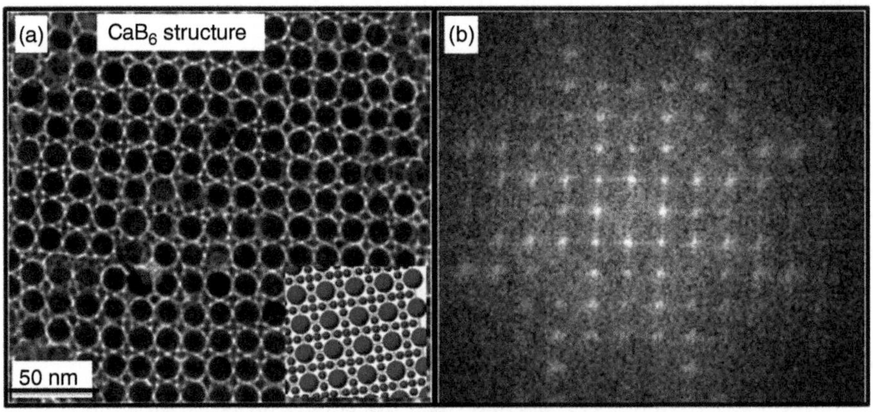

Figure 7.13 TEM image of binary superlattices formed with $A = 11.9$ nm coated with oleylamine and $B = 4.0$ nm coated with dodecanethiol: (a) TEM image of CaB_6-type binary nanoparticles superlattices with (100) planes parallel to the substrate and (b) corresponding FFT pattern to panel (a). *Source:* Jingjing Wei 2015, p. 137/American Chemical Society.

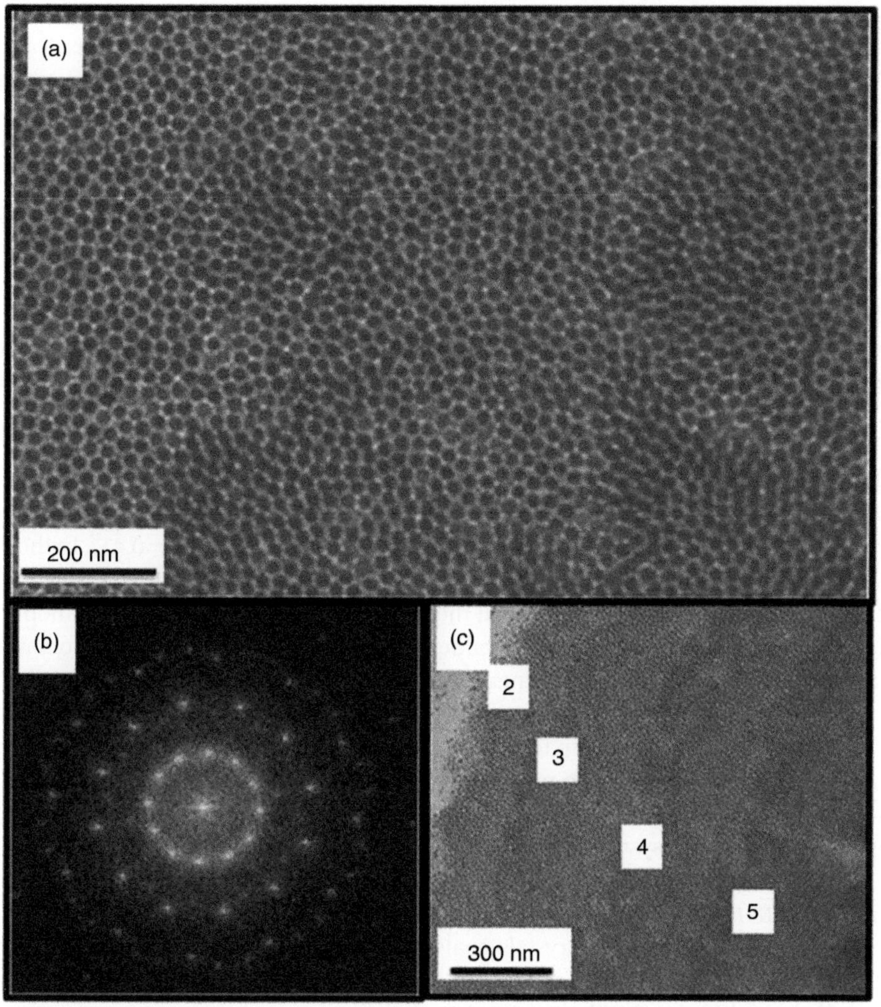

Figure 7.14 TEM image of binary superlattices formed with 11.9-nm Ag@OAM and 4.0 nm Ag@DDT binary mixtures: (a) TEM image of quasicrystal order, (b) corresponding FFT pattern to panel (d), and (c) TEM image of large-scale quasicrystal with layer-by-layer structures (the numbers correspond to the different layers of quasicrystal order). *Source:* Jingjing Wei 2015, p. 137/American Chemical Society.

cluster being surrounded by large nanoparticles packed in cubic structure. This fits perfectly with the corresponding (100) plane of CaB_6 crystalline phase (inset in Figure 7.13a).

As observed in Figure 7.12, the quasicrystal domains are characterized by a crystalline structure having octahedral shape with six small Ag nanoparticles separated by large Ag nanoparticles in a random manner (Figure 7.14a). The corresponding FFT pattern shows 12-fold symmetry, revealing its dodecagonal quasicrystal order (Figure 7.14b). Figure 7.14c shows that the quasicrystal superlattices grow through a layer-by-layer process.

According to theoretical calculations based on the hard-sphere model, at this size ratio (γ_{eff} =0.50), AlB_2 is predicted, whereas the other structures such as NaCl, Cu_3Au, and $NaZn_{13}$ are produced (Figure 7.11c). Furthermore, when the γ_{eff} value increases to 0.56 and 0.61, another unstable and unpredicted phase of Cu_3Au structure with a relatively low packing density formed is obtained. Similarly, at large γ_{eff} values (~0.75), the expected $MgZn_2$ structure is not produced, and only an amorphous film is observed. These data clearly show that the binary structures produced with two different coating agents markedly differ from the hard-sphere model (Figure 7.11b,c). Furthermore, the phase diagrams in Figure 7.11a,c show marked changes compared with data acquired when using binary 3D superlattices of nanoparticles bearing the same coating agents for both small and large components. Hence, the use of two different coating agents for small and large nanoparticles induces the formation of various structures with low packing densities.

A question arises: Will such changes be related to the type of nanoparticles used? To answer this, Ag nanoparticles is replaced by Au both coated by dodecanethiol while keeping the same average diameter (4.0 nm) for Au nanoparticles. All other experimental conditions remained unchanged ($[Au]_S/[Ag]_L = 4$; $\gamma_{eff} = 0.41$). In those conditions, CaB_6 and dodecahedral quasicrystal structures are clearly observed (Figure 7.15). They are consistent with the results obtained from Ag to Au nanoparticle binary system with a similar γ_{eff} value. However, the size of the crystal domains of both CaB_6 and dodecahedral quasicrystal structures produced is smaller than those with Ag counterparts. This can be attributed to stronger interactions between Au atoms and SH– head group than those between SH– and Ag atoms. From these data, one can assume that the production of quasicrystal order in binary superlattices relies on the presence of small and large nanoparticles coated by two different ligands in the binary systems.

To confirm such claim, similar experiments are extended to other types of nanoparticles differing by their sizes and coating agents.

Let us here consider nanocrystals coated with similar alkyl chains. Alkyl chain lengths of oleic acid/OAM are similar (C_{18}) with a double bond at the C_9 position for both ligands differing by the interactions between the coating agent polar head groups and the surface atoms of the nanoparticles. Oleic acid is covalently attached to Co or Fe atoms [27], whereas OAM is weakly associated to the Ag surface atoms [31]. With 12-nm $CoFe_2O_4$ nanoparticles coated with oleic acid and 3.7-nm Ag nanoparticles with OAM, Figure 7.16a shows a periodic (3^2.4.3.4) Archimedean tiling. The corresponding FFT pattern (Figure 7.16a, inset) indicates a fourfold rotational symmetry as already observed previously [22, 32]. In addition, dodecagonal quasicrystal structures and AlB_2-type 3D superlattices are also observed (Figure 7.16b).

This clearly shows that coating agent interactions, both in terms of bond strength of head groups/surface atoms and the length of the alkyl chains, are important in the quasicrystal formation. Similar data with the formation of either CaB_6 or at large γ_{eff} value (~0.75) structure in binary superlattices were obtained with either Fe_2O_3, Fe_3O_4, or PBs coated with oleic acid and Au or Pd coated with dodecanethiol [5, 14, 22]. This corroborates our findings stating that, to produce either CaB_6 or at large γ_{eff}

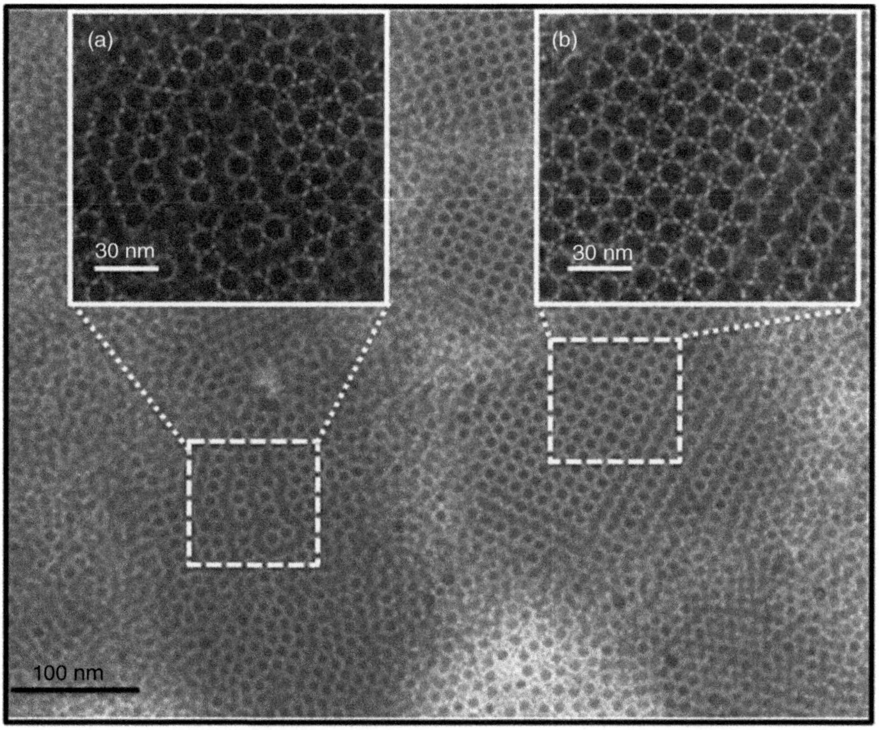

Figure 7.15 Low magnification of TEM image of quasicrystal and CaB_6-type superlattices from Au–Ag binary mixtures produced with $[Au]_S/[Ag]_L = 4$ (11.9-nm Ag coated with oleylamine and 4-nm Au coated with dodecanethiol). Insets (a) and (b): high magnification of the area of the panel. (a) Quasicrystals and (b) CaB_6. *Source:* Jingjing Wei 2015, p. 137/ American Chemical Society.

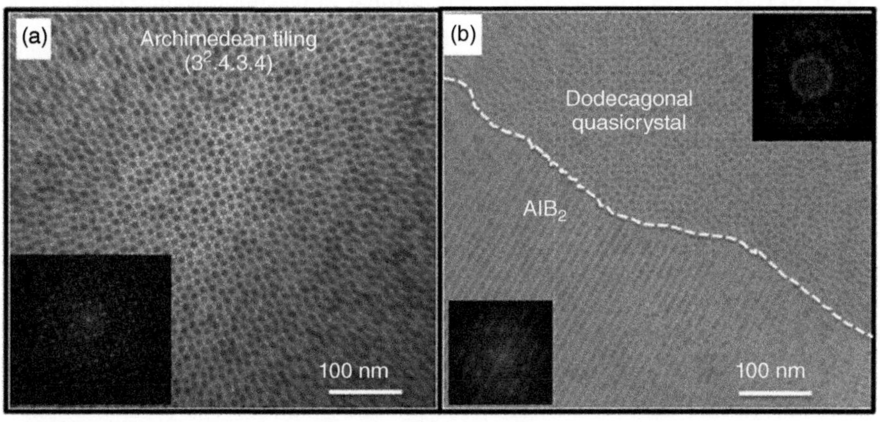

Figure 7.16 (a) and (b) TEM images of the binary $Co_AFe_2O_4$/Ag 3D superlattices formed with 12.0-nm $CoFe_2O_4$ and 3.7-nm Ag nanoparticles coated with oleic acid and oleylamine, respectively, with $[Co_AFe_2O_4]/[Ag] = 1/4$. *Source:* Jingjing Wei 2015, p. 137/American Chemical Society.

value (~0.75) structures in binary superlattices, the small and large nanoparticles must be coated by two different coating agents.

Similarly, with 12-nm $Co_AFe_2O_4$ nanoparticles coated with oleic acid and 4-nm Ag nanoparticles coated with dodecanethiol, CaB_6-type and dodecagonal quasicrystal structures are observed. Hence, in both cases, CaB_6-type and dodecagonal quasicrystal structures are obtained instead of the NaCl phase predicted from the hard-sphere model at this size ratio. Note that the length of the ligand used for the formation of large γ_{eff} value (~0.75) structure is different from C_{12} dodecanethiol to C_{18} oleic acid. While we stress that the ligand length difference is not required for the formation of quasicrystal structure even though the formation of large γ_{eff} value (~0.75) structures are observed while using C_{12} dodecanethiol to C_{18} oleic acid, they were also present in the case of 12-nm $CoFe_2O_4$ and 3.7-nm Ag coated with oleic acid and OAM, respectively.

The above results unambiguously show that coating agents can play a pivotal role in the formation of binary 3D superlattices. With similar interactions between alkyl chains and between head polar groups/surface atoms, the resulting binary 3D superlattices fit well with the phase diagrams based on hard-sphere model. However, divergences emerge when two different coating agents are used, and the resulting binary structures, such as the quasicrystal phase, cannot be predicted from the same phase diagram.

Several other factors could be involved in the formation of these binary 3D superlattices. Moreover, the formation of quasicrystals and CaB_6 structures by the addition of large amount of surfactant [11, 33] has been observed. This was explained by the change, even in nonpolar solvents [5], of the surface electrical charges (either positive or negative) of the nanoparticles. All the data described in this chapter are obtained after washing the nanoparticles several times before deposition process. We know that the localized surface plasmon resonance (LSPR) is highly sensitive to Ag nanoparticle surface and consequently to the washing process. From that we identify, in the limit of the technique, the number of washing needed to keep the surface free of molecules involved in the syntheses. This permits to conclude that, in the present conditions, the formation of CaB_6 and quasicrystalline structures cannot be attributed to Coulombic interactions; van der Waals forces, as described previously [8], cannot be retained because these structures are produced with a rather large variety of nanoparticles differing by their Hamaker constant. The flexibility and the specific interactions of the ligand head groups with the nanocrystal surface atoms possibly play a major role. Previously [34], it was pointed out that deformability of the coating layer is attributed to the change in local coordination environment inducing a change in the packing mode of binary nanocrystal superlattices. However, here, the softness and the crystalline structure of oleic acid and OAM are similar, whereas the binary structures shown in Figures 7.8c and 7.16 markedly differ with the transition from NaCl-type structure to a large variety of structures such as quasicrystal, Archimedean tiling, and AlB_2 type, respectively. From that, the softness of the organic layer alone cannot be retained. All the experiments are done on TEM grid, indicating that the role of substrate is negligible. However, the role of evaporation time seems to be the key parameter. By changing the solvent used to disperse nanoparticles and keeping all the other experimental conditions constant, the evaporation time can be tuned. By replacing toluene with hexane, the evaporation time drops from four hours to five minutes. Let us mix 7.2-nm Ag nanoparticles coated with OAM to 4-nm Ag nanoparticles coated with either OAM or dodecanethiol and dispersed in hexane. At the end of the evaporation, whatever the coating agent of small Ag nanoparticles, same NaCl-type structures, as predicted, are produced. This can be explained in terms of ligand exchange processes during the superlattice growth processes and consequently of ligand–ligand interactions. Adsorption and desorption processes differ when the nanoparticles are coated with different types of ligands, and consequently, the ligand exchange takes place during the crystallization. Hence, the formation of binary superlattices with lower packing density may be favored kinetically because the required energetic penalty is smaller than that of a denser structure. The final structure strongly depends on the strength of attachment between the ligands and nanocrystal surface atoms. Hence, during solvent evaporation, one can assume that ligand exchange processes take place as illustrated in Scheme 7.1. Complete evaporation of the carrier solvent provides sufficient time for this ligand exchange process to occur in the colloidal solution. During the formation of nanocrystal superlattices, the system must pay a large energetic penalty to enable the ligand exchange process, which can affect the thermodynamics of the assembly of binary nanocrystal superlattices during solvent evaporation. Thus, the formation of a superlattice with lower packing density may be favored kinetically because the required energetic penalty is smaller than that of a denser structure. In other words, a binary structure is likely to be in a kinetically trapped state at a local free energy minimum in spite of its lower packing density.

The data presented here with Ag/Ag and Co_A/Co_A binary 3D superlattices seem to be highly consistent. However, several major questions remain open: (i) How can you explain that Co/Ag and Co/Au binary 3D superlattices of Co_A nanoparticles

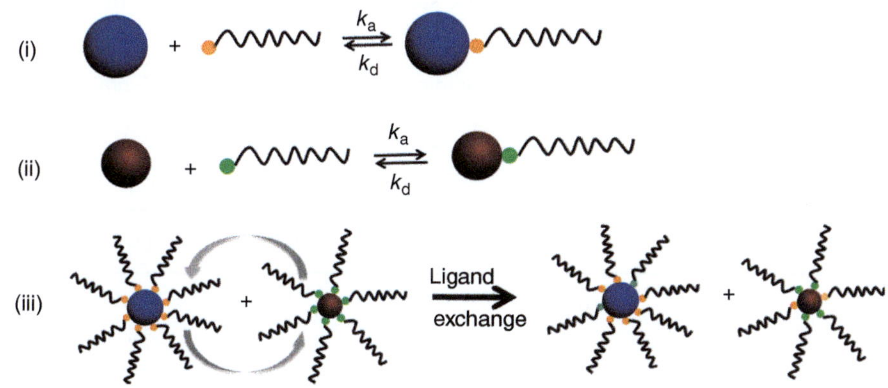

Scheme 7.1 (i) and (ii) The equilibriums of the bonding of nanocrystal ligand with different sizes and different coating agents by adsorption and desorption kinetic constants, k_a and k_d. (iii) Ligand exchange between nanocrystals with different surface coatings. *Source:* Wei et al. [17], scheme 1 (p. E)/American Chemical Society.

coated with oleic acid and Ag or Au [4] nanocrystals coated with dodecanethiol the binary 3D superlattices obtained agree with the hard-sphere model? (ii) Why does the concentration ratio between small and large nanoparticles is a key parameter for a good agreement with the hard-sphere model (see below)? [20].

7.2.2 Relative Concentration of the Small and Large Nanoparticles on the Binary 3D Superlattices

The prerequisite for applications is to evaluate and to improve the stability of such binary system. One of the major parameters is related to the relative number of small nanoparticles compared with the large ones involved during the deposition process [20]. This is well demonstrated by using binary systems composed of Co_A (7.2 nm) and Ag (4 nm) nanoparticles ($\gamma_{eff} = 0.59$) at various $[Co_A]/[Ag]$ concentrations (Table 7.1, Figures 7.2–7.6). For $[Co_A]/[Ag] = 1$, the domain sizes, d_s, of AlB_2-type binary superlattices are rather small (100 nm $< d_s <$ few μm), whereas large domains (up to 10 μm) of AlB_2-type binary superlattices are produced on increasing the Ag nanoparticles by a factor of 2, i.e. $[Co_A]/[Ag] = 1/2$. A further increase of Ag nanoparticle concentration to $[Co_A]/[Ag] = 1/4$ induces the formation of 30% of AuCu-type binary structures even though the AlB_2-type remains predominant (70%). Moreover, a further increase of Ag nanoparticle concentration to $[Co_A]/[Ag] = 1/8$ leads to $NaZn_{13}$-type lattices. On decreasing the $[Co_A]/[Ag]$ ratio to 1/12, single-component superlattices of Ag nanoparticles take over at the expense of $NaZn_{13}$-type structures, together with small domains of $NaZn_{13}$-type structures. Hence, on decreasing the concentration ratio $[Co_A]/[Ag]$, a structural transition from hexagonal AlB_2 to cubic $NaZn_{13}$, undergoing the metastable cubic AuCu-type structure, takes place at 25 °C with $\gamma = 0.54$ or 0.59 (Figure 7.17 and Table 7.1).

This clearly shows that the entropic force drives both AlB_2- and $NaZn_{13}$-type structures with a primitive lattice transition from the hexagonal to the cubic one. Hence, the entropic force has the power and rivals interparticle interactions between nanoparticles [7, 11, 12, 35]. To break the stable hexagonal symmetry, metastable

Table 7.1 Various binary superlattices self-assembled from 7.2-nm Co and 4.0-nm Ag nanoparticles ($\gamma = 0.59$).

	[Co]/[Ag]				
T_d (°C)	1	1/2	1/4	1/8	1/12
25	AlB_2	AlB_2	AuCu AlB_2	$NaZn_{13}$	$NaZn_{13}$ Ag
65	AlB_2	AlB_2	AuCu	AuCu $AuCu_3$ Ag	AuCu $AuCu_3$ Ag

Source: Yang et al. [20], table 1 (p. 2154)/American Chemical Society.

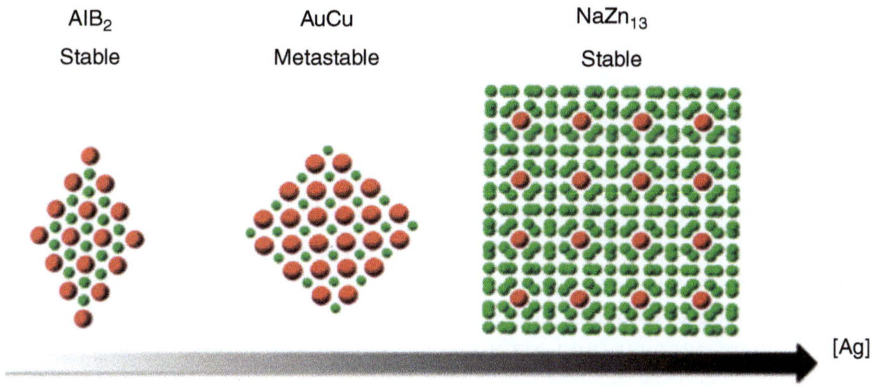

Figure 7.17 Structure transition with the change of the nanoparticle concentration ratio $[Co_A]/[Ag]$ and from a thermodynamically stable phase AlB_2-type to another stable phase $NaZn_{13}$-type binary structures observed. *Source:* Yang et al. [20]. Copyright 2015 American Chemical Society.

cubic AuCu-type structure appears. This clearly shows some discrepancies of the hard-sphere model. According to other data [5] on semiconductor binary superlattices, AuCu- and $AuCu_3$-type structures, predicted by the hard-sphere model, are considered as unstable phases, whereas, here, with size ratio of Co_A/Ag in the range of $0.49 \leq \gamma \leq 0.59$, these structures are produced. This could be explained by ligand exchange process even though here the ligand–surface atom interactions are rather strong. This allows us to conclude that the energy contribution induced by the ligand exchange processes, as already observed, cannot be neglected.

7.2.3 Temperature Effect

7.2.3.1 Influence on the Binary Phase Diagram

From the data already developed above, we know that γ value, coating agents, evaporation time, and nanocrystal concentration ratio are major parameters to control the binary crystalline structures. However, other parameters such as substrate temperature could play a role [20]. Similar experiments, as described above (7.2-nm Co_A and 4.0-nm Ag nanoparticles, $\gamma = 0.59$), are performed at 65 °C instead of 25 °C. Table 7.1 shows major changes when the amount of small nanoparticles increases, i.e. $[Co_A]/[Ag]$ decreases. For $1/2 \leq [Co_A]/[Ag] \leq 1$, the binary superlattices remain stable with the formation of AlB_2-type structures. The presence of AuCu-type binary 3D superlattices, with a packing density of ~0.61, is formed at $[Co_A]/[Ag] = 1/4$ and 25 °C. On increasing the temperature to 65 °C (Table 7.1), the AlB_2-type structure disappears, and only the AuCu-type structure remains due to an increase in the energetic contribution. At $[Co_A]/[Ag] = 1/8$ and 65 °C, AuCu- and $AuCu_3$-type structures appear instead of only $NaZn_{13}$-type structure at 25 °C. On a whole, at $\gamma = 0.59$, by changing the temperature from 25 to 65 °C, four types of Co_A/Ag binary structures, such as AuCu-, AlB_2-, $AuCu_3$-, and $NaZn_{13}$-type structures, are observed. Bodnarchuk et al. [5] reported similar results claiming that a low substrate temperature favors the formation of small nanoparticle assemblies, while the high-temperature phase, such as AuCu-type or $AuCu_3$-type structure, is formed.

7.2.3.2 Structural Transformation of Co_AAu_{13} Subjected to High Temperature

Let us consider 7.2-nm Co_A and 3.7-nm Au nanoparticles coated with oleic acid (C_{18}) and dodecanethiol (C_{12}), respectively, and dispersed in toluene. The relative nanoparticle concentration was kept constant and equal to $[Co_A]/[Au] = 1/10$. Figure 7.18a shows the TEM image of the $NaZn_{13}$-type (Co_AAu_{13}) binary nanoparticle superlattice, oriented along a ⟨100⟩ axis. The corresponding FFT pattern (inset Figure 7.18a) yields to a square symmetry and confirms the cubic lattice of the structure [7]. Note

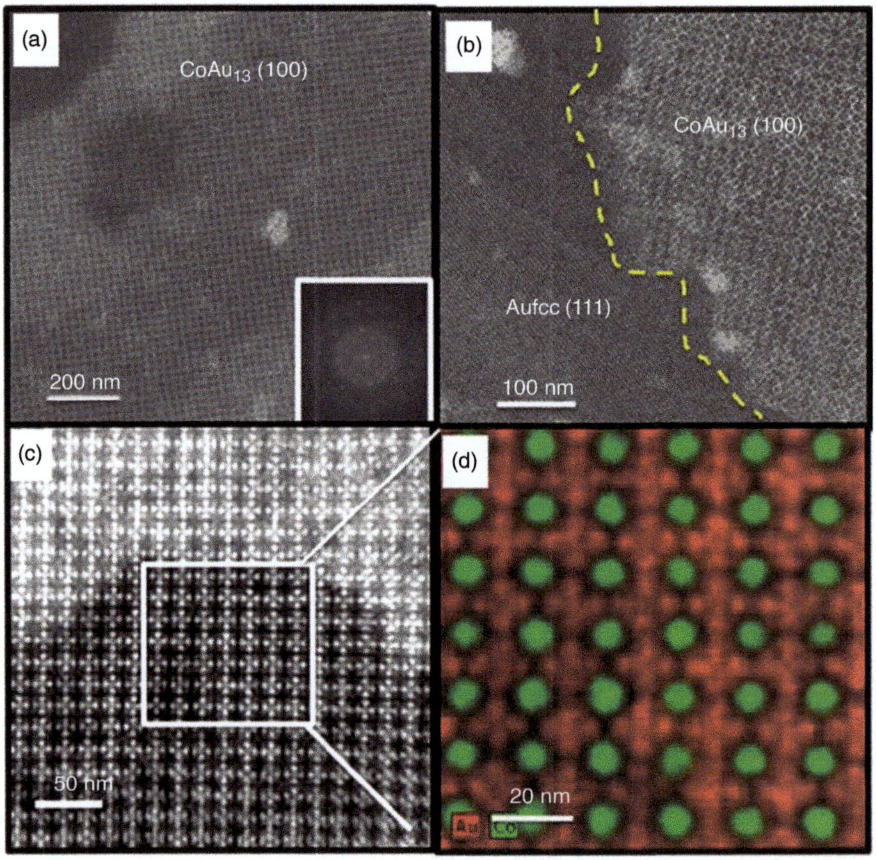

Figure 7.18 (a) TEM image of $NaZn_{13}$-type $CoAu_{13}$ binary nanoparticle superlattices. Inset: the FFT of the corresponding structure. (b) TEM image of the crystal boundary of single-component Au superlattices and $CoAu_{13}$ binary nanoparticle superlattices. (c) HAADF-STEM image of the $CoAu_{13}$ binary nanoparticle superlattice. (d) Superimposition of the elemental maps acquired from the region indicated by white square shown in (c). *Source:* Altantzis et al. [4]/American Chemical Society.

that both single-component Au nanoparticle assemblies and AlB_2-type structures can be observed in some regions, which cannot be avoided during the formation of $NaZn_{13}$-type binary nanoparticle superlattices with a large excess of small nanoparticles (Figure 7.18b). From the high-angle annular dark-field scanning transmission electron microscopy (HAADF-STEM) image (Figure 7.18c), it is clear that nanoparticles are highly ordered with the formation of $CoAu_{13}$ superlattices. The brighter and darker features in the image correspond to Au ($Z = 79$) and Co ($Z = 29$) nanoparticles, respectively. Au nanoparticles are periodically ordered in a cubic framework, whereas Co_A nanoparticles are located at the centers of the Au cubic framework, creating a simple cubic lattice. The superposition of the two chemical maps of a binary nanoparticle superlattice with Co_AAu_{13} structure is shown in Figure 7.18d.

The Co_AAu_{13} sample is subjected to thermal annealing at 200 °C for five minutes inside the TEM column [4]. The $CoAu_{13}$ binary nanoparticle superlattice can no longer be observed. The smaller 3.7-nm Au nanoparticles fuse into larger Au nanoparticles, whereas the 7.2-nm Co nanoparticles remain stable (Figure 7.19a–c). From the STEM-EDS (scanning transmission electron microscopy-energy dispersive X-ray spectroscopy) analysis, it can be concluded that the Co nanoparticles maintain their initial ordering into a simple cubic sublattice within the superlattice. The size of the Au nanoparticles evolved from 3.7 to 9.8 nm.

Hence, in $CoAu_{13}$ binary 3D superlattices, large Co nanoparticles form a simple cubic unit cell, which is filled with 13 Au nanoparticles, yielding an icosahedral nanoparticle cluster (Figure 7.20).

The HAADF-STEM images (Figure 7.21a) show that the fused Au nanoparticles are faceted. According to the theoretical prediction [36], the 3.7-nm Au nanoparticles partially fused, forming in this way adatoms at the nanoparticle surfaces. A partial or total fusion of the Au nanoparticles takes place with the appearance of either inhomogeneous or highly homogeneous contrast. The presence of twinning in the Au nanoparticles after fusion (Figure 7.21b) reveals their polycrystallinity. Hence, by heating at 200 °C for five minutes, the structure of binary 3D superlattices transits from $CoAu_{13}$ to CoAu with Co and Au nanocrystal sizes of 7.2 and 9.8 nm, respectively.

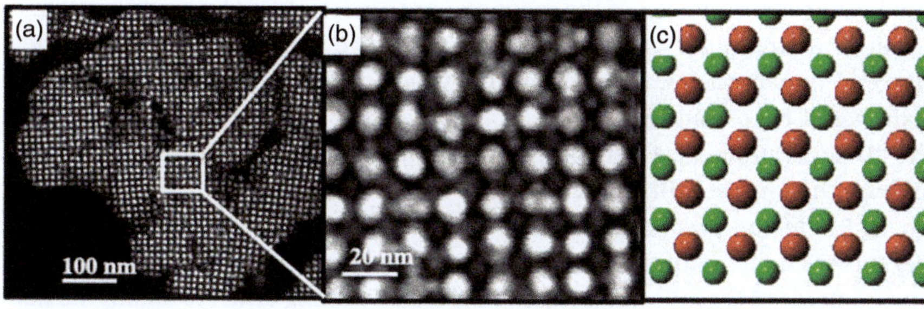

Figure 7.19 HAADF-STEM image of non-irradiated $CoAu_{13}$ binary nanoparticle superlattices, heat treated in-situ in the TEM column at 200 °C for 5 min; (b) magnified image of the region indicated by the red square in (a); (c) structure model. *Source:* Altantzis et al. [4]/American Chemical Society.

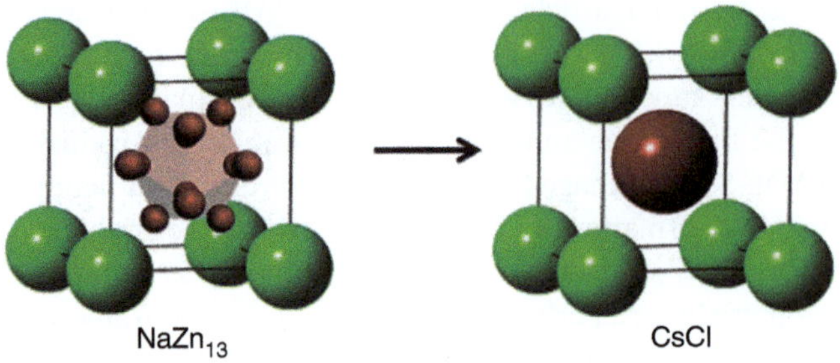

Figure 7.20 Schematic illustration of the structural transition from $NaZn_{13}$-type to CsCl-type binary nanoparticle superlattices. *Source:* Based on Wei et al. [18].

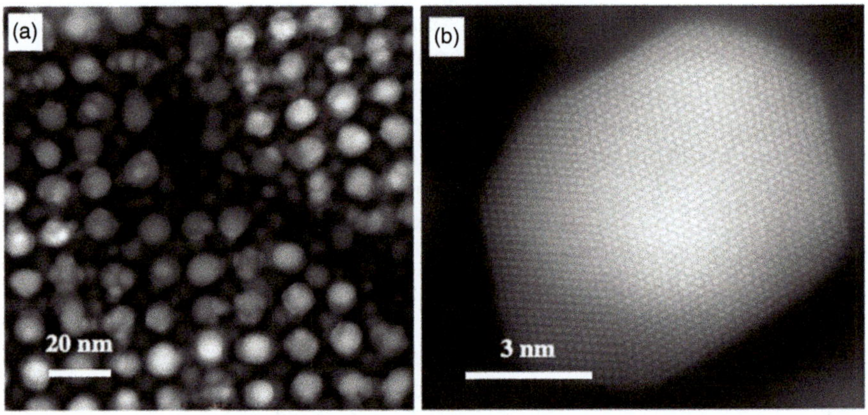

Figure 7.21 (a) High-magnification HAADF-STEM image of the Au nanoparticles within the CsCl-type binary nanoparticle superlattices. (b) High-resolution HAADF-STEM image of the Au particle. A twin defect is also present, revealing the polycrystallinity of the formed particles after fusion. *Source:* Altantzis et al. [4]/American Chemical Society.

7.2.3.3 Influence of the Magnetic Properties on Co/Ag Binary Systems

Let us look at the influence of the magnetic properties of Co/Ag binary system on the final structure(s). As we already mentioned above and in Chapter 1, by annealing the Co_A nanoparticles, their structures evolve from amorphous (CO_A) to single-domain hcp (Co_{HCP}) nanocrystals.

The 3D superlattice magnetic property as a function of temperature is measured as follows: the sample is first demagnetized at room temperature. Then, it is cooled down to 5 K in the absence of an external magnetic field. The zero-field-cooled (ZFC) curve is then monitored by heating the sample from 5 to 350 K in the presence of a probe field (10 G), followed by their successive cooling down to obtain the field-cooled (FC) curve. At low temperatures, the thermal fluctuations are smaller than the internal magnetic energy barriers, magnetizations are frozen along the easy axes of nanoparticles, and the system is in a blocked state. By increasing the temperature, the thermal fluctuations increase, and the magnetization of individual particles can

Table 7.2 Co nanoparticle self-assembled in fcc 3D superlattices: blocking temperature of amorphous (Co_A) and hcp (Co_{HCP}) Co nanoparticles differing by their nanoparticle diameter.

Crystalline structure	Co_A	Co_A	CO_{HCP}	Co_A	Co_{HCP}
Size (nm)	4.6	7.2	7.2	9.2	9.3
Blocking temperature (K)	50	185	250	250	325

Source: Based on Yang et al. [21].

align themselves, gradually, along the applied probe field. The increasing of the magnetization continues on the ZFC curve up to a maximum known to be the average blocking temperature of the sample, labeled as T_b. Above T_b, the nanoparticles are submitted to thermal fluctuations transiting them toward a superparamagnetic phase. In strongly interacting superparamagnetic particle systems, the blocking temperature may increase by the coupling between nanoparticles [37] because the dipole–dipole interaction can be considered as an extra energy barrier adding to the anisotropy energy, which consequently hinders thermal fluctuations of the spins. The energy of the dipole–dipole interaction between two magnetic moments separated by a distance can be calculated [38, 39]. Here, we measure the blocking temperature of Co nanoparticles self-assembled in fcc 3D superlattices. Table 7.2 shows that the blocking temperature of Co 3D superlattices increases with the Co nanoparticle size and nanocrystallinity (from amorphous to single-domain hcp phase).

From Table 7.2, we know that fcc 3D superlattices of Co nanoparticles differing by their sizes and nanocrystallinity (amorphous or hcp) are in the superparamagnetic state at 25 °C for 7.2 nm differing by their nanocrystallinity and for amorphous 9.2-nm Co_A nanoparticles. Conversely, 9.3-nm Co_{HCP} nanocrystals are ferromagnetic at 25 °C and reach superparamagnetic regime at 52 °C.

With 7.2-nm Co_A and 4.0-nm Ag nanoparticles ($\gamma \neq 0.59$), AlB_2-type structure is produced as described above and shown in Figures 7.3 and 7.22a,b. By replacing amorphous Co nanoparticles with their hcp counterparts, still superparamagnetic at room temperature (Table 7.2), Figure 7.22c,d shows similar AlB_2 3D superlattices [21]. This clearly indicates that the change of the crystalline structure from amorphous to hcp crystalline structure of superparamagnetic Co nanoparticles does not play any role on the AlB_2 final crystalline structure. With 9.2-nm Co_A and 4-nm Ag nanoparticles ($\gamma = 0.49$), NaCl-type structure is obtained (Figure 7.22e,f). These results clearly indicate that the nanocrystallinity of 7.2-nm Co nanoparticles does not play any role during the growth of binary 3D superlattices at 25 °C, for which both the amorphous- and hcp-phase Co nanoparticles are in a superparamagnetic state [21].

By replacing 9.2-nm Co_A nanoparticles with their hcp counterpart, keeping the same experimental condition, marked changes in the binary systems are observed. Figure 7.23b shows TEM images of the asymmetric ordering of the nanoparticles. The FFT pattern (inset Figure 7.23a) with 12-fold symmetry confirms the dodecagonal quasicrystalline ordering [5, 15, 20]. In a careful study of the TEM grid, small fractions of $CoAg_{13}$ ($NaZn_{13}$-type structures) are also observed in Figure 7.5b. In the

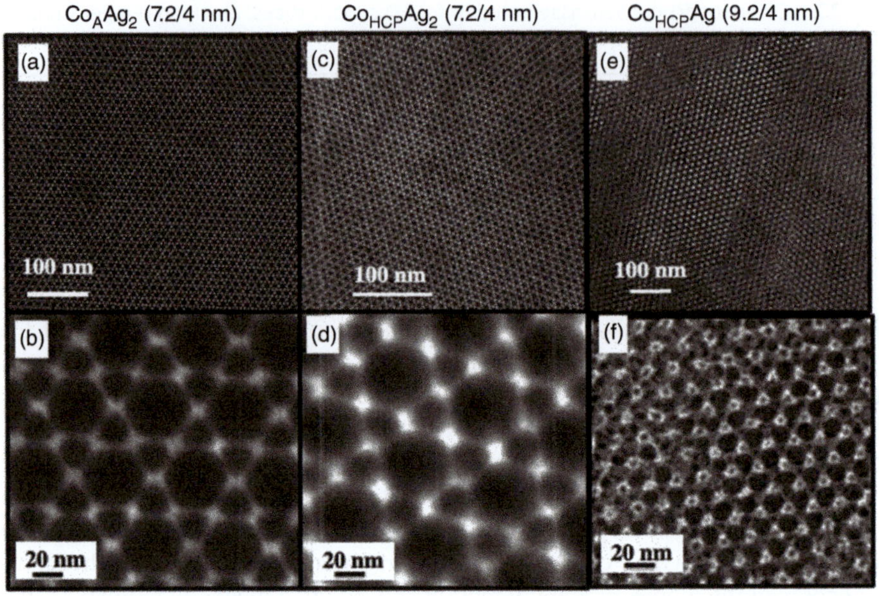

Figure 7.22 TEM and HRTEM images of binary nanocrystal superlattices assembled from Co nanoparticles differing by their crystallinity and 4-nm Ag nanoparticles: (a, b) 7.2-nm CO_A/Ag, (c, d) 7.2-nm CO_{HCP}, and (e, f) 9.2-nm Co_A nanoparticles. (a–d) ALB_2-type structures and (e, f) NaCl-type binary systems. [Co]/[Ag] = 1/2. *Source:* Zhijie Yang 2015, p. 137/ American Chemical Society.

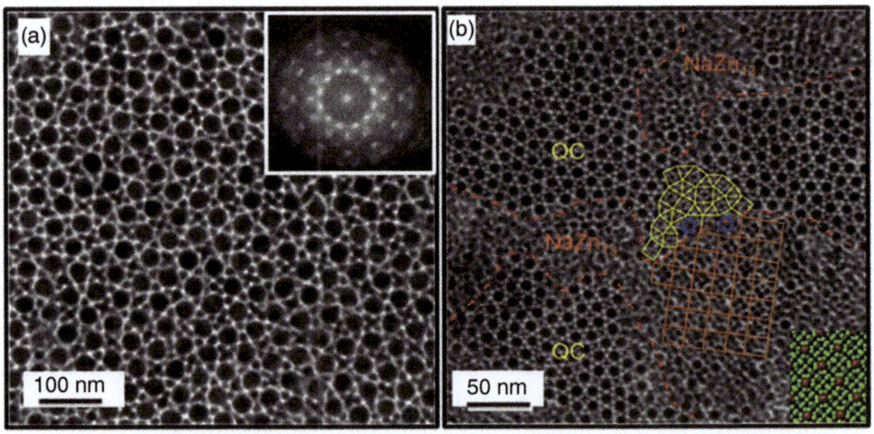

Figure 7.23 (a) TEM images of quasicrystalline ordering assembled from 9.3-nm Co_{HCP} nanoparticles and 4-nm Ag nanoparticles (inset) FFT pattern of the quasicrystalline superlattice. (b) Transition from the quasicrystalline phase to the $NaZn_{13}$-type binary superlattice in the presence of the point defects (cycles). *Source:* Zhijie Yang 2015, p. 137/ American Chemical Society.

Table 7.3 3D superlattices of binary systems differing by their compositions and nanocrystallinities.

Materials	Co_ACo_A	Co_AAg	Co_AAg	Co_AAg	Co_{HCP}/Ag
Crystalline structure	Co_ACo_{A2}	Co_AAg_2	Co_AAg_{13}	Co_AAg	QC $Co_{HCP}Ag_6$ $Co_{HCP}Ag_{13}$
Size (nm)	7.2/4.6	7.2/4	7.2/4	9.2/4	9.3/4
Blocking temperature (K)	120	166	145	80	300

Source: Based on Yang et al. [21].

$CoAg_{13}$-type structure, a small 9.3-nm Co nanocrystal is surrounded by 12 other 4.0-nm Ag nanoparticles, each at the vertices of a slightly distorted icosahedron, and these icosahedral particle clusters are inserted into the cubic lattices of large 9.3-nm Co nanoparticles as observed previously [3]. Hence, the straight lines connecting the centers of Co_{HCP} nanoparticles in the $CoAg_{13}$ structures result in periodic squares (Figure 7.4b).

Note that the small domains with $Co_{HCP}Ag_{13}$ structures are isolated from the quasicrystal domains with smooth interfaces, which is not frequently discerned. At the interface between quasicrystals and $CoAg_{13}$ 3D superlattices, point defects are commonly observed (Figure 7.23). The data described above clearly show an abrupt difference in the binary superlattices with magnetic state. With thick binary systems, the blocking temperature of 9.3-nm Co 3D superlattices is 325 K [21]. Due to the fact that the distance between Co nanoparticles in binary superlattices (5-nm) increases compared with what is observed with single component (3 nm) [40], the blocking temperature is slightly smaller (300 K) for binary than for single-component 3D superlattices (Table 7.3).

Nevertheless, with Co_A/Ag 3D superlattices, the system is superparamagnetic at room temperature, whereas it is ferromagnetic with Co_{HCP}/Ag structure. The same experiments, as described above, are performed at 338 K instead of 298 °C. In this case, both amorphous and hcp 9.3-nm Co nanoparticles are in the superparamagnetic state. A mixed colloidal solution of Co (amorphous or hcp) and Ag nanoparticles is heated to 338 K and deposited at this temperature. For the 9.2-nm Co_A nanoparticles, Co_AAg_3 ($AuCu_3$-type) and Co_AAg_2 (AlB_2-type) binary nanocrystal superlattices (Figure 7.24a,c) are the dominant structures instead of the NaCl type at 298 K. For 9.3-nm hcp Co nanoparticles, a $Co_{HCP}Ag_3$ ($AuCu_3$-type) structure (Figure 7.24b) is the main structure with small domains of $Co_{HCP}Ag_2$ (AlB_2 type), as shown in Figure 7.24d. Note that small domains (not shown here) of $Co_{HCP}Ag_4$ (AB_4-type structures) are obtained. It needs to be pointed out that neither the quasicrystalline order nor the $CoAg_{13}$ ($NaZn_{13}$-type) structure is observed after careful TEM analysis. Hence, the loss of the ferromagnetic state in 9.3-nm hcp Co nanocrystal by increasing the temperature above T_b would result in the driving force in the formation of binary structures of the same magnitude, comparing the hcp phase with Co_A nanoparticles. At this stage, it is demonstrated that Co nanoparticle

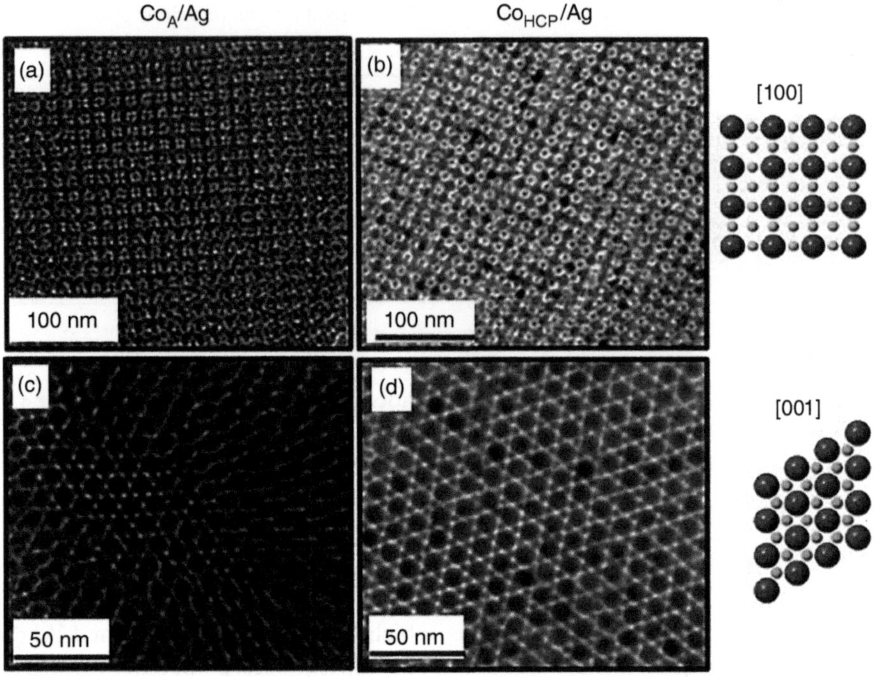

Figure 7.24 TEM images of binary nanocrystal superlattices assembled from 9.2-nm Co_A nanoparticles/9.3-nm Co_{HCP} nanocrystals and 4-nm Ag nanoparticles: (a) Co_AAg_3, (b) $Co_{HCP}Ag_3$, (c) Co_AAg, and (d) $Co_{HCP}Ag$ binary structures. *Source:* Zhijie Yang 2015, p. 137/ American Chemical Society.

magnetic state (superpara- or ferromagnetic) plays a crucial role during the growth of binary 3D superlattices. Hence, in the superparamagnetic regime of Co nanoparticles, the binary structures remain quasi-unaltered with the change of nanocrystallinity of Co nanoparticles, whereas quasicrystalline ordering is observed with ferromagnetic Co nanoparticles.

7.2.4 Unexpected Behavior Induced by Mixing Small and Large Nanoparticles

The crystalline structures of nanoparticles determine the final superstructures: hence, by mixing two colloidal solutions of 8.4-nm *single-domain* Co (ε-phase) nanoparticles coated with oleic acid and 5.4-nm single-domain Au nanoparticles coated with dodecanethiol at the end of the solvent evaporation, Au nanoparticles are concentrated into polyhedral (i.e. truncated tetrahedron, distorted square, or rhombus-like) 3D superlattices embedded in 3D superlattice film of Coε nanoparticles [41]. High-resolution scanning electron microscopy (HRSEM) images of polyhedral nanocrystal superlattice surfaces exhibit complex periodical pattern characteristic of vicinal surfaces. The crystallographic planes that build the terraces differ from one polyhedral 3D superlattice to another. By replacing *single-domain*

Au nanoparticles with their *polycrystalline* counterparts, keeping single-domain Coε nanoparticles, segregation is observed between Au and Co 3D superlattices with low energy surface. Obviously, the influence of nanocrystallinity during their superlattice formation plays a major role.

7.3 Solvent-Mediated Crystallization of Nanocrystal 3D Assemblies of Silver Nanocrystals: Unexpected Superlattice Ripening

Solvent–ligand interactions in colloidal nanocrystals are of significant importance as they can be used to modulate the way they pack into superlattices [18]. We demonstrate that the crystal structures of the nanocrystal superlattices made of 2.2-nm Ag nanocrystals can be controlled by using different carrier solvents. Specifically, the superlattices structures are tuned from body-centered cubic (bcc) to fcc when varying solvents from hexane to tetrachloroethylene (TCE). Furthermore, by simultaneously annealing these two samples at different temperatures, bcc structures originating from hexane solutions are dominated by simple coalescence mechanism while fcc structure stemming from TCE solutions undergoes Ostwald ripening process that can produce a variety of binary nanocrystal superlattices such as NaCl, AlB_2, $NaZn_{13}$, and $MgZn_2$, the formation of those structures being well explained by a pure entropy-driven process. This is believed to be due to variations in the ligand coverage ratio of the nanocrystals in different solvents that are changing the superlattice structure stability. Those findings provide insights into the solvent-mediated nanocrystal superlattices and the Ostwald ripening process in nanocrystal superlattices.

7.4 Collective Properties Involved in Self-Assemblies of Binary 3D Superlattices

Unfortunately, less effort has been devoted to studies on the collective physical properties of these binary systems [15, 19, 23]. Small magnetic nanoparticles act like "super" spin and, in a 3D superlattice due to dipolar interactions, tend to align their magnetic moments in the direction of the local magnetic field. In either Fe_3O_4/Fe_3O_4 or Fe_3O_4/FePt binary systems, collective magnetic dipolar interactions were observed. However, the role of small magnetic nanoparticles within magnetic binary 3D superlattices in the collective magnetic coupling is still unknown and needs to be further studied.

Let us consider 7.2-nm Co_A as large nanoparticles and either 4-nm Ag or 4.6-nm Co_A as small nanoparticles. Figures 7.3 and 7.6 show the formation of long-range (with domain sizes up to 5 μm) binary systems with AlB_2- and $MgZn_2$-type 3D superlattices for $Co_A(Ag)_2$ and $Co_A(Co_A)_2$, respectively. The interparticle distances between large Co_A nanoparticles are 4.8 and 4.2 nm, respectively. The position of the observed peak in the ZFC curve corresponds to the average magnetic blocking temperature (T_b) of the nanoparticles. Very surprisingly, Table 7.3 and Figure 7.25 show that the blocking temperatures are lower for $Co_A(Co_A)_2$ (121 K) than $Co_A(Ag)_2$ (166 K).

The present data clearly show that, due to their nonmagnetic behavior, Ag nanoparticles act as a spacer to tune the interparticle distance. Hence, the dipolar coupling between Co nanoparticles controls, as expected, the magnetic properties of the 3D superlattices through modulation of the interparticle distance. The average distances between 7.2-nm Co_A nanoparticles in the Co_ACo_{A2} (4.2-nm) and Co_AAg_2 (4.8-nm) binary systems are larger than that determined for the single-component 7.2-nm Co_A 3D superlattices (3.4 nm). This could explain the decrease in the blocking temperature of binary 3D superlattices compared with single-component 3D superlattices (Tables 7.2 and 7.3). However, the blocking temperature of Co_ACo_{A2} binary 3D superlattices is smaller than that of Co_AAg_2 ones (Figure 7.25). Here, the interparticle distance cannot be the major factor controlling the blocking temperature value (through the dipole–dipole interaction) because these two binary 3D superlattices have close Co_A/Co_A interparticle distances (4.2 and 4.8 nm, respectively). The major difference between these two binary 3D superlattices lies in the fact that the small nanoparticles are magnetic or not. Consequently, this shows that other parameters such as the coexistence of two magnetic sublattices with different nanoparticle sizes can affect the magnitude and symmetry of the magnetic energy barriers and reduce the blocking temperature. Let us consider Co_AAg_{13} produced by the same nanoparticle sizes (7.2 and 4 nm, respectively) as shown in Figure 7.5. Here, the interparticle distance between Co_A nanoparticles is 8.9 nm. Figure 7.25 shows the ZFC/FC magnetization curves of Co_AAg_{13} binary 3D superlattices. The

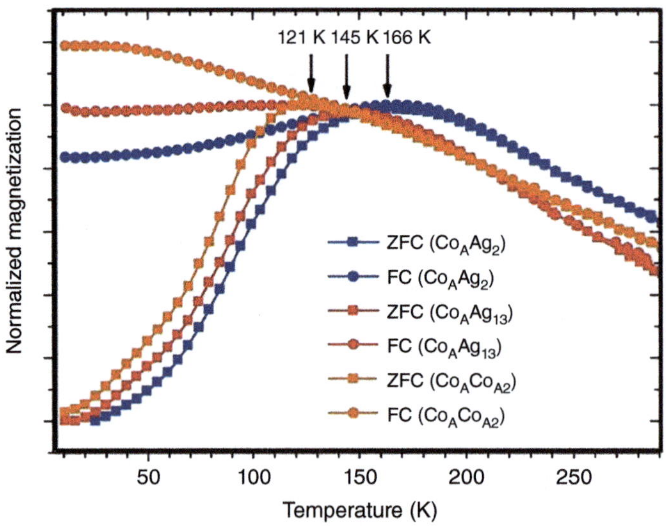

Figure 7.25 Zero-field-cooled (ZFC) and field-cooled (FC) magnetizations versus temperature curves measured at 10 G for $CoCo_2$ (orange) and $CoAg_2$ (blue) and $CoAg_{13}$ binary 3D superlattices (red). *Source:* Yang et al. [21], figure 4b (p. 4911)/John Wiley & Sons.

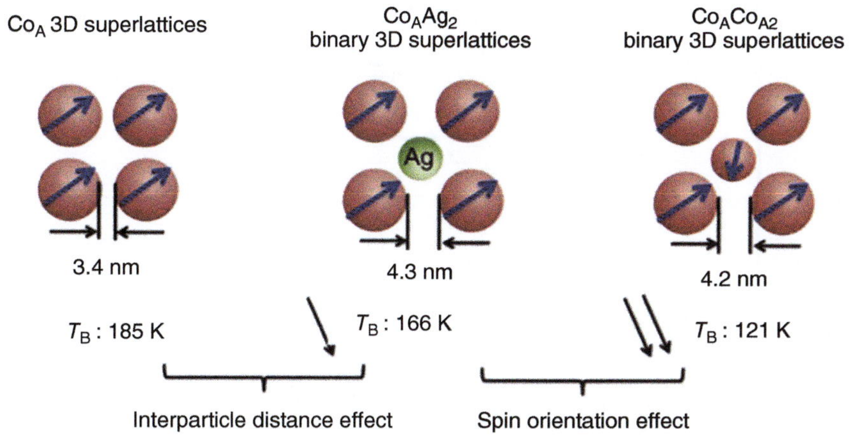

Scheme 7.2 Various crystal models of binary nanoparticles superlattices. (a) NaCl, (b) CuAu, (c) AlB_2, (d) $MgZn_2$, (e) Cu_3Au, (f) Fe_4C, (g) CaB_6, (h) $NaZn_{13}$ types. *Source:* Yang et al. [21], scheme 1 (p. 4912)/John Wiley & Sons.

blocking temperature of Co_AAg_{13} 3D superlattices (145 K) is higher than that of Co_ACo_{A2} binary 3D superlattices (121 K), whereas the interparticle distance between large (7.2 nm) Co_A nanoparticles is smaller (4.2 nm). The rather low blocking temperature and the smoothly increase in the FC curve on decreasing temperature observed with Co_ACo_{A2} 3D superlattices clearly indicate a decrease in the magnetic interactions compared with binary systems produced with nonmagnetic small (Ag) associated to the large Co_A nanoparticles. Hence, Co_ACo_{A2} 3D superlattices are characterized by weaker dipolar interactions than their counterpart as Co_AAg_2 and Co_AAg_{13} 3D superlattices. These data clearly show that the magnetic dipolar interactions in 7.2-nm Co_A 3D superlattices are tunable by engineering the Co_A–Co_A interparticle distance with the formation of Co_A/Ag binary 3D superlattices. For the Co_AAg_{13} binary 3D superlattices, a larger interparticle distance between Co_A nanoparticles results in reduced dipolar interactions, leading to a lower blocking temperature of the sample. Furthermore, a decrease in the blocking temperature is observed for AlB_2 binary 3D superlattices by replacing 4.6-nm Co nanoparticles with nonmagnetic Ag nanoparticles. This suggests that in the Co_ACo_{A2} binary 3D superlattices, the spin directions between the large and small Co_A nanoparticles differ. The dipolar interactions in the system are reduced as shown in Scheme 7.2.

7.5 Conclusion

In this chapter, we showed that by mixing two uniform nanoparticles differing by their diameters, binary crystalline structures are produced. The larger nanoparticles are self-assembled. The smaller ones fill into the interstices of the larger. These structures of silica microparticles exist in Mother Nature in the most stable phases. Here, we self-assemble two types of metal nanoparticles having a very low size

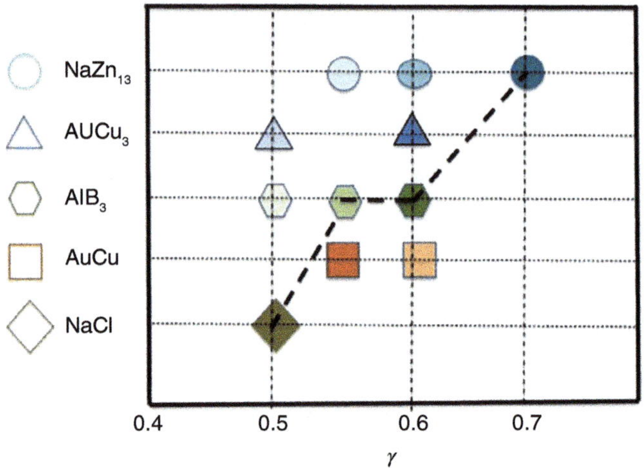

Figure 7.26 Various crystalline structures of binary systems obtained. The dashed line corresponds to the agreement between experimental data and hard-sphere model. *Source:* Yang et al. [20], figure 6 (p. 2155)/American Chemical Society.

distribution and highly cleaned to get rid of impurities in the experimental limits. Various binary crystalline phases (Figure 7.26) are produced. They mainly depend on the relative ratio between large and small particle sizes, γ, as predicted by a hard-sphere model. In some experimental conditions, a good agreement between the calculated model and experiments is obtained. However, a rather large number of discrepancies can be pointed out. The relative concentration between small and large nanoparticles, coating agent, temperature, crystalline structure of the nanoparticles, and magnetic properties play major roles. Note that some discrepancies cannot be explained yet. Furthermore, unexpected highly stable 3D superlattice surfaces can be produced as obtained in bulk atomic crystals. The latter are unstable. Finally, we described some intrinsic properties when small magnetic nanoparticles reduce the dipolar interactions between the larger ones.

References

1 Eldridge, M.D., Madden, P.A., and Frenkel, D. (1993). Entropy-driven formation of a superlattice in a hard-sphere binary mixture. *Nature* 365: 35–37.
2 Ohara, P.C., Leff, D.V., Heath, J.R., and Gelbart, W.M. (1995). Crystallization of opals from polydisperse nanoparticles. *Phys. Rev. Lett.* 75: 3466–3469.
3 Shevchenko, E.V., Talapin, D.V., O'Brien, S., and Murray, S.B. (2005). Polymorphism in AB_{13} nanoparticle superlattices: an example of semiconductor–metal metamaterials. *J. Am. Chem. Soc.* 127: 8741–8747.
4 Altantzis, T., Yang, Z., Bals, S. et al. (2016). Thermal stability of $CoAu_{13}$ binary nanoparticle superlattices under the electron beam. *Chem. Mater.* 28: 716–719.

5 Bodnarchuk, M.I., Kovalenko, M.V., Heiss, W., and Talapin, D.V. (2010). Energetic and entropic contributions to self-assembly of binary nanocrystal superlattices: temperature as the structure-directing factor. *J. Am. Chem. Soc.* 132: 11967–11977.

6 Boneschanscher, M.P., Evers, W.H., Qi, W.J. et al. (2013). Electron tomography resolves a novel crystal structure in a binary nanocrystal superlattice. *Nano Lett.* 13: 1312.

7 Chen, J., Ye, X., and Murray, C.B. (2010). Systematic electron crystallographic studies of self-assembled binary nanocrystal superlattices. *ACS Nano* 4: 2374–2381.

8 Evers, W.H., Nijs, B.D., Filion, L. et al. (2010). Entropy-driven formation of binary semiconductor-nanocrystal superlattices. *Nano Lett.* 10: 4235–4241.

9 Kang, Y., Ye, X., Chen, J. et al. (2013). Engineering catalytic contacts and thermal stability: gold/iron oxide binary nanocrystal superlattices for CO oxidation. *J. Am. Chem. Soc.* 135: 1499–1505.

10 Redl, F.X., Cho, K.S., Murray, C.B., and O'Brien, S. (2003). Three-dimensional binary superlattices of magnetic nanocrystals and semiconductor quantum dots. *Nature* 423: 968.

11 Shevchenko, E.V., Talapin, D.V., Kotov, N.A. et al. (2006). Structural diversity in binary nanoparticle superlattices. *Nature* 439: 55–59.

12 Shevchenko, E.V., Talapin, D.V., Murray, S.B., and O'Brien, S. (2006). Structural characterization of self-assembled multifunctional binary nanoparticle superlattices. *J. Am. Chem. Soc.* 128: 3620–3637.

13 Shevchenko, E.V., Ringler, M., Schwemer, A. et al. (2008). Self-assembled binary superlattices of CdSe and au nanocrystals and their fluorescence properties. *J. Am. Chem. Soc.* 130: 3274–3275.

14 Talapin, D.V., Shevchenko, E.V., Bodnarchuk, M.I. et al. (2009). Quasicrystalline order in self-assembled binary nanoparticle superlattices. *Nature* 461: 964–967.

15 Urban, J.J., Talapin, D.V., Shevchenko, E.V. et al. (2007). Synergism in binary nanocrystal superlattices leads to enhanced p-type conductivity in self-assembled $PbTe/Ag_2Te$ thin films. *Nat. Mater.* 6: 115–121.

16 Vanmaekelbergh, D. (2011). Self-assembly of colloidal nanocrystals as route to novel classes of nanostructured materials. *Nano Today* 6: 419.

17 Wei, J., Schaeffer, N., and Pileni, M.P. (2015). Ligand exchange governs the crystal structures in binary nanocrystal superlattices. *J. Am. Chem. Soc.* 137: 14773–14784.

18 Wei, J., Schaeffer, N., and Pileni, M.P. (2016). Solvent-mediated crystallization of nanocrystal 3D assemblies of silver nanocrystals: unexpected superlattice ripening. *Chem. Mater.* 28: 293–302.

19 Yang, Z., Wei, J., Bonville, P., and Pileni, M.P. (2015). Beyond entropy: magnetic forces induce formation of quasicrystalline structure in binary nanocrystal superlattices. *J. Am. Chem. Soc.* 137: 4487.

20 Yang, Z., Wei, J., and Pileni, M.P. (2015). Metal–metal binary nanoparticle superlattices: a case study of mixing Co and Ag nanoparticles. *Chem. Mater.* 27: 2152–2157.

21 Yang, Z., Wei, J., Bonville, P., and Pileni, M.P. (2015). Engineering the magnetic dipolar interactions in 3D binary supracrystals via mesoscale alloying. *Adv. Funct. Mater.* 25: 4908.

22 Dong, A., Chen, J., Ye, X. et al. (2011). Enhanced thermal stability and magnetic properties in NaCl-type FePt-MnO binary nanocrystal superlattices. *J. Am. Chem. Soc.* 133: 13296–13299.

23 Ye, X., Chen, J., Diroll, B.T., and Murray, C.B. (2013). Tunable plasmonic coupling in self-assembled binary nanocrystal superlattices studied by correlated optical microspectrophotometry and electron microscopy. *Nano Lett.* 13: 1291.

24 Xue, Y., Li, X., Li, H., and Zhang, W. (2014). Quantifying thiol–gold interactions towards the efficient strength control. *Nat. Commun.* 5: 4348.

25 Jin, R., Zeng, C., Zhou, M., and Chen, Y. (2016). Atomically precise colloidal metal nanoclusters and nanoparticles: fundamentals and opportunities. *Chem. Rev.* 116: 10346–10413.

26 Grönbeck, H., Curioni, A., and Andreoni, W. (2000). Why gold nanoparticles self-assemble on sulfhydryl (SH) monolayers? *J. Am. Chem. Soc.* 122: 3839–3842.

27 Legrand, J., Petit, C., Bazin, D., and Pileni, M.P. (2000). Collective effect on magnetic properties of 2D superlattices of nanosized cobalt. *Appl. Surf. Sci.* 164: 186–192.

28 Hiramatsu, H. and Osterloh, F.E. (2004). A simple large-scale synthesis of nearly monodisperse gold and silver nanoparticles with adjustable sizes and with exchangeable surfactants. *Chem. Mater.* 16: 2509–2511.

29 Jun, Y.W., Seo, J.W., and Cheon, J. (2008). Nanoscaling laws of magnetic nanoparticles and their applicabilities in biomedical sciences. *Acc. Chem. Res.* 41: 179.

30 Smith, D.K., Goodfellow, B., Smilgies, D.M., and Korgel, B.A. (2009). Self-assembled simple hexagonal AB_2 binary nanocrystal superlattices: SEM, GISAXS, and defects. *J. Am. Chem. Soc.* 131: 3281–3290.

31 Wei, J., Schaeffer, N., and Pileni, M.P. (2015). Surface plasmon resonance properties of silver nanoparticles differing in size and coating agent ordered in 3D supracrystals. *Chem. Mater.* 28: 293–302.

32 Talapin, D.V., Lee, J.S., Kovalenko, M.V., and Shevchenko, E.V. (2010). Prospects of colloidal nanoparticles for electronic and optoelectronic applications. *Chem. Rev.* 110: 389–458.

33 Sun, Z., Luo, Z., and Fang, J. (2010). Assembling nonspherical 2D binary nanoparticle superlattices by opposite electrical charges: the role of Coulomb forces. *ACS Nano* 4: 1821–1828.

34 Boles, M.A. and Talapin, D.V. (2015). Many-body effects in nanocrystal superlattices: departure from sphere packing explains stability of binary phases. *J. Am. Chem. Soc.* 137: 4494–4502.

35 Dong, A., Chen, J., Vora, P.M. et al. (2010). Binary nanocrystal superlattice membranes self-assembled at the liquid–air interface. *Nature* 466: 474–477.

36 Shim, J.H., Lee, B.J., and Cho, Y.W. (2002). Thermal stability of unsupported gold nanoparticle: a molecular dynamics study. *Surf. Sci.* 512: 262–268.

37 Gross, A.F., Diehl, M.R., Beverly, K.C. et al. (2003). Controlling magnetic coupling between cobalt nanoparticles through nanoscale confinement in hexagonal mesoporous silica. *J. Phys. Chem. B* 107: 5475.

38 García-Otero, J., Porto, M., Rivas, J., and Bunde, A. (2000). Influence of dipolar interaction on magnetic properties of ultrafine ferromagnetic particles. *Phys. Rev. Lett.* 84: 167.

39 Du, J., Zhang, B., Zheng, R.K., and Zhang, X.X. (2007). Memory effect and spin-glass-like behavior in Co–Ag granular films. *Phys. Rev. B* 75: 014415.

40 Yang, Z., Cavalier, M., Walls, M. et al. (2012). A phase-solution annealing strategy to control the cobalt nanocrystal anisotropy: structural and magnetic investigations. *J. Phys. Chem. C* 116: 15723–15730.

41 Goubet, N., Yang, J., Albouy, P.A., and Pileni, M.P. (2014). Spontaneous formation of high-index planes in gold single domain nanocrystal superlattices. *Nano Lett.* 14: 6632–6638.

8

Analogy Between 3D Lattices and Atomic Crystals: Crystalline Structure

Here, we would like to answer the following general question: Does an analogy exist between atoms and nanocrystals when they are ordered in crystalline structure? To answer this, we have to take into account the difference in scale between atoms and nanocrystals and consequently in crystalline array scales. Also, we will share with the reader some unexpected data that could permit to demonstrate a universal behavior in crystal growth processes.

8.1 Atomic Crystals, Shaped 3D Lattices, and Minerals

Here we concentrate on analogy between atomic and nanoparticles crystals [1, 2]. We know that the arrangement in fcc structure is a dense packing of tetrahedral subunits. In fcc arrangements, *single* crystals are cuboctahedra (Figure 8.1a) and truncated octahedra resulting from different growth rates between the {111} and {100} facets. Multiply-twinned particles (MTPs) like decahedra (Figure 8.1b) and icosahedra (Figure 8.1c) are formed by assembling 5 and 20 *single*-crystalline fcc tetrahedral subunits, respectively. At the earliest stages of a solid's growth, the atoms reorganize into a completely new structure with a preferred symmetry to form clusters. Further growth takes place by adding layers of atoms to this frozen core. Layered growth imposes certain restrictions on the outer symmetry or morphology of the clusters. This was demonstrated for clusters having large binding energies and resistance to the evaporation of atoms [3]. A rather large number of metallic clusters are produced during the last three decades under chemical vacuum deposition (CVD) [4, 5]. Figure 8.1d–f shows copper clusters produced by ultravacuum technique with shapes similar to the scheme (Figure 8.1a–c). By using reverse micelles (Chapter 1), similar structures are produced with formation of cuboctahedral (Figure 8.1d), decahedral (Figure 8.1e), and icosahedral (Figure 8.1f) silver atomic nanocrystals [1].

In Chapters 3 and 14, it is shown that either at the end of the solvent evaporation of Au nanocrystals dispersed in toluene or by keeping Au nanocrystals under saturated atmosphere, a large number of fcc 3D lattices with well-defined morphologies

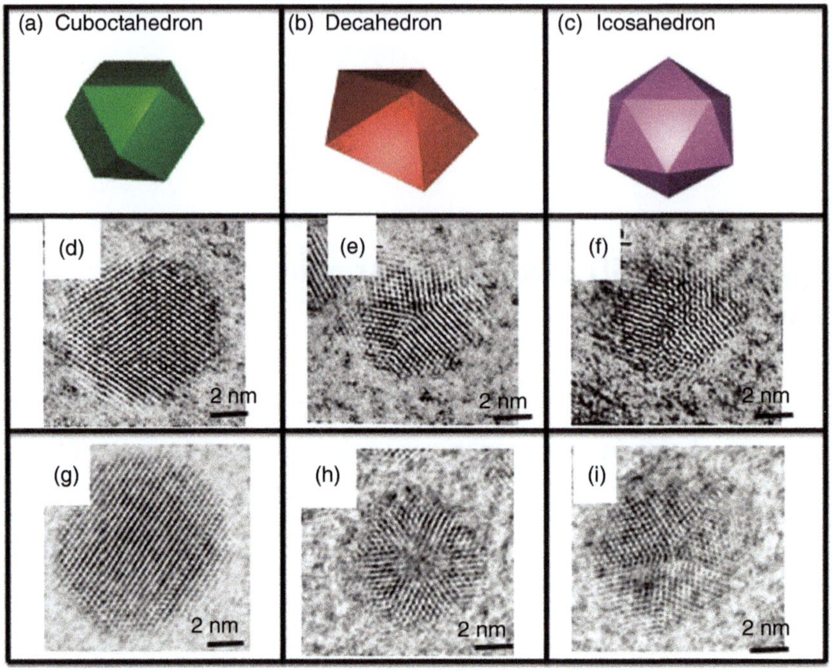

Figure 8.1 Scheme of cuboctahedral (a), decahedral (b), and icosahedral (c) atomic assemblies. (d, e, f) TEM (Transmission Electron Microscopy) images of copper clusters produced by chemical vapor deposition having cuboctahedral (d), decahedral (e), and icosahedral (f) shapes. (g, h, i) TEM images of silver nanocrystals produced from water-in-oil droplets called reverse micelles with cuboctahedral (g), decahedral (h), and icosahedral (i) shapes. *Source:* Pileni [1]/American Chemical Society.

are produced. Icosahedral shaped 3D lattices terminated by (111) planes are obtained (Figure 8.2a). This morphology corresponds to MTP crystals like those obtained at the nanometer scale for nanoparticles called MTP. Octahedral *single* Au 3D lattice of 5-nm nanocrystals, with threefold symmetry (Figure 8.2b), is also produced [6]. Similar shapes have been produced by others [7, 8]. To our knowledge, a perfect decahedral 3D lattice has never been observed. This is probably due to the strong constraints involved in the formation of such aggregate morphology. From these results, it is concluded that nanocrystals retain the crystalline morphology of fcc clusters with tetrahedral subunits.

Other nanocrystal shapes with threefold symmetry like triangular nanocrystals, nanodisks, and nanoprisms are produced from modified fcc tetrahedral precursors to give triangular lamellar particles. At the nanometer scale, several groups produced triangular nanocrystals by either chemical reduction of metal salts, photoinduced aggregation of small nanoparticle seeds, or ultrahigh vacuum (UHV) with the epitaxial growth of metal particles on substrates [9–13]. These nanodisks are fcc *single* crystals with flat surfaces orientated in the {111} direction as shown by HRTEM (High-resolution transmission electron microscopy) and the corresponding Fourier transforms (FT) (Figure 8.3a,b and inset). Another way to produce *single* triangular Ag nanocrystals is to anneal at 50 °C, Ag nanocrystals ordered in 3D fcc superlattices.

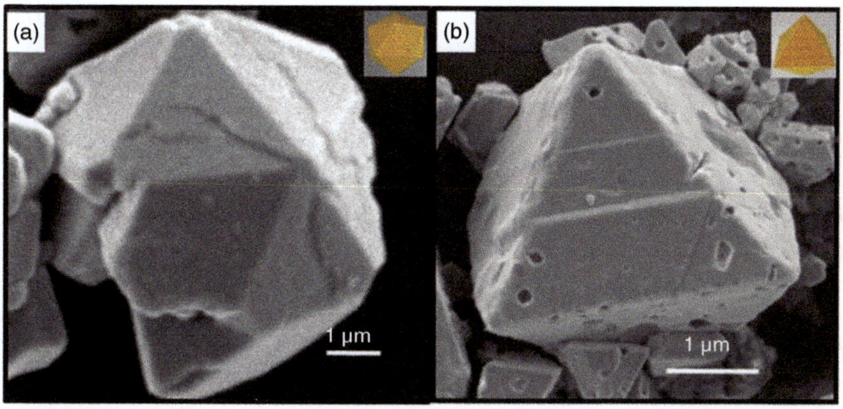

Figure 8.2 SEM (Scanning Electron Microscopy) images of icosahedral (a) and octahedral (b) 3D superlattices of 5 nm Au nanocrystals. *Source:* Goubet and Pileni [2]/American Chemical Society.

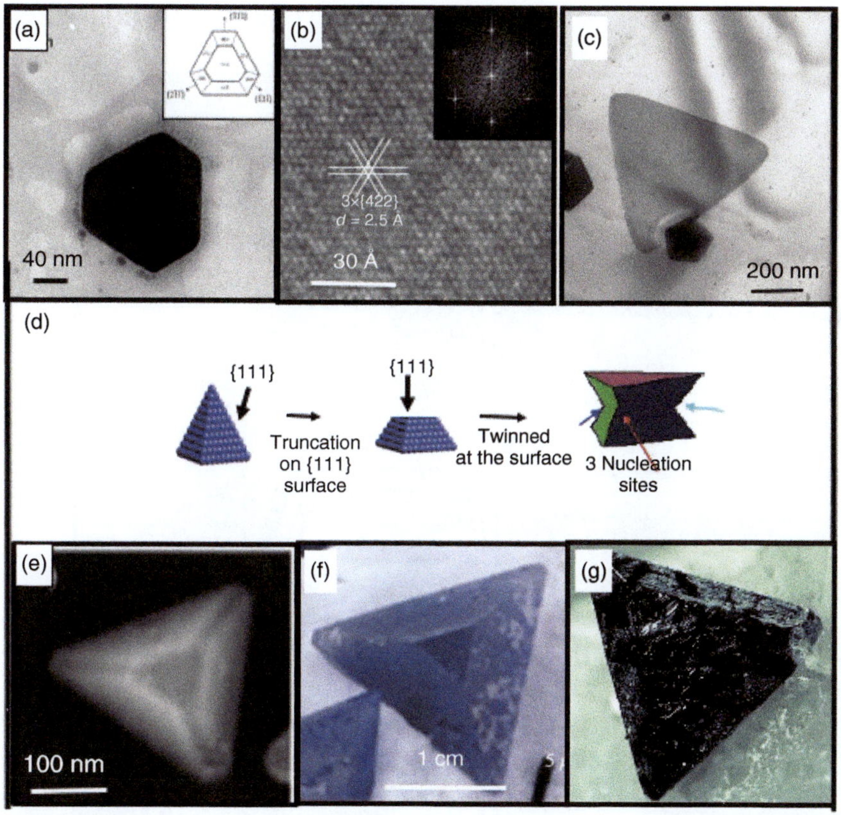

Figure 8.3 TEM (a) and HRTEM (b) images of Ag nanodisks produced from synthesis in reverse micelles in presence of an excess of hydrazine [10]. (c)TEM image of Ag nanotriangle produced by annealing at 50 °C of an assembly of 5-nm Ag nanocrystals deposited on HOPG (Highly Oriented Pyrolytic Graphite) [14]. (d) Schematic of the growth mechanism from nucleus to nanodisk. (e) SEM image of truncated tetrahedral Au nanocrystal [15]. (f and g) Truncated tetrahedral mineral: benitoite (f) and tetrahedrite (g). *Source:* Goubet and Pileni [2]/American Chemical Society.

Note that the size of triangular Ag crystals, from a few tens to a few hundred nanometers, is controlled by the nanocrystal ordering scale (Figure 8.3c) [14].

Figure 8.3d shows that a model is adopted by the scientific community to explain the formation of triangular particles: two regular fcc tetrahedral clusters truncated on a {111} surface adhere to each other, leading to a bitetrahedral cluster with three active sites for accelerated growth, maintaining the overall threefold symmetry in the final nanocrystal [9, 16]. To our knowledge, at the sub-nanometer scale, the formation of either tetrahedron and their truncated forms or bi-tetrahedral clusters has never been observed. However, truncated tetrahedral Au nanocrystals with rather large dimensions (210 ± 20 nm) (Figure 8.3c) were synthesized [15]. In nature, a rather large number of truncated tetrahedral minerals such as tetrahedrite and benitoite (Figures 8.3f,g) are found.

Similarly, to what observed with atomic crystals (Figure 8.3e), Figure 8.4a shows truncated tetrahedral 3D lattices of 5-nm Au nanocrystals. This confirms that threefold symmetry can be produced both with atoms and nanocrystals. Figure 8.4b shows fcc 3D lattices with the bipyramid morphology of two truncated tetrahedral *single* 3D lattices with a *single* twin plane along the {111} surface. Such a bipyramid

Figure 8.4 (a and b) SEM images of truncated tetrahedral (a) and bitetrahedral (b) 3D lattices of 5-nm Au nanocrystals. (c) Spinel mineral (https://www.irocks.com). (d) Polarized light microscopy image of plagioclase minerals obtained from https://www.flickr.com/photos/zeissmicro/7300492584. *Source:* (a) Pileni (2019)/The Chemical Society of Japan and (b–d) Goubet and Pileni [2]/American Chemical Society.

having a size in the micrometer range is similar to that in the sketch of the model (Figure 8.3d) proposed as an intermediate in the nanodisk production (Figure 8.3a,c). Hence by using nanocrystals, it is possible to image intermediated structure. This is probably due to a kinetic process slow enough with nanomaterials and too fast at the atomic level. This claim is also supported by the fact that similar intermediate structures are produced by mother nature. Figure 8.4c shows minerals like spinels with a similar morphology (bipyramid) as that observed with nanocrystals assembled in 3D lattices (Figure 8.4b) and as assumed at the nanoscale (Figure 8.3d). To our knowledge, such intermediates (truncated tetrahedrons and bipyramids) have never been observed at the nanoscale (below 10 nm).

By using reverse micelles (Chapter 1), in the presence of a large amount of salt (NaCl) [17], the formation of large truncated decahedra with fivefold symmetry has been observed (Figure 8.5a). Figure 8.5b shows a very high atomic ordered. It results

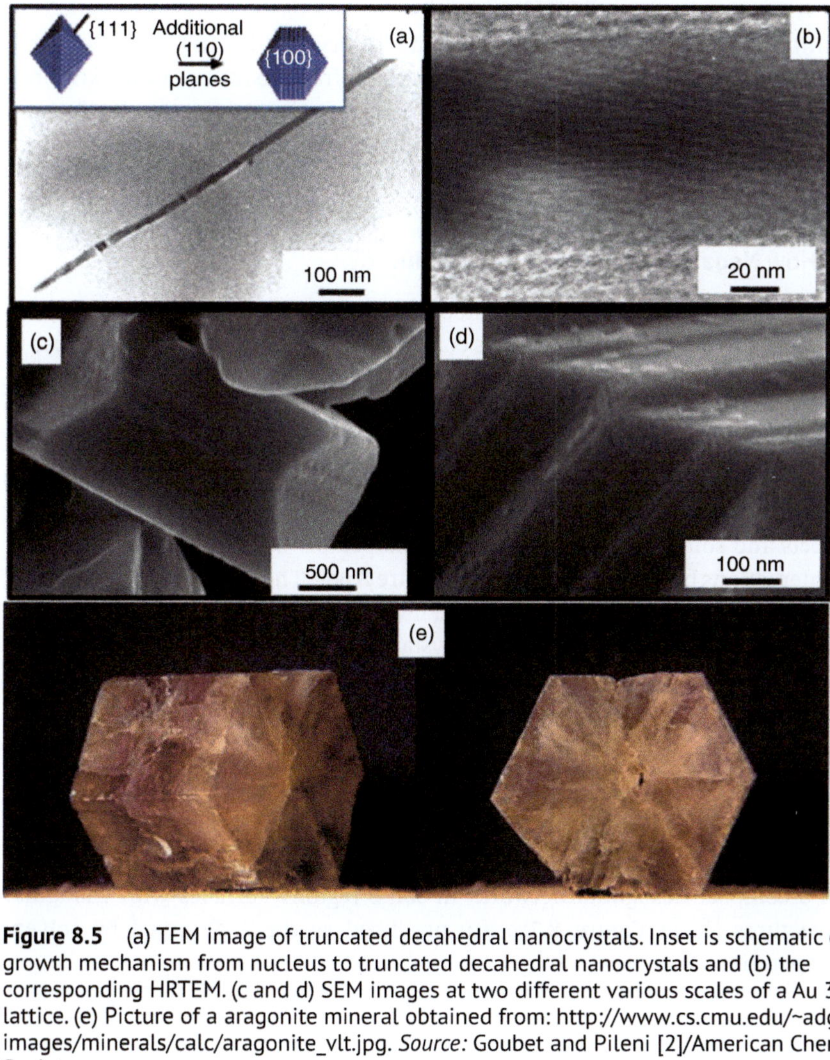

Figure 8.5 (a) TEM image of truncated decahedral nanocrystals. Inset is schematic of the growth mechanism from nucleus to truncated decahedral nanocrystals and (b) the corresponding HRTEM. (c and d) SEM images at two different various scales of a Au 3D lattice. (e) Picture of a aragonite mineral obtained from: http://www.cs.cmu.edu/~adg/images/minerals/calc/aragonite_vlt.jpg. *Source:* Goubet and Pileni [2]/American Chemical Society.

of additional intermediate {100} in regular decahedral nanocrystals with a preferential growth of the {111} facets compared with the growth {100} facets. With 3D lattice multiple parallel twins on {111}, truncated tetrahedral facets are observed with Au 3D lattices (Figure 8.5c,d) as found in plagioclase minerals (Figure 8.4d). Very surprisingly, minerals such as aragonites with similar shapes are characterized by additional {100} plane like nanorods (Figure 8.5e). From these examples, it is obvious that the crystal growth processes are universal feature even though the bonding strength involved to form 3D lattices from nanocrystals is weak compared with atomic crystals (bulk and/or nanocrystals). This opens the door for finding new physical and chemical properties with the obvious possible analogies between atoms and nanocrystals.

8.2 Negative 3D Lattices

By definition, negative crystals are birefringent crystals having cavities within the crystal bounded by the crystal faces with the refractive index of the extraordinary ray less than that of the ordinary ray. Hence, the voids take the form of ordinary crystal surfaces, because they are directly related to the lattice of the solid crystal.

8.2.1 How Negative 3D Lattices Are Produced?

In Chapter 4, we describe two simultaneous growth processes. Here, we will concentrate on 5-nm *single* Au nanocrystals coated with dodecanethiol and dispersed in toluene [6, 18]. In short, keeping the colloidal solution under toluene saturation, during seven days, interfacial film and precipitate-shaped 3D lattices are formed when some Au nanocrystals remain in the colloidal solution (Figure 8.6a). The formation of interfacial film 3D lattice is due to layer-by-layer growth process (Figure 8.6b). Such film is characterized by highly homogeneous surface with very few defects and some holes. The precipitate-shaped 3D lattices are due to van der Waals interactions between nanocrystals (Figure 8.6d). The crystalline structures of these 3D lattices are fcc (Figure 8.6c,e), and their building blocks *single* domains. In solution, some polycrystalline Au nanocrystals remain. Because dodecanethiol is part of the reactants used during the nanocrystal synthesis, even though at the end of the process the nanocrystals washed several times, free dodecanethiol molecules are kept physiosorbed to the coating agents.

A more careful observation of the interfacial film shows two distinct domains in the film morphology (Figure 8.7):

(i) Figure 8.7a,b shows zone with highly flat surface of Au nanocrystals ordered in a compact hexagonal network. In some regions, terraces with triangular structures (Inset Figure 8.7b) are produced as those observed by solvent evaporation.

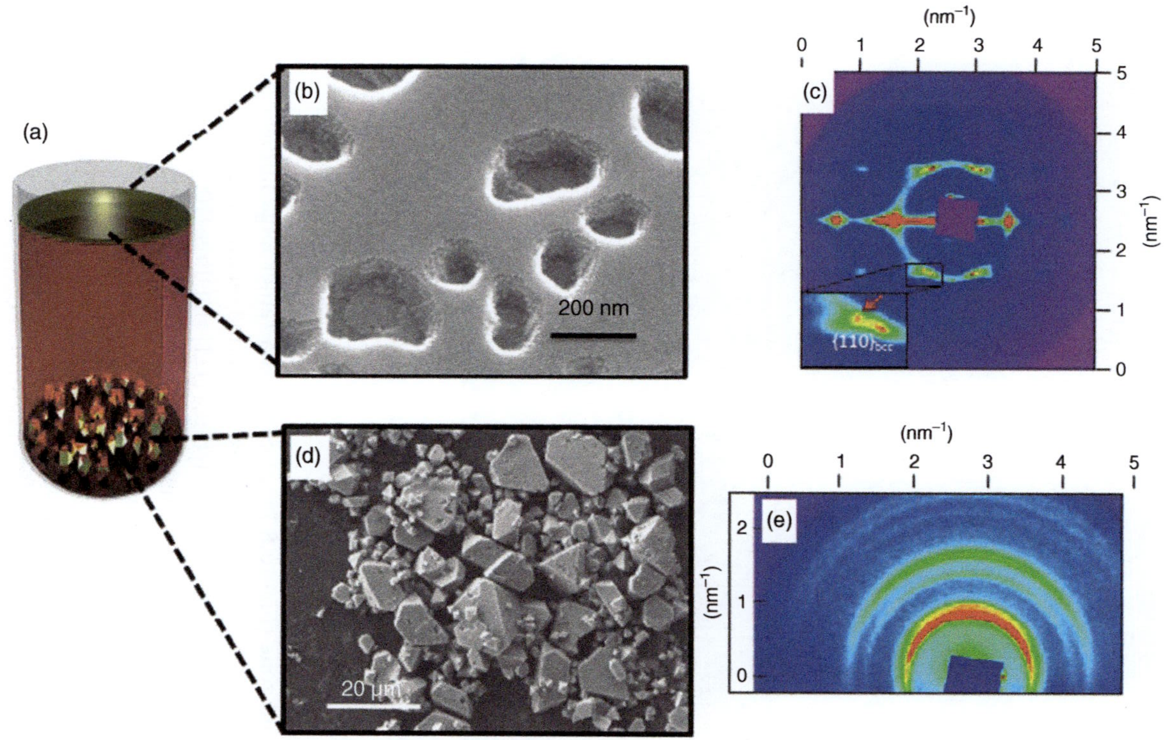

Figure 8.6 (a) Scheme of the beaker containing a colloidal solution of 5.1 nm dodecanethiol-coated Au nanocrystals dispersed in toluene and maintained under toluene during seven days. (b) Titled SEM image of interfacial 3D lattices containing well-defined holes and (c) the corresponding X-ray diffraction pattern of the interfacial structure with a mixture of fcc and bcc crystalline lattices. (d) SEM image of precipitate 3D lattices and (e) the corresponding SAXS diffraction shown in an fcc structure. *Source:* (b) Goubet and Pileni [19]/Springer Nature.

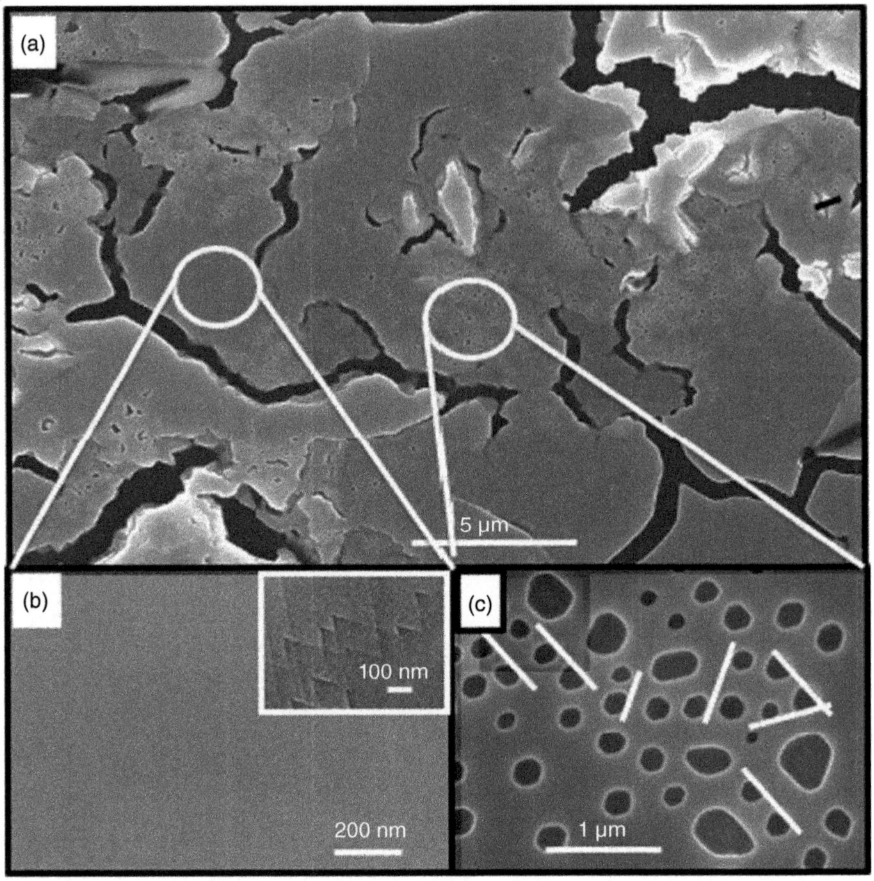

Figure 8.7 (a) Overall SEM image of interfacial 3D lattice film of 5.1-nm Au nanocrystals coated with dodecanethiol. (b) Enhancement of the SEM image of the homogeneous film shown in (a). (c) Enhancement of the SEM image of the film with well-defined holes. *Source:* Pileni (2019)/The Chemical Society of Japan.

(ii) The second zone (Figure 8.7a, c) contains a rather large number and well-defined holes. The inside faceted walls are tilted relative to the basal superlattice surface and characterized with three {111} and three {100} crystallographic planes (Figure 8.7c). Even though the size of the holes differs, similar crystallographic patterns are observed.

The fast Fourier transform (FFT) patterns deduced from HRSEM (High Resolution Scanning Electron Microscopy) images of various regions show that the orientation of the FFT spots remains "quasi" unchanged with only 5° as maximum disorientation indicating a coherence length up to 20 μm in the homogeneous domain of the film. Figure 8.8 shows that the coherence length in both regions (with and without holes) reaches more than several dozen of micrometers.

The HRSEM image around a negative 3D lattice [19] (Figure 8.9a,b) shows successive ordered domains of expanded (zones 1) and compact networks (zones 2)

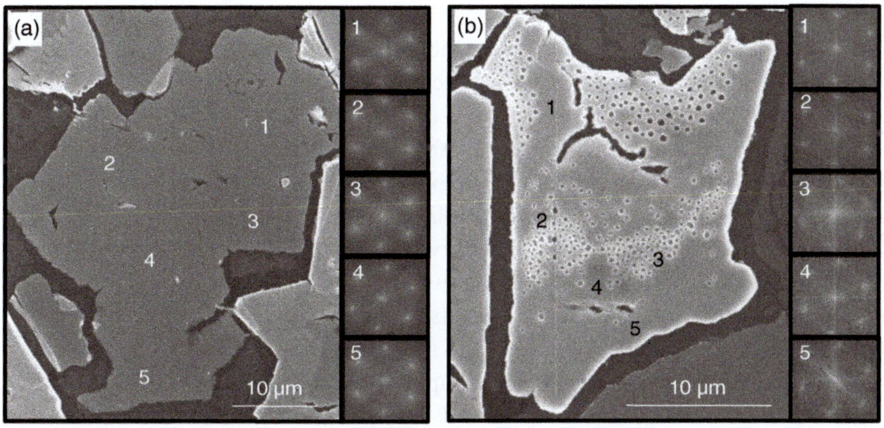

Figure 8.8 SEM image of two domains of interfacial 3D lattice film and corresponding the Fourier transform over all the SEM image (a) homogeneous film and (b) holes films. *Source:* Goubet and Pileni [19]/Springer Nature.

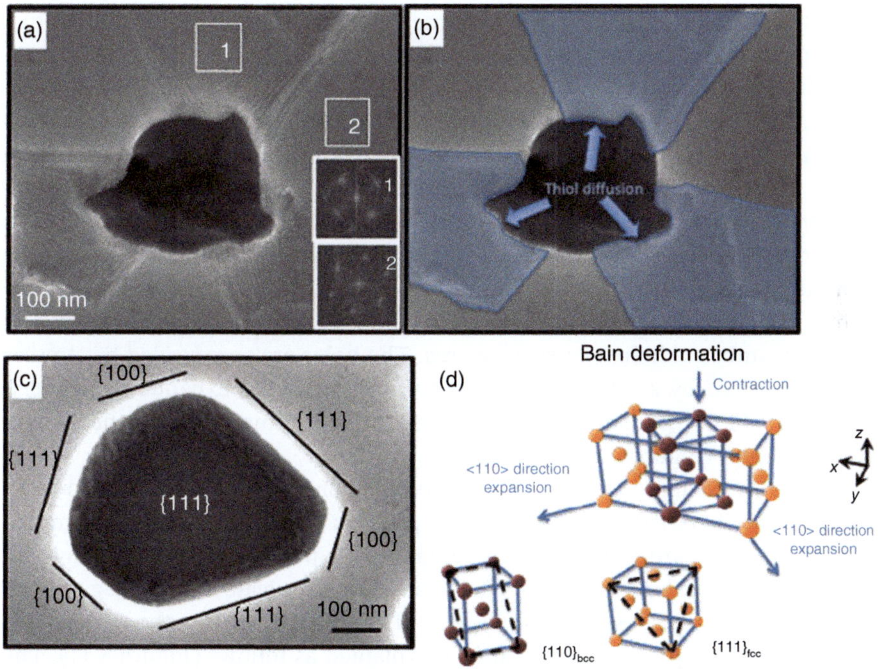

Figure 8.9 (a) HRSEM image of the nanocrystal network deformation induced by inclusion. (b) Superimposed expanded regions, highlighted in blue, due to the thiol diffusion on the image (a). (c) The facets of a hole. (d) Scheme of the Bain transformation and the corresponding {111} plane in the face-centered cubic unit cell and {110} plane in the body-centered cubic unit cell after deformation. *Source:* Goubet and Pileni [19]/ Springer Nature.

repeated three times around the hole as symmetry center. The FFT patterns (insets Figure 8.9a), corresponding to zone 1, exhibit distorted twofold symmetry, whereas the unexpanded regions (zone 2) are characterized by threefold symmetry, i.e. compact assembly. Such stresses, exclusively observed near the negative 3D lattices, induce a transition from threefold to twofold symmetry on the same basal plane. The greatest expanded surfaces of the nanocrystal superlattices are linked with the {100} inner surface of the hole, whereas the compact ones are linked to the {111} inner plane. The presence of planes with various compacities $\{111\}_{fcc}$ and $\{110\}_{bcc}$ parallel to the same superlattice surface, without well-distinguished delimitations between them like crystallographic facets (Figure 8.9c), is confirmed by SAXS (Small Angle X Ray Scattering) implying a structure transition.

One can conclude that the resulting shapes of the holes are images of truncated octahedrons. Hence, the interfacial film is composed of coexisting domains with distorted threefold and twofold symmetries as supported by the Fourier-transformed patterns of areas 1 and 2. The diffraction profile shows planes indexed as fcc and bcc structures (Figure 8.6c). From the literature, we know that the addition of organic molecules to a colloidal solution of nanocrystals leads to bcc superlattices, whereas without addition the superlattices exhibit fcc structure. The major differences between these data and those presented here are related to the fact that two structural arrangements (fcc and bcc) are observed on the same nanocrystal superlattice film. From chemical analysis of dried Au nanocrystal powder, we know that they are associated to a large excess of free dodecanethiol molecules [20]. By analogy to data obtained by Nagaoka et al. [21], we assume that the fcc/bcc transition is due to the presence of dodecanethiol molecules within the superlattice. To support such claim, keeping the same experimental conditions as above, a large amount of free dodecanethiol is added to the initial Au colloidal solution. At the end of the crystal growth process (seven days), the interfacial film is still characterized by a mixture of fcc and bcc structures with a surface markedly covered by a very large number of holes per surface area. The holes are surrounded by stresses and characterized by a rather low size distribution and the expanded and unexpanded networks with distorted two- and threefold symmetries with the hole as symmetry center (Figure 8.10).

However, their shapes are not well faceted. Regions without holes as shown in Figure 8.8a cannot be detected. The HRSEM pictures and FFT patterns confirm that the interfacial films are nanocrystals ordered in 3D superlattices with the same orientation along several micrometers. These data clearly indicate that the negative 3D lattices are due to the presence of dodecanethiol molecules during the nanocrystal superlattice growth process. This could be explained as follows (Figure 8.11). The high surface tension at air–toluene interface favors self-assemblies of free $C_{12}H_{25}SH$ molecules and nanocrystals coated with $C_{12}H_{25}SH$ to form a Gibbs monolayer [22]. A drop of the surface tension follows this process. Due to rather strong van der Waals interactions between Au nanocrystals compared with free $C_{12}H_{25}SH$ molecules, the nanocrystals tend to self-assemble at the interface and produce a 3D lattice film with the emergence of concentrated dodecanethiol zones trapped during the nanocrystal superlattice growth. Hence, fcc 3D lattice film of Au nanocrystals is

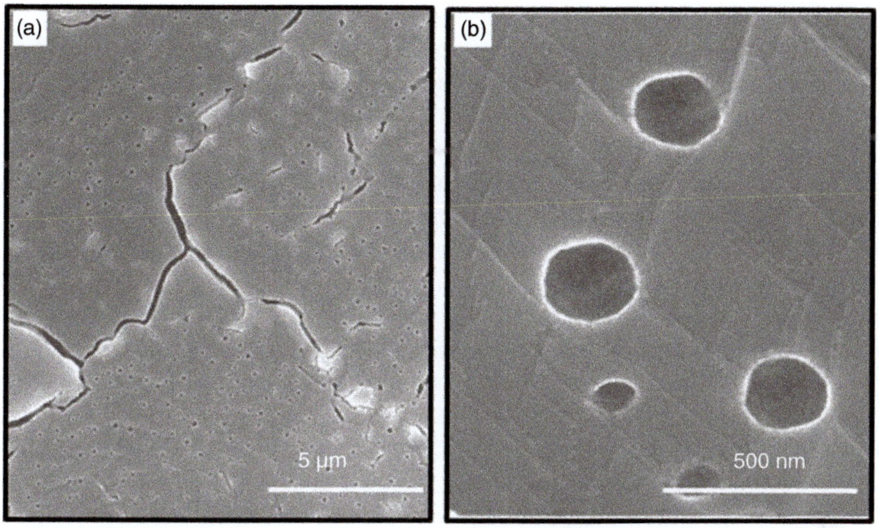

Figure 8.10 SEM images of interfacial supracrystal film produce with an excess of dodecanethiol at various scales. *Source:* Goubet and Pileni [19]/Springer Nature.

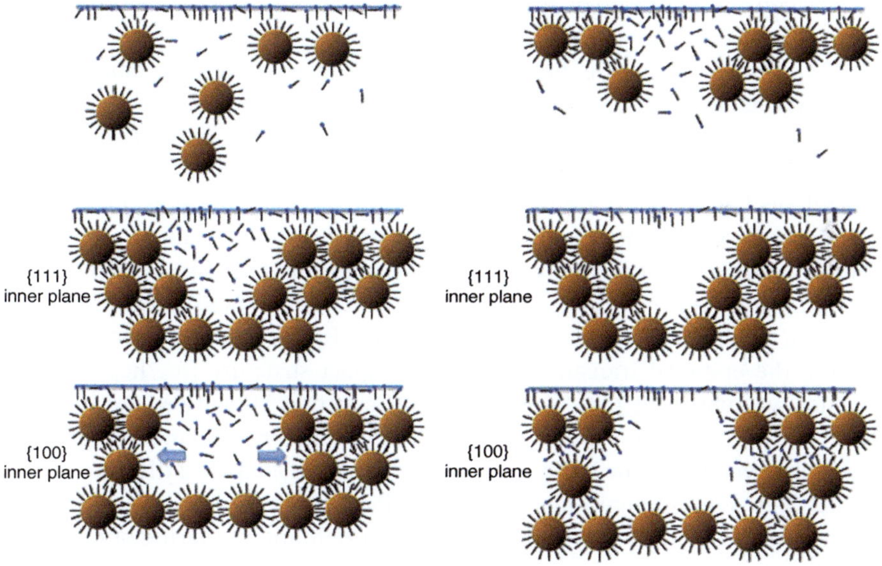

Figure 8.11 Mechanism of negative 3D lattice formation. *Source:* Goubet and Pileni [19] and Goubet et al. [6].

produced with inclusions due to the presence of $C_{12}H_{25}SH$ molecules. This explains the formation of negative 3D lattices with {111} and {100} as inner facets. The strong affinity between the free $C_{12}H_{25}SH$ involved in the concentrated zones and chemisorbed at the Au nanocrystal surfaces makes possible diffusion of dodecanethiol molecules inside the nanocrystal arrays through the nanocrystal gaps of the {100} planes, whereas the {111} planes remain compact.

Moreover, the sphere diameter inscribed inside the octahedral sites of the fcc nanocrystal superlattice is estimated around 3 nm. By filling this gap, the insertion of solvated $C_{12}H_{25}SH$ molecules induces a swelling of the nanocrystal superlattice in certain directions. Dodecanethiol diffusion is anisotropic and characterized by a Bain deformation (Figure 8.9d). This is possible since the interfacial film is in contact with the solvent and the alkyl chains of "free" and the coated $C_{12}H_{25}SH$ molecules are solvated. It reflects the intrinsic motion of solvated $C_{12}H_{25}SH$ molecules through the superlattices. The concentrated solvated thiol zones at the toluene interface induce constraints similar to the mechanical ones through the diffusion of the organic molecules inside the superlattice. Such mechanism is confirmed by the marked increase in the number of the holes in the presence of large amount of free $C_{12}H_{25}SH$ molecules to the colloidal solution. No uniform surface is observed, and the full surface of interfacial 3D lattices contains holes. We have to note that the structure of the negative 3D lattices is not as well defined and their facets are screened as compared with that observed in the absence of a large excess of $C_{12}H_{25}SH$ molecules. This could be explained by the fact that diffusion of $C_{12}H_{25}SH$ along the {100} direction is not exclusive with the appearance of smaller amounts of $C_{12}H_{25}SH$ along the {111} direction. Recently, a structural transformation of the orientational ordering of nanocrystals within the superlattices was observed controlling the reversible nanocrystal superlattice structure through the solvent–ligand interactions with a transition from fcc to bcc to the overall superlattice [23]. Furthermore, this process is reversible due to the high ability of the solvent molecules to move from the superlattices to outside and vice versa. Here, solvated $C_{12}H_{25}SH$ molecules migrate into the superlattices and remain trapped there during the drying process. The size of solvated $C_{12}H_{25}SH$ molecules and consequently its boiling point are responsible for the fact that the molecules remain in the superlattices. This makes the Bain transitions irreversible, and consequently, the fingerprint of diffusion of molecules through the superlattices can be observed *a posteriori* in a dry system. Similar deformation is usually observed by applying a mechanical constrains on the material. By shearing a di-block polymer solution, the epitaxial transition between compact structure (fcc/hcp) and bcc structure was observed [24]. Note that such negative 3D lattices are observed when the interfacial films are composed of *single*-domain Au nanocrystals. They are not observed when the building blocks of the interfacial 3D lattices are a mixture between *single*-domain and polycrystalline phases as obtained with larger nanocrystals (between 6 and 8 nm) [6, 25]. The orientation ordering of nanocrystals in the superlattice seems to be a key parameter for the Bain deformation that could be assimilated to another way to induce phase transition in nanocrystal superlattice, compared with the usual cited parameter such as the ratio between coating agent and metallic core length.

From these data, it is concluded that organic matter like thiols at the air–solvent interface in a concentrated zone induces polyhedral inclusions, called negative 3D lattices. This takes place when a strong affinity exists between thiol freely diffusing in solution and the coating agent of nanocrystals.

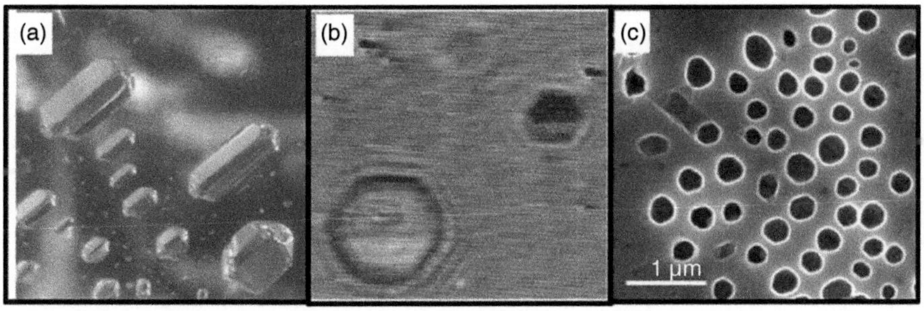

Figure 8.12 (a) Bubble inclusion in quartz, (b) STM (Scanning Tunneling Microscope) image of sub-surface Ar bubble in $Al_{99}Cu_1(111)$, and (c) negative 3D lattices of 5-nm Au nanocrystals coated with dodecanethiol and grown under thermodynamic process. *Source:* Schmid et al. [26]/American Physical Society.

8.2.2 Analogy Between Negative 3D Lattices, Atomic Crystals, and Minerals

Generally speaking, negative crystals are a specific kind of crystal foreign material (inclusion) trapped within a host crystal. These inclusions can be solid, liquid, or gaseous. They can become trapped within the crystal either during its growth or once the host crystal has formed. Negative atomic crystals are found in a very large number of crystals and/or gems. Their presence in gems drops their values. As an example, Figure 8.12a shows air bubble in quartz. This is due to the defects during the crystal growth process. It also can be homemade by bombarding aluminum surfaces with Ar ion gas bubbles between the surface and the subsurface formed with a depth of a few layers on Al(111) (Figure 8.12b) [26]. Here, negative crystals are produced by molecule inclusion during the 3D lattice growth processes (Figure 8.12c). Similarly, the Bain deformation was obtained in atomic metal crystalline thin films with stress induced by epitaxial growth [27]. In these cases, there are external mechanical constrains, whereas in the present work, none of them are involved. Here, inclusion of organic matter within the nanocrystal superlattice results in the Bain transformation. This clearly shows that, as already reported [28], Bain transformation could be also applied to nanocrystal superlattice even though the interactions are far away different from the "classical" crystals.

8.3 Vicinal Surface of 3D Lattices

Foreign nanocrystals on superlattice, doping, and impurities have strong influence on the nanocrystal superlattice nucleation and growth [8, 14, 29–31]. From our knowledge, these overall methods produce 3D lattices characterized by low energy surfaces. None of them show high-index plane surfaces with steeped and kinked vicinal surfaces extended on microscopic scale.

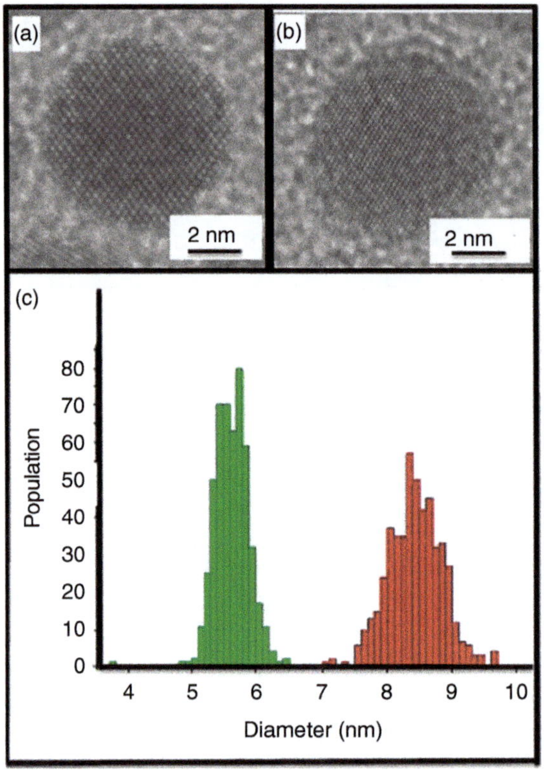

Figure 8.13 (a and b) HRTEM images of *single*-domain Au (a) and Co (b), respectively. (c) Histograms of Au (red) and Co (green) nanocrystals.

Let us now consider a mixture of *single*-domain Au and Co nanocrystals. Two colloidal solutions of *single-domain* Au and Co (ε phase) nanocrystals coated with dodecanethiol and oleic acid, respectively, are dispersed in toluene and deposited separately on TEM grid. Figure 8.13a,b confirm they are both *single* domains. From the TEM images, the deduced diameter and size distribution of both nanocrystals (Figure 8.13c) show 5.6 and 8.4 nm, respectively, with a very low size distribution (<8%).

The two colloidal solutions are mixed and the solvent is slowly evaporated. On the bottom of the vessel is the silicon substrate previously placed. The SEM image obtained at the end of the evaluation shows segregation of the nanomaterials (Figure 8.14a). The chemical map in Figure 8.14b shows that Au nanocrystals are concentrated into polyhedral 3D lattices (green) embedded in a film of Co(ε) nanocrystals (red).

Figure 8.15 shows HRSEM pictures of polyhedral Au nanocrystal superlattice surfaces exhibiting complex periodical patterns. These observations are characteristic for vicinal surfaces. The crystallographic planes that build the terraces can be different from one polyhedral 3D lattice to the other ones. The major observed vicinal surfaces are based on {111} terraces (Figure 8.15). However, few surfaces are also observed with less compact {100} and {110} terraces that are usually unstable in

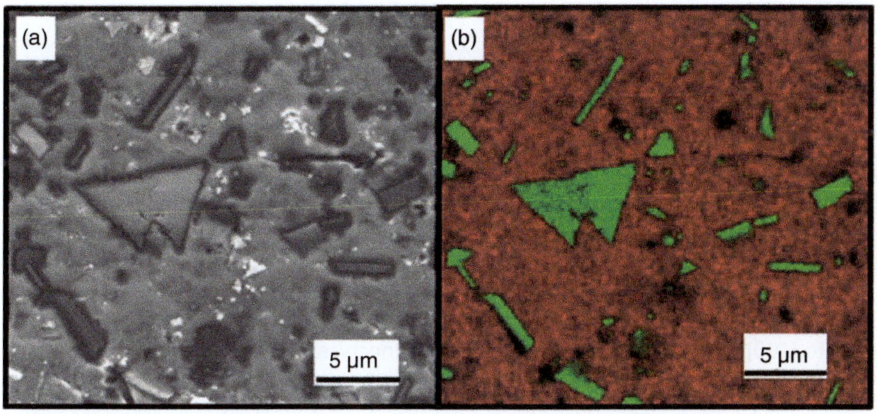

Figure 8.14 (a) SEM image of the mixture of the two colloidal solutions of Au and Co *single*-domain nanocrystals. (b) Energy-dispersive X-ray chemical map showing *single*-domain Au nanocrystal superlattice (green) within the *single*-domain Co(ε) nanocrystal film (red). The map is made by using the Mα_1 X-ray line for Au and Kα_1 for Co(ε). *Source:* Goubet et al. [14]/American Chemical Society.

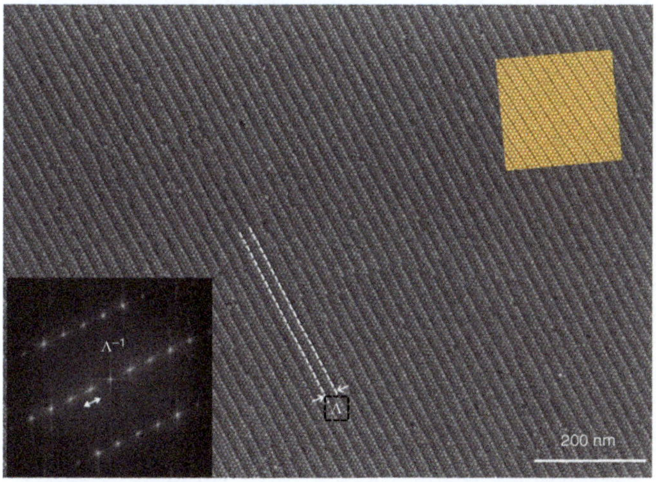

Figure 8.15 HRSEM pictures of the {221} nanocrystal superlattice plane that is vicinal to {111}. Insets: FFT pattern corresponding HRSEM picture and modeled plane. *Source:* Pileni (2019)/The Chemical Society of Japan.

atomic system (Figure 8.16). Parallel linear steps correspond to one nanocrystal in height (Figures 8.15 and 8.16b). Their periodicity is kept on micrometer range (Figure 8.15) that is confirmed by the image Fourier transform with lines of sharp reflections (insets Figures 8.15 and 8.16). Given its $\{hkl\}$ indexation, such a surface can be described by the $\{h_1k_1l_1\}$ and $\{h_2k_2l_2\}$ Miller indices for terraces and step microfacets, respectively (Figure 8.17a). Figure 8.16a shows 3D lattice surfaces that are kinked instead of stepped. At the difference of stepped surfaces, the sharp reflections of the Fourier transform are no more organized in parallel lines (Figure 8.16a). This is due to the presence of a second type of step microfacet with index $\{h_3k_3l_3\}$

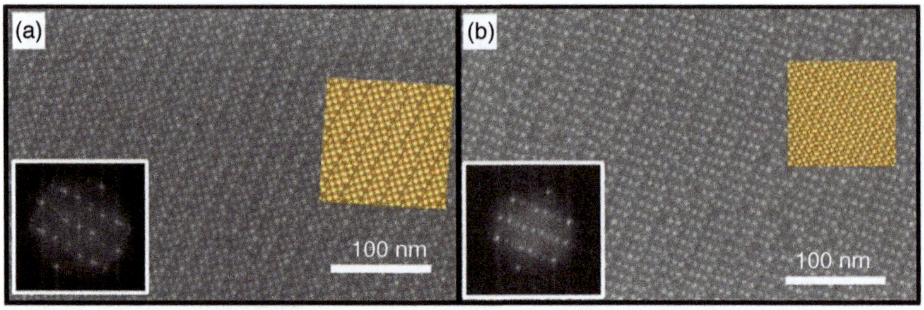

Figure 8.16 (a) HRSEM pictures of the {35 9 1} nanocrystal superlattice plane that is vicinal to {100}. (b) HRSEM pictures of the {430} nanocrystal superlattice plane that is vicinal to {110}. Each FFT patterns and modeled planes are insets in the corresponding HRSEM pictures. *Source:* Pileni (2019)/The Chemical Society of Japan.

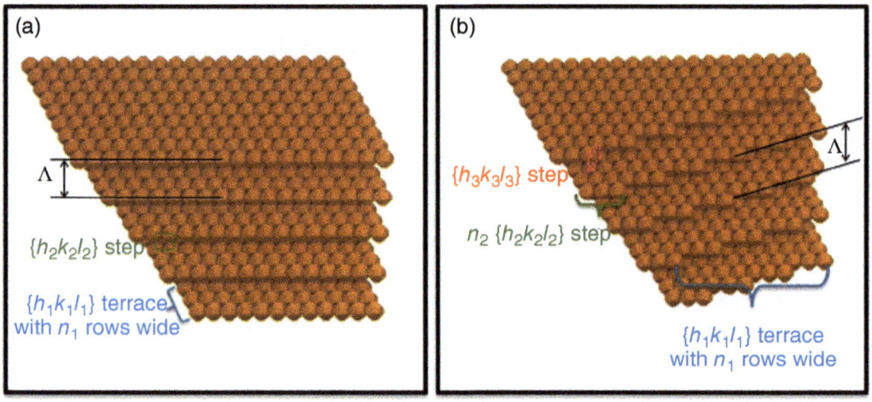

Figure 8.17 (a) Description of the stepped surface notation by their microfacets. (b) Description of the notation for kinked surfaces by their microfacets. *Source:* Goubet et al. [14]. Copyright © 2014, American Chemical Society.

(inset Figure 8.17b). A third number is introduced for the vicinal surface notation as the (*hkl*) vector is now decomposed on three sets of indices. In this way, the kinked nanocrystal superlattice surfaces are characterized by their chirality, as observed with atomic surfaces [32]. The density of low-coordinated nanocrystals is higher in the case of chiral surfaces than the achiral ones. One has to notice that some surfaces could not be indexed due to their complex symmetry. (*hkl*) vicinal surfaces containing $(h_1k_1l_1)$ terraces are usually produced by introducing a small convenient miscut angle α between the reference $(h_1k_1l_1)$ plane and the actual plane of cut. Here, due to the spontaneous formation of vicinal 3D lattice surfaces, we prefer to assign this parameter to a misorientation angle. In order to check if the Au 3D superlattices grow in direct contact with the silicon substrate or on a film of Co nanocrystals, the sample has been cleaved. HRSEM pictures of the edge show the systematic presence of an underlying film of Co nanocrystals. (Figure 8.18) reveals that the underlying Co nanocrystals are well ordered into fcc 3D superlattices. This strongly suggests that Co nanocrystal superlattice template is necessary to induce vicinal surfaces.

Figure 8.18 (a) SEM image of cleaved nanocrystal superlattice sample with high-index plane and (b) zoom-in of the interface between Au and Co nanocrystal superlattices. The upper and bottom inset FFT patterns correspond to the Au and Co nanocrystal assemblies, respectively. *Source:* Goubet et al. [14]/American Chemical Society.

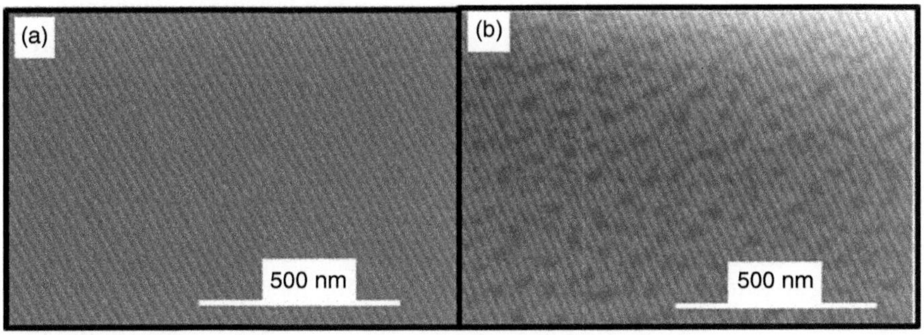

Figure 8.19 High-index nanocrystal superlattice plane before (a) and after (b) exposition to nitric acid. *Source:* Goubet et al. [14]/American Chemical Society.

At this point of the study, we need to identify which factors induce the formation of such vicinal surfaces that usually do not form spontaneously in crystals made of atoms. Indeed, by treating the sample with nitric acid, the Co nanocrystals can be dissolved while Au 3D lattice vicinal surfaces remain intact (Figure 8.19). This shows that Au nanocrystal superlattices with vicinal surfaces are stable. In the *absence of Co nanocrystals*, Au 3D lattices are characterized by low-index facets as observed previously (Chapter 3). This clearly indicates that Co nanocrystals are needed to produce vicinal surfaces. Note that when pure Au nanocrystal solution is deposited in the presence of a small excess of oleic acid, only compact superlattice planes are observed. Replacing 5.6-nm *single*-domain Au nanocrystals by the corresponding polycrystalline one, the nanocrystal superlattices are also material segregated, but the Au film reveals only low-index surfaces (Figure 8.20). No polyhedral

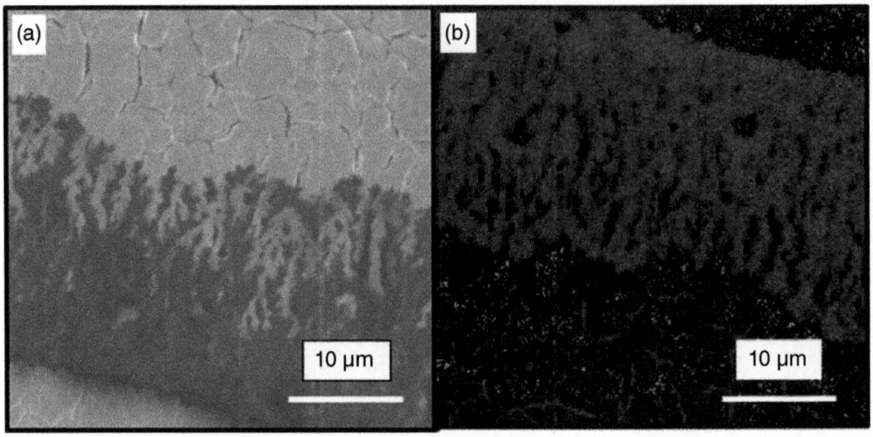

Figure 8.20 (a) SEM image of the mixture of the two colloidal solutions of Au and Co nanocrystals. The crystalline structure of Au nanocrystals is polycrystals, whereas Co nanocrystals with it remains e-phase. (b) Energy-dispersive X-ray chemical map showing *single*-domain Au nanocrystal superlattice (green) within the *single*-domain Co(ε) nanocrystal film (red). The map is made by using the $M\alpha_1$ X-ray line for Au and $K\alpha_1$ for Co(ε). *Source:* Goubet et al. [14]/American Chemical Society.

shapes with high symmetry surfaces are observed. Figure 8.20b shows that the upper surface of Au nanocrystal film is made of compact low-index planes and presents disorder at their boundaries with Co nanocrystal film. The slight variation in size distribution (from 5% to 6%) cannot explain such difference in self-assembly behavior. From these data, it is concluded that Au polyhedral 3D lattice with vicinal surfaces can be produced only if Au nanocrystals are in *single*-crystalline phase. Hence, the nanocrystallinity of nanocrystals involved in such process play a key role.

To explain such unusual growth process, we investigate the formation of Au 3D lattices differing by their nanocrystallinities [29]. The self-assembly process is followed using synchrotron-based *in situ* GISAXS (grazing-incidence small-angle scattering) and GIWAXS (grazing-incidence wide-angle scattering) techniques simultaneously at beamline D1 at the Cornell High Energy Synchrotron Source (CHESS). A drop of colloidal solution, small enough to keep its geometry with a convex meniscus, is enclosed in an evaporation chamber saturated with toluene solvent vapor to slow down the evaporation process. GISAXS and GIWAXS patterns of *single*-domain Au nanocrystals were recorded simultaneously, in the presence and absence of Co(ε) nanocrystals during the evaporation process in toluene atmosphere and a light inert gas flow.

In the absence of Co(ε) nanocrystals, after 10 minutes of evaporation, the GISAXS pattern of *single*-domain Au nanocrystals' colloidal solution became structured with the appearance of many well-defined Bragg reflections. The indexation reveals that *single*-domain Au nanocrystal superlattices correspond to a low-symmetry bct lattice film. Note that no evolution of the lattice parameters is observed during the solvent evaporation. Small arcs in the wide-angle scattering appear, corresponding to the scattering of the {111} and {200} Au atomic lattice planes of individual nanocrystals. The presence of directional scattering indicates a coherent alignment

Table 8.1 Au 3D lattice parameters differing by their nanocrystallinity and *single*-domain Au nanocrystals and polycrystalline Au nanocrystals, deduced from GISAXS patterns in the various experimental conditions.

	Toluene atmosphere		Hexane atmosphere
	Without Co(ε)	**With Co(ε)**	**With Co(ε)**
Single-domain Au nanocrystals	$a_{compressed\_fcc} = 12.3\,nm$ $c_{compressed\_fcc} = 11.2\,nm$	$a_{fcc} = 12.3\,nm$	$a_{fcc} = 12.4\,nm$
	$\delta_{pp} = 3.3\,nm$	$\delta_{pp} = 3.3\,nm$	$\delta_{pp} = 3.4\,nm$
Polycrystalline Au nanocrystals	$a_{hcp} = 8.2\,nm$ $c_{hcp} = 13.5\,nm$	$a_{hcp} = 8.2\,nm$ $c_{hcp} = 13.4\,nm$	$a_{hcp} = 8.2\,nm$ $c_{hcp} = 13.2\,nm$
	$\delta_{pp} = 2.6\,nm$	$\delta_{pp} = 2.6\,nm$	$\delta_{pp} = 2.5\,nm$

Let us admit an error of 0.1 nm without influence of Co(ε) nanocrystals.

of the atomic lattice planes for neighboring *single*-domain Au nanocrystals. Table 8.1 shows that the interparticle distance ($\delta_{pp} = 3.3\,nm$) is larger than that obtained at zero toluene vapor tension ($\delta_{pp} = 1.7\,nm$) [33]. This is due to the swelling of superlattices with toluene molecules induced by the high solvent vapor pressure, as it was previously reported [34]. A coherent alignment of the atomic lattice planes for neighboring *single*-domain Au nanocrystals is deduced from the arcs. A spot corresponds to a slightly tilted configuration by an angle of 8.1° of the nanocrystals as shown schematically in Figure 8.21a. Similar tilted configurations were recently

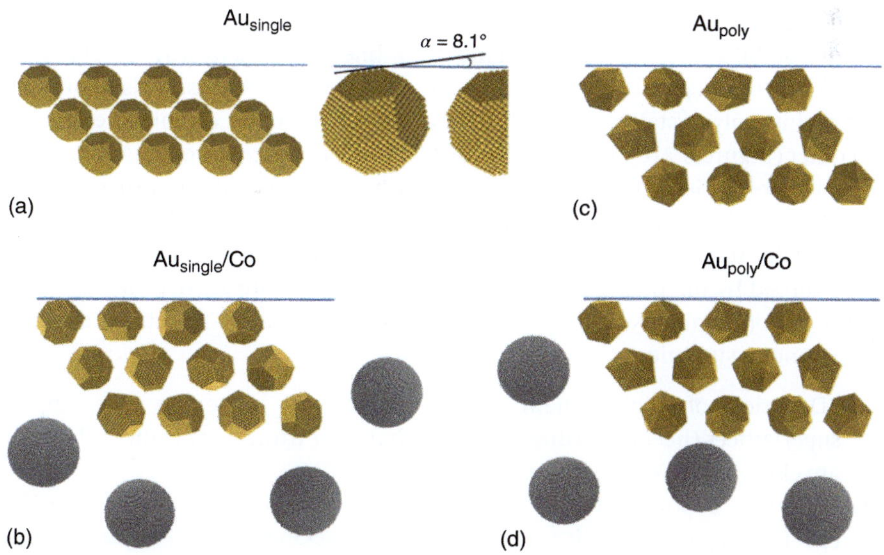

Figure 8.21 Schematized representation of (a, b) *single*-domain and (c, d) polycrystalline Au nanocrystal superlattices without (a and c) and with (b and d) Co(ε) nanocrystals. For clarity, the capping ligands are not represented. *Source:* Based on Goubet et al. [14].

observed in the case of PbS truncated octahedron nanocrystals with rearrangement during evaporation [34]. However, in the present case, only a slight reconfiguration is observed in the further course of evaporation, leading to a broadening and fusion of the two {200} peaks into a broader one, corresponding to coaxial alignment of the nanocrystals with the superlattice.

In the presence of Co(ε) nanocrystals, the GISAXS pattern can be indexed with a pure fcc lattice. The deduced δ_{pp} remains unchanged (Table 8.1). The GIWAXS pattern exhibits homogenous rings indicating a loss of *single*-domain Au nanocrystals' orientational ordering within the superlattice compared with what is observed in the absence of Co(ε) nanocrystals, and the arc corresponding to the {200} atomic *single*-domain Au lattice plane oriented along the [100] superlattice axis disappeared. Hence, the assembly of *single*-domain Au nanocrystals is sensitive to the presence of Co(ε) nanocrystals that lead to a change of lattice symmetry from bct to fcc with a loss of orientational order and of the 8.1° tilted configuration by of nanocrystals (Figure. 8.21b).

Similar experiments as described above are performed by replacing single domain by polycrystalline Au nanocrystals. In the absence of *Co(ε)* nanocrystals, polycrystalline Au nanocrystals start to self-assemble into a superlattice thin film at 18.3 minutes. The indexing of the GISAXS pattern reveals that the polycrystalline Au nanocrystals are assembled into an hcp lattice with δ_{pp} value (2.6 nm) larger than that deduced for zero toluene vapor pressure (δ_{pp} = 2.2 nm). Note that *single*-domain Au nanocrystals are more sensitive to the swelling effect than polycrystalline Au nanocrystals (Table 8.1). This indicates a larger flexibility and ductility of *single*-domain Au nanocrystals than 3D lattices of polycrystalline Au nanocrystals. The diffuse lines along the q_z direction are characteristic for the presence of large amount of stacking faults along the [001] directions. Even with the presence of defects in the superlattice stacking sequence (. . .A/B/A/B. . .), the positions of the scattering maxima along the q_z lines allow us to attribute the overall packing to a hcp lattice rather than fcc one. The GIWAXS signals show that neighboring polycrystalline nanocrystals are not coherently oriented relative to each other (Figure 8.21c). *In the presence of Co(ε) nanocrystals,* the GISAXS pattern shows the formation of an oriented film with a hcp crystalline structure. The GIWAXS signals do not show any changes compared with the absence of Co(ε) nanocrystals, and consequently, 3D lattices of polycrystalline Au nanocrystals are not perturbed by the presence of Co(ε) nanocrystals (Figure 8.21d). Hence, *in the absence of Co(ε)* nanocrystals, the Au nanocrystal superlattice structures depends on the nanocrystallinity:

(i) 3D lattices of *single*-domain Au nanocrystals are related to non-compact superlattices (bct or bcc) due to the fact that their truncated octahedral shapes are close to the Wigner–Seitz cell of the bcc lattice.
(ii) 3D lattices of polycrystalline Au nanocrystals show compact superlattices (fcc (111) or hcp (001)) with a high density of stacking faults. The nanocrystal

diameters of *single*-domain and polycrystalline Au nanocrystals deduced from GISAXS and TEM measurements perfectly agree.

(iii) Very surprisingly, in the presence of Co(ε) nanocrystals, no differences in superlattice structures of polycrystalline nanocrystals (hcp) and interparticle distances (δ_{pp} = 2.6 nm) are observed (Table 8.1). We note that the changes of the Au 3D lattice structure induced by the presence of Co(ε) nanocrystals are as follows:

(i) A crystalline structure transition from bct to fcc keeping same interparticle distance (δ_{pp} = 3.3 nm).
(ii) A loss of the nanocrystal orientational order of nanocrystal assemblies.
(iii) A loss of the nanocrystal alignment from a disorientation of 8.1°.

To explain such pronounced transition from low energy to vicinal surfaces of Au 3D lattices, we have to consider atomic crystals. In the bulk phase [35–38], the majority of metals have either fcc, bcc, or hcp structure. The different unit cells lead to a difference in physical properties: fcc lattices are usually soft and "ductile" that means they can be bent and shaped easily. Moreover, bcc lattices are less ductile but stronger while hcp lattices are usually brittle. It is well known that by cutting a bulk *single* atomic crystal with a small deviation from a compact plane, a vicinal surface is obtained with well-defined steps and terraces (Figure 8.22) [39].

By analogy between atomic crystals and 3D lattices [2], we could expect a similar behavior between these crystalline phases. The formation of the vicinal surface could be attributed to a similar cutting process with a deviation of α = 8.1° with a loss of nanocrystal orientational ordering. The kinked nanocrystal superlattice surfaces are characterized by their chirality, as observed with atomic surfaces [32]. Once again, analogy between atomic crystals [39] and 3D lattices can be pointed out. Vicinal surfaces of 3D lattices are formed and are produced from bulk atomic crystals. At the opposite to atomic vicinal surfaces, the ones produced with Au

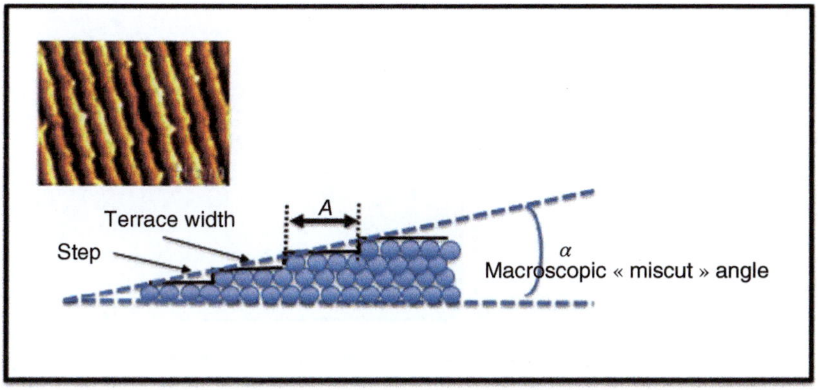

Figure 8.22 Scheme of the cut of a bulk *single* atomic crystals. *Source:* Based on Smilgies et al. [29].

nanocrystals are highly stable. To prove such claim, the sample is treated with nitric acid, and the Co(ε) nanocrystals are dissolved, while Au 3D lattice vicinal surfaces remain intact. In the future, such high-energy 3D lattice surfaces could open a new area with surface intrinsic properties.

8.4 Quasi-3D Lattices

A quasicrystal is an ordered structure without any periodicity. This as any other crystal continuously fills at available space with a lack in the translational symmetry. That means the shifted copy never matches exactly with its original. The diffraction patterns show sharp peaks and permit to identify the quasicrystal.

From Chapter 7, we know that the solvent evaporation process of a mixture of two colloidal solutions containing either small (S) or large (L) hydrophobic nanocrystals dispersed in toluene induces the formation of binary 3D lattices and, in some specific conditions, quasi-"3D lattices" are produced. Similarly, in nature, precious opals consist of silica spheres of uniform size (micrometer sizes) ordered or not in crystalline structures [40–44]. Brazilian gem opals are a regularly ordered array of silica spheres of two different uniformed sizes mixed together. Long-range ordering produces several different types of regular structures (Figure 8.23).

Mineral quasicrystals have been discovered. Icosahedrite ($Al_{63}Cu_{24}Fe_{13}$) (Figure 8.24a) is characterized by an icosahedral symmetry with a specific diffraction pattern (Figure 8.24b). It offers the evidence for the existence of natural quasicrystals [46]. Furthermore, a rather large number of meteorites with quasicrystalline structure have been discovered (Figure 8.24c,d) [45].

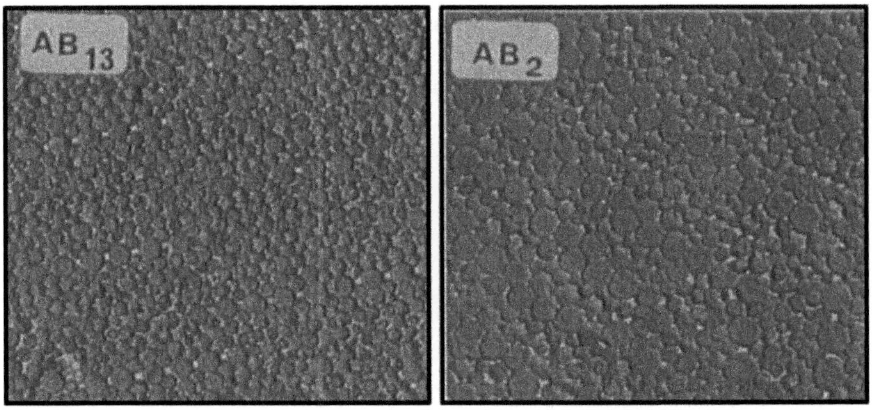

Figure 8.23 Electron micrographs of replicas of chips of various structures. *Source:* Murray and Sanders [44].

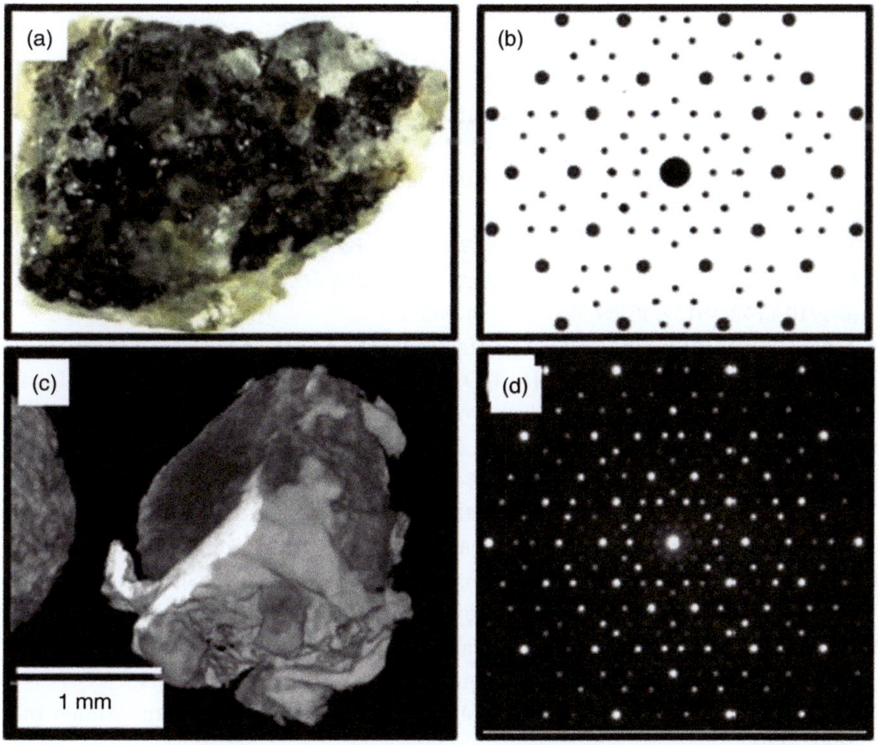

Figure 8.24 Photographs (a, c) and diffraction patterns (b, d) of natural quasicrystals. Icosahedrite (a, b) and Khatyrka meteorite (c, d) with specific composition $Al_{63}Cu_{24}Fe_{13}$ and $Al_{71}Ni_{24}Fe_5$, respectively. *Source:* (a) Paul J. Steinhardt (2012), p. 9/IOP Publishing, (b) Materialscientist/Wikipedia Commons, (c) Bindi et al. [45]/Springer Nature.

8.5 Conclusions

As natural crystals, negative and quasi-3D lattices are produced. Unexpected data concerns the surface of 3D lattices with the formation of high-index planes in the Au *single*-domain superlattices. Vicinal surfaces of 3D lattices are formed, and they are produced from bulk atomic crystals. In contrast to atomic vicinal surfaces, the ones produced with Au nanocrystals are highly stable. To prove this, the sample is treated with nitric acid, and the Co(ε) nanocrystals are dissolved, whereas Au 3D lattice vicinal surfaces remain intact. In the future, high-energy 3D lattice surfaces could open a new area with surface intrinsic properties.

References

1 Pileni, M.P. (2007). Control of the size and shape of inorganic nanocrystals at various scales from nano to macrodomains. *J. Phys. Chem. C* 111: 9019–9038.

2 Goubet, N. and Pileni, M.P. (2011). Analogy between atoms in a nanocrystal and nanocrystals in a supracrystal: is it real or just a highly probable speculation? *J. Phys. Chem. Lett.* 2: 1024–1031.

3 Markov, I. (2016). *Crystal Growth for Beginners: Fundamentals of Nucleation, Crystal Growth and Epitaxy*, 3e. Singapore: World Scientific https://doi.org/10.1142/10127. ISBN 978-981-3143-85-2.

4 Trager, F. and zu Putlitz, G. (1986). *Metal Clusters*. Springer-Verlag.

5 Kirkland, A.I., Jefferson, D.A., Duff, D.G. et al. (1993). Structural studies of trigonal lamellar particles of gold and silver. *Proc. R. Soc. London, Ser. A* 440: 589–609.

6 Goubet, N., Portalès, H., Yan, C. et al. (2012). Simultaneous growths of gold colloidal crystals. *J. Am. Chem. Soc.* 134: 3714–3719.

7 Compton, O.C. and Osterloh, F.E. (2007). Evolution of size and shape in the colloidal crystallization of gold nanoparticles. *J. Am. Chem. Soc.* 129: 7793–7798.

8 Rupich, S.M., Shevchenko, E.V., Bodnarchuk, M.I. et al. (2010). Size-dependent multiple twinning in nanocrystal superlattices. *J. Am. Chem. Soc.* 132: 289–296.

9 Salzemann, C., Urban, J., Lisiecki, I., and Pileni, M.P. (2005). Characterization and growth process of copper nanodisks. *Adv. Funct. Mater.* 15: 1277–1284.

10 Maillard, M., Giorgio, S., and Pileni, M.P. (2002). Silver nanodisks. *Adv. Mater.* 14: 1084–1086.

11 Jin, R., Cao, Y.W., Hao, E. et al. (2002). Controlling anisotropic nanoparticle growth through plasmon excitation. *Nature* 425: 487.

12 Ino, S. and Ogawa, S.J. (1967). Multiply twinned particles at earlier stages of gold film formation on alkalihalide crystals. *J. Phys. Soc. Jpn.* 22: 1365–1374.

13 Chapon, C., Granjeaud, S., Humbert, A., and Henry, C.R. (2001). Structure and morphology of nanometer-sized Pd clusters grown at high temperature on natural graphite single crystals. *Eur. Phys. J. AP* 13: 23–30.

14 Goubet, N., Yang, P., Albouy, P.A., and Pileni, M.P. (2014). Spontaneous formation of high-index planes in Au single domain nanocrystal superlattices. *Nano Lett.* 14: 6632–6638.

15 Kim, F., Connon, S., Song, H. et al. (2004). Platonic gold nanocrystals. *Angew. Chem. Int. Ed.* 43: 3673–3677.

16 Lofton, C. and Sigmund, W. (2005). Mechanisms controlling crystal habits of gold and silver colloids. *Adv. Funct. Mater.* 15: 1197–1208.

17 Lisiecki, I., Filankembo, A., Sack-Kongehl, H. et al. (2000). Structural investigations of copper nano-rods by HRTEM. *Phys. Rev. B* 61 (4968): 1–7.

18 Courty, A., Henry, A.I., Goubet, N., and Pileni, M.P. (2007). Large triangular single crystals formed by mild annealing of self-organized silver nanocrystals. *Nat. Mater.* 6: 900–907.

19 Goubet, N. and Pileni, M.P. (2014). Negative supracrystals inducing a FCC–BCC transition in a gold nanocrystal superlattice. *Nano Res.* 7: 171–179.

20 Wan, Y.F., Goubet, N., Albouy, P.A., and Pileni, M.P. (2013). Hierarchy in Au nanocrystal ordering in supracrystal: II. The control of inter-nanocrystal distances. *Langmuir* 29: 13576–13581.

21 Nagaoka, Y., Chen, O., Wang, Z., and Cao, Y.C. (2012). Structural control of nanocrystal superlattices using organic guest molecules. *J. Am. Chem. Soc.* 134: 2868–2871.

22 Hostetler, M.J., Templeton, A.C., and Murray, R.W. (1999). Dynamics of place-exchange reactions on monolayer-protected gold cluster molecules. *Langmuir* 15: 3782–3789.

23 Bain, C.D., Troughton, E.B., Tao, Y.T. et al. (1989). Formation of monolayer films by the spontaneous assembly of organic thiols from solution onto gold. *J. Am. Chem. Soc.* 111: 321–335.

24 Bang, J. and Lodge, T.P. (2003). Mechanisms and epitaxial relationships between close-packed and bcc lattices in block copolymer solutions. *J. Phys. Chem. B* 107: 12071–12081.

25 Portalès, H., Goubet, N., Sirotki, S. et al. (2012). Crystallinity segregation upon selective self- assembling of gold colloidal single nanocrystals. *Nano Lett.* 12: 5292–5298.

26 Schmid, M., Hebenstreit, W., Varga, P., and Crampin, S. (1996). Quantum wells and electron interference phenomena in Al due to subsurface noble gas bubbles. *Phys. Rev. Lett.* 76: 2298–2301.

27 Roldan Cuenya, B., Doi, M., Löbus, S. et al. (2001). Observation of the fcc-to-bcc Bain transformation in epitaxial Fe ultrathin films on $Cu_3Au(001)$. *Surf. Sci.* 493: 338–360.

28 Bian, K., Choi, J.J., Kaushik, A. et al. (2011). Shape-anisotropy driven symmetry transformations in nanocrystal superlattice polymorphs. *ACS Nano* 5: 2815–2823.

29 Smilgies, D.M., Li, R., and Pileni, M.P. (2018). Au nanocrystal superlattices: nanocrystallinity, vicinal surfaces, and growth processes. *Nanoscale* 10: 15371.

30 Cargnello, M., Johnston-Peck, A.C., Diroll, B.T. et al. (2015). Substitutional doping in nanocrystal superlattices. *Nature* 524: 450–453.

31 Shenton, W., Pum, D., Sleytr, U.B., and Mann, S. (1997). Synthesis of cadmium sulphide superlattices using self-assembled bacterial S-layers. *Nature* 389: 585–587

32 Jenkins, S.J. and Pratt, S. (2007). Beyond the surface atlas: a roadmap and gazetteer for surface symmetry and structure. *J. Surf. Sci. Rep.* 62: 373–429.

33 Goubet, N., Richardi, J., Albouy, P.A., and Pileni, M.P. (2011). Which forces do control the supracrystal nucleation in organic media? *Adv. Funct. Mater.* 21: 2693–2704.

34 Weidman, M.C., Smilgies, D.M., and Tisdale, W.A. (2016). Kinetics of the self-assembly of nanocrystal superlattices measured by real-time in situ X-ray scattering. *Nat. Mater.* 15: 775–781.

35 Tegenkamp, C. (2009). Vicinal surfaces for functional nanostructures. *J. Phys. Condens. Matter* 21: 013002.

36 Denk, P. (2000). Vicinal surfaces. PhD thesis. Ludwig-Maximilians-Universität München.

37 Schwennicke, C., Wanf, X.S., Einstein, T.L., and William, E. (1998). Evolution of surface morphology of vicinal Si(111) surfaces after aluminum deposition. *Surf. Sci.* 418: 22–31.

38 Desjonquères, M.C., Spanjaard, D., Barreteau, C., and Raouafi, F. (2002). Stability of metal vicinal surfaces revisited. *Phys. Rev. Lett.* 88: 056104.

39 (2003). Special issue on vicinal surfaces. *J. Phys. Condens. Matter* 15: 47.

40 Sanders, J.V. (1968). Diffraction light from opals. *Acta Crystallogr. A* 24: 427–434.

41 Sanders, J.V. (1964). Colours of precious opals. *Nature* 204: 1151–1153.

42 Sanders, J.V. and Murray, M.J. (1978). Ordered arrangements of spheres of two different sizes in opal. *Nature* 201–203.

43 Sanders, J.V. (1980). Close-packed structures of spheres of two different sizes I. Observations on natural opal. *Philos. Mag. A* 4: 705–720.

44 Murray, M.J. and Sanders, J.V. (1980). Close-packed structures of spheres of two different sizes II. The packing densities of likely arrangements. *Philos. Mag. A* 42: 721–740.

45 Bindi, L., Yao, N., Lin, C. et al. (2015). Natural quasicrystal with decagonal symmetry. *Sci. Rep.* 5: 9111.

46 Bindi, L., Steinhardt, P., Yao, N., and Lu, P.J. (2009). Natural quasicrystals. *Science* 324: 1306–1309.

9

Analogy Between 3D Superlattices and Atomic Crystals: Physical Properties

The analogy between 3D superlattices and atomic crystals related to the growth mechanism of crystals shown in Chapter 8 makes it possible to explore new topics in chemistry and physics. Now we can ask the following question: Are there any analogies in the physical properties of materials composed of atoms with those of nanocrystals? To answer this question, we have first study 3D superlattice physical properties and then look if some similitude with atomic crystals could be pointed out. For this purpose, one of the ways is to study physical properties of both 3D superlattice and disordered assemblies produced from the same batch of nanocrystals. Very few data are valuable in the literature. Nevertheless, a number of these can be pointed out.

9.1 Magnetic Properties

Some of the magnetic properties of 3D superlattices are similar to what observed with magnetic atomic crystals [1, 2]. By controlling the deposition conditions, it is possible to produce films of either disordered or 3D fcc superlattices of Co nanocrystals (Chapter 6). It has already been shown that there is a difference in the magnetic behavior when Co nanocrystals are self-assembled in 1D and 2D hexagonal networks compared with the isolated nanocrystals [3, 4]. More recently, we have reported differences in the magnetic behavior of fcc 3D superlattices of Co nanocrystals compared with disordered 3D assemblies of the same nanocrystals, arising from the narrow distribution of energy barriers in the 3D superlattice. From an application point of view, these systems are already interesting for potential use in magnetic data storage. For both study of collective property and also future applications, it is desirable to use nanocrystals with a high magnetic anisotropy. This can be achieved by annealing the nanocrystals, which induces the crystallographic transition amorphous to hcp structures. This has already been demonstrated for cobalt nanocrystals either isolated or in 2D assemblies [5]. Improving the crystalline structure of Co nanocrystals in 3D self-assemblies by *in situ* annealing is not trivial as in many systems (e.g. FePt [6], CoPt [7]); this leads to an undesirable nanocrystal coalescence.

Metal Nano 3D Superlattices: Synthesis, Properties, and Applications, First Edition.
Marie-Paule Pileni. © 2023 WILEY-VCH GmbH. Published 2023 by WILEY-VCH GmbH.

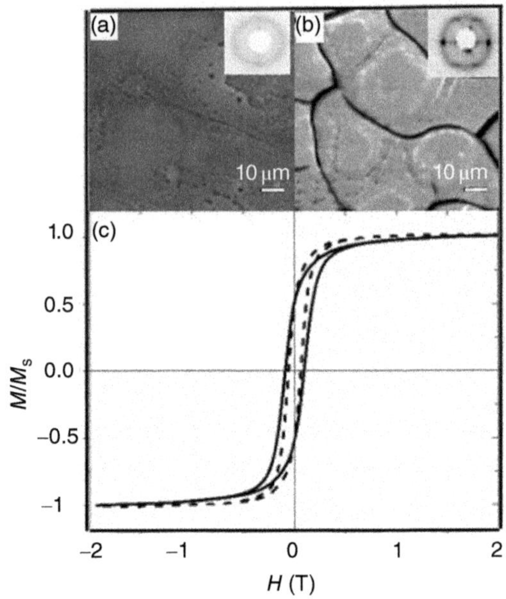

Figure 9.1 SEM image of fcc 3D superlattices (a) and amorphous aggregates (b) of 7-nm cobalt nanocrystals. Inset: X-ray diffraction patterns (c) M/M_s versus H curves of disordered (dashed line) and ordered (solid line) 3D assemblies. *Source:* (a, b) Lisiecki et al. (2007) [1]/ American Chemical Society.

Let us compare the magnetic response of disordered aggregates (Figure 9.1a) with that of fcc 3D superlattices (Figure 9.1b) produced from the same batch of nanocrystals. Figure 9.1c shows the magnetization versus field curves for the ordered and disordered 3D cobalt nanocrystal assemblies. For the 3D superlattices, the approach to saturation is more gradual than for the disordered aggregates.

This difference is explained by considering the model of amorphous ferromagnets proposed by Chudnovsky [8]. In this model, there is only short-range structural order that leads to a local random anisotropy with a short-range correlation length. The randomness of this local anisotropy leads to a very small uniform anisotropy in the sample, referred to as coherent anisotropy. The model predicts an approach to saturation in these materials given by $1/H^{1/2}$ (where H is the applied magnetic field) that has been confirmed experimentally [9]. It is extended to disordered nanocrystallized films [10] where a $1/H^{1/2}$ approach to saturation is also observed. In addition, by increasing the anisotropy, a deviation from $1/H^{1/2}$ to $1/H^2$ on approaching saturation is observed that would apply in the case of a perfect uniaxial system. By analogy to what is observed in Figure 9.1c, it is concluded that in the disordered aggregates of Co nanocrystals, the direction of anisotropy randomly fluctuates from one magnetic nanocrystal to another and leads to an extremely small uniform local anisotropy with a square root magnetization law in approaching saturation. For Co 3D superlattices, due to long-range mesoscopic order, the total anisotropy is rather large with inter-nanocrystal coupling energies, leading to a smoother magnetization curve on approaching saturation. This clearly shows the magnetic properties of ordered and disordered aggregates of nanocrystals behave similarly to those of highly crystallized and amorphous materials. Similarly, as predicted, the coercive field increases for 3D superlattices (900 Oe) compared with disordered aggregates (600 Oe). Again, as above, this agrees with a change in the local anisotropy with the nanocrystal ordering.

From this comparison between the approach to saturation magnetization and the hysteresis loop of either amorphous material or highly crystallized 3D superlattices, it is reasonable to conclude that there is a similarity in the magnetic behavior between disordered aggregates or highly ordered 3D superlattices as in amorphous and highly crystalline materials of atoms. To go further, to a first approximation, we could assume that the magnetic behavior of atoms and nanocrystals is similar in a given order. This clearly shows that 3D superlattices behave similarly as bulk materials.

9.2 Coherent Longitudinal Acoustic Phonons in Small 3D Superlattices

3D fcc 3D superlattices of polycrystalline of 7.1-nm Co nanocrystals coated with dodecanoic acid are characterized by a center-to-center nanocrystal distance of 10.1 nm and consequently the average nanocrystal spacing of 3 nm. The 3D superlattice thickness varies from a few tens of nanometers to few micrometers.

Picosecond acoustic technique enables to generate and detect ultrasounds up to several hundred GHz frequency range [10, 11]. As above, the 800-nm pump excites the Co 3D superlattices deposited on either HOPG (Highly Ordered Pyrolytic Graphite) or a silicon wafer at quasi-normal angle of incidence ($\vartheta_1 = 6°$). The corresponding 2D $\Delta T/T$ maps shows that, for any probe wavelength, the oscillation patterns are anharmonics with a slight reduction of the frequency as function of time (Figure 9.2). These dynamic oscillations for HOPG and silicon substrates are perfectly superimposed indicating an independent behavior with the substrate. A linear dependence of the oscillation periods with the probe wavelengths is observed for various external incident angles.

Figure 9.3a shows similar oscillations at various incident angles. Whatever the incident angle, the period linearly depends on the probe wavelengths. From the slope of such linear variation at various incidence angles, the index of refraction and the speed of sound are 1.26 ± 0.03 and $1235 \pm 12\,\text{m}\,\text{s}^{-1}$, respectively. The rather low value for the speed of sound is attributed to the large mass of the nanocrystals and the rather weak interactions between the interdigitized dodecanoic chains (such interdigitation is controlled by van der Waals interactions). The rather low absorption of the 800-nm pump, occurring on the micrometer scale (i.e. much longer than the acoustic wavelength) by the 3D superlattices, confirms that there is no localized pump absorption in the sample but only in the substrate, which is considered as transducer. The oscillation period and consequently the speed of sound markedly depend on the sample temperature (Figure 9.3b). From 435 to 110 K, the whole pattern of oscillations shrinks by nearly a factor of 5 that is almost proportional to the temperature. This change in period is fully reversible, as verified by successively operating several cycles and observing no permanent modification of the sample.

Neglecting a potential change of the index of refraction with temperature, this highlights the role of the coating agents as springs connecting the nanocrystals. The speed of sound evolves from $870\,\text{m}\,\text{s}^{-1}$ at 435 K to more than $3500\,\text{m}\,\text{s}^{-1}$ at 110 K with a clear change of behavior around 250 K. Figure 9.4 shows two linear fits with

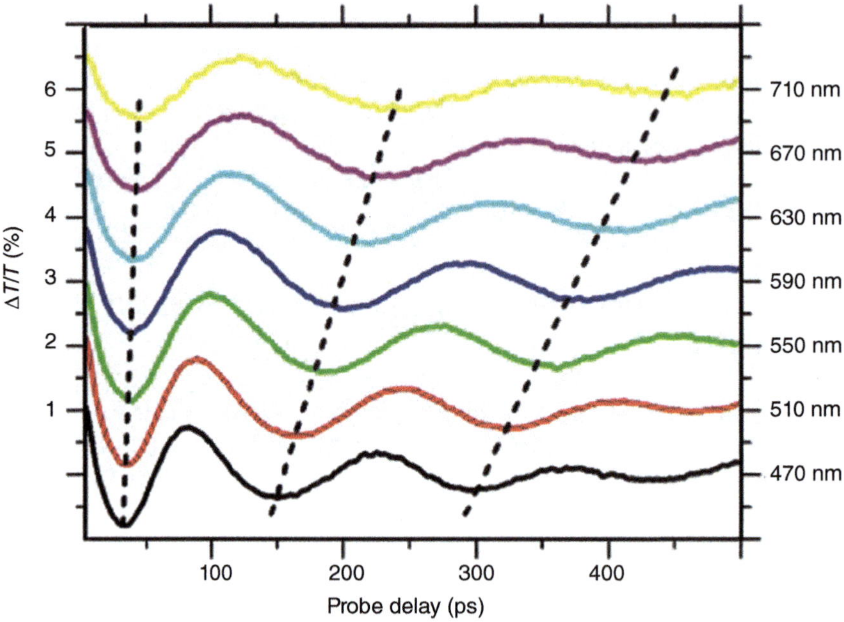

Figure 9.2 2D maps showing the $\Delta T/T$ signal plotted after the subtraction of the slowly varying background signal for 7-nm Co 3D superlattices deposited on HOPG. Dashed lines are a guide to the eye for highlighting the variable period of oscillation with wavelength. *Source:* Lisiecki et al. (2013) [12]/American Chemical Society.

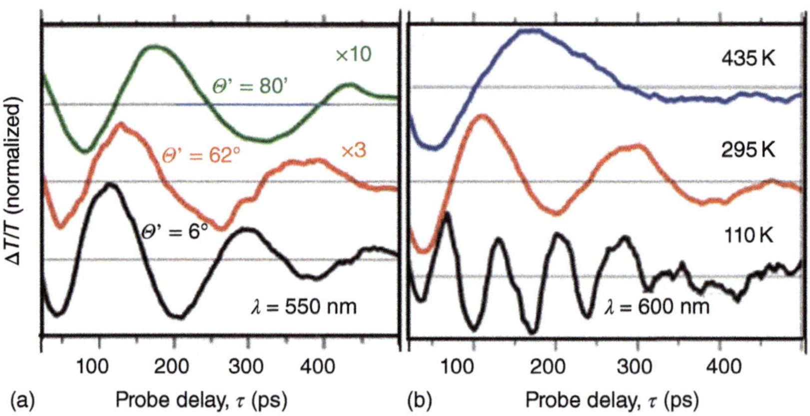

Figure 9.3 (a) $\Delta T/T$ dynamics as a function of incident angles (a) recorded at $\lambda_{pr} = 550\,nm$ and of sample temperature (b) for the same probe wavelength $\lambda_{\pi r} = 600\,nm$. *Source:* Lisiecki et al. (2013) [12]/American Chemical Society.

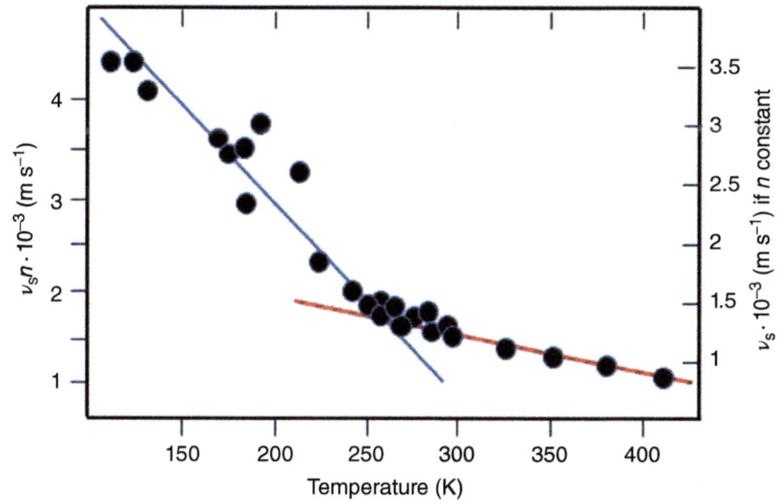

Figure 9.4 On the left: the product between the speed of sound and the index of refraction as a function of temperature. On the right axis: the speed of sound reported in case no change of the index of refraction occurs with temperature. *Source:* Lisiecki et al. (2013) [12]/American Chemical Society.

different slopes and intercepts that can describe the data. Note that this strong variation is at odds with the behavior of acoustic waves in ordinary crystals for which the speed of sound displays only weak temperature dependence. This is a very peculiar effect of this supracrystalline material that has never been observed in normal crystals, where both the speed of sound and the index of refraction are almost insensitive of temperature. This temperature dependence of sound velocity is reminiscent of that observed in glass-forming materials in the supercooled state between the crystalline melting (T_f) and glass transition (T_g) temperatures [13]. The data presented here could be explained as follows: first pump pulse at 800 nm, focused on the sample and absorbed by the transducer (HOPG or silicon substrate), induces an "instantaneous" thermal expansion that impulsively creates a traveling compression wave (i.e. a phonon wavepacket with broad spectrum) that moves in the sample with the longitudinal speed of sound. Such localized strain wave results in a moving discontinuity in the material index of refraction, corresponding to a mirror able to partially reflect a second delayed probe pulse with broadband spectrum in the visible (covering the 460–720 nm wavelength range). From the qualitative behavior of v_s and n relatively to the density, it is deduced that the increase of $v_s \cdot n$ at low temperature in our sample is mainly due to an increase of the rigidity of the dodecanoic network: the lower the temperature, the higher their stiffness giving rise to an increase of the speed of sound. The crossing point at 250 K can be assigned to a temperature transition, attributed to a phase transition of the alkyl chain configuration from ordered to disordered [14]. This temperature dependence corresponds to two different regimes of the rigidity of the bonds between the interdigitized alkyl chains. Molecular dynamics simulations of a similar system (dodecanethiol self-assembled monolayers on gold nanocrystals) [15] show that at low temperatures (up to 200 K), the passivating molecules organize into oriented molecular bundles with ordered parallel alignment. For higher temperatures, intramolecular conformational defects (e.g. reduction of both the percentage of molecules with trans configuration and of their end-to-end length) induce disorder in the coating agents, which starts at the bundles' outer boundaries and then with increasing temperature propagates inward toward the nanoparticle surface. At around 290 K, melting of the molecular bundles occurs, resulting in a uniform (random) orientation of the molecules around the nanocrystals. Alkyl chain dynamics are studied in a similar system [16] and are found to evolve continuously with the increase in temperature: at low temperature, only a small portion of each chain is dynamically active, and only at 400 K, the whole chain is involved. We note that a second phase transition of the alkyl chains from the trans configuration to a liquid-like phase takes place at 340 K. No such transition is observed in our data, probably because it has no effect on the rigidity.

These data clearly show propagation of longitudinal acoustic wavepackets [12] in a self-assembled 3D superlattice film made of highly ordered nanocrystals linked by alkyl chains, in analogy with atomic crystals. Because of the large mass of the nanocrystals and the low stiffness of the connecting alkyl chains, the oscillation period turns out to be quite long, that is, on the order of 200 ps at room temperature. In contrast with ordinary solids, the speed of sound in these artificial materials is strongly dependent on temperature due to the change in the dynamic properties of the dodecanoic lattices.

9.3 Breathing Modes

Silver (or gold) nanoparticles markedly adsorb light by exciting electronic surface plasma dipolar oscillations. This resonance induces intense Raman scattering by the vibration of the nanoparticles. The low-frequency Raman spectra of these materials always show a very intense peak close to the Rayleigh line, which is attributed to the excitation of the quadrupolar vibration mode of the nanoparticles via the plasmon–phonon interaction. For aggregates of Ag to Au nanocrystals, the quadrupolar modes appear as sharp intense lines. For spherical nanocrystals with sizes larger than ~1 nm, the cluster vibrations are described by modeling the nanocrystal with a continuum nanosphere of a diameter D equal to the size of the nanocrystal [17] and using the longitudinal, v_l, and transversal, v_t, sound velocities of Ag bulk. The vibrations are characterized by the quantum numbers n and l, and the vibrational frequencies are given by the following equation:

$$\nu = \frac{S_{ln} v_t}{D} \tag{9.1}$$

where S_{ln} depends on the ratio v_l/v_t.

Let us deposit on a substrate, by evaporation of a colloidal solution, 5-nm Ag nanocrystals with low size distribution. The substrate temperature controls, during the solvent evaporation process, the nanocrystal assemblies from disordered film

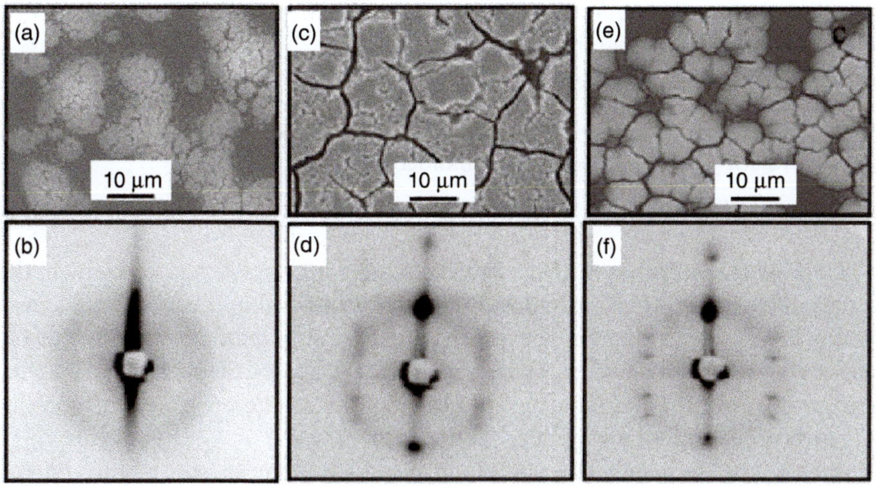

Figure 9.5 SEM images (a, c, and e) and corresponding X-ray diffraction pattern (b, d, f) of 5-nm Ag nanocrystals deposited on a substrate. The substrate temperatures are 10, 22, and 30 °C respectively. *Source:* Courty et al. [18].

(Figure 9.5a,b) considered as amorphous film, small Ag fcc 3D superlattices (Figure 9.5c,d), and finally long-range fcc 3D superlattices (Figure 9.5e,f).

As observed with similar systems [19], the Stokes/anti Stokes Raman spectra of these samples differing by their ordering from amorphous to highly crystallized 3D superlattices show quadrupolar modes appearing as two sharp intense lines. The Stokes/anti-Stokes spectra of both "large and small" fcc 3D superlattices are shifted two lower frequencies than amorphous film (Figure 9.6a,b).

Let us first consider the data obtained with amorphous film (Ag nanocrystals are randomly dispersed). Raman scattering intensity, $I(\nu)$, at the frequency, ν, is

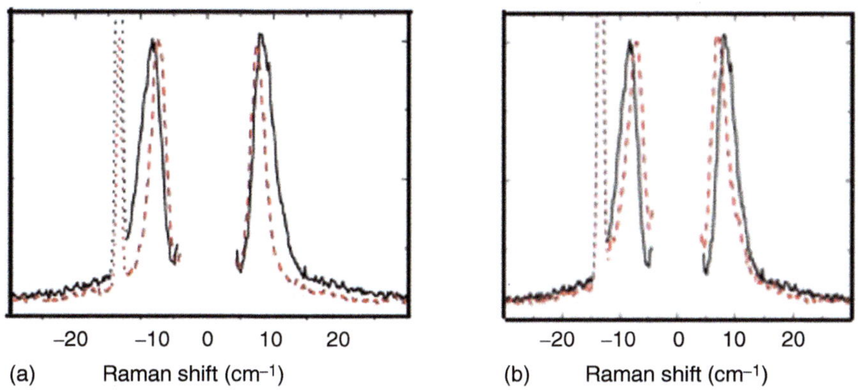

Figure 9.6 Comparison of stokes-anti stokes Raman spectra of amorphous film (black) with (a) small (b) large 3D superlattices. *Source:* Courty et al. (2005) [20]/Springer Nature.

proportional to the frequency distribution of the nanocrystals, $F(\nu)$. From Eq. (9.1), this is related to the inverse size distribution, $F(D^{-1})$:

$$F(\nu)\mathrm{d}\nu = F(D^{-1})\mathrm{d}D^{-1} \tag{9.2}$$

Therefore, for a random assembly of nanocrystals:

$$I(\nu) \propto F(\nu) \tag{9.3}$$

This proportionality is verified with film of Ag nanocrystals, in which the nanocrystals are disordered assembled. Figure 9.7 shows that the line shape agrees with the inverse size histogram because the intensity of Raman scattering frequency is inversely proportional to the nanocrystal diameter [20]. This confirms data published previously [19] and indicates intra-nanoparticle coherence, i.e. the atoms in the nanocrystals coherently vibrate.

The major difference of the Stokes/anti-Stokes Raman spectra of the ordered films compared with their amorphous counterpart is a shift toward low frequencies (Figure 9.6a,b) and to the quadrupolar lines: with "large" 3D superlattices, the quadrupolar lines are perfectly superimposed to the amorphous film (Figure 9.8a), whereas with "small" 3D superlattices (size$\leq \lambda/10$), the quadrupolar line is narrower than the amorphous film (Figure 9.8b).

Such changes in the Raman scattering spectra can be attributed to the following [21]:

(i) *The effect of the electric field (Lorentz field)*: the electromagnetic field, induced on each nanocrystal by the vibration-fluctuating electric dipoles of the neighboring nanocrystals organized in fcc 3D superlattices, changes the nanocrystal polarizability fluctuation.

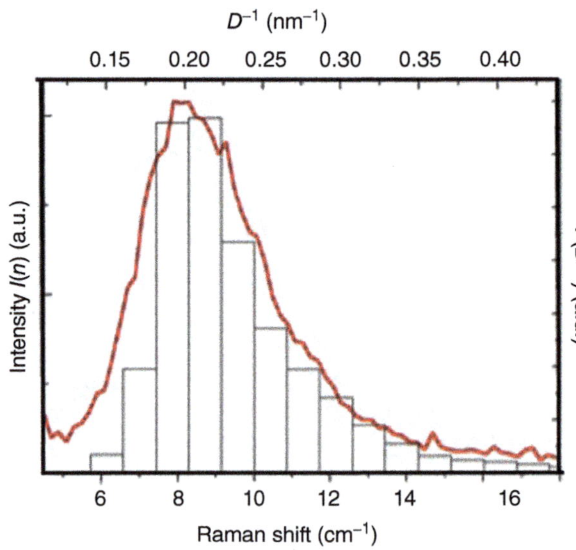

Figure 9.7 Comparison of the Stokes line shape of amorphous film of Ag nanocrystals with their inverse diameter distribution (D^{-1}). *Source:* Courty et al. (2005) [20]/ Springer Nature.

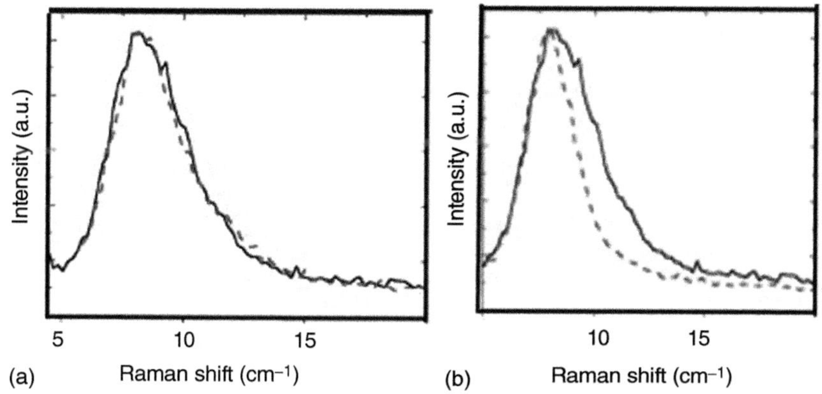

Figure 9.8 (a) Stokes–anti-Stokes Raman spectra of amorphous film (solid line) and large 3D superlattices (dashed line). (b) Stokes–anti-Stokes Raman spectra of amorphous film (solid line) and "small" 3D superlattices (dashed line). In both cases, there is superimposition of both line shapes (same symbols) after horizontal shifting. *Source:* Courty et al. (2005) [20]/Springer Nature.

(ii) *Vibrational coherence between nanocrystals in the 3D superlattice*: the van der Waals bonding between thiol chains is sufficient to establish a correlation between the vibrating nanocrystals, so that they vibrate coherently in a 3D superlattice.

Two features appear related to the 3D superlattice size [22]:

(1) When fcc 3D superlattices are small enough (size $< \lambda/10$, where λ is the laser excitation wavelength) and characterized by a rather large number of defects with slightly diffused spots in the small-angle X-ray diffraction pattern (Figure 9.5d), the light is scattered by *stationary modes*. The active vibration modes as those of a molecule are localized and determined only by their symmetry. In consequence, the selection rule for the phonon momentum vector does not limit the Raman active vibration mode number, so the intensity of Raman scattering from a 3D superlattice $I_{sc}(\nu)$ is proportional to the square of that of nanocrystal amorphous aggregates $I(\nu)$ as

$$I(\nu) \propto L^2(\nu)F^2(\nu) \propto L^2(\nu)I^2(\nu) \tag{9.4}$$

where ν, $I(\nu)$, and $L(\nu)$ are the frequency, the Raman scattering intensity for a disordered aggregate (without *inter*-nanocrystals), and the Lorentz field factor, respectively. The 3D superlattice Stokes and anti-Stokes $I_{sc}(\nu)$ profiles are horizontally shifted and vertically rescaled to match the peak maxima. The line profile is given by the square of that corresponding to a disordered arrangement of nanocrystals. The narrowed peak of the "small" 3D superlattices has the same profile as the square of the non-narrowed peak of the "disordered" aggregates (Figure 9.9). These data clearly indicate inter-nanocrystal coherence inside fcc 3D superlattices [20]. Hence, nanocrystals in a 3D superlattice behave as atoms in a nanocrystal. These coherences could explain the change

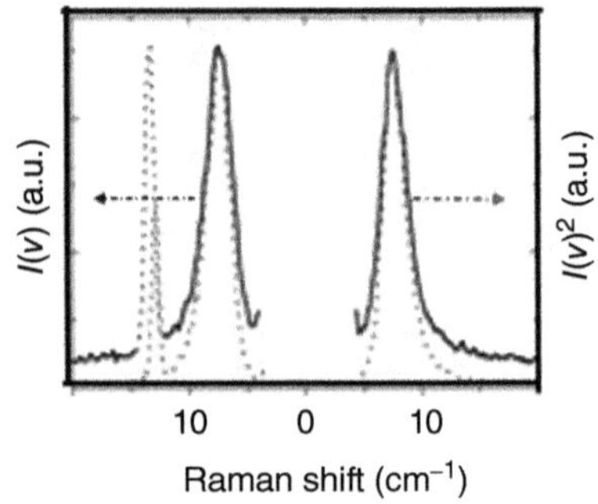

Figure 9.9 Square of Stokes and anti-Stokes intensity profile intensity for a small 3D superlattice (dashed line) and intensity profile of disordered aggregates (solid line). *Source:* Courty et al. (2005) [19]/Springer Nature.

in the transport properties observed previously with silver nanocrystal self-organizations [22].

(2) *With large-size 3D superlattices (size≥λ)*: Figure 9.8a shows a perfect superimposition of the Stokes and anti-Stokes spectra of disordered aggregates and large 3D superlattices. This is attributed to the fact that the light is scattered by *propagating modes* taking into account the volume of the 3D superlattice proportional to the number of nanocrystals used as building blocks. From mathematical calculation, show that the 3D superlattice Raman intensity $I_{sc}(\nu)$ is

$$I_{sc}(\nu) \propto L^2(\nu) I(\nu) \tag{9.5}$$

This clearly explains the data shown in Figure 9.8a where the quadrupolar lines of "large" 3D superlattices perfectly superimposed to that of amorphous film.

Instead of using dry Ag 3D superlattices deposited on a substrate, let us consider the water-dispersed Au "clustered" structure (Chapter 14) with average size of 196 ± 86 nm (around 14 layers) characterized by a large peak centered around 700 nm [23]. The energy flow upon light irradiation is investigated by pump–probe experiments. At long time scales, the differential transmission ($\Delta T/T$) spectra, investigated by pump–probe experiments and recorded as a function of time and of the probe wavelength (λ), and a mechanical oscillation period are observed (Figure 9.10). This period, attributed to the breathing mode of the whole assembly, is around of 300 ps [24]. The calculated period is given by $T = 2D/v_s$ where D is the average diameter (198 nm ± 84 nm) of the assembly and v_s is the speed of sound inside the 3D superlattice. With Co 3D superlattices, we found (see above) that the speed of sound in the 3D superlattice is 1235 ± 12 m s^{-1}. If we assume the speed of sound remains similar with Au compared with Co 3D superlattices, the resulting oscillation period of the Au "clustered" structure is 236 ± 77 ps consistent with experimental data. These results provide a clear indication from which the coherent vibration of nanocrystals in a 3D superlattice can be directly observed by using water-soluble hybrids, whereas an indirect process is needed for dried 3D superlattices.

From these data, it is concluded that nanocrystals vibrate coherently in a 3D superlattice as atoms in a nanocrystal.

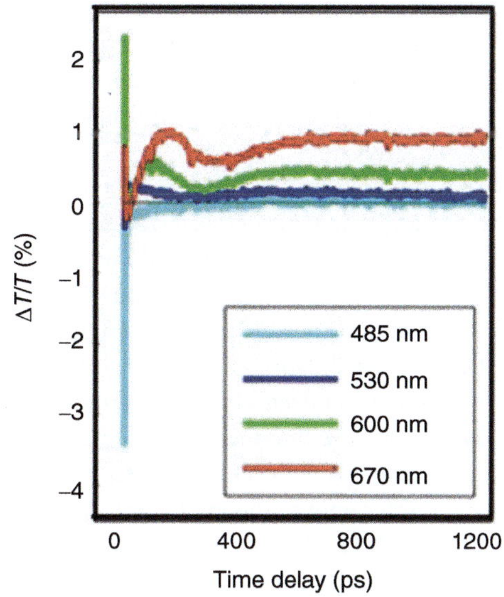

Figure 9.10 Dynamics of the $\Delta T/T$ at selected probe wavelengths on nanometer time scales of energy flow in the Au clustered structures. *Source:* Mazzanti et al. (2019) [24]/National Academy of Science.

9.4 Conclusions

From these examples, the door is open for finding new physical and chemical properties with the obvious possible analogies between atoms and nanocrystals. Evidently, the strength involved to form 3D superlattices from nanocrystals is not the same as that to produce bulk and/or nanocrystals from atoms. Nevertheless, this could be a new concept to go further in the paradigm concerning physical and chemical properties of matter at nano- and micrometer scales.

References

1 Lisiecki, I., Parker, D., Salzemann, C., and Pileni, M.P. (2007). Face-centered cubic supra-crystals and disordered three-dimensional assemblies of 7.5 nm cobalt nanocrystals: influence of the mesoscopic ordering on the magnetic properties. *Chem. Mater.* 19: 4030–4036.

2 Parker, D., Lisiecki, I., Salzemann, C. et al. (2007). Emergence of new collective properties of cobalt nanocrystals ordered in fcc supracrystals: II, Magnetic investigation. *J. Phys. Chem. C* 111: 12632–12638.

3 Petit, C. and Pileni, M.P. (1988). Synthesis in situ of cadmium sulfite synthetized in reverse micelles and hydrocarbon gels. *J. Phys. Chem.* 92: 2282–2286.

4 Petit, C., Lixon, P., and Pileni, M.P. (1990). Synthesis in situ of cadmium sulfate semiconductor: 2. Effect of functionalized surfactant. *J. Phys. Chem.* 94: 1598–1603.

5 Petit, C., Wang, Z.L., and Pileni, M.P. (2005). Seven nanometer HCP cobalt nanocrystals for high temperature magnetic applications through a novel annealing process. *J. Phys. Chem. B* 109: 15309–15316.

6 Zeng, H., Sun, S., Sandstrom, R.L., and Murray, C.B. (2003). Synthesis and magnetic properties of CoPy nanoparticles. *J. Magn. Magn. Mater.* 256: 227.

7 Yu, A.C.C., Mizuno, M., Sasaki, Y. et al. (2002). Structural characteristics and magnetic properties of chemically synthesized CoPt nanoparticles. *Appl. Phys. Lett.* 81: 3768.

8 Chudnovsky, E.M. (1989). Dependence of the magnetization law on structural disorder in amorphous ferromagnets. *J. Magn. Magn. Mater.* 79: 127–130.

9 Filippi, J., Amaral, V.S., and Barbara, B. (1991). High-field magnetization curve of random-anisotropy amorphous magnet: observation of a crossover and link to structural short-range order. *Phys. Rev. B* 44: 2842.

10 Thomas, L., Tuaillon, J., Perez, J.P. et al. (1995). Approach to saturation in nanocrystallized films of iron and nickel. *J. Magn. Magn. Mater.* 140–144: 437–438.

11 Gusev, V., Lomonosov, A.M., Ruello, P. et al. (2011). Depth-profiling of elastic and optical inhomogeneities in transparent materials by picosecond ultrasonic interferometry: theory. *J. Appl. Phys.* 110: 124908.

12 Lisiecki, I., Polli, D., Yan, C. et al. (2013). Coherent longitudinal acoustic phonons in three-dimensional supracrystals of cobalt nanocrystals. *Nano Lett.* 13: 4914–4919.

13 Durvasula, L.N. and Gammon, R.W. (1979). Brillouin scattering from shear waves in amorphous polycarbonate. *J. Appl. Phys.* 50: 4339–4344.

14 Lee, K.W., Lee, C.E., Choiand, J.Y., and Kim, Y. (2007). High intensity photoluminescence of microcrystalline $CsPbBr_3$ films: evidence for enhanced stimulated emission at room temperature. *J. Curr. Appl. Phys.* 7: 31–33.

15 Luedtke, W.D. and Landman, U. (1998). Structure and thermodynamics of self-assembled monolayers on gold nanocrystallites. *J. Phys. Chem. B* 102: 6566–6572.

16 Mukhopadhyay, R., Mitra, S., Johnson, M. et al. (2007). Evolution of the alkyl-chain dynamics in monolayer-protected gold clusters. *Phys. Rev. B* 75: 075414.

17 Duval, E., Boukenter, A., and Champagnon, B. (1986). Vibration eigenmodes and size of microcrystallites in glass: observation by very-low-frequency Raman scattering. *Phys. Rev. Lett.* 56: 2052–2055.

18 Courty, A., Albouy, P.A., Mermet, A. et al. (2005). Intrinsic vibrational coherence in face centered cubic supracrystals: Raman scattering measurements. *J. Phys. Chem. B* 109: 21159–21166.

19 Portales, H., Saviot, L., Duval, E. et al. (2001). Resonant Raman scattering by breathing modes of metal nanoparticles. *J. Chem. Phys.* 115 (8): 3444–3447.

20 Courty, A., Mermet, A., Albouy, P.A. et al. (2005). Vibrational coherence of self-organized silver nanocrystals in f.c.c. supra-crystals. *Nat. Mater.* 4: 395–398.

21 Duval, E., Mermet, A., Courty, A. et al. (2005). Coherence effects on Raman scattering from self-organized Ag nanocrystals: theory. *Phys. Rev. B* 72: 085439.

22 Taleb, A., Silly, F., Gusev, A.O. et al. (2000). Electron transport properties of nanocrystals: isolated or in "supra" crystalline phases. *Adv. Mater.* 12: 633–637.

23 Yang, N., Deeb, C., Pelouard, J.L. et al. (2017). Water-dispersed hydrophobic Au nanocrystal assemblies with a plasmon fingerprint. *ACS Nano* 11: 7797–7806.

24 Mazzanti, A., Yang, Z., Silva, M.G. et al. (2019). Light–heat conversion dynamics in highly diversified water-dispersed hydrophobic nanocrystal assemblies. *Proc. Natl. Acad. Sci. U.S.A.* 116: 8161–8166.

10

3D Superlattice Stability

In this chapter,we will discuss the stability of 2D and 3D superlattices. For this, we will consider the temperature of the support, the influence of the edge and the coating.

10.1 Influence of Temperature

The stability of 3D superlattices markedly depends on temperature [1].

In Chapter 11, we describe an intrinsic properties produced by subjection decanethiol silver nanocrystals self-assembled in 2D and 3D superlattices on a substrate (amorphous carbon or highly orientated pyrolytic graphite [HOPG]) to annealing process during eight days at 50 °C.

With amorphous carbon as substrate, many of the particles are regular multiply-twinned particles (MTPs). In addition, a small amount of flat triangular particles with more or less rounded edges are obtained. The High Resolution Transmission Electron Microscopy (HRTEM) images of flat particles show that the crystallinity of the particles is very high even at their edges with an fcc structure.

With HOPG instead of amorphous carbon, large triangular crystals are produced. The Transmission Electron Microscopy (TEM) enhancement shows that these triangular particles are equilateral, flat, and more or less truncated at the edges and that they are aligned on the substrate. Among 1000 particles, it has been found that there are 29% triangles, 3% hexagons, 6% decahedra, 19% faceted particles corresponding to pieces of triangles or hexagons, and 43% un-faceted particles with irregular. From the bright- and dark-field images, the triangle-shaped particles show uniform contrast indicating a high crystallinity and low thickness of the triangular particles. Several triangular particles are aligned on the HOPG substrate, and two epitaxial orientations of the triangles are observed. It is shown that the single fcc silver triangles are formed from a progressive coalescence of self-organized spherical 5-nm silver nanocrystals.

Metal Nano 3D Superlattices: Synthesis, Properties, and Applications, First Edition.

Hence, the three characteristic stages of the triangular single-crystal formation mechanism are as follows:

(i) Formation of silver nanocrystal arrays resulting from the evaporation of the nanocrystal solution dispersed in decane at 50 °C during 20 hours.
(ii) Annealing of the nanocrystal arrays induces the coalescence of neighboring nanocrystals by a partial desorption of the oxidized thiols in disulfide. The nanocrystal ordering favors their coalescence by reducing the interparticle distance. The only source of energy is brought by the annealing temperature. A small amount of extra energy is sufficient to initiate the coalescence process.
(iii) A recrystallization process of the coalesced nanocrystals leads to the formation of larger triangular crystals, free of twin planes. This last process can be accompanied by a fusion of small triangles into larger one. On HOPG, the silver triangular single crystals are epitaxially oriented. As the diffusion of silver atoms on an HOPG surface is faster than on amorphous carbon, the silver particles recrystallize more easily on HOPG than on amorphous carbon.

10.2 Edging Process

10.2.1 Ag Nanocrystals

Let us consider 5-nm silver nanocrystals produced from reverse micelles, coated with dodecanethiol and dispersed in hexane. A colloidal drop is deposited on a copper TEM grid covered with either amorphous carbon or HOPG. As expected [2], the nanocrystals deposited on amorphous carbon are ordered over a very short range (Figure 10.1a,b). In contrast, the same nanocrystals self-assemble over a very large scale in mono- and bilayers on HOPG (Figure 10.1c,d and inset).

Let us consider the ordering produced by deposition of a drop of Ag nanocrystal colloidal solution on HOPG substrate. Figure 10.2a shows three areas observed: two monolayers marked as 1 and 2 and a bilayer marked as 3 [3–5].

A careful comparison of monolayer 1 and 2 morphologies shows that one of the major difference is obtained from the fast Fourier transform (FFT): monolayer 1 FFT (Figure 10.2b) shows well-defined spots indicating a high level of nanocrystals ordering. In contrast, the monolayer 2 FFT (Figure 10.2b) shows a distorted hexagon with diffuse spots, indicating a lower level of ordering and thus the presence of some defects. Stacking faults, grain boundaries, and/or lack of nanocrystals in the hexagonal network break periodicity and generate distortion. If the hexagons are slightly locally distorted, dodecanethiol chains emanating from surfaces of two adjacent nanocrystals do not exactly face each other, meaning that the number of chains involved in the interdigitation is lower than for an undistorted hexagon [3, 6–10]. This difference in the level of nanocrystal ordering is probably due to surface defects of HOPG itself. The cleavage induces bundles, steps, and/or islands at the surface with some large defect-free monodomains of about 1 μm^2 that is in the range of the monolayer area [10]. The two monolayers are distinguished also by the distances separating nanocrystals from edge to edge ($d_{\text{p-p}}$) in the networks. The average distance $\delta_{\text{p-p}}$ measured in the three directions of the hexagonal network of

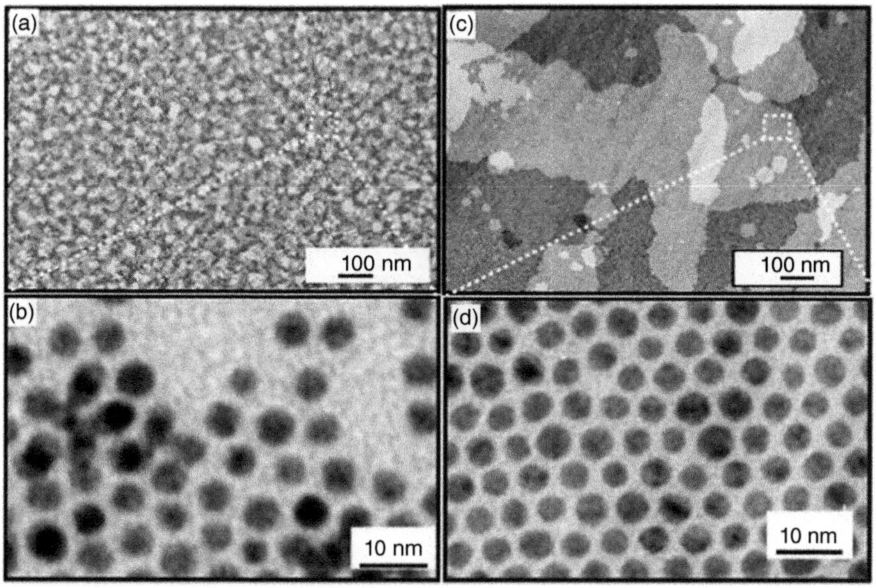

Figure 10.1 TEM images at low magnification (a, c) and higher magnification (b, d) of dodecanethiol-coated Ag nanocrystals deposited on an amorphous carbon (a, b) and on a highly oriented pyrolytic graphite (HOPG) substrate (c, d). *Source:* Klecha et al. [5]/American Chemical Society.

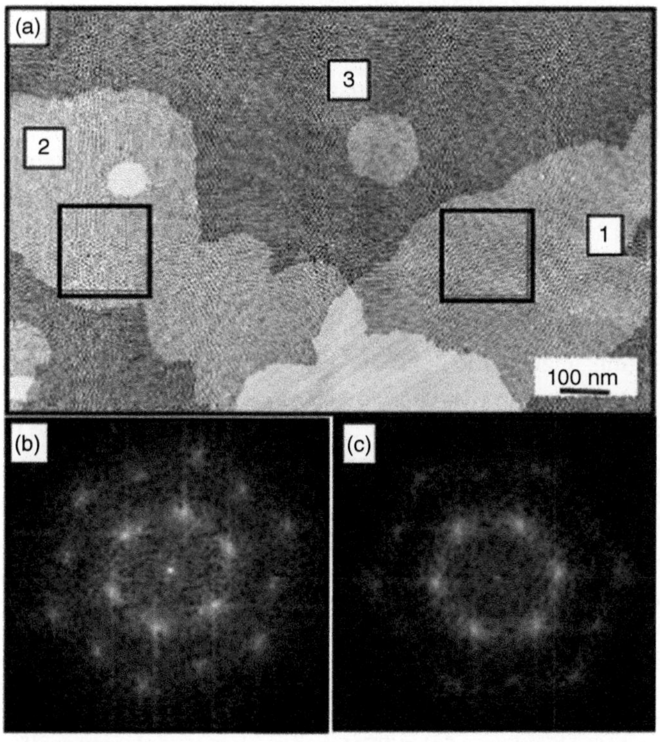

Figure 10.2 TEM images of silver nanocrystals deposited on HOPG substrate (a). Fast Fourier transform of monolayer 1 (b) and monolayer 2 (c). *Source:* Klecha et al. [5]/American Chemical Society.

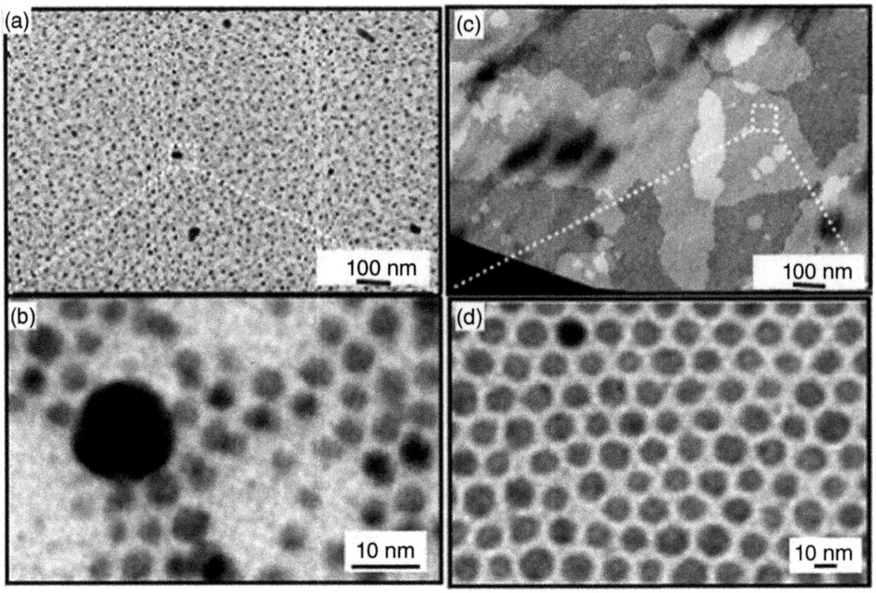

Figure 10.3 TEM images at low magnification (a, c) and higher magnification (b, d) of dodecanethiol-coated Ag nanocrystals after oxygen plasma exposure. Ag nanocrystals are deposited on an amorphous carbon (a, b) and on a highly oriented pyrolytic graphite (HOPG) substrate (a, d). Same areas of TEM grid as shown in Figure 10.1 after exposure to oxygen plasma for 5 s at 5 W. *Source:* Klecha et al. [5]/American Chemical Society.

monolayer 1 is shorter ($\delta_{p\text{-}p}$ $=2.1\pm0.1$ nm) than that measured in monolayer 2 ($\delta_{p\text{-}p}$ $=2.3\pm0.1$ nm), indicating in this case a difference in the number of carbon atoms contributing to interdigitation.

The samples shown in Figure 10.1 are loaded into the RIE apparatus and etched with a 5-W oxygen plasma power during 5 s. After exposure, the TEM images of the two samples (one the same zone as shown in Figure 10.1) markedly differ. On amorphous carbon, a large number of coalesced objects are observed over the entire surface of the TEM grid (Figure 10.3a,b), whereas, on HOPG, the plasma treatment leaves the assemblies essentially identical with no changes in either the nanocrystal integrities or their ordering. Figure 10.3c,d shows that the hexagonal pattern is intact, as if the oxygen plasma has no effect on the nanocrystal deposit.

Note in Figure 10.3c that dark shadows appear on the TEM grid; they are due to the HOPG sheet structure. These different behaviors seem to be related to the range of nanocrystals ordering.

To confirm, the sample shown in Figure 10.2 is subjected to a higher plasma oxygen power. After etching (10 W, 5 s), the TEM pattern (Figure 10.4a) markedly differs from that before etching (Figure 10.3a) and from that obtained at lower plasma power (5 W, 5 s) (Figure 10.3c).

Hence, the TEM images show three nanocrystal orderings:

(i) Monolayer 1 appears unchanged by the oxygen plasma (Figure 10.4a). However, a higher magnification of the TEM image shows coalescence of a few nanocrystals (Figure 10.5a).

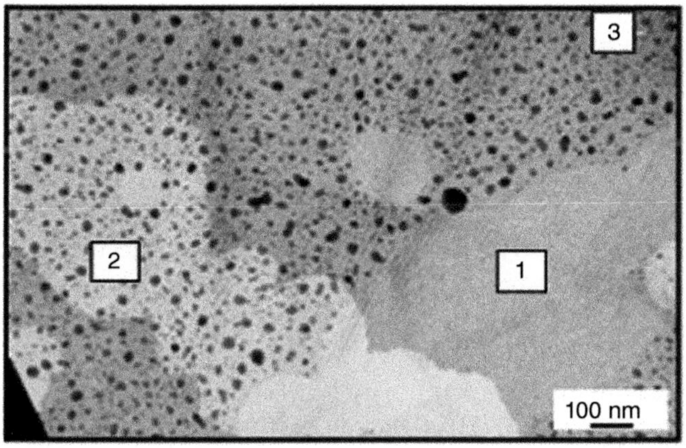

Figure 10.4 TEM images of silver nanocrystals deposited on HOPG substrate after oxygen plasma exposure (5 s, 10 W). Monolayers are marked 1 and 2; bilayer is marked 3.
Source: Pileni (2011)/Royal Society of Chemistry.

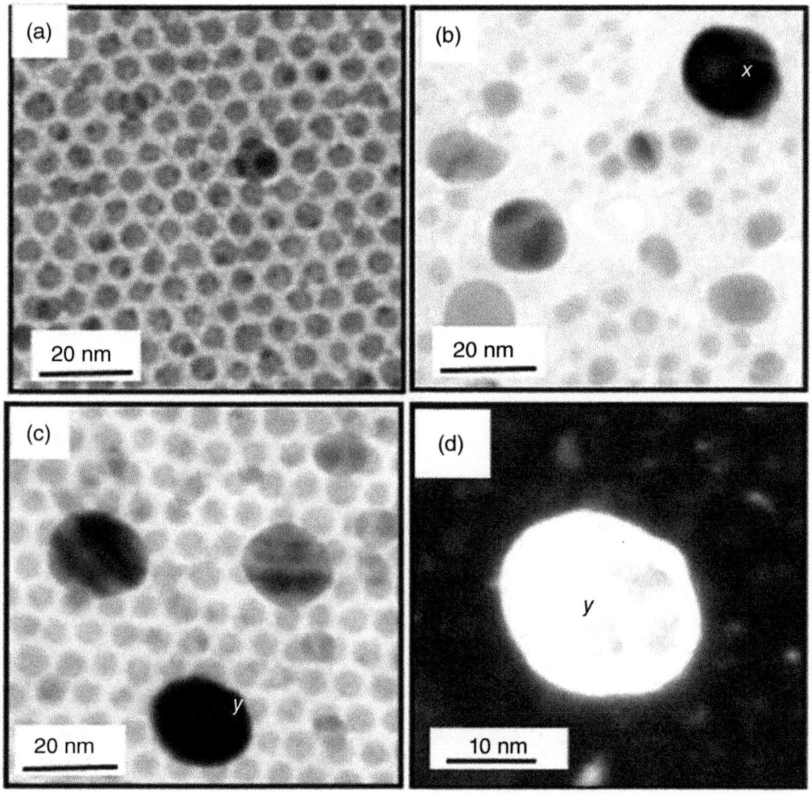

Figure 10.5 TEM images of silver nanocrystals corresponding of monolayers marked as 1 (a), 2 (b), and 3 (c). Dark-field images of coalesced object of the monolayer 2 after plasma (d).
Source: Klecha et al. [5]/American Chemical Society.

(ii) Monolayer 2 (Figure 10.5b) is totally destroyed with coalescence of a very large number of nanocrystals forming large particles (Figure 10.5b).
(iii) Bilayer 3 shows coalescence of nanocrystals forming the top layer (Figures 10.4 and 10.5c), whereas the bottom layer is slightly modified as observed in monolayer 1.

It thus appears that the level of ordering distinguishes the two monolayers (1 and 2) is the key parameter in the nanocrystal stability that provides protection from coalescence induced by oxygen plasma. When coalescence occurs (i.e. for monolayer 2 and bilayer 3), the sizes and size distributions (Figure 10.6) of large coalesced objects are similar. Hence, the nanocrystal coalescence takes place when they are deposited either on the bare substrate itself or on a nanocrystal monolayer.

The bottom layer of the bilayer behaves as monolayer 1 (slightly damage), and coalescence occurs laterally (i.e. parallel to the substrate). This clearly rules out the particle–substrate interactions as a parameter in the mechanism of coalescence and in the stability of the monolayer 1. High magnifications of monolayer 2 (Figure 10.5b) and bilayer 3 (Figure 10.5c) after etching show coalesced (Figure 10.5c) with a spheroidal shape. The corresponding dark-field image (Figure 10.5d) shows very bright particles, indicating diffracting crystals and consequently well-crystallized objects with the formation of single crystals. Figure 10.7a shows an image of a crystal with well-defined lattice spacing (Figure 10.7b) of 2.34 Å (determined from the profile shown in Figure 10.7c) corresponding to the (111) lattice spacing of the fcc Ag crystal (δ_{theor} = 2.36 Å). The fringes extend across the entire object right up to the edges, confirming a well-crystallized object. This coalesced single crystal (Figure 10.7a) is very thin as the underlying nanocrystals are observable. Electronic diffraction and energy-dispersive X-ray spectrometry (EDX) do not reveal the formation of silver oxide either for coalesced objects or for the intact nanocrystals.

To confirm the absence of silver oxide although silver nanocrystals are subjected to oxygen plasma, EELS measurements have been made. Figure 10.8a shows a

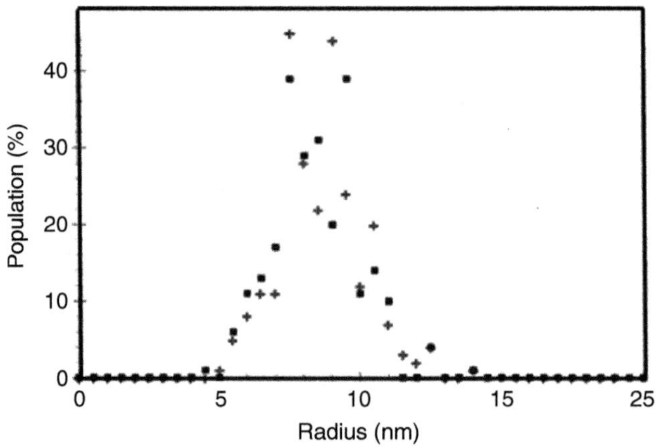

Figure 10.6 Size distribution of the coalesced objects of monolayer 2 (•) and of the bilayer 3 (+) produced after oxygen plasma exposure (5 s, 10 W). *Source:* Based on Klecha et al. [4].

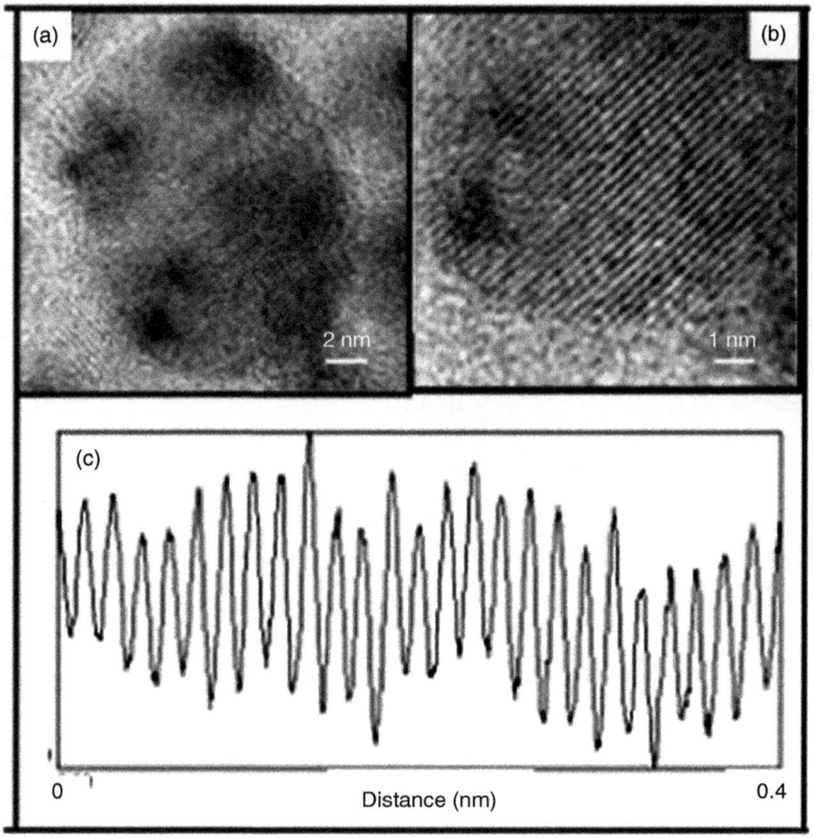

Figure 10.7 (a) High-resolution TEM images of a single monocrystal on bilayer 3 after oxygen plasma exposure. (b) Higher magnification of an edge and (c) profile. *Source:* (b and c) Klecha et al. [5]/American Chemical Society.

high-angle annular dark-field image (HAADF) after plasma exposure (5 s, 10 W) of monolayer 2. Such images are formed using electrons scattered beyond the typical Bragg angles and are dominated by Z-contrast rather than diffraction contrast. The technique is highly sensitive to variations in the atomic number Z of the atoms in the sample. This implies that all the bright objects in this image contain silver. The same area of monolayer 2 observed by classical bright-field TEM (inset Figure 10.8a) shows an opposite contrast (the BF image will exhibit diffraction effects). The spectrum, in the energy-loss range containing the K edges of carbon and oxygen of the bare substrate (zone z'), shows that oxygen has a very low but nevertheless clearly detectable intensity indicating its presence on the bare substrate (Figure 10.8b). On zone z corresponding to the middle of a coalesced particle (Figure 10.8a), the spectrum exhibits a broadband at the $M_{4,5}$ edge of silver (Figure 10.8c). Moreover, at the K-edge of oxygen, a very small signal may be present, but the intensity is certainly much lower than on the substrate. The substrate is still detected confirming formation of a very thin coalesced particle. The EELS (Electron Energy Loss Spectroscopy) elemental maps of the coalesced particle (z) are generated using the spectrum-image technique [11] and

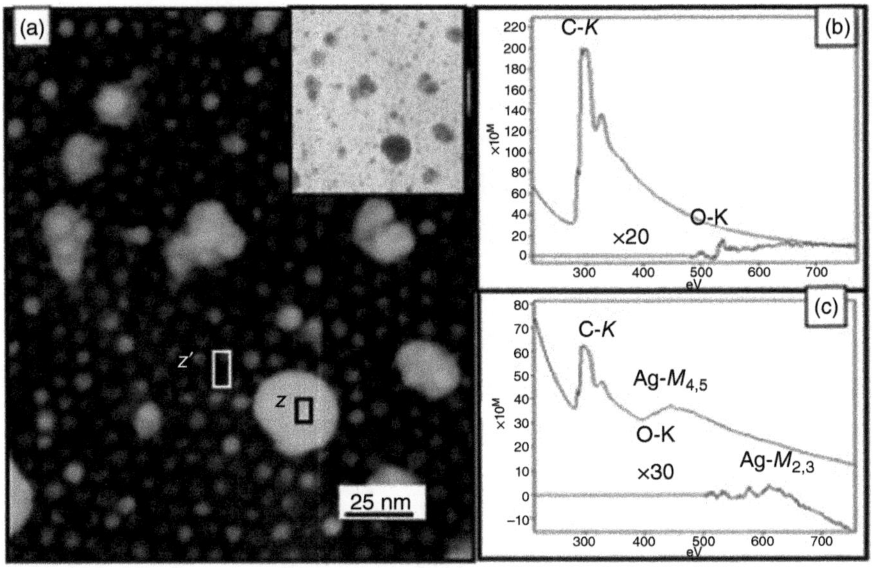

Figure 10.8 (a) High-angle annular dark-field (HAADF) image after oxygen plasma exposure (5 s, 10 W) of monolayer 2, corresponding bright-field image in inset. (b) Electron energy-loss spectrum of the substrate marked *z′* in (a) and (c) EELS of a coalesced object marked *z* in (a). *Source:* Klecha et al. [5]/American Chemical Society.

the standard Egerton background subtraction method [12] with 30- and 40-eV background fitting and signal integration windows, respectively.

Hence, from EELS measurements, the following is observed:

(i) The carbon map shows the bare substrate as bright, but the area covered by the coalesced particle is darker, although there is presumably still just as much carbon present in this area. This is because the silver particle scatters electrons strongly to high angles outside the acceptance angle of the spectrometer, thus reducing the total signal at all energies, including those of the carbon signal.

(ii) The silver map shows a strong signal everywhere, whereas the distribution of oxygen is homogeneous on the substrate and is somewhat reduced on the coalesced object, as for carbon.

(iii) The presence of this oxygen signal everywhere means that it is difficult to put a lower limit on the amount of silver oxide present. However, no bright shell corresponding to a surface oxidation is observable. Layers up to 3 nm, when present, are readily detected by this technique [13, 14], and isolated silver oxide particles are known to show a clear, easily detected oxygen signal. From these analyses, it is concluded that the coalesced particles are pure silver.

Figure 10.9a,b shows HRTEM images of the same three particles before and after plasma treatment, respectively. After exposure (Figure 10.9b), the atomic planes of three MTP nanocrystals tend to align, with planes maintaining adhesion of the three partially coalesced silver nanocrystals. This implies a reorganization of the atoms [11, 15] and the first stage of the larger particle formation.

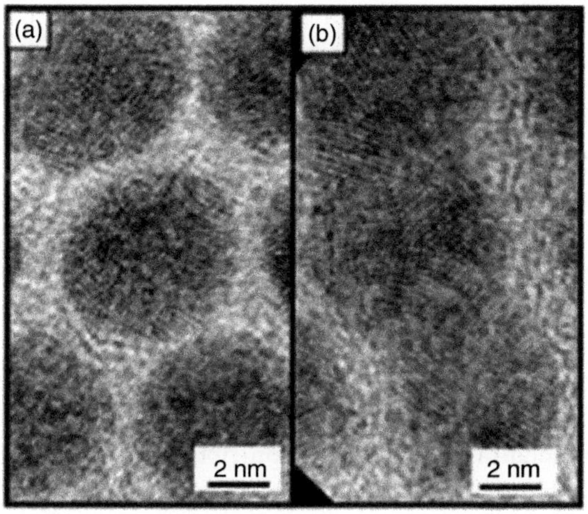

Figure 10.9 High-resolution TEM images of nanocrystals of monolayer 1 (a) before and (b) after oxygen plasma exposure (5 s, 10 W). *Source:* Klecha et al. [5]/American Chemical Society.

To understand the mechanism, we have to consider the effects of oxygen plasma and atomic oxygen onto alkanethiol self-assembled monolayer (SAM)-coated flat metallic surfaces. Reactive ion etching technology is a combination of physical etching and chemical etching. At the substrate surface, materials can be removed by both ion bombardment and chemical reaction. A variety of activated species including radicals, ions, free electrons, and gas molecules are employed, especially in the case of oxygen plasma, to decompose the organic protecting groups. The ability of oxygen to etch the hydrocarbon chains is related to its ability to form linkages such as R–(C=O)–R that can then react further via decarboxylation to produce volatile carbon-containing species, resulting in the formation of a steady-state etch front [16–20]. Thus, when SAM-coated metallic surfaces are subjected to a 5-W oxygen plasma during 5 s, they are 80% destroyed [17, 19]. However, no damages are observed when silver nanocrystals self-assembled over a long range (i.e. deposited on HOPG) subjected to the same treatment. Dodecanethiol chains covering the silver nanocrystals are certainly damaged, but not enough to involve coalescence. This clearly indicates that alkanethiols adsorbed on nanocrystals do not have analogous sensitivity to the plasma exposure as when adsorbed on a flat surface. More, the interdigitation of the surfactant chains in the case of the nanocrystals slows down the progression of the etch front. This explains the results on amorphous carbon at 5 W where the nanocrystals are ordered over a short range and on HOPG with a higher plasma power (10 W). We cannot totally dismiss a mild heating as a plasma is applied. However, the exposure time is very short (5 s), and the different behaviors for two monolayers (1 and 2) lying on a same substrate permit us to neglect the heating contribution. Monolayer 1 that has the higher level of ordering (i.e. greater alkanethiol chains interdigitated) is quite intact, whereas monolayer 2 that has less number of interdigitated chains is destroyed. Later, the behavior of the coating agent adsorbed on silver nanocrystals is close to that of SAM. The reactivity of the species from the oxygen plasma is initiated in proximity to defect sites (distorted hexagons). Then, the inflow is trapped in the interdigitated dodecanethiol chains and reacts with them to form carbon-containing oxygen functional groups [16]. Hence, the desorbing hydrocarbons take up oxygen ions and oxidized protected silver from oxidation. This is supported by experiments on gold nanoparticles. When similar oxygen plasma treatment is applied to uncoated gold nanoparticles ranging in sizes from 1 to 8 nm, oxide derivative is formed, whereas gold-55 clusters stabilized by a ligand shell remain in their reduced form [21].

Similarly, no oxidation takes place when organized gold nanoparticles stabilized by a ligand are subjected to a low-temperature oxygen plasma [22]. Nevertheless, we cannot exclude either a change in the electronic states due to the ordering or a strong interaction between the surfactant molecules [23], contributing to the inertness of the nanocrystals from oxidation. The complete formation from atomic diffusion of the nanocrystals shown in Figure 10.9b, to larger particles (Figure 10.5b,c), cannot be followed step by step, and the equipment used does not allow changing the oxygen plasma power progressively from 5 W (no effect) to 10 W (large particles). However, the atomic diffusion and coalescence mechanism involved in the formation of large crystals from monolayer 2 or bilayer 3 are similar to that observed with monolayer 1. When interdigitation is not strong enough, the coating agent tends to be removed physically or chemically by the reactive species of the oxygen plasma. Then, the high mobility of surface atoms favors the recrystallization reducing their proportion and the surface energy and involving formation either of single crystals or of the polycrystalline phase. Similar mechanism has been observed by heating an assembly of 5-nm silver nanocrystals [1] and on recrystallization of MTP particles into single crystals observed under UHV (Ultra High Vaccum) [24].

Hence, an intrinsic property of 2D self-organizations is due to the ordering: self-assembled silver nanocrystals are not sensitive to a low-level oxygen plasma exposure (5 W, 5 s) when ordered over a long range (on HOPG), whereas they coalesce under the same treatment when ordered on short range (on amorphous carbon).

10.2.2 Au Nanocrystals

From Chapters 1 and 2, we know the synthesis of Au nanocrystals, produced from inorganic chemistry [25], is a mixture of single-domain and polycrystalline Au nanocrystals. Furthermore, from thermodynamic equilibrium, it is possible to select separately single-domain and polycrystalline Au nanocrystals. As mentioned, an interface supracrystal film and shape 3D superlattices are produced at the air–solvent interface and in the bottom of the beaker, respectively, whereas some Au nanocrystals are still dispersed in solution. It is shown in Chapter 2 that the building blocks of the 3D superlattices produced single-domain nanocrystals, whereas those remaining in solution are polycrystals. 3D superlattices are easily dispersed in hexane. Hence, a drop of colloidal solution of either single or MTP nanocrystals is deposited on TEM grid. As expected, well-defined 2D superlattices are obtained

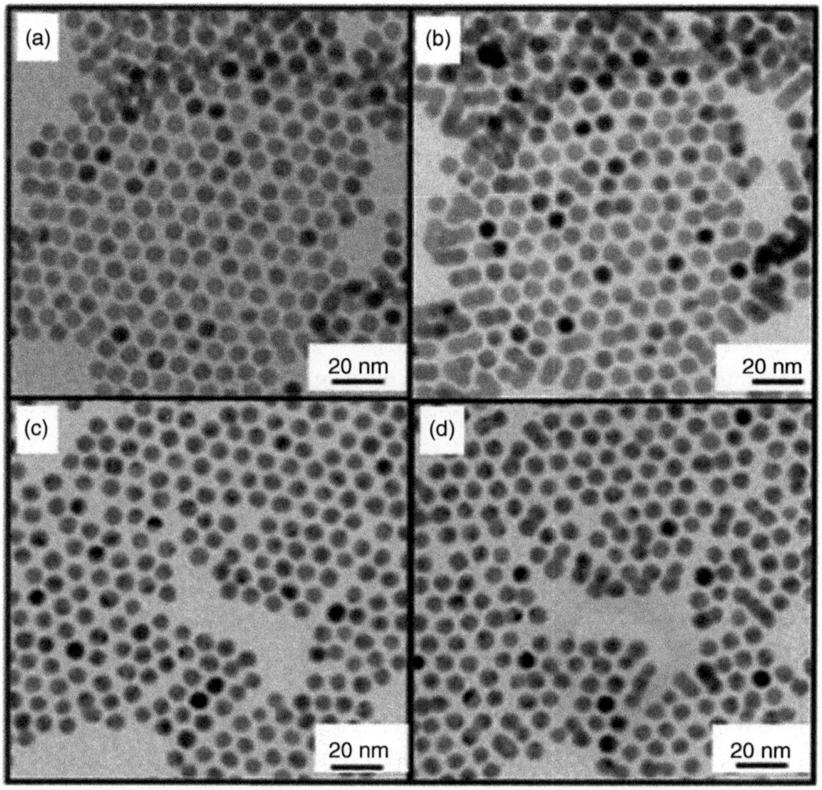

Figure 10.10 TEM images of dodecanethiol-coated Au nanocrystals deposited on an amorphous carbon substrate before (a, c) and after (b, d) oxygen plasma exposure. Same areas after exposure to oxygen plasma for 15 s at 5 W: single-domain Au nanocrystals (a, b) and polycrystalline phase (c, d).

(Figure 10.10a,c). To evaluate the stability of such self-assemblies, the two samples are loaded into the RIE apparatus and etched with a 5-W oxygen plasma power during a given time. After 5 and 10 s' etching, no detectable changes are observed. However, after 15 s, Figure 10.10b,d shows that in both case at the edge of the monolayers, nanocrystals tend to coalesce, whereas when the nanocrystals are well ordered in hexagonal network, they remain quite unchanged. By comparison, it seems that single-domain Au nanocrystals are slightly less stable than their polycrystalline counterparts.

10.2.3 Influence of the Coating Agent

As described in Chapter 1, we know that it is possible to produce nanocrystals having similar diameter from various precursors.

Let us first compare the stability of around 6-nm Au nanocrystals coated with dodecanethiol and produced from various precursors and procedures. Three procedures are used:

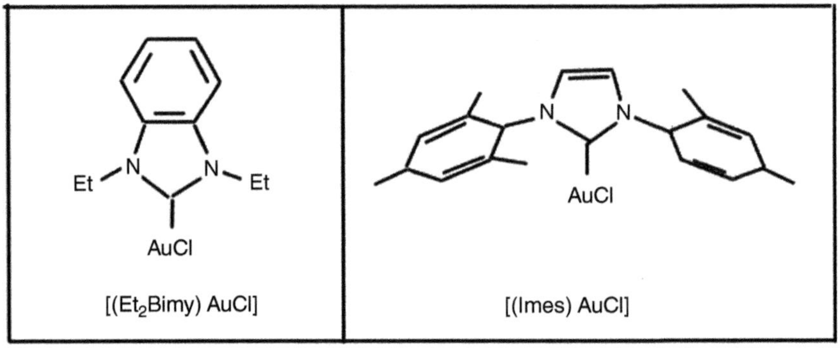

Figure 10.11 [Au(NHC)Cl] precursors used for the preparation of Au nanocrystals coated with dodecanethiol.

(1) *Reverse micelles* [26, 27] tetrachloroauric acid [$HAuCl_4$] is used as precursor.
(2) *Inorganic chemical reaction*: chloro(triphenylphosphine)gold [(PPh_3)AuCl] is used as precursor.
(3) *Thermal decomposition* [28, 29]: the precursors are shown in Figure 10.11.

Figure 10.12 shows Au nanocrystals (around 6 nm) coated with dodecanethiol produced from various methods and with various precursors. They self-assemble in 2D superlattice (Figure 10.12a–d) and after (Figure 10.12e–h) oxygen plasma exposure (80 s, 10 W). Obviously, Figure 10.12e–h clearly shows the formation of large proportion of coalesced objects. After etching, the nanocrystals cannot be redispersed because their coating agents are lost. This first series of experiments [30–32]

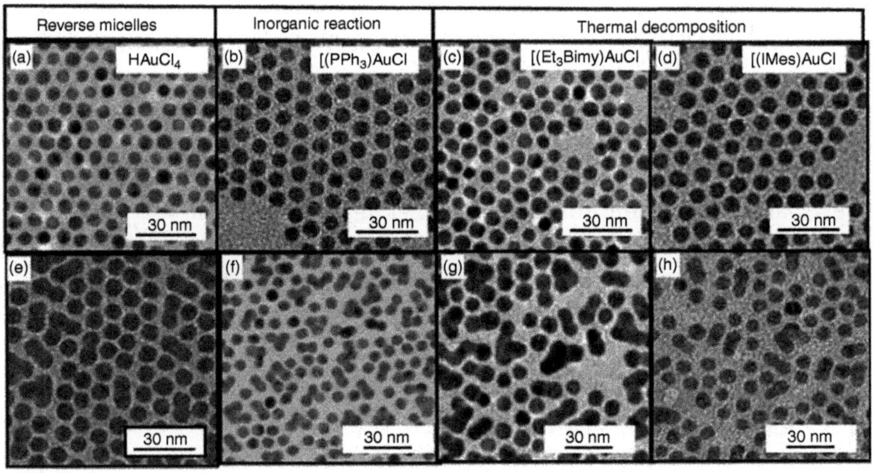

Figure 10.12 TEM images of 6-nm Au nanocrystals produced from various precursors deposited on an amorphous carbon-covered TEM grid before (a, b, c, d) oxygen plasma exposure and same areas after (e, f, g, h) exposure to oxygen plasma for 80 s at 10 W. The precursors are [$HAuCl_4$] (a, e), [(PPh_3AuCl] (b, f), [Et_2 Bimy)AuCl] (c, g), and [I Mes)AuCl] (d, h). *Source:* Ling et al. [32]. Copyright © 2015, American Chemical Society.

demonstrate that all dodecanethiol-coated Au nanocrystals produced from various precursors are similarly affected by oxygen plasma treatment. Similar data are obtained with 2.6-nm Au nanocrystals indicating that their resistance against aggregation when arranged as 2D networks is independent of the nanocrystal sizes. In addition, the method to prepare the nanocrystals does not have any influence on the stability. This suggests that the stability of Au nanocrystals coated with dodecanethiol is strongly related to the nature of the ligand. Hence, the dodecanethiol-coated Au nanocrystals coalesce at 80 s of exposure, suggesting that dodecanethiol ligands on the surface had partially been desorbed and induced agglomeration of the nanocrystals.

Instead of using dodecanethiol as ligand, let us consider various N-heterocyclic carbenes (NHCs) as new ligands. NHCs emerge as a major class of ligands in organic chemistry [29, 33–35]. Figure 10.13 shows a series of NHC-coated Au nanocrystals with various ligands. For simplicity, let us name them A, B, and C: {[(NHC^AH^+) $AuCl_4^-$], [(NHC^BH^+)$AuCl_4^-$], [(NHC^CH^+)$AuCl_4$]}.

The major difference between the various NHC ligands used in this study is related to their steric hindrance. Hence, three different families of NHC ligands (**NHC^{A-C}**) bearing long alkyl chains at different positions (Figure 10.13) stabilize Au nanocrystals. Here, we mention the major differences of this series:

(1) NHC^A is characterized by aliphatic chains with different lengths (C_7) in the position 5 of the aromatic ring. The small ethyl substituents are positioned at the nitrogen atoms. Similar data were produced with alkyl chains (C_9).
(2) NHC^B is characterized by the long aliphatic chains placed both on the nitrogen atoms and at position 5 of the aromatic ring.
(3) With NHC^C, the long alkyl chains are only present on the nitrogen atoms.

NHC^B has the largest volume among these three NHC ligands, whereas NHC^A had the smallest volume ($NHC^B > NHC^C > NHC^A$). In Chapter 1 [30], it is shown that the

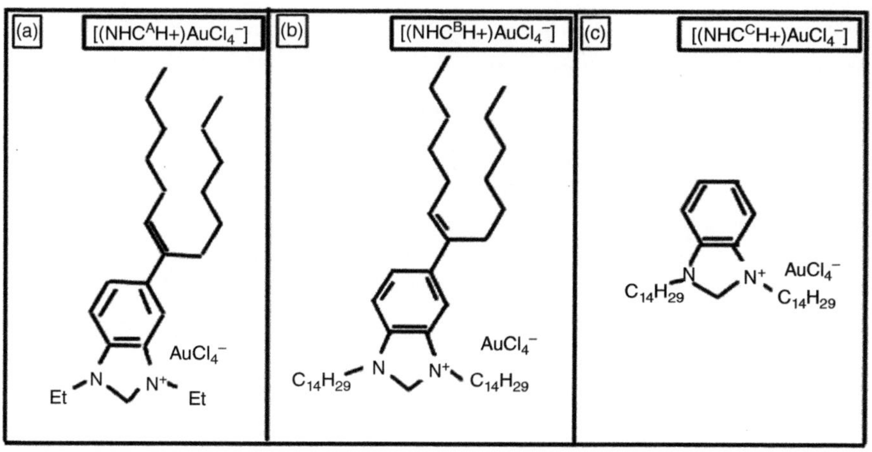

Figure 10.13 Benzimidazolium chloroaurate salts used for the preparation of NHC-coated Au nanocrystals (a) [(NHC^AH^+)$AuCl_4^-$], (b) [(NHC^AH^+)$AuCl_4^-$], and (c) [(NHC^AH^+)$AuCl_4$]}.

steric hindrance of these NHCs had an effect on nanocrystal size, likely related to the volume of the NHC. The NHC-stabilized Au nanocrystals are dispersed in toluene and deposited on copper TEM grids. As previously observed, the TEM images used as references show that NHC^A Au nanocrystals (Figure 10.14a) exhibit an average diameter of 5.7 nm. Similar size is obtained by replacing the C_7 alkyl chain by C_9 (5.9 nm). NHC^C Au nanocrystals (Figure 10.14b) exhibit slightly smaller average sizes (4.9 nm). With NHC^C Au nanocrystals (Figure 10.14c), the average diameter is 4 nm. After TEM control, the samples are loaded into the RIE apparatus and submitted to oxygen plasma for 80 s. At this exposure time, in contrast to Au nanocrystals coated with dodecanethiol (DTT), no changes in either the nanocrystal integrity or their 2D assembly are detected by TEM for all NHC-based Au nanocrystals. The exposure time is then increased to 120 s. Figure 10.14d–f shows that the structures are not affected by the oxidative treatment. This demonstrates that NHC-coated Au nanocrystals are resistant to oxygen plasma treatment for at least 120 s.

Note that it has been impossible to subject the grid because the carbon film on copper TEM grids is strongly affected at this exposure time and totally broken down at an exposure time exceeding 180 s.

The influence of pure oxygen on Au nanocrystals coated either with dodecanethiol or NHCs studied by NMR (Nuclear Magnetic Resonance) shows that the exposure to oxygen induces the oxidation of a significant amount of dodecanethiol into its corresponding disulfide. By comparison, the oxygen treatment of NHC-coated Au

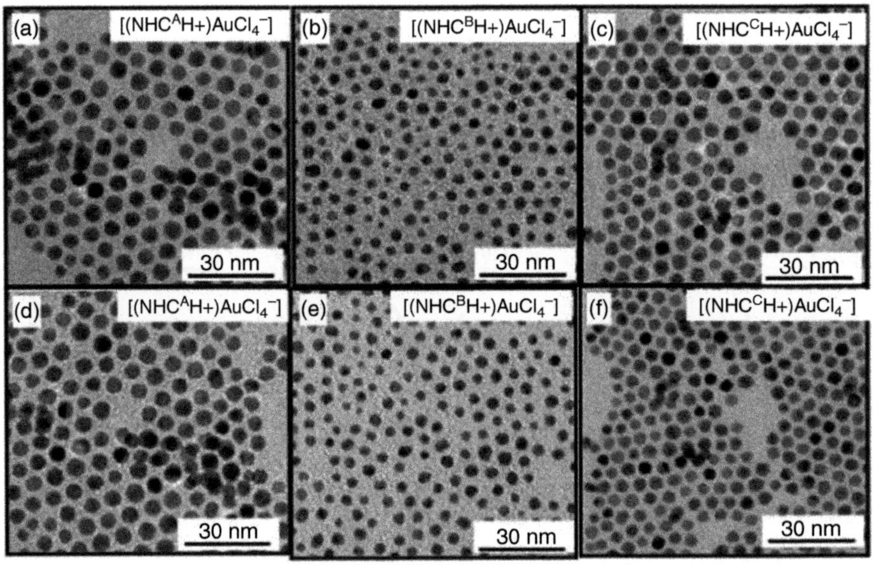

Figure 10.14 TEM images of NHC^AAu nanocrystals (a), NHC^BAu nanocrystals (b), and NHC^CAu nanocrystals (c) deposited on an amorphous carbon-covered TEM grid before oxygen plasma exposure (a, b, c). Same areas after exposure (d, e, f) to oxygen plasma for 120 s at 10 W: (a, d) NHC^AAu nanocrystals, (b, e) NHC^BAu nanocrystals, and (c, f) NHC^CAu nanocrystals. *Source:* Ling et al. [32]. Copyright © 2015, American Chemical Society.

nanocrystals induces the formation of lower amounts of ligand-derived species that might be related to the increased oxygen stability of the NHC–Au unit. Thus, although NHC ligands bound to the surface of nanocrystals are not completely insensitive to the effect of oxygen, they show more resistance to oxygen than thiols.

10.3 Solvent-Mediated Crystallization of Nanocrystal 3D Assemblies

Let us consider two colloidal solutions of 2.2-nm Ag nanocrystals coated with oleylamine (C_{18}–NH_2) and dispersed either in hexane or tetrachloroethylene (TCE) [36]. For the deposition, few drops of the colloidal solutions were placed on the substrates, and 3D films are produced (Figure 10.15a,c). The crystalline structures of these 3D superlattices depend on the solvent used to disperse nanocrystals: with hexane as solvent, the small-angle X-ray scattering (SAXS) pattern of the films (Figure 10.15b) is specific to a body-centered cubic (bcc) structure, whereas when TCE is used as solvent, face-centered cubic (fcc) crystalline structure is produced (Figure 10.15d). This is further confirmed by the TEM images (Figure 10.15ac), which reveal a change in the 2.2-nm Ag nanocrystals ordering with typical (110) and (111) crystal plane parallel to the substrate for bcc and fcc structures, respectively. The lattice fringes corresponding to (110) and (220) crystal planes of the bcc and fcc superlattices are 4.2 and 2.8 nm, and the edge-to-edge interparticle distances (δ_{pp}), calculated

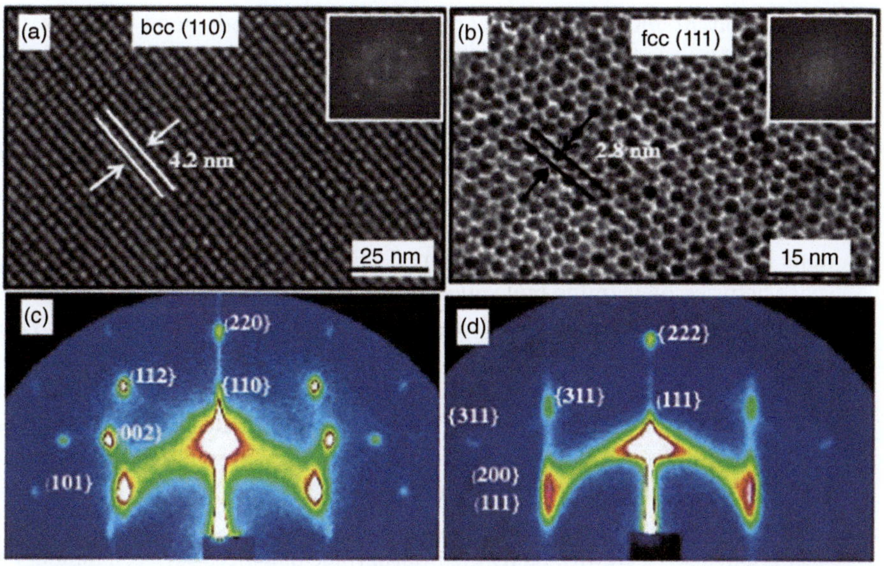

Figure 10.15 (a, c) TEM images of superlattices assembled with 2.2-nm Ag nanocrystals from different solvents: hexane (a) and TCE (c) and (b, d) small-angle X-ray diffraction pattern of superlattices assembled with 2.2-nm Ag nanocrystals from solvent hexane (b) or solvent TCE (d). *Source:* (a–d) Wei et al. [36]/American Chemical Society.

from the SAXS patterns, are 3.0 and 3.5 nm, respectively. This indicates a change in the degree of interdigitation between nanocrystals with the solvent used during superlattice growth.

These data clearly demonstrate that the memory on the 3D superlattice fabrication is retained with a strong influence on their crystalline structures. Such change in the crystalline structure with the solvent used to disperse nanocrystals is relatively surprising and opens a new mean of predicting the crystalline structure of this type of assemblies. With hexane used as solvent, the observed results agree with the phenomenological approach proposed [37] where the softness of the ligand shell can be parameterized by the ratio (χ) of the ligand length (L) to the particle radius (R) (i.e. $\chi = L/R$). When the ratio $\chi < 0.7$, the nanocrystals contact as hard spheres and form close-packed assemblies with fcc symmetry. By opposition, if the ratio $\chi > 0.7$, the interdigitation between ligands becomes long ranged and can interact directly with chains belonging to its second nearest neighbors [38], favoring the formation of more open bcc structure with lower packing fraction. Here, the 2.2-nm Ag nanocrystals are coated with oleylamine whose chain length is around 1.8 nm. Hence, the softness ratio of the nanocrystals is calculated to be very large with $\chi = 1.6$. This large softness ratio favors the formation of bcc structure in these conditions and can account to the formation of bcc structure when using hexane as the carrier solvent. However, the influence of the softness fails to explain observation of fcc structure when TCE is used as a solvent, even though the softness ratio of $\chi = 1.6$ is expected for the same coating agent and inorganic core. Thus, other parameters have to be taken into account to predict such crystalline structure:

(1) *Ligand–solvent interactions*: hexane and TCE are both good solvents for nanocrystals coated by organics like oleylamine. There are stronger interactions between solvent and oleylamine molecules when nanocrystals are dispersed in TCE because of the presence of double bonds present in both species, compared with ligand–solvent interactions when using hexane. Then, some of the oleylamine molecules are taken off from the nanocrystals into the solvent, reducing the ligand coverage of nanocrystals. Moreover, due to the interactions between the double bonds, more TCE molecules richly insert into the space between oleylamine molecules, preventing the interdigitation of ligands between the nearest nanocrystals during the formation of superlattices. This can be confirmed by the EDX mapping of the superlattices obtained from TCE, which shows that the Cl element is distributed in the superlattices. In such case, a ligand anchored to a given core can hardly extend to and interact directly with the chain belonging to it is second nearest neighbors, consequently giving rise to the fcc crystal structure in superlattices, as shown in Scheme 10.1.

    This can also be confirmed by the interparticle distances between nanocrystals within the superlattices as mentioned above. The calculated chain length of oleylamine is 1.8 nm. When using TCE as solvent, the interparticle distance is $\delta_{pp} = 3.5$, which is almost double value of the length of oleylamine molecule, indicating that the interdigitation between nanocrystals within fcc

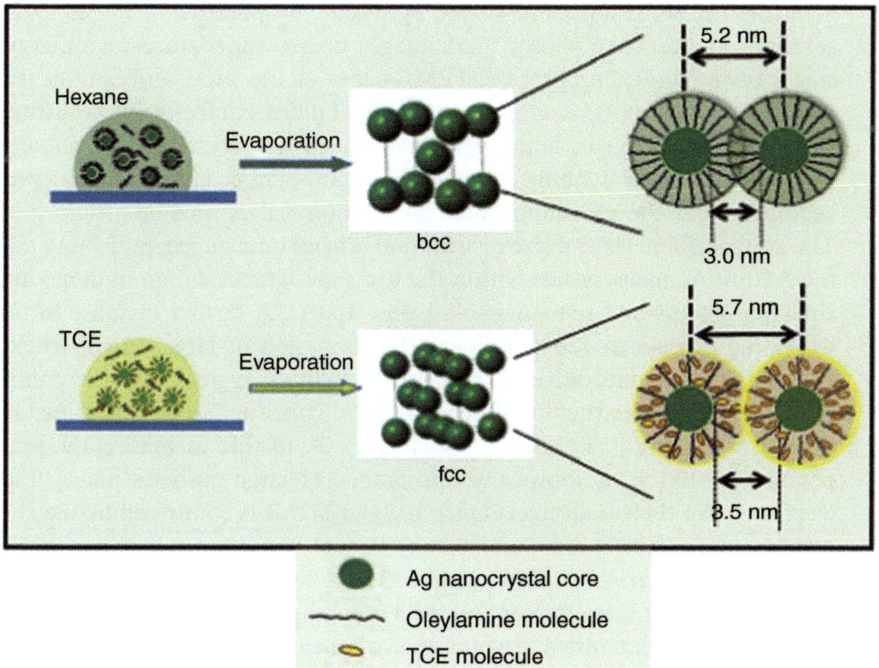

Scheme 10.1 The superlattice growth mechanism of 2.2-nm oleylamine-coated Ag nanocrystals dispersed in hexane or TCE solvents. *Source:* Wei et al. [36]/American Chemical Society.

superlattices is much weaker than that obtained from hexane ($\delta_{pp} = 3.0$ nm). Furthermore, the trapping of the guest molecules into the nanocrystal superlattices can induce the structure transition from fcc to bcc [39]. However, none of the guest molecules have been introduced into our system.

(2) *Influence of halides*: according to a recent study [40], the stability of dodecanethiol-capped Au nanocrystals is significantly reduced when halides are present because the halide ions can exchange the thiolates on Au nanocrystal surface. Hence, it can be assumed here that TCE molecules take the place of some oleylamine ligands, reducing the ligand coverage of 2.2-nm silver nanocrystals dispersed in TCE. Furthermore, as mentioned above, the small TCE molecules could be incorporated into the space between ligands, decorating the soft shell as a mixture of oleylamine and TCE molecules. During evaporation of the carrier solvent, the interdigitation between ligands is largely reduced, consequently giving rise to fcc structures.

These two samples (bcc and fcc 3D superlattices of 2.2-nm Ag nanocrystals) are placed in a furnace equipped with a programmable heating system and annealed simultaneously. The temperature was controlled from 25 to 130 °C with a heating rate of 1 °C per minute:

* With the bcc superlattices produced by hexane evaporation of the colloidal solution, Figure 10.16 shows TEM images of bcc superlattices at different stages of the annealing process. TEM images of the superlattices annealed between 25 and 100 °C exhibit a (110) crystal plane parallel to the substrate (Figure 10.16a–d). The lattice fringe of this (110) crystal plane remains unchanged (around 4.2 nm). By redispersing the sample after annealing process in hexane, the remaining colloidal solution is deposited on a TEM grid. The average diameter and size distribution remain unchanged, indicating that the 2.2-nm Ag nanocrystals within the bcc superlattices do not undergo any ripening process and remain stable below 100 °C. A further increase in the annealing process to 120 °C induces the formation of larger nanoparticles (Figure 10.16e,f), indicative of nanocrystal melting, aggregation, and consequently a partial destruction of bcc structure. This was further confirmed by the corresponding FFT (inset Figure 10.16f). By further increasing the temperature to 130 °C, it shows the appearance of large particles and a total destruction of their ordering (Figure 10.16g,h). This is confirmed by the corresponding FFT (inset in Figure 10.16h). In such condition, it is impossible to redisperse the thermal-treated samples. The average diameter is then determined by a direct measurement on the TEM images. Figure 10.17 shows that bcc superlattices submitted to a progressive annealing process remain stable at below 120 °C, further increased above this threshold.
* With superlattices produced from nanocrystals dispersed in TCE and subjected to the same treatment as described above, the corresponding TEM images (Figure 10.18a–c) and their respective FFT (inset Figure 10.18a) show that the superlattices remain stable in the temperature ranging from 25 to 70 °C with the lattice fringe of (220) crystal plane (Figure 10.18b). Upon temperature increase of above 70 °C, new structures emerge coexisting with the fcc film.

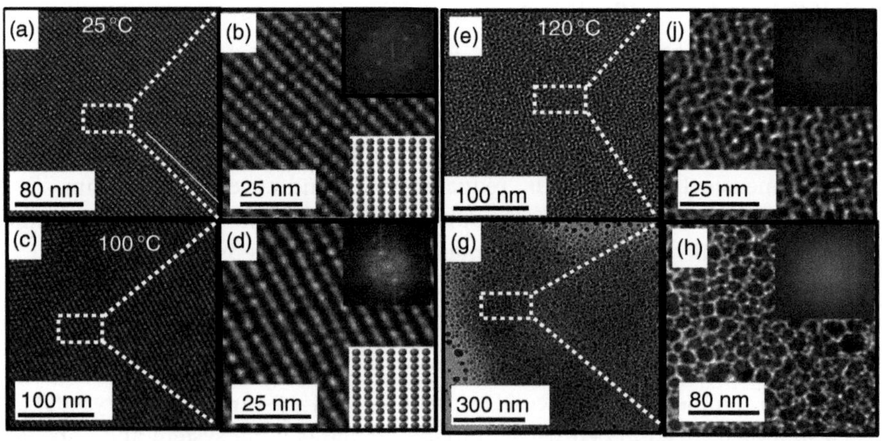

Figure 10.16 TEM images and magnified TEM images of the bcc superlattices produced from hexane evaporation of 2.2-nm Ag nanocrystal colloidal solution and annealed at different stages: (a, b) at 25 °C, (c, d) at 100 °C, (e, f) at 120 °C, and (g, h) at 130 °C. (i, j) Inset up is the corresponding FFT; inset down is the model of 110 crystal plane of bcc superlattice. *Source:* (a, b, d, e, f) Wei et al. [36]/American Chemical Society.

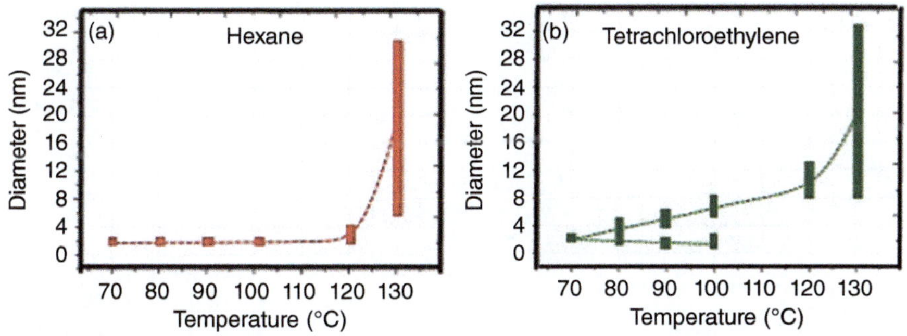

Figure 10.17 Diameter range of nanoparticles during the annealing process at different temperatures produced from 2.2-nm Ag nanocrystals deposited on a substrate. The as-nanocrystals were dispersed in hexane (a) or in TCE (b). *Source:* Wei et al. [36]/American Chemical Society.

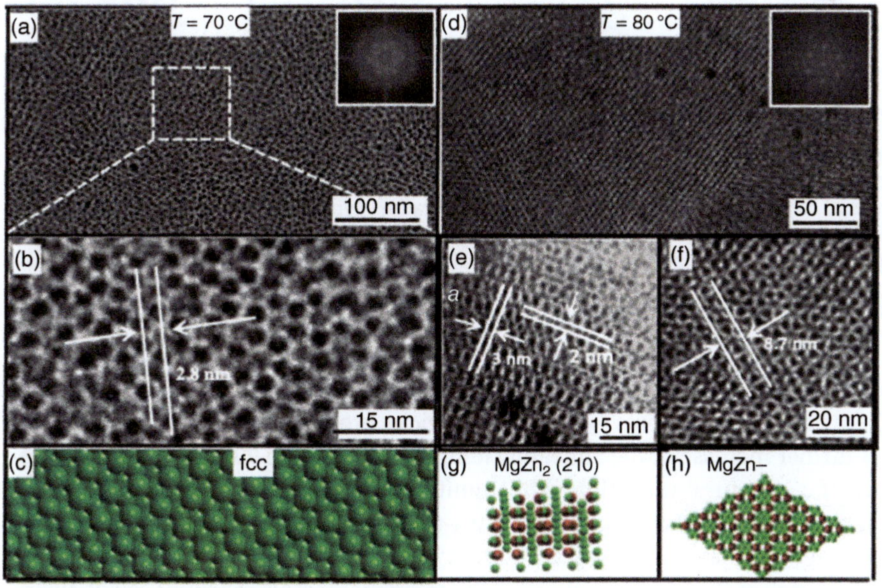

Figure 10.18 (a, b) TEM images of the fcc superlattices produced from TCE annealed to 70 °C. (c) is the corresponding (111) crystal plane model of (b). (d, e, f) TEM images of the fcc superlattices produced from TCE annealed to 80 °C. (g, h) Model of (210) crystal plane and (001) crystal plane of $MgZn_2$-type binary superlattices. *Source:* (a, b, d, e, f) Wei et al. [36]/American Chemical Society.

At 80 °C, Figure 10.18d–f shows that the superlattices are composed of two distinct nanocrystal sizes with the appearance of a bimodal distribution with average nanocrystal diameter values of 2 and 3 nm, respectively. The crystalline structure is well known and called binary nanocrystal superlattices. A careful characterization of various crystal planes parallel to the substrate of such structure (Figure 10.18e,f) shows that it matches well a structure of atomic solid $MgZn_2$ characterized by four

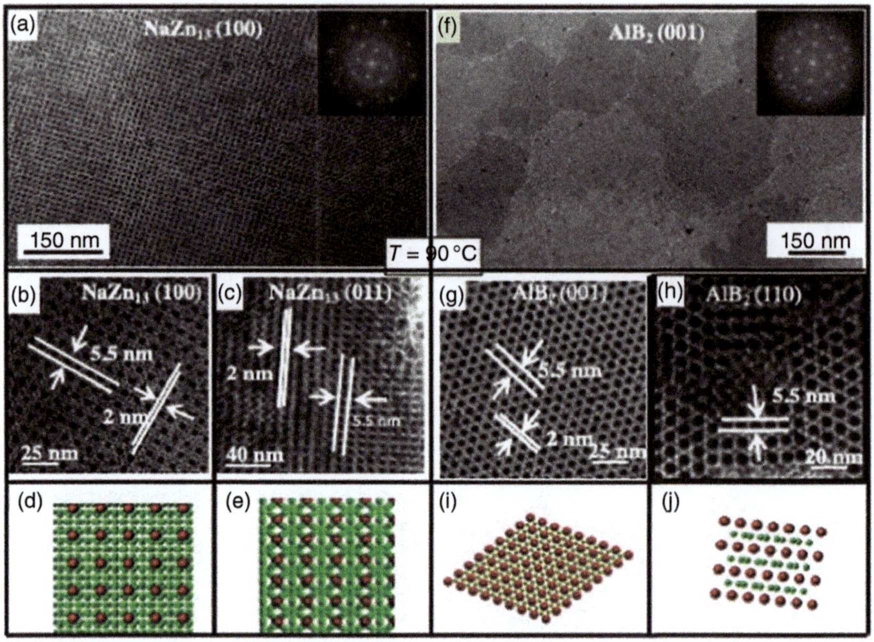

Figure 10.19 TEM image and magnified TEM image of films produced from 2.2-nm Ag nanocrystals dispersed in TCE characterized by fcc superlattices and annealed at 90 °C. Two types of binary structures are observed: (a, b, c) $NaZn_{13}$-type binary nanocrystal superlattices with various crystal planes parallel to the substrate and (f, g, h) AlB_2-type binary nanocrystal superlattices with various crystal planes parallel to the substrate. (d, e, i, j) Crystal plane models correspond to (b, c, g, h); inset images in panels (a) and (f) are the corresponding FFT patterns. *Source:* (a, b, f, g, h) Wei et al. [36]/American Chemical Society.

molecular units per unit cell with the smaller Zn atoms ordered in tetrahedra [41, 42], where the larger Mg atoms fall into the vacancies provided by the Zn tetrahedrons.

A further increase of the annealing temperature of up to 90 °C shows the appearance of a structural transition with simultaneous formation of other two kinds of binary nanocrystal superlattices: $NaZn_{13}$-type [27] and AlB_2-type [43] structures. Figure 10.19a shows the typical (100) crystal plane of $NaZn_{13}$-type superlattices with the size of the crystal domains in the micrometer scale. The unit cell of $NaZn_{13}$ structure contains 104 small spheres and 8 large spheres, which can be viewed as an icosahedron consisting of 13 small spheres inside simple cubic lattices of large spheres. Figure 10.19 shows the typical (001) crystal plane of AlB_2-type superlattices that are based on a hexagonal unit cell and consisting of hexagonally ordered large nanocrystals, together with small nanocrystals inserted into the vacancies between the large nanocrystal layers. The crystal domain sizes of this structure are up to 2 μm. Figure 10.19b,c,g,h shows (100), (011) crystal planes of $NaZn_{13}$-type superlattices and (001), (110) crystal planes of AlB_2-type superlattices, respectively. From these, an average diameter of nanocrystals in the binary system of 5.5 nm for large nanocrystals and 2 nm for small ones can be deduced. TEM images and their corresponding size histograms of the nanocrystals after redispersion further confirm the

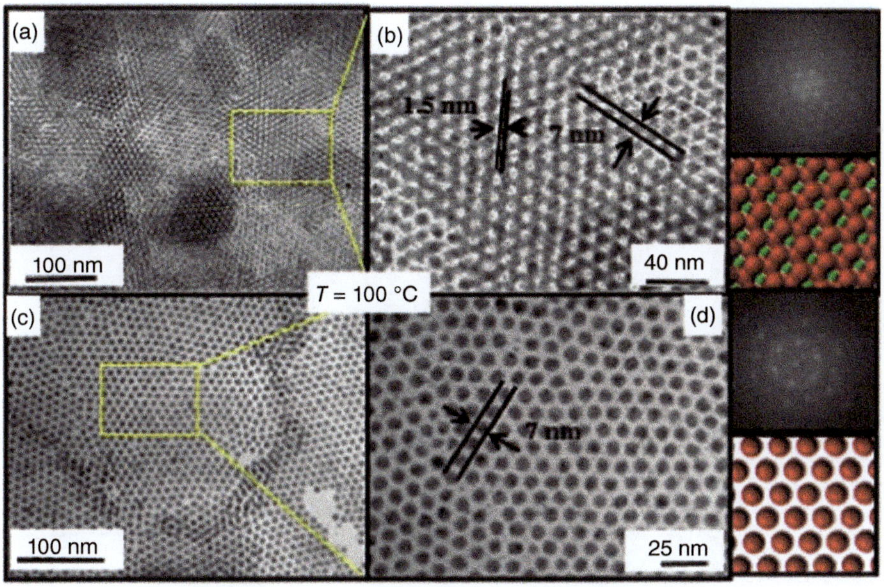

Figure 10.20 TEM image and magnified TEM image of annealing results at 100 °C of films produced from 2.2-nm Ag nanocrystals dispersed in TCE characterized by fcc superlattices. (a, b) NaCl-type binary nanocrystal superlattices with (111) crystal planes parallel to the substrate and (c, d) one- to two-layer nanocrystals with hexagonal packing mode. Right-side images of panels (b) and (d) are crystal plane models and the corresponding FFT patterns. *Source:* Wei et al. [36]/American Chemical Society.

bimodal distribution and average diameters of 2 and 5.5 nm, respectively. A further temperature increase to 100 °C induces another structural transition with the appearance of NaCl-type [44] binary superlattices in place of the $NaZn_{13}$ and AlB_2-type superlattices that were observed at lower temperatures.

Figure 10.20a,b shows the typical (111) crystal plane of NaCl-type binary superlattices. The average diameters for the two types of nanocrystals are approximately 7 and 1.5 nm, respectively. It is pointed out here that the larger nanocrystals (7-nm) self-assembled in both NaCl-type binary system and single-component close-packed hexagonal monolayer, whereas the original superlattices (2.2-nm Ag nanocrystals) disappear. The number of layers ordered into close-packed hexagonal pattern drops to one or two, as confirmed by the FFT pattern (Figure 10.20c,d), and the interparticle distance (3.0 nm) decreases slightly compared with that of the superlattices prior to the treatment (3.5 nm). No structural transitions could be observed above 120 °C with binary nanocrystal superlattices disappearing and a few layers of 10-nm Ag nanocrystals ordered in compact hexagonal network appearing at higher temperatures (Figure 10.21a,b).

Hence, from room temperature to 70 °C, the 2.2-nm fcc Ag superlattices produced from TCE solution keep their integrity and remain stable. From 70 to 100 °C, a progressive disappearance of the 2.2-nm Ag superlattices and the appearance of new structures with a simultaneous highly controlled smooth increase and decrease in

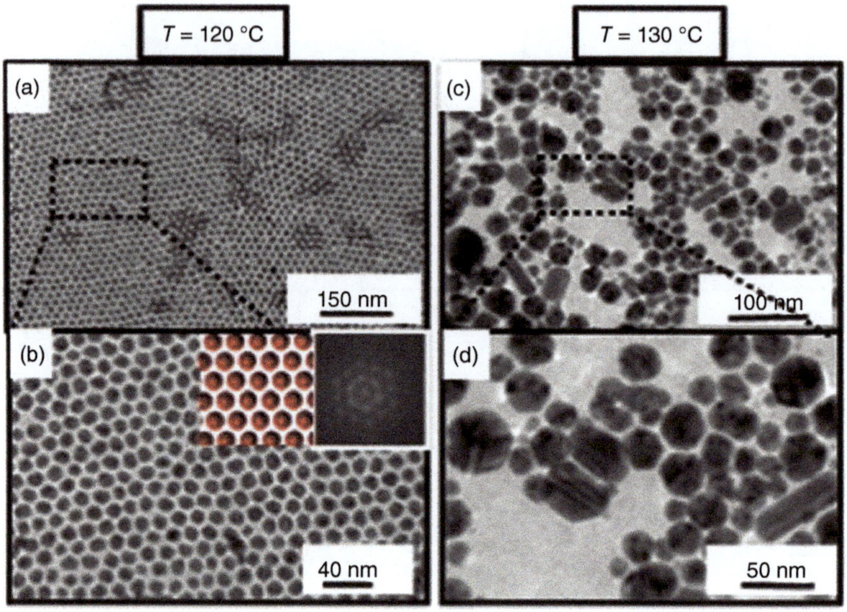

Figure 10.21 TEM image and magnified TEM image of annealing results at 120 °C (a, b) and 130 °C (c, d) of films produced from 2.2-nm Ag nanocrystals dispersed in TCE characterized by fcc superlattices. Inset image in panel (b) is the corresponding FFT. *Source:* Wei et al. [36]/American Chemical Society.

the Ag nanocrystal diameters occur, both characterized by a relatively low size distribution. Such processes induce various structural transitions due to the mass transport between smaller and larger nanoparticles to reach a total disappearance of the small nanoparticles with the formation of homogeneous single-component superlattices. Figure 10.17b shows a simultaneous progressive increase and decrease of the large and small nanocrystals, respectively, to reach large nanoparticles with a rather low size distribution. At 130 °C, very large nanocrystals with various shapes such as plates and polyhedrons are obtained (Figure 10.21c,d).

For temperatures ranging from room temperature to 120 °C, the structural transitions are quantified by taking into account the ratio of effective diameter γ_{eff} [45], defined as $\gamma_{eff} = D_{eff\text{-}small}/D_{eff\text{-}large}$, where D_{eff} is the sum of metal core diameter and twice the thickness of the organic ligand. Here, it is rather difficult to determine the shell thickness of the organic ligands. In first approximation, the shell thickness of the ligands is assumed to be similar to that measured previously [36] for oleylamine-coated Ag nanocrystals (~1.3 nm).

For the annealing temperatures ranging from 70 to 100 °C, the structure of superlattices evolves from fcc to $MgZn_2$ to AlB_2 to $NaZn_{13}$ to NaCl structure. This corresponds to an approximate change of γ_{eff} from 1.0 to 0.82 to 0.51 to 0.51 to 0.43, approximately. Theoretical predictions [46] based on the calculation of free energy of entropy contribution pointed out that $MgZn_2$, $NaZn_{13}$, AlB_2, and NaCl are stable in the range of $0.76 \leq \gamma_{eff}^{theo} \leq 0.84$, $0.54 \leq \gamma_{eff}^{theo} \leq 0.625$, $0.45 \leq \gamma_{eff}^{theo} \leq 0.61$, and

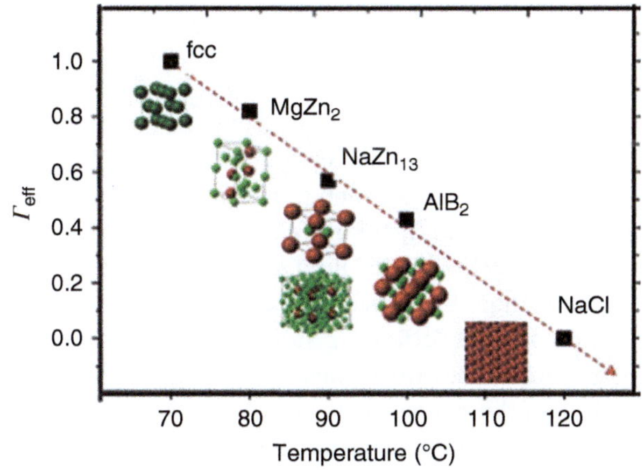

Figure 10.22 The effective diameter ratio (γ_{eff}) of nanocrystals in binary superlattices appearing in the annealing process at different temperatures and the corresponding theoretical predictions γ_{eff}^{theo} based on the calculation of entropy contribution (color area). *Source:* Wei et al. [36]/American Chemical Society.

$0.414 \leq \gamma_{eff}^{theo} \leq 0.45$, respectively. Assuming the estimation of the shell length described above, the various γ_{eff} values corresponding to the binary structures produced and compared with γ_{eff}^{theo} as a function of temperature can be given, as depicted in Figure 10.22. A good agreement with hard-sphere model prediction is observed.

These data unambiguously show that, during the annealing process, the nanocrystals are locally confined within the superlattices and ripening process occurs between the neighboring nanocrystals within the superlattices. Based on simple hard-sphere packing rules, self-assembly of colloidal nanocrystals is typically driven by the changes in the free-volume entropy combined with weak attractive interparticle forces. The gain in free volume entropy is related to the packing density (space-filling fraction p). Generally, it favors structures with the highest packing fraction [47]. During this process, the entropic variation is positive and reaches its maximum when the ordering has the highest packing density. In other words, entropy is the main driving force for forming binary structures during annealing process, similar to the assemblies of the colloidal balls in the absence of organic coatings. This is highly consistent with data shown in Chapter 7 and others [48] describing binary nanocrystal mixtures with the same kind of ligand packing into structures predicted by the hard-sphere model. Regardless of the solvent's nature, the 3D superlattice is stable ranging from 25 to 70 °C, whereas the nanocrystals dispersed in TCE coalesce at 60 °C. This confirms a stability increase of nanocrystals self-assembled in 3D superlattices compared with the same nanoparticles dispersed in a solvent. However, the crystalline structures of the 3D superlattices markedly change when nanocrystals are initially dispersed in TCE or in hexane during the annealing process. These changes can also be argued to be related to the ligand coverage of nanocrystals in the superlattices. Here, a tentative mechanism is proposed: as discussed above, the nanocrystal surface is poorly covered with a mixture of oleylamine and TCE molecules when using TCE as a solvent. Upon progressive heating, the Ag surface atoms that are not bound to a ligand can move freely between nanocrystals, enabling typical Ostwald ripening. Furthermore, it is worth noting here that the space between the ligands is fully occupied by TCE molecules (Scheme 10.1), which can promote the mobility of surface Ag atoms and enhance the mass transfer between the neighboring nanocrystals. However, it is widely accepted that Ostwald ripening is a common fluid phenomenon where the mass transfer occurs easily through the fluid medium. Here, the small Ag nanocrystal core are embedded in a liquid-like medium composed of the ligands and solvent, hence promoting the mobility of surface Ag atoms. When all the atoms involved in the formation of small nanoparticles are dissolved, only large nanocrystals forming the original network maintain the nanocrystals in compact hexagonal network. At 130 °C, regardless of the solvent's nature, oleylamine molecules are desorbed from the surface, and the nanoparticles coalesce. Besides, the structural stability of nanocrystal superlattices differs from fcc to bcc structure, as reported recently [32], and a higher structural stability in the bcc superlattice than the fcc superlattice was observed. Hence, the inherent structural stability of superlattices can also probably play a role during the heat treatment of the different superlattices from fcc to bcc in our case.

10.4 Conclusions

The fcc triangular single silver crystals are obtained, under atmospheric pressure at 50 °C, by annealing the 2D/3D silver nanocrystal superlattices. The triangle size depends on the initial nanocrystal ordering range of the substrate. The crystalline growth properties of silver nanocrystals can be tailored by controlling their organization. Furthermore, on HOPG, an epitaxial orientation of the triangles is observed.

Dodecanethiol-coated silver nanocrystals self-assembled in 2D are subjected to oxygen plasma using the reactive ion etching process. The investigation by transmission electronic microscopy of the same areas before and after exposure allows one to distinguish two behaviors under plasma treatment, which are determined by the level of ordering of the 2D organizations. Higher-ordered self-assemblies remain unchanged, while less-ordered organizations coalesce into larger nanocrystals with spheroidal shapes that could be single crystals. There is no oxidation of the silver either on the large coalesced nanocrystals or on the stable self-assemblies.

It is shown that NHC-stabilized Au nanocrystals exhibit considerably greater resistance to oxygen plasma than dodecanethiol-protected Au nanocrystals. In addition, NHC ligands bound to the surface of Au nanocrystals are significantly more resistant to O_2-induced transformations than dodecanethiol. Exposure to oxygen is a major cause of degradation for nanocrystals and more particularly for thiol-coated Au nanocrystals. Therefore, the results presented here clearly reinforce our vision of NHC ligands as a promising tool for developing reliable and safe applications in the field of nanomaterials.

It is shown that 2.2-nm Ag nanocrystals are used as building blocks to grow nanocrystal superlattices and that the carrier solvent plays a determining role in the packing way of nanocrystals. Significant ligand–ligand interdigitation between engorging nanocrystals should also be taken into account when hexane is used as the solvent, which results in bcc structure as predicted by Landman theory. However, drastic changes appear in the case of TCE dispersions; a mixed coverage of ligand and solvent and a well-solvated organic shell with the insertion of TCE molecules lead to not only a reduced ligand–ligand interdigitation but also the formation of fcc structure with high packing density. Furthermore, by simultaneously annealing these two samples at various temperatures, significant changes appear. For bcc structures originated from the hexane suspension evaporation, a simple coalescence mechanism dominates. However, for fcc structures originated from the TCE solutions, Ostwald ripening process takes place and enables the temperature-dependent production of a variety of binary nanocrystal superlattices, such as NaCl, AlB_2, $NaZn_{13}$, and $MgZn_2$. This can be well explained by an entropy-driven process. Those findings provide insights into the solvent-mediated nanocrystal superlattices and the Ostwald ripening process in nanocrystal superlattices.

References

1 Courty, A., Henry, I., Goubet, N., and Pileni, M.P. (2007). Large triangular single crystals formed by mild annealing of self-organized silver nanocrystals. *Nat. Mater.* 6: 900–907.

2 Motte, L., Courty, A., Ngo, A. et al. (2005). Self assemblies of nanoparticles. *Nanocrystals Forming Mesoscopic Structures* (ed. M.P. Pileni). Weinheim: Wiley-VCH, Chapter 1.

3 Klecha, E., Arfaoui, I., Richardi, J. et al. (2011). Does oxygen plasma can be considered as a general technique to reveal nanocrystals ordering on various scale? *Phys. Chem. Chem. Phys.* 13: 2953–2962.

4 Klecha, E., Ingert, D., and Pileni, M.P. (2010). How the level of ordering of 2D nanocrystals superlattices is controlled by their deposition mode? *J. Phys. Chem. Lett.* 1: 1616–1622.

5 Klecha, E., Ingert, D., and Pileni, M.P. (2009). Immunity of coated self-organized nanocrystals against strong oxidation: a new intrinsic property due to nanocrystal ordering. *Langmuir* 25: 2824–2830.

6 Harfenist, S.A., Wang, Z.L., Alvarez, M.M. et al. (1996). Highly oriented molecular Ag nanocrystal arrays. *J. Phys. Chem.* 100: 13904–13910.

7 Wang, Z.L. (1998). Structural analysis of self-assembling nanocrystal superlattices. *Adv. Mater.* 1: 13–30.

8 Courty, A., Lisiecki, I., and Pileni, M.P. (2002). Vibration of self-organized silver nanocrystals. *J. Chem. Phys.* 116: 8074–8078.

9 Wang, Z.L., Harfenist, S.A., Whetten, R.L. et al. (1998). Bundling and interdigitation of adsorbed thiolate groups in self-assembled nanocrystal superlattices. *J. Phys. Chem. B* 102: 3068.

10 Donnet, J.B. and Custodero, E. (1994). Caractérisation des défauts de surface d'un graphite pyrolytique à forte orientation par microscopie à effet tunnel. *C. R. Acad. Sci. Paris* 319: 17.

11 Jeanguillaume, C. and Colliex, C. (1989). Spectrum-image: the next step in EELS digital acquisition and processing. *Ultramicroscopy* 28: 252.

12 Egerton, R.F. (1996). *Electron Energy-Loss Spectroscopy in the Electron Microscope*, 2e. New York: Plenum Press.

13 Lisiecki, I., Walls, M., Parker, D., and Pileni, M.P. (2008). 2D self-organization of core/shell Co_{hcp}/Co nanocrystals. *Langmuir* 24: 4295–4299.

14 Duhamel, C., Champion, Y., Tencé, M., and Walls, M. (2005). Synthesis of controlled-chemistry ultrafine Fe_xNi_{1-x}. *J. Alloys Comp.* 393: 204.

15 Doraiswamy, N., Jayaram, G., and Marks, L.D. (1995). Unusual island structures in Ag growth on Si(100)-(2×1). *Phys. Rev. B* 51: 10167.

16 Wagner, A.J., Wolfe, G.M., and Fairbrother, D.H.J. (2004). Atomic oxygen reactions with semifluorinated and *n*-alkanethiolate self-assembled monolayers. *J. Chem. Phys.* 120: 3799.

17 Raiber, K., Terfort, A., Benndorf, C. et al. (2005). Removal of self-assembled monolayers of alkanethiolates on gold by plasma cleaning. *Surf. Sci.* 595: 56.

18 Xue, C.Y. and Yang, K.L. (2007). Chemical modification of inert organic monolayers with oxygen plasma for biosensor applications. *Langmuir* 23: 5835.

19 Bain, C.D., Throughton, E.B., Tao, Y.T. et al. (1989). Formation of monolayer films by the spontaneous assembly of organic thiols from solution onto gold. *J. Am. Chem. Soc.* 111 (1): 321–335.

20 Torres, J., Perry, C.C., Bransfield, S.J., and Fairbrother, D.H. (2002). Radical reactions with organic thin films: chemical interaction of atomic oxygen with an X-ray modified self-assembled monolayer. *J. Phys. Chem. B* 106: 6265.

21 Boyen, H.G., Kästle, C., Weigl, F. et al. (2002). Oxidation-resistant gold-55 clusters. *Science* 297: 1533.

22 Onoue, S., Junhui, H., and Kunitake, T. (2006). Fabrication of gold nanosheet and nanowire by oxygen plasma induced fusion of densely arrayed nanoparticles. *Chem. Lett.* 35: 214.

23 Mueggenburg, K.E., Lin, X.M., Goldsmith, R.H., and Jaeger, H.M. (2007). Elastic membrane of closed packed nanoparticles. *Nat. Mater.* 6: 656.

24 Allpress, J.G. and Sanders, J.V. (1967). The structure and orientation of crystals in deposits of metals on mica. *Surf. Sci.* 7: 1–25.

25 Zheng, N., Fan, J., and Stucky, G.D. (2006). One-step one-phase synthesis of monodisperse noble-metallic nanoparticles and their colloidal crystals. *J. Am. Chem. Soc.* 128: 6550–6551.

26 Pileni, M.P. (1993). Reverse micelles: a microreactors. *J. Phys. Chem.* 97: 6961–6974.

27 Salzemann, C., Zhai, W., Goubet, N., and Pileni, M.P. (2010). How to tune the Au inter-nanocrystal distance in 2D self-ordered superlattices? *J. Phys. Chem. Lett.* 1: 149–154.

28 Sun, S.H., Murray, C.B., Weller, D. et al. (2000). Monodisperse FePt nanoparticles and ferromagnetic FePt nanocrystal superlattices. *Science* 287: 1989.

29 Hahn, F.E. and Jahnke, M.C. (2008). Heterocyclic carbenes: synthesis and coordination chemistry. *Angew. Chem. Int. Ed.* 47: 3122.
30 Ling, X., Roland, S., and Pileni, M.P. (2015). Supracrystals of N-heterocyclic carbene-coated Au nanocrystals. *Chem. Mater.* 27: 414.
31 Roland, S., Ling, X., and Pileni, M.P. (2016). N-Heterocyclic carbene ligands for Au nanocrystal stabilization and three-dimensional self-assembly. *Langmuir* 32: 7683–7696.
32 Ling, X., Schaeffer, N., Roland, S., and M.P.P. (2015). Superior oxygen stability of N-heterocyclic carbene-coated Au nanocrystals – comparison with dodecanethiol. *Langmuir* 31: 12873–12882.
33 Bourissou, D., Guerret, O., Gabbai, F.P., and Bertrand, G. (2000). Stable carbenes. *Chem. Rev.* 100: 39.
34 Herrmann, W.A. (2002). N-Heterocyclic carbenes: a new concept in organometallic catalysis. *Angew. Chem. Int. Ed.* 41: 1290.
35 Hopkinson, M.N., Richter, C., Schedler, M., and Glorius, F. (2014). An overview of N-heterocyclic carbenes. *Nature* 510: 485.
36 Wei, J., Schaeffer, N., and Pileni, M.P. (2016). Solvent-mediated crystallization of nanocrystal 3D assemblies of silver nanocrystals: unexpected superlattice ripening. *Chem. Mater.* 28: 293–302.
37 Korgel, B.A., Zaccheroni, N., and Fitzmaurice, D. (1999). Melting transition of a quantum dot solid: collective interactions influence the thermally-induced order-disorder transition of a silver nanocrystal superlattice. *J. Am. Chem. Soc.* 121: 3533–3534.
38 Landman, U. and Luedtke, W.D. (2004). Small is different: energetic, structural, thermal, and mechanical properties of passivated nanocluster assemblies. *Faraday Discuss.* 125: 1–22.
39 Nagaka, Y., Chen, O., Wang, Z.W., and Cao, C. (2012). Structural control of nanocrystal superlattices using organic guest molecules. *J. Am. Chem. Soc.* 134: 2868–2871.
40 Yu, Y., Goodfellow, B.W., Rasch, M.R. et al. (2015). Role of halides in the ordered structure transitions of heated gold nanocrystal superlattices. *Langmuir* 31: 6924–6932.
41 Shevchenko, E.V., Talapin, D.V., Kotov, N.A. et al. (2006). Structural diversity in binary nanoparticle superlattices. *Nature* 439: 55–59.
42 Shevchenko, E.V., Talapin, D.V., Murray, C.B., and O'Brien, S. (2006). Structural characterization of self-assembled multifunctional binary nanoparticle superlattices. *J. Am. Chem. Soc.* 128: 3620–3637.
43 Redl, F.X., Cho, K.S., Murray, C.B., and O'Brien, S. (2003). Three-dimensional binary superlattices of magnetic nanocrystals and semiconductor quantum dots. *Nature* 423: 968–971.
44 Parthe, E.Z. (1961). Space filling of crystal structures: a contribution to the graphical presentation of geometrical relationships in simple crystal structures. *Z. Kristallogr. – Cryst. Mater.* 115: 52–79.
45 Chen, J., Ye, X., and Murray, C.B. (2010). Systematic electron crystallographic studies of self-assembled binary nanocrystal superlattices. *ACS Nano* 4: 2374–2381.

46 Evers, W.H., Nijs, B.D., Filion, L. et al. (2010). Entropy-driven formation of binary semiconductor-nanocrystal superlattices. *Nano Lett.* 10: 4235–4241.
47 Bodnarchuk, M.I., Kovalenko, M.V., Heiss, W., and Talapin, D.V. (2010). Energetic and entropic contributions to self-assembly of binary nanocrystal superlattices: temperature as the structure-directing factor. *J. Am. Chem. Soc.* 132: 11967–11977.
48 Bian, K.F., Wang, Z.W., and Hanrath, T. (2012). Comparing the structural stability of PbS nanocrystals assembled in fcc and bcc superlattice allotropes. *J. Am. Chem. Soc.* 134: 1078.

11

Intrinsic Properties Related Due to the Self-Assemblies of Nanocrystals

11.1 Epitaxial Crystal Growth as a Result of the Nanocrystal Ordering

Very surprinsing we demonstrate it that an epitaxial crystal growth process is related to the nanocrystal ordering [1, 2].

Ag nanocrystals of 5 nm coated with decanethiol are dispersed in decane. The colloidal solution is deposited on either amorphous carbon or highly oriented pyrolytic graphite (HOPG). Because of the high boiling point of decane, 20-hour annealing at 50 °C is required to evaporate all the solvent. For each substrate, one is extracted from the oven and kept at room temperature, whereas the second one is maintained for eight days at 50 °C. At the end of *decane* evaporation times (20 hours), one grid is extracted from the oven, whereas the second one remains in it for eight days. At the end of each process, the transmission electron microscopy (TEM) grid is kept at room temperature.

After 20-hours annealing on amorphous carbon, local 2D nanocrystal organizations (Figure 11.1a) are observed by TEM. This is confirmed by the corresponding power spectrum (inset in Figure 11.1a) with one diffuse ring. Multilayer nanocrystal assemblies (3D) are also observed in Figure 11.1b and lead give one ring with diffuse spots corresponding to the first order (inset Figure 11.1b). This indicates that the Ag nanocrystals self-organize over short range, in a few layers. On HOPG as a substrate, the domains of ordered nanocrystals, observed after 20 hours of evaporation, are very large compared with those obtained on amorphous carbon (Figure 11.1c,d and insets). The power spectrum in the inset of Figure 11.1c is typical of 2D hexagonal networks, and the inset of Figure 11.1d shows spots corresponding to (220) plane reflections of a fcc 3D arrangement oriented in the [111] direction. Note that the well-defined second order confirms a large domain ordering. From Figure 11.1, it is concluded that the ordered nanocrystal domains observed after 20 hours of evaporation are very large compared with those obtained on amorphous carbon.

After eight days' annealing, the two samples markedly differ from those obtained after 20-hour annealing, with the formation of particles with various ill-defined shapes (Figures 11.2a and 11.3a).

Metal Nano 3D Superlattices: Synthesis, Properties, and Applications, First Edition.

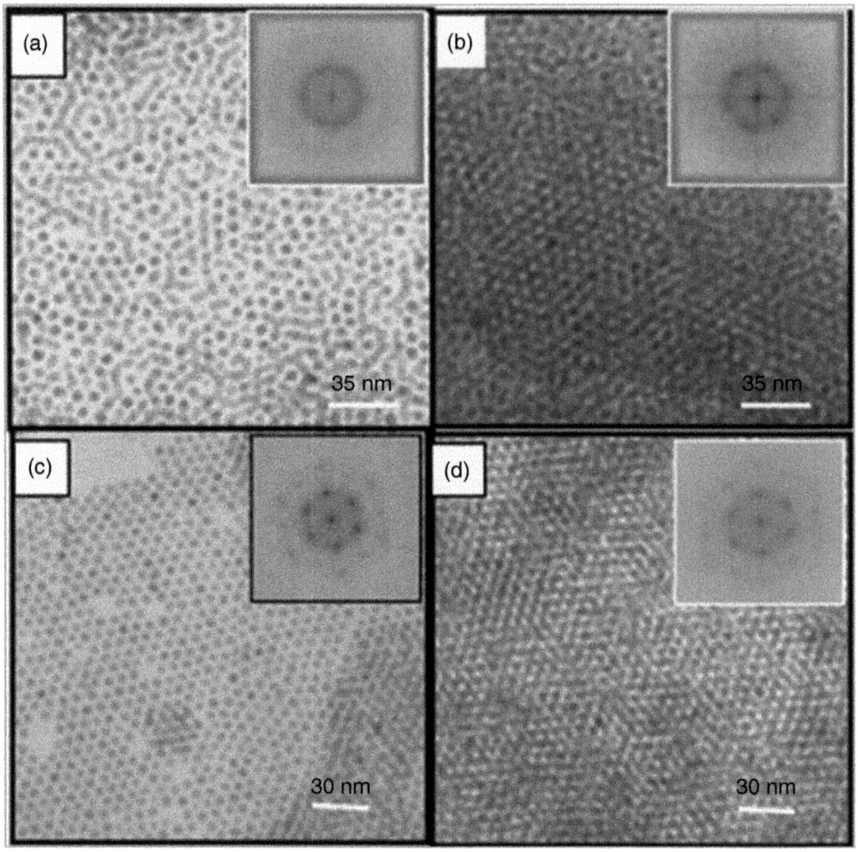

Figure 11.1 2D and 3D 5-nm Ag nanocrystal organizations on amorphous carbon (a, b) and HOPG (c, d). 2D (a, c) and 3D (b, d) nanocrystal assemblies obtained at the end of solvent evaporation time (20 hours). Insets: the corresponding power spectrum. *Source:* Courty et al. [1]/Springer Nature.

With amorphous carbon as substrate, Figure 11.2a shows a large variety of shapes. Many of the particles are regular multiply-twinned particles (MTPs), shown in Figure 11.2b. In addition, we observe a small amount of flat, triangular particles with more or less rounded edges (Figure 11.2a). High-resolution transmission electron microscopy (HRTEM) images of flat particles (Figure 11.2c) show that the crystallinity of the particles is very high (Figure 11.2d) with an fcc structure. The power spectrum obtained for the triangular particles (Figure 11.2e) shows three families of lattice plane with a lattice distance of 0.250 nm. The angle between the planes is 60. The measured distance corresponds to 1/3(422) reflections that are normally forbidden. It has been observed previously that the 1/3(422) reflections are [3–5] present on very thin particles or films of fcc structure crystals. These reflections appear when the number of (111) planes is not congruent with three or when two or more defects on the (111) planes are present. To confirm the high crystallinity of the triangular particles and their low thickness, bright-field (Figure 11.2f) and dark-field

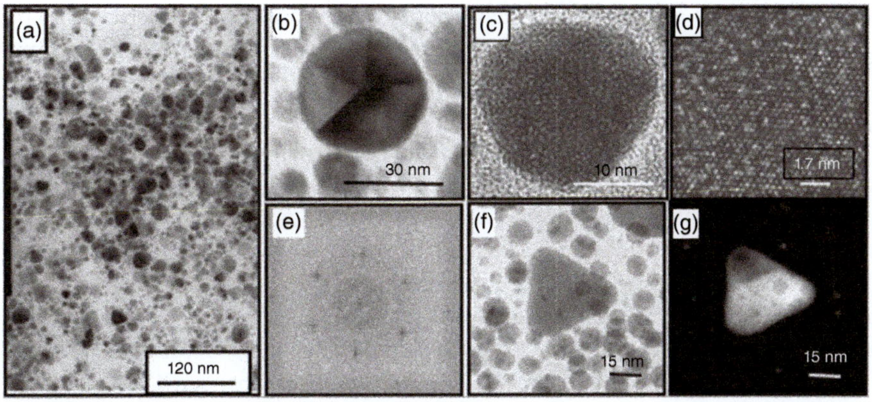

Figure 11.2 (a) TEM image of aggregates produced by annealing for eight days of self assemblies of 5 nm silver nanocrystals deposited on amorphous carbon. (b) High-magnification image of a selected particle. (c) A low-magnification image of the particle. (d) HRTEM image of an isolated triangular particle. (e) The power spectrum of the image in (d). (e, f) Bright-field image of a triangular single crystal (e) and the corresponding dark-field image (f). *Source:* (a–g) Courty et al. [1]/Springer Nature.

(Figure 11.2g) TEM images show a triangular-shaped particle with uniform contrast with very bright faceted particle, indicating a strongly diffracting crystal and consequently a well-crystallized object.

With HOPG instead of amorphous carbon, large triangular crystals are produced after 8-day annealing (Figure 11.3a). The TEM enhancement shows that these

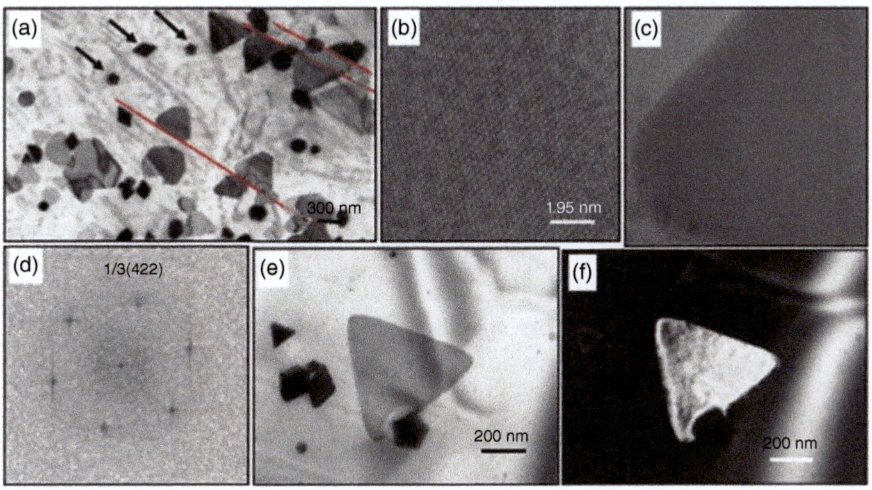

Figure 11.3 (a) TEM image of aggregates produced by annealing for eight days of self assemblies of 5 nm silver nanocrystals deposited on HOPG. The arrows indicate the decahedral particles. The lines show the alignment of the triangular particles on the substrate. (b) HRTEM image of one Ag triangular particle. (c) A low-magnification image of the particle. (d) The power spectrum of the HRTEM image. (e) Bright-field and (f) dark-()field images of the Ag particles in (a). *Source:* Courty et al. [1]/Springer Nature.

triangular particles are equilateral, flat (more or less truncated at the edges), and aligned on the substrate. Since the experiments are performed under the same experimental conditions as those used for amorphous carbon, we deduce the change in the triangle size, and these preferential orientations are induced by the HOPG substrate. The largest sizes (defined as the diameter of the circle surrounding each particle) are around 940 nm, and the smallest sizes around 15 nm. A few decahedral particles with maximum sizes around 200 nm with pentagonal profiles (see particles with arrow in Figure 11.3a) and hexangular particles with more or less rounded edges are also obtained. In addition, the TEM image shows particles with irregular shapes corresponding to pieces of triangles or hexagons. Among 1000 particles, it has been found that there are 29% triangles, 3% hexagons, 6% decahedra, 19% faceted particles corresponding to pieces of triangles or hexagons, and 43% un-faceted particles with irregular shapes. In this analysis, the spherical nanocrystals used at the beginning of the deposition process that remain on the substrate are not taken into account. The HRTEM pattern (Figure 11.3b) shows that the crystallinity of the particles is very high even at the edges (Figure 11.3c). Furthermore, the power spectrum performed on several triangular particles produced from various samples shows three families of lattice planes (Figure 11.3d) with lattice distances of 0.248, 0.251, and 0.265 nm, respectively. The angles between the planes are 62°, 62°, and 56°. Although the angles and the lattice distances are different from 60° and 0.250 nm, respectively, the spots are attributed to the 1/3(422) reflection. The distortion in the lattice fringes observed in the power spectrum is explained by the fact that the crystal is not strictly in the Bragg condition and there is a tilt of 2.7°. This disorientation does not preclude the visualization of lattice fringes in small (or thin) particles. From this result, we deduce that the basal plane of triangular particles is (111). Their lateral facets are probably (111)- or (100)-type facets [6, 7]. The bright- and dark-field TEM images (Figure 11.3e,f) support high crystallinity and low thickness of the triangular particles. They are clearly more faceted on HOPG than on amorphous carbon.

At this stage, we ask: What is the difference between the final triangular particles produced either on amorphous carbon or on HOPG? Let us consider the crystal produced by using amorphous carbon: Figure 11.2f,g shows the bright- and dark-field images of crystals produced by annealing nanocrystal self-assembled on amorphous carbon. On the bright-field image (Figure 11.2f), we observe a triangular-shaped particle with uniform contrast. The corresponding dark-field image (Figure 11.3e) shows a very bright faceted particle indicating a strongly diffracting crystal and consequently a well-crystallized object. Similarly, the bright- and dark-field images (Figure 11.3e,f) confirm the high crystallinity and the low thickness of the triangular particles. They are clearly more facetted on HOPG than on amorphous carbon. *The single-domain triangular crystals are larger by almost 1 order of magnitude with HOPG instead of amorphous carbon as a substrate.* The triangular single crystals are stable for over a year without coalescence.

By energy-dispersive X-ray spectroscopy (EDX), performed on several triangular single crystals, the oxygen is not always detectable. Statistically the amount of oxygen is very small ($I_O/I_{Ag} \leq 2\%$). The presence of carbon and sulfur, in the EDX spectrum, indicates that decanethiol chains remain attached to the triangle and prevent its entire oxidation. A careful observation, by HRTEM, shows that the triangle edges are well defined. Nevertheless, sometimes, it is observed a surrounding thin layer of amorphous matter. This thin layer could be attributed to either the presence of thiols remaining attached to the triangles as observed in particular focus conditions [6] or a thin amorphous Ag oxide layer produced by a few thiols desorption as a result of their oxidation in disulfide [8]. The desorption of thiols as a result of the heating at 50 °C is less. Nevertheless, from EDX measurements, we deduce that only a small amount appears to be oxidized. We are not surprised by the fact that the Ag particles are not totally oxidized even after exposure to air for several days. Indeed, it has been shown recently that bulk Ag, even under a beam containing hyperthermal oxygen, shows an amorphous layer of Ag oxide, which contains only 3–5 atom% of oxygen [9]. In conclusion, whatever the substrate is (amorphous carbon or HOPG), single-domain triangular Ag particles are produced. A thin film protects these particles from oxidation. The major difference is related between triangular particles produced on amorphous carbon, or HOPG is related to the triangular size. On HOPG, the size is larger by a factor of 10 compared with amorphous carbon.

As described above, several triangular particles are aligned on the substrate (see red lines in Figure 11.3a). To determine the existence of epitaxial orientation, electron diffraction has been performed on several isolated triangular particles. Two epitaxial orientations of the Ag triangles are observed. The main epitaxial orientation, which concerns 70% of the total number of Ag single crystals, is defined by $[110]_{Ag} \parallel [110]_{HOPG}$ and $(111)_{Ag} \parallel (001)_{HOPG}$ (Figure 11.4a,b). The second epitaxial orientation, which concerns 20% of the Ag single crystals, is defined by $[112]_{Ag} \parallel [11.0]_{HOPG}$ and $(111)_{Ag} \parallel (001)_{HOPG}$ (Figure 11.4c). The remaining 10% of triangles show no preferential orientation with HOPG. For the main epitaxial orientation, the perfect alignment between the (220) Ag planes and (100) graphite planes is exceptional (Figure 11.4a), and in most of the electron diffraction patterns, a small angular misorientation β with an average value of 3° is measured (Figure 11.4b). For the secondary epitaxial orientation, we also observe an angular misorientation of 3° between the 1/3(422) planes of Ag and the (110) planes of graphite (Figure 11.4c). The Ag atoms of the first layer of the triangular particle are thus not strictly at equivalent sites on the graphite substrate. The angular misorientations are higher than those obtained in ultrahigh-vacuum (UHV) systems [10]. This discrepancy could be explained by the HOPG surface preparation. It is cleaved in air without further treatment, and thus, we cannot assure a perfectly clean surface. The mismatch could also be due to the fact that the triangle size (between 300 and 900 nm) is larger than those usually obtained by UHV experiments (≤100 nm) [10, 11]. Figure 11.4d,e gives schematic representations of the azimuthal orientations of Ag particles on HOPG for both epitaxial orientations. We explain both orientations by taking into account the lattice parameter misfit between the Ag crystal and the substrate. From the diffraction patterns, the mean lattice parameter value of the Ag particles is very close to the bulk Ag value ($a = 0.408$ nm). The distance between two nearest neighbor Ag atoms of the (111) basal plane of a particle is thus close to 0.289 nm. For a complete accommodation of the particle on the substrate, this distance is expected to be 0.283 nm for particles in the main

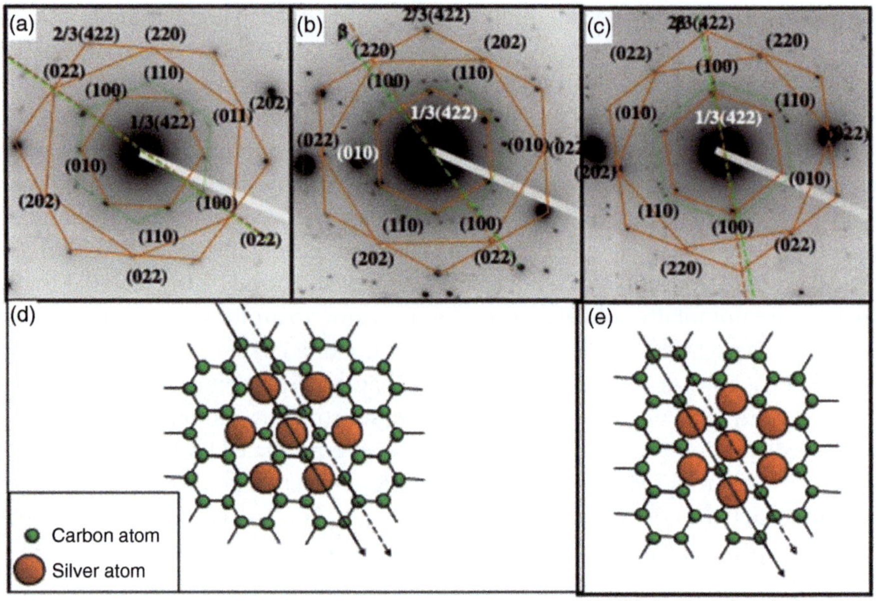

Figure 11.4 Electron diffraction patterns of triangular Ag particles atomic crystals on HOPG. (a–c) Electron diffraction patterns for different isolated triangular particles in the main (a, b) and secondary (c) epitaxial orientations. The dotted and solid hexagons indicate the contributions of the HOPG substrate and the Ag crystal, respectively. The angular misorientation β is represented by the angle between the orange and the green dotted lines; its values are 0° (a) and 3° (b–e). Schematic representations of the azimuthal orientation of Ag particle in the main (d) and secondary (e) epitaxial orientations. *Source:* (a–c) Courty et al. [1]/Springer Nature.

epitaxial orientation and 0.245 nm for the particles in the secondary epitaxial orientation. Thus, the corresponding lattice parameter misfits are 2% and 15%, respectively, which clearly favors the main epitaxial orientation. Since epitaxy occurs on the HOPG substrate, we suppose that, during coalescence, crystals align with the substrate in an orientation, which minimizes the energy of their interface with the substrate. This implies that the particles are in contact with the substrate. We know from scanning tunneling microscopy (STM) measurements that the alkyl chains are oriented along the substrate [12], reducing the distance between the nanocrystals and the substrate. In addition, this is also supported by the fact that at low temperature, the alkyl chains behave as a solid with a rigid configuration, whereas in increasing the temperature to 50 °C, the chains are in a liquid-like state. Consequently, the chains are more mobile on the substrate, and the particle–substrate distance decreases. This favors the nanocrystal epitaxial orientation on HOPG even after a partial thiol desorption.

To determine the formation mechanism of the Ag single triangular crystals, a kinetic study is proposed with nanocrystals deposited in HOPG: six samples, prepared simultaneously, are deposited in the oven (50 °C). Each sample is removed after a given time and observed by TEM. After three hours of annealing (Figure 11.5a), as above, small ordered domains formed by the stacking of a few nanocrystal layers are observed. After one day of annealing (Figure 11.5b), some

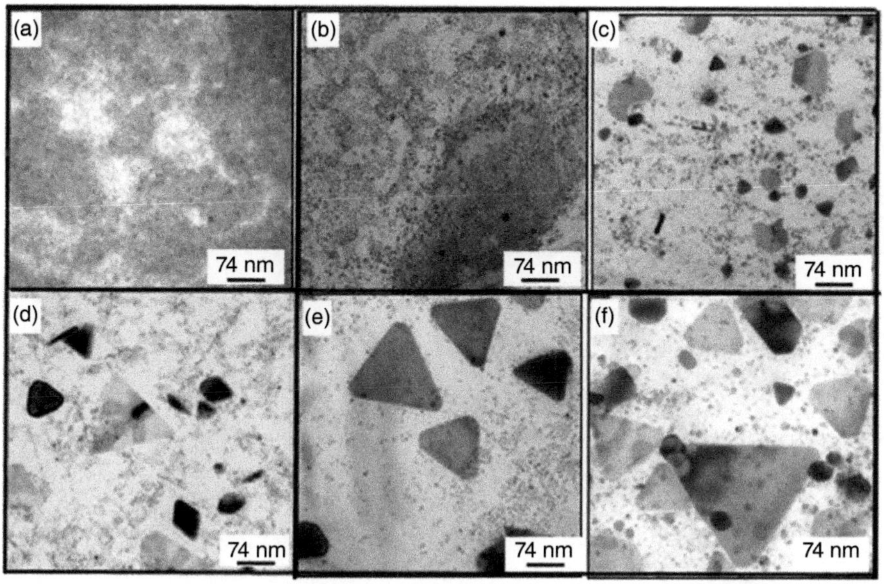

Figure 11.5 TEM images at various annealing times of 5 nm Ag nanocrystals deposited on HOPG: A progressive formation of triangular Ag single-crystal. (a–f) After an annealing time of three hours (a), one day (b), two days (c), three days (d), six days (e), and eight days (f). *Source:* Courty et al. [1]/Springer Nature.

nanocrystals coalesce to reach a diameter of around 20 nm. The initial organization is partially removed. After two days of annealing (Figure 11.5c), the coalesced Ag nanocrystals keep on growing to form triangles or hexagons with defects, and small particles still remain on the substrate. After three days of annealing (Figure 11.5d), the size of the coalesced particles with a more regular shape increases and reaches a maximum at around 150 nm. After six days (Figure 11.4e) and eight days (Figure 11.5f) of annealing, the size of the triangular particles still continues to increase. Note that particles smaller than 10 nm are still observed on the TEM grid. Furthermore, the single fcc triangular particles are stable for almost a year. From this step-by-step observation (from Figure 11.5a–f), we deduce that the single Ag triangles are formed from a progressive coalescence of self-organized spherical 5-nm Ag nanocrystals.

This is in agreement with HRTEM experiments performed on samples annealed for six days: Figure 11.6 shows partial recrystallization of the Ag nanocrystals during coalescence. Thus, in agreement with the previous results, not all the nanocrystals are not totally coalesced after six-day annealing.

Let us follow the coalescence at the beginning of the process. Figure 11.6a shows two icosahedral nanocrystals in contact. We observe a thin region between the two nanocrystals, where the atomic planes of both particles tend to align. This implies a movement of the atoms to reorganize. This movement is favored by the heating, inducing the coalescence process.

On the same TEM grid, we observe that twin planes maintain the adhesion of partially coalesced Ag nanocrystals (Figure 11.6b). These planes have to move and reorientate to form a new particle. Figure 11.6c shows plane defects in the newly

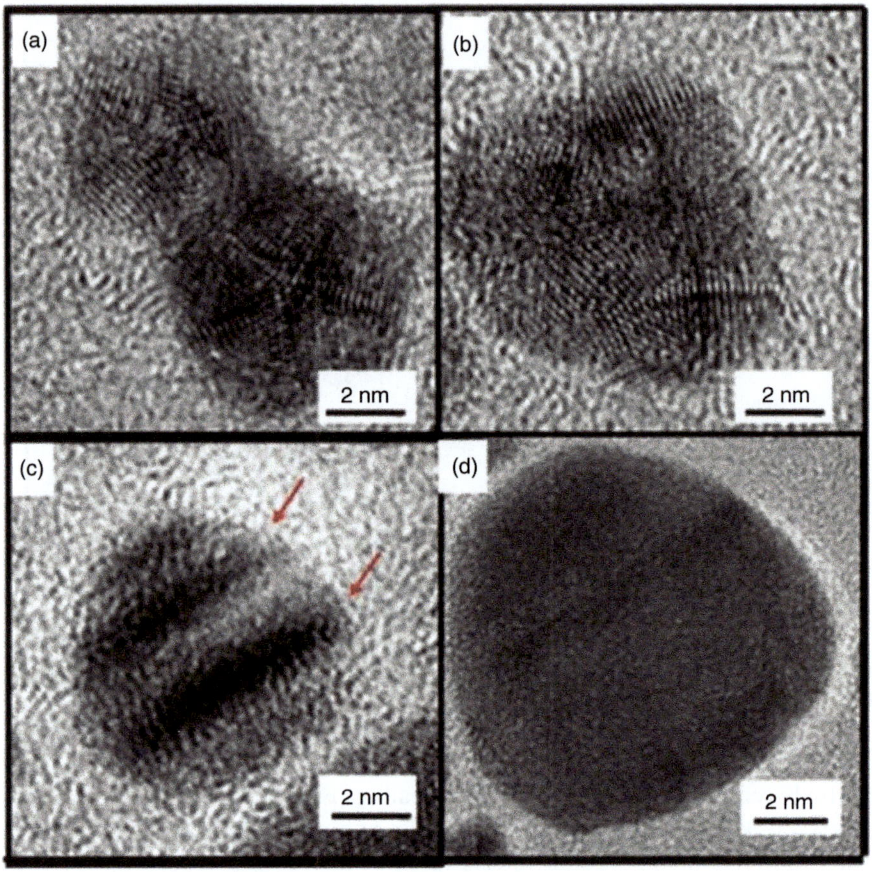

Figure 11.6 HRTEM images of coalesced Ag nanocrystals deposited HOPG: (a) Neck formation between two icosahedral particles, (b) coalesced particles with remaining structural stress (twin boundaries), (c) stacking faults, marked by the arrows in a new particle formed by coalescence, and (d) a newly formed particle, totally relaxed without structural defects. *Source:* Courty et al. [1]/Springer Nature.

formed particles (see arrows). Moreover, Figure 11.6d shows a totally relaxed spherical particle, which no longer has structural defects. This implies a complete atomic rearrangement and a removal of twin planes.

Since most of the initial nanocrystals are MTP, the results obtained by HRTEM and TEM show clearly that the crystalline structure and shape of the new crystals can bear no relation to that of its precursor. The recrystallization of MTP into single crystals, observed under UHV, is attributed to the high mobility of the surface atoms. Note that this process is energetically more favorable as it reduces the proportion of surface atoms and thus the surface energy. We know also that Ag MTP can be dissolved by the presence of oxygen combined with chloride. As we have no chloride in the initial colloidal solution, this process cannot be taken into account to explain the crystal growth mechanism described above. Furthermore, these results show clearly that all the initial self-organized nanocrystals do not start to coalesce at the same annealing time. This progressive coalescence explains the size evolution of the triangular single crystals during the annealing process. That does not exclude fusion of small triangles into larger ones. This is consistent with data observed elsewhere from which nanoprisms are produced by light irradiation of the colloidal solution. Note that here all the experiments are carried out in the dark. Hence in both cases, the formation of single triangular crystals can be obtained when the system is submitted to an external energy, which can be either light or mild annealing. However, we should point out that the two mechanisms are not the same. In our case, coalescence and recrystallization take place on a substrate, whereas in the other, photoconversion and fusion of nanocrystals occur in solution. Moreover, epitaxial orientations of the triangles are observed. From this, the three characteristics stages of the triangular single-crystal formation mechanism can be identified:

(i) Formation of Ag nanocrystal arrays resulting from the evaporation of the nanocrystal solution dispersed in decane at 50 °C during 20 hours.
(ii) Annealing of the nanocrystal arrays induces the coalescence of neighboring nanocrystals by a partial desorption of the oxidized thiols in disulfide. The nanocrystal ordering favors their coalescence by reducing the interparticle distance. The only source of energy is brought by the annealing temperature. A small amount of extra energy is sufficient to initiate the coalescence process.
(iii) A recrystallization process of the coalesced nanocrystals leads to the formation of larger triangular crystals, free of twin planes. This last process can be accompanied by a fusion of small triangles into larger one. On HOPG, the Ag triangular single crystals are epitaxially oriented. As the diffusion of Ag atoms on a HOPG surface is faster than on amorphous carbon, the Ag particles recrystallize more easily on HOPG than on amorphous carbon. This is consistent with the observation of more facetted particles on HOPG.

This mechanism is illustrated in Figure 11.7 where we have schematically drawn the different stages of triangular single-crystal formation.

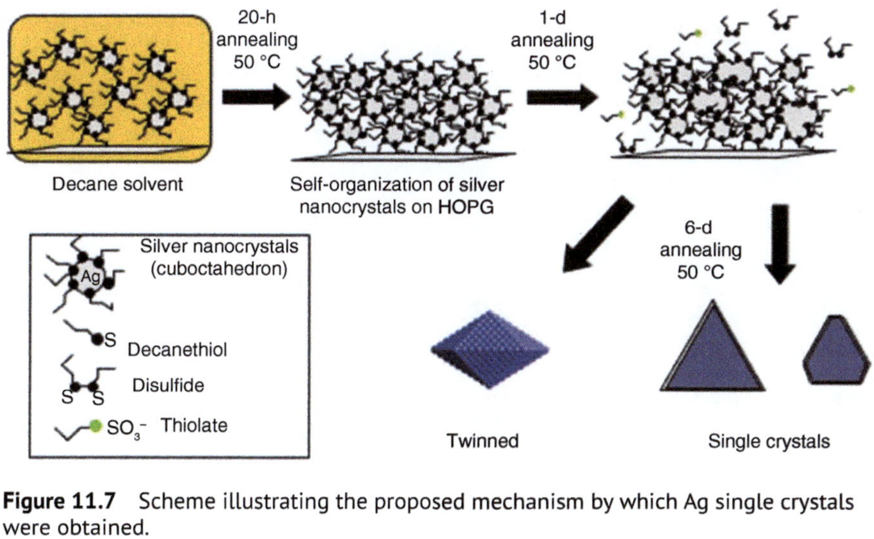

Figure 11.7 Scheme illustrating the proposed mechanism by which Ag single crystals were obtained.

Different stages of triangular single-crystal formation are observed: immediately after the solvent evaporation, the nanocrystals start to coalesce. After two annealing days, the nanocrystals continue growing to form triangles or hexagons with defects. In increasing the annealing, the size of coalesced particles increases with a more and more triangular shape. Note that particles smaller than 10 nm are still observed on the TEM grid. These triangular particles, formed by a progressive coalescence of self-organized spherical 5-nm Ag nanocrystals, are single-domain fcc structure. Furthermore, they are stable for almost a year. One open question remains: What is the driving force to produce single-domain triangular particles instead of aggregates?

11.2 Unexpected Electronic Properties of Micrometer-Thick 3D Superlattices of Au Nanocrystals

Usually the sample studied by scanning tunneling image and spectroscopy the sample are very thin. Here highly thick 3D superlattices are observed [13].

As shown in Chapter 4 by keeping a colloidal solution of Au nanocrystals [14, 15] under solvent saturation in seven days, two simultaneous 3D superlattice growth processes take place: at the air–toluene interface, well-defined layer-by-layer-thick 3D superlattice films (Figure 11.8a), called interfacial 3D superlattices, are formed. In

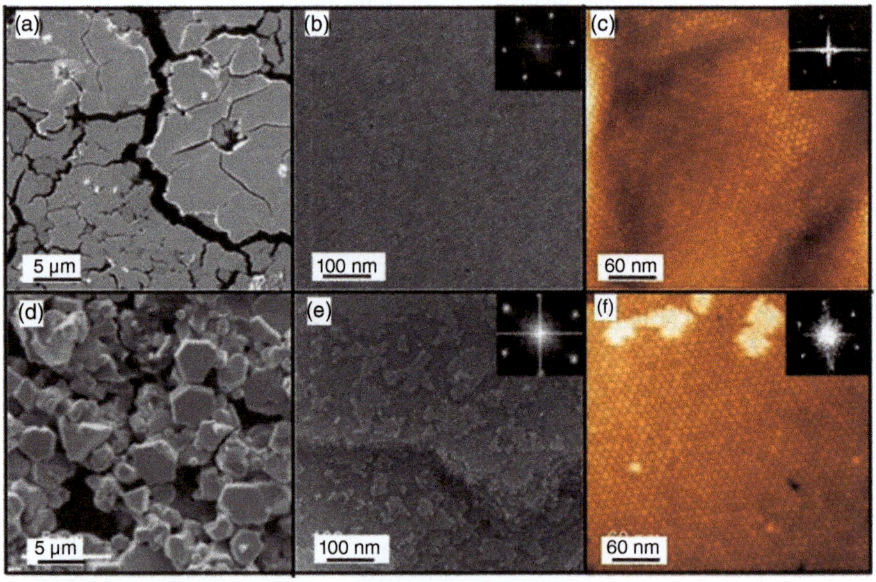

Figure 11.8 Typical scanning electron microscope (a, d), field-emission scanning electron microscope (b, e), and scanning tunneling microscope topographic (c, f) images for interfacial (a–c) and precipitated (d–f) 3D superlattices. The topographic STM images are obtained at 5 K with a bias voltage of 2.5 V and a current set point of 50 pA in (c) and 5 V and 150 pA in (f). Insets in (b), (c), (e), and (f) show the corresponding fast Fourier transform patterns. *Source:* Yang et al. [13]/American Chemical Society.

the meantime, a homogeneous growth takes place in solution with the formation of randomly oriented, well-defined, thicker 3D superlattices with different polyhedral shapes (Figure 11.8d), called precipitated 3D superlattices. Whatever the growth mechanisms are, the 3D superlattices have an fcc structure, which is determined by the small-angle X-ray diffraction technique. Field-emission scanning electron microscopy (FESEM) studies (Figure 11.8b,e) reveal a well-defined hexagonal arrangement of nanocrystals on the surface of both interfacial and precipitated 3D superlattices as illustrated by the fast Fourier transform (FFT) pattern shown in Figure 11.8b,e insets, respectively. This hexagonal arrangement corresponds to the {111} surface of the fcc structure. The thicknesses of interfacial and precipitated 3D superlattices are typically 300 nm and 5 μm, respectively. Then, the interfacial and precipitated 3D superlattices are composed of about 40 and 700 nanocrystal layers, respectively.

Contrary to what is expected, these thick 3D superlattices are conductive enough to be investigated by using STM/scanning tunneling spectroscopy (STS). All the STM/STS experiments are carried out with a chemically etched tungsten STM tip at 5 K on a commercial setup (Omicron Nanotechnology) under UHV conditions (base pressure $< 10^{-11}$ mbar). Data analysis and image processing were performed using Mathematica and the WSxM program [16].

Typical topographic STM images of interfacial and precipitated 3D superlattices are shown in Figure 11.8c,f, respectively. These STM images show a clear hexagonal arrangement of nanocrystals on the surface of the 3D superlattices. The corresponding FFT patterns (Figure 11.8c,f insets) show well-defined hexagonal characteristics similar to those of FESEM images (Figure 11.8b,e insets). Especially, for the precipitated 3D superlattices, the defects such as vacancies, adsorbed nanocrystals, and small pieces of stacking layers shown in the STM image (Figure 11.8f) appear clearly in the FESEM image (Figure 11.8e) of a precipitated 3D superlattice. It is worth noting that the topographic STM image of the interfacial 3D superlattice shown in Figure 11.1c is obtained with a bias voltage of 2.5 V. However, the thicker precipitated 3D superlattice is imaged at a higher bias voltage, e.g. 5 V (Figure 11.8f), and some STM images have been taken up to 9 V. Such a high voltage suggests that the potential drop between the STM tip and the first nanocrystal layer of the 3D superlattices is not equal to the applied voltage but is instead markedly reduced due to the highly resistive nature of the films. Indeed, the nanocrystals are coated with thiol molecules, which play the role of a dielectric barrier. Hence, the current passes across the supracrystalline network via tunneling between nanocrystals. Such a process confers a very high resistivity to the 3D superlattice [17, 18] Of course, the voltage drop through the film increases when the number of nanocrystal layers increases explaining why imaging of the very thick precipitated 3D superlattices necessitates higher voltages than the much thinner interfacial 3D superlattices.

Local spectroscopic measurements are performed to gain some insight into the transport mechanism involved. Simultaneously with the topographic STM image, *I–V* curves in a grid of 200 × 200 points are recorded with an open feedback loop in the scanning tunneling spectroscopic mode [19].

All experiments are repeated several times with different STM tips and different spots on the sample, and the results are representative and consistent with the more comprehensive dataset. *I–V* curves show a symmetric feature about zero bias

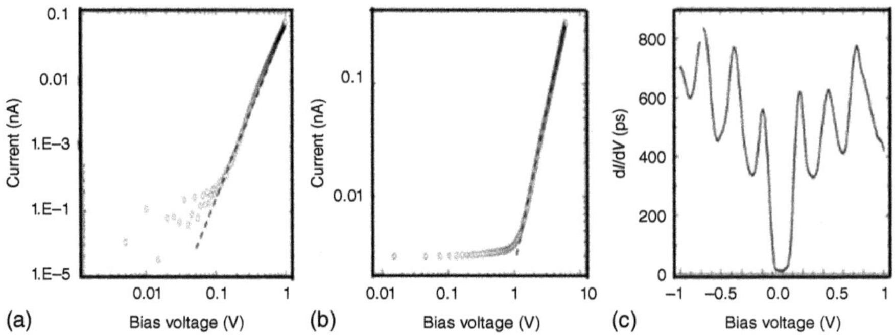

Figure 11.9 Log–log plot of the average *I*–*V* curves on the surface of interfacial (a) and precipitated (b) 3D superlattices. From several measurements, the scaling exponent (ζ) is about 2.7–3.4 and 2.8–3.0 for interfacial and precipitated 3D superlattices, respectively. (c) The Coulomb staircase in the d*I*/d*V* curve (averaged from 126 individual curves) of isolated Au nanocrystals adsorbed on the STM tip measured at 5 K. *Source:* Yang et al. [13], figure 2 (p. 2053)/American Chemical Society.

voltage. Additionally, the current is blocked when the bias voltage is lower than a threshold voltage (V_T). However, the current starts to flow with a bias voltage of above V_T with the opening of conduction channels through the 3D superlattice [20]. Figure 11.9a,b shows the *I*–*V* curves plotted in log–log coordinates. The increase in the current with the applied voltage reveals power-law dependence, i.e. $I \propto (V - V_T)^{\zeta}$. The scaling exponent ζ is related to the number of accessible current paths through the 3D superlattices, depending on the 3D superlattice dimensionality [21]. For 1D and 2D arrays, ζ is predicted to be ~1 and 5/3, respectively [22]. The scaling exponent is expected to increase for 3D arrays due to the greater numbers of current paths even though this has not been theoretically demonstrated [21]. Here, from several measurements, ζ is determined to be 2.7–3.4 for interfacial 3D superlattices, whereas for precipitated 3D superlattices, it is 2.8–3.0. The fact that the scaling exponents are comparable for the 3D superlattices having ~40 or ~700 nanocrystal layers seems to indicate that interfacial and precipitated 3D superlattices have similar transport mechanisms.

We carried out spectroscopic mappings (Figure 11.10) to measure the spatial variations in the electronic properties of the 3D superlattices. Conductance maps, $\mathrm{d}I/\mathrm{d}V(\vec{r},V)$, were derived numerically from the *I*–*V* data on each spectroscopic grid point at selected values of the bias voltage. Figure 11.10a,b shows a conductance map at 0.45 V for the interfacial 3D superlattice and a map at 3 V for the precipitated one. The corresponding topographic STM images are shown in Figure 11.10c,d insets. The conductance maps show light spatial variations at the nanocrystal scale as indicated in Figure 11.10a,b by the red and blue circles. Some nanocrystals can be clearly identified by their unique conductance. The d*I*/d*V* curves averaged from the regions delimited by the red and blue circles are illustrated by the red and blue curves in Figure 11.10c,d. For comparison, dashed lines in the plot illustrate the d*I*/d*V* curves averaged overall the surface. It is clear that from the d*I*/d*V* curves measured above, a given nanocrystal exhibits a slight additional staircase-like

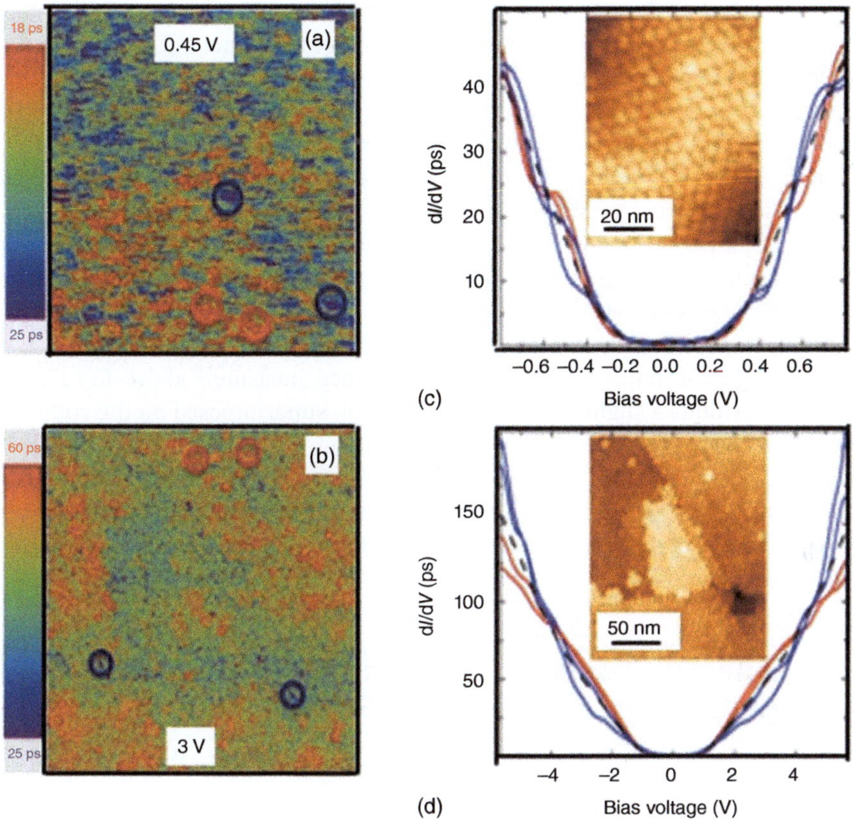

Figure 11.10 The conductance map at 0.45 V of interfacial (a) and at 3 V of precipitated (b) 3D superlattices. The fingerprint of isolated nanocrystals, i.e. staircase-like structures, is found in the dI/dV curves (c, d) averaged from particular regions indicated by the red and blue circles on conductance maps (a, b). For both interfacial and precipitated 3D superlattices, the dI/dV curves averaged overall the surface are shown by dashed lines in (c, d). The insets in (c) and (d) show the topographic STM images corresponding to the conductance maps (a) and (b), respectively. These STM topographic images are obtained at 5 K with bias voltages and current set points at 2.3 V, 130 pA, and 5 V, 200 pA, respectively. *Source:* Yang et al. [13]/American Chemical Society.

modulation (red or blue lines) superimposed on the collective tunneling conductance background $dI/dV \propto (V - V_T)^{\zeta-1}$ (dashed lines). This modulation can be easily associated with a Coulomb staircase through an isolated nanocrystal. Figure 11.9c shows the dI/dV curve of a single nanocrystal adsorbed on the STM tip; the blocking energy of 250 meV is comparable with the 300 meV observed in the interfacial 3D superlattice as shown in Figure 11.10c.

Then, the collective electronic properties of 3D superlattices are associated with the fingerprint of its isolated component nanocrystals. For the precipitated 3D superlattices, faint Coulomb oscillations are also possibly observed, but their characteristic energy is of the order of 2 V instead of 250 mV for an isolated particle (Figure 11.10d). This is attributed to the large voltage drop through the precipitated

3D superlattice. In this case, the voltage between the STM tip and the first particle layer is so reduced compared with the applied voltage that it needs higher voltage to observe the Coulomb oscillations. Except for these small Coulomb oscillations, the conductance maps are quite homogeneous. In particular, the underlying $\propto(V-V_T)^{\zeta-1}$ conductance background is very homogeneous. This homogeneity is attributed to both the high degree of nanocrystal ordering within the 3D superlattices and the low nanocrystal size distribution.

Hence, 3D superlattices of 5-nm single-domain Au nanocrystals with thickness up to 5 μm can be surprisingly well imaged by STM. The average I–V curves clearly show a power-law dependence ($I \propto (V-V_T)^{\zeta}$) for the bias voltage above a threshold voltage (V_T) with a scaling exponent $\zeta \sim 3$ that indicates a 3D multiple paths tunneling through the nanocrystals. The conductance measured above for a given nanocrystal exhibits a slight additional modulation superimposed on the collective conductance background that mimics the Coulomb staircase structure normally associated with an isolated nanocrystal. The observation of collective properties with the fingerprint of isolated nanocrystals makes it possible to benefit from the unique properties of nanocrystals while avoiding their size-related potential risks in the future applications.

11.3 Collective Magnetic Properties of Co Nanocrystals Self-Assembled in 3D Superlattices

Co-based materials have attracted tremendous attention due to their unique magnetic, electronic, and catalytic properties [23–25].

A general feature characterizing single magnetic nanocrystals is their superparamagnetic behavior. The magnetocrystalline anisotropy energy in axial symmetry depends mainly on the particle volume, V, and its anisotropy constant, K. In the superparamagnetic regime, the anisotropy energy barrier, $E_b = K\,V$, is usually of the same magnitude as the thermal energy. Then, the magnetization vector fluctuates among the easy directions of magnetization. This process is called superparamagnetic relaxation. When a magnetic material is subjected to an increasing magnetic field, the spins within the material are aligned with the field. Its magnetization increases and reaches a maximum value called the saturation magnetization, M_s, as the magnitude of the magnetic field decreases and the total magnetization decreases. In ferromagnets, a residual magnetic moment remains at zero field. The value of the magnetization, at zero field, is called the remanent magnetization, M_r. If the magnetic sample is cooled at low temperature and then a magnetic field is applied, a progressive increase in the magnetization is observed to reach a maximum called blocking temperature, T_b. At higher T_b value, the magnetization of the particle does not depend on temperature, and the susceptibility falls. This is called zero-field-cooled (ZFC) magnetization.

11.3.1 Influence of Nanocrystal Ordering on the Magnetic Properties

Collective magnetic properties due to the nanocrystal ordering are observed [26, 30].

Let us consider 7-nm amorphous Co nanoparticles coated with dodecanoic acid. We know that by controlling the substrate temperature ($18 < T < 45\,°C$) during the colloidal solution evaporation process, from the same batch of nanocrystals, it is possible to produce either amorphous film or 3D superlattices [26]. Here, we use amorphous Co nanoparticles and compare the magnetic properties of the either ordered (fcc) or disordered films.

To measure the field-cooled (FC) magnetization versus temperature, a field of 20 Oe is applied at 300 K before cooling the sample to 5 K and subsequently measuring the magnetization from 5 to 300 K. Figure 11.11 shows the ZFC curve normalized to the blocking temperature (T) versus the temperature curves of both the ordered (red) and disordered (black) 3D superlattices of 7-nm Co assemblies. The ZFC peak is significantly narrower for 3D superlattice film. The width of the ZFC peak is related to the distribution of energy barriers, E_b, in the system: a larger distribution gives a broader peak. The energy barriers involved are the anisotropy energy ($E_a = k_a V$ where k_a is the anisotropy constant and V is the nanoparticle volume) and the dipole–dipole interaction energy (E_{dd}), which varies with the particle distance [28]. Because the ordered and disordered samples are made with the same

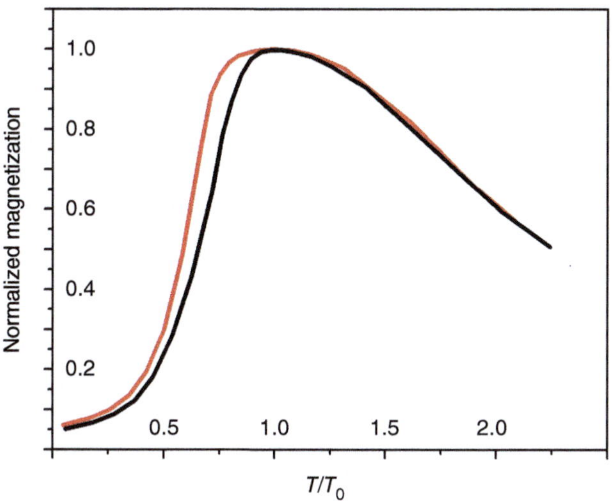

Figure 11.11 Normalized to the blocking temperature and the ZFC curve of amorphous aggregates (red) and 3D superlattices (black), corresponding to these configurations calculated for 8-nm Co nanocrystals and 0.125 as an effective coupling constant. *Source:* Lisiecki et al. [27], figure 3c (p. 4033)/American Chemical Society.

batch of nanocrystals, the change in width of the ZFC peak is not a consequence of a change in the size dispersion of nanoparticles between the two samples or a difference in anisotropy. We therefore explain the difference in the distribution of E_b by the change in the structural environment of the Co nanoparticles. As pointed out, dipolar forces have a strong directional dependence, and consequently, dipolar interactions in the assembly should be sensitive to the detailed geometrical arrangement of the nanoparticles. In the 3D superlattices, the fcc domains have long coherence lengths, and, therefore, the geometric environment of the nanoparticles is fairly uniform. In the disordered 3D aggregates, we have a mixture of many very small fcc domains characterized by a short coherence length and disordered domains with an irregular stacking periodicity. Therefore, we expect the distribution of E_{dd} (and hence E_b) in the 3D superlattice to be lower than in the disordered sample, leading to the observed narrowing of the ZFC peak. We acknowledge that this effect of order is fairly subtle. However, the data are highly reproducible.

In Chapter 10, it is shown that the change of the magnetization versus field curves for the ordered and disordered samples at 5 K is attributed to the change of the flipping of the spins requiring higher fields when the nanocrystals are ordered in a long-range superlattice compared with the disordered system [27, 29]. Furthermore, the more gradual approach to saturation for ordered assemblies than disordered assemblies is explained in terms of fluctuation of the anisotropy direction that is random with disordered aggregates and highly ordered with fcc Co 3D superlattices.

11.3.2 Influence of Co Nanocrystallinity on fcc 3D Superlattices

In Chapters 1 and 2, we know that the internal crystallinity called nanocrystallinity of rather uniform Co nanoparticles can be improved by annealing process. The average diameter remains the same (within the accuracy limit of the measurements) as those of the native nanocrystals, i.e. 7.5 ± 0.4 nm. The electron diffraction pattern taken from a population of several hundred nanocrystals shows rings corresponding to pure hexagonal close-packed (hcp) Co without the signature of Co oxide.

Figure 11.12 shows the SEM images and the corresponding diffraction for Co 3D superlattices before and after annealing at 350 K. It is clearly shown that the crystalline structure of 3D superlattices remains similar (fcc) before and after annealing. However, the nanocrystallinity of the building blocks evolves from *amorphous* phase to hcp structure.

Here, we would like to show the role of nanocrystallinity on the magnetic properties. The magnetization versus applied field curves of Co nanoparticles in their amorphous and hcp states, deposited on HOPG substrate (Figure 11.13a), shows an increase in saturation magnetization. The curves of ZFC magnetization versus temperature show that annealing the sample at 350 °C induces a drastic increase in τ_B from 100 to 280 K. This confirms the crystallographic transition from Co polycrystals to hcp nanocrystals. This leads to an increase in the anisotropy of the nanocrystals, which in turn gives an increase in the energy barriers and hence in τ_B. As shown in Figure 11.13b, the ZFC curves, normalized to τ_B of the samples in the native and annealed states, superimpose. This result shows that, after annealing at 350 °C, the conversion to a single-domain hcp structure is complete and no coalescence between

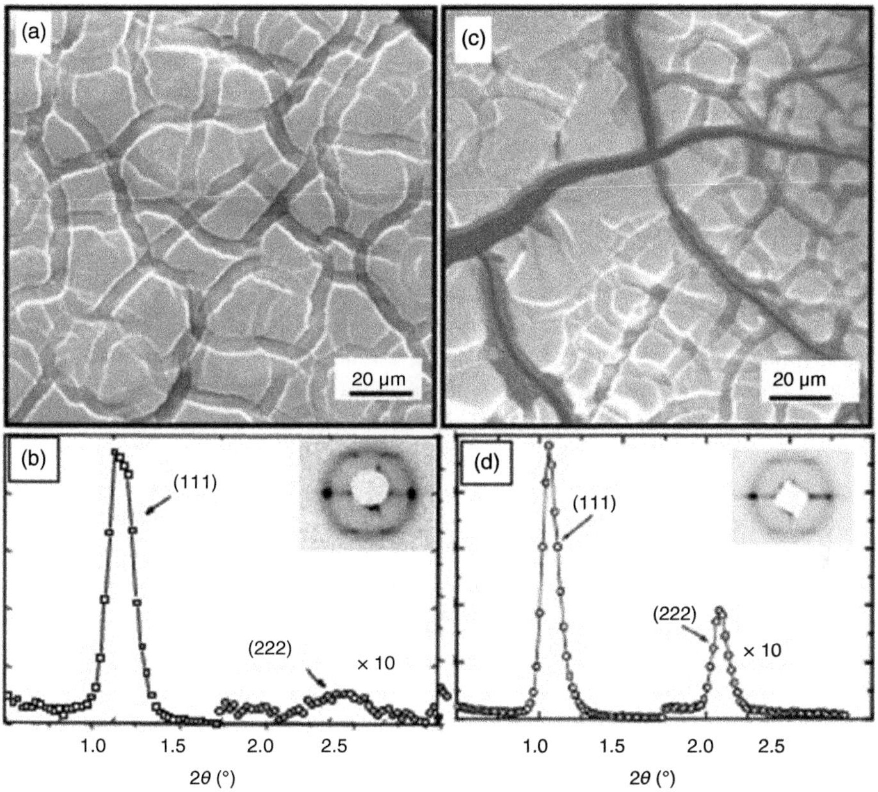

Figure 11.12 (a, b) SEM images of a 3D superlattice and (c, d) the corresponding diffractogram of 7.5-nm Co nanocrystals not annealed (a, b) and annealed at 350 °C (c, d). *Source:* (a, c) Lisiecki et al. [27]/American Chemical Society.

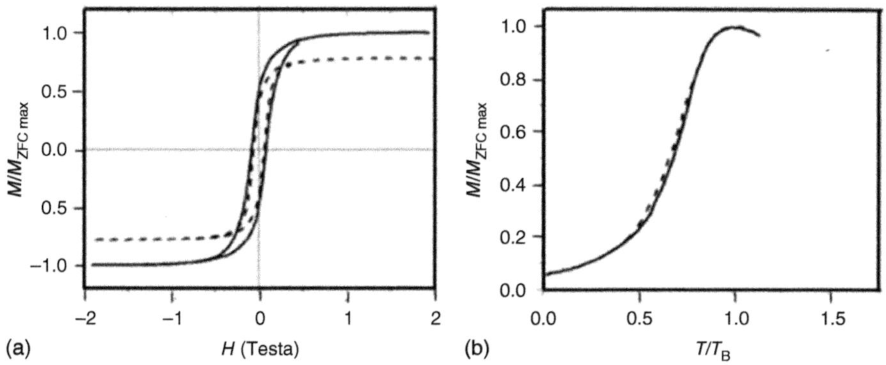

Figure 11.13 (a) Magnetization versus field curves at 5 K of a supracrystalline sample, native (dashed line) and annealed at 350 °C (solid line). (b) ZFC magnetization versus T/T_b curves for the 3D superlattice of amorphous (dashed line) and hcp (solid line) nanocrystals. *Source:* Based on Lisiecki et al. [27].

nanocrystals occurs. The absence of a peak at around 8 K in the ZFC curves confirms the absence of oxidation of the metallic nanomaterial before and after the annealing treatment. Additionally, these results show that the ZFC measurements could be used as a powerful tool to probe the particle crystallinity [29].

11.3.3 Magnetic Properties of Single-Domain ε-Phase Co Nanocrystals at Various Interactions Scales

11.3.3.1 Magnetic Properties of Co (ε-Phase) Nanocrystals Dispersed in PMMA

As mentioned in Chapter 1, at the nanoscale, Co nanocrystals have three nearly isoenergetic crystal structures: hcp, fcc, and epsilon (ε). Bulk Co displays two stable phases of hcp and fcc at ambient pressure [30]. Due to its relatively smaller magnetocrystalline anisotropy constant ($K_c \approx 1.5 \times 10^5 \text{J m}^3$), ε-structure is widely known as a softer magnetic material than the hcp and fcc phases. Hence, at room temperature, the thermal fluctuations ($k_B T$) are much larger than the internal magnetic energy ($K_c V$, V being the volume of the nanocrystal) of an individual ε-Co nanocrystal up to several tens of nanometers in diameter. In this condition, the nanocrystal presents a superparamagnetic phase with zero average magnetization that has been shown to be a crucial property enabling the elaboration of its self-assembled 2D ordered thin films in the absence of internal stray fields [31]. Because of their low magnetic anisotropy, these 2D superlattices have attracted a considerable attention for promising technological applications, which is reinforced after the observation of spin-dependent electron transport [21]. Most of the previous reports are focused on the 2D superlattice of ε-Co nanoparticles [31–34]. Here, we compare single-domain ε-Co nanocrystals coated with oleic acid in either not or highly interacting states. In the first case, Co nanocrystals are dispersed in PMMA [poly(methyl methacrylate)] matrix. In highly interacting state, films of nanocrystals either disordered or ordered in fcc 3D superlattices are produced. This study is performed for various ε-Co nanocrystal diameters from 7.6 to 9.2 nm. The ZFC/FC curves are performed from 5 to 350 K, and the sample is subjected to a field of 100 Oe. Figure 11.14 shows the ZFC/FC curves of isolated ε-Co nanocrystals dispersed in PMMA and differing by their average diameters. An increase in the blocking temperature with increasing their diameters is observed (Table 11.1). The linear increase in the blocking temperature with increasing the nanocrystal volumes follows the well-known equation for noninteracting nanoparticles: $T_b \approx KV/k_B \ln(\tau_0/\tau_m)$, where τ_0, τ_m, K, and V are the inverse frequency by which the magnetic moments overpass the anisotropy barriers, the experimental measuring time, the magnetocrystalline anisotropy constant, and the volume of the magnetic nanoparticle, respectively.

Table 11.1 shows, as expected, an increase in the blocking temperature with increasing the nanocrystal diameter. Because of the low size distribution and the high sphericity of ε-Co nanocrystals, in first approximation, the magnetocrystalline anisotropy and the magnetization distribution are neglected. Furthermore, it is assumed that the anisotropy constants of nanocrystals differing by their sizes remain unchanged in such size range (7.6 to 9.2 nm).

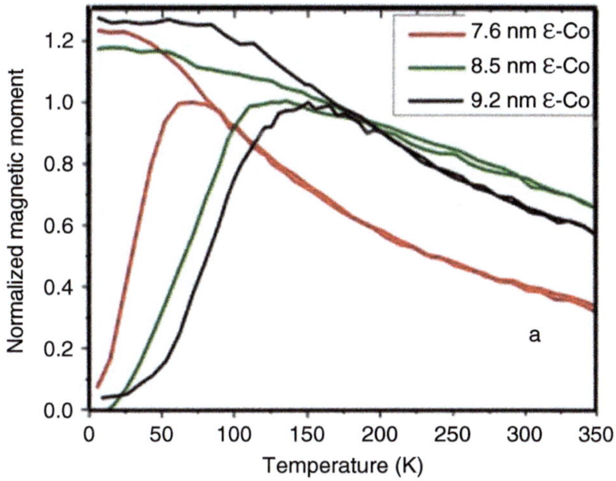

Figure 11.14 ZFC/FC thermal evolution of the magnetic moments in an ensemble of isolated ε-Co nanocrystals embedded in PMMA (H = 100 Oe). The three curves correspond to the 7.6-nm (black), 8.5-nm (red), and 9.2-nm (blue) nanocrystals and are normalized to their peak value at blocking temperature. *Source:* Yang et al. [35], figure 6 (p. 8)/ IOP Publishing Ltd.

Table 11.1 Different characteristic temperatures and the respective ΔT corresponding to the samples in two ordered and disordered series, deduced from magnetization ZFC/FC curves measured via an in-plane 100 Oe magnetic field.

Size (nm)	Experimental conditions	T_p (K)	T_{irr} (K)	T_{pm} (K)	$\Delta T = (T_{irr} - T_p)$ (K)	$\Delta T' = (T_p^{supra} - T_p^{Disord})$ (K)
7.6	Isolated	75	—	—	—	—
	Supracrystals	133	144	235	11	6
	Amorphous film	128	143	235	16	
8.5	Isolated	115	—	—	—	—
	Supracrystals	160	170	278	10	16
	Amorphous film	136	190	285	44	
9.2	Isolated	150	—	—	—	—
	Supracrystals	236	255	288	19	21
	Amorphous film	215	269	300	44	

Source: Yang et al. [35], table 1 (p. 8)/IOP Publishing Ltd.

11.3.3.2 Magnetic Properties of Co (ε-Phase) Nanocrystals Deposited on Substrate

Let us compare the magnetic properties of films either disordered or ordered in fcc 3D superlattices. Figure 11.15a shows a typical magnetic thermal evolution measured for both ordered and disordered assemblies of ε-Co nanocrystals. It shows marked changes in the ZFC/FC curve compared with what is observed when the ε-Co nanocrystals are isolated in PMMA (Figure 11.11). For any ε-Co nanocrystal size, the following is observed:

The FC curves markedly change when nanocrystals form films compared with that obtained when they are isolated in PMMA matrix. It decreases with decreasing

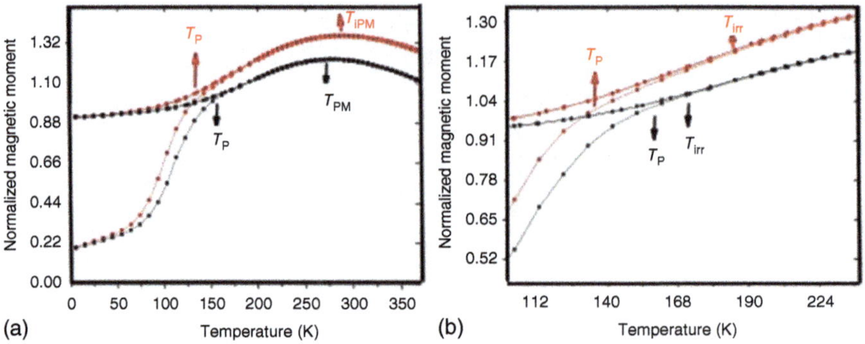

Figure 11.15 The comparison between the ZFC/FC measurements of the 3D superlattice (black squares) and the disordered sample (red circles) of 8.5-nm ε-Co nanocrystals. The temperature range is from 5 to 350 K (a) and from 100 to 225 K (b). The temperatures corresponding to irreversibility and the ZFC peaks are marked with arrows. (H = 100 Oe). *Source:* Yang et al. [35], figure 7 (p. 9)/IOP Publishing Ltd.

temperature to reach a minimum, whereas it continuously increases for isolated nanocrystals. This feature was previously reported as the signature of the presence of large interparticle magnetic interactions, for instance, in spin glass (SG) systems [36–39].

(i) With film of nanocrystals, the ZFC curves are very broad compared with isolated nanocrystals. The broadening of the ZFC/FC curves is followed by a decrease of the blocking temperature as previously observed in the case of interacting assemblies of magnetic nanoparticles [40].

A careful study of ZFC curves obtained with $H = 100\,\mathrm{Oe}$ as applied field reveals three characteristic temperatures (Figure 11.15a,b):

(a) T_p corresponds to the maximum of the magnetization reached on increasing the temperature (Figure 11.15a,b).
(b) T_{irr} corresponds to the temperature at which the FC and ZFC curves split (Figure 11.15b).
(c) T_{pm} corresponds to a maximum in the ZFC curve above the blocking temperature (Figure 11.15b).

T$_p$ corresponds to temperature at which $k_{\mathrm{B}}T_p \approx E_{\mathrm{total}}$. As expected, T_p increases with increasing the nanocrystal volume and consequently the magnetic anisotropy energy. Table 11.1 shows that *Tp* temperatures are larger than those observed for the isolated nanocrystals. Such feature is attributed to the presence of magnetic interactions between nanocrystals [31, 32]. For a given nanocrystal size, T_p is larger for ordered 3D superlattices than the disordered films. This is attributed to the different contributions of the interparticle magnetic interaction energies, E_{inter}, to the total magnetic energy, E_{total}, of each of these two systems. Table 11.1 shows that the temperature difference, $T_p{}^{\mathrm{supra}} - T_p{}^{\mathrm{dis}}$, increases with increasing the particle diameter. This reveals that, while the nanoparticle volume fraction remains unchanged, the nanocrystal ordering induces an increase in the magnetic interactions probably due to the spatial and orientational ordered dipolar interactions compared with disordered films. This is partially supported by the fact that the blocking temperature increases with degree of ordering: with 2D superlattices ordered in compact in hexagonal network of 9-nm ε-Co is 165 K [41], whereas it is 235 K for thick 3D ordered 3D superlattices. Figure 11.15b shows the irreversibility temperature (T_{irr}) below which the magnetic moments have a bistable state, depending on the conditions under which they approach the point. In a perfectly uniform system of nanoparticles, the irreversibility point and the blocking temperature are one unique point. Such splitting between T_{irr} and T_p ($\Delta T = T_{irr} - T_p$) was observed previously [42] and explained a large energy distribution in a system due to a large nanoparticle size distribution: T_p was attributed to the average blocking temperature and T_{irr} to the blocking temperature of the largest particles. Here, because of the very low size distribution ($\sigma = 7\%$), this is ruled out. However, such feature could be explained in terms of the formation of the aggregates present in the film. In fact, even though the 3D superlattices are highly ordered, at the edges, the ordering drops and forms small disordered aggregates. Similarly, with disordered film, because of the low size distribution of nanocrystals, the formation of locally ordered aggregates cannot be excluded. Such small discrepancies in the nanocrystal ordering induce some change in the interparticle distances and consequently in the dipolar magnetic interactions that explain the differences in T_p and T_{irr}. As in disordered system, the nanoparticles have more liberty of positioning, and ΔT is more pronounced for disordered than ordered films. Moreover, because of increasing interparticle interactions, ΔT increases with the nanocrystal size (Table 11.1).

At higher temperatures ($T_{pM} > T_p \approx T_{irr}$), another characteristic peak is observed in the ZFC curve measured with 100 Oe as applied probe field (Figure 11.15). This feature has previously observed also in other interacting nanoparticle assemblies [31, 43] and did not attracted attention. As already observed with T_p and T_{irr}, T_{pM} increases with increasing the nanocrystal diameter and consequently is also attributed to an increase in the magnetic anisotropy energy. The peak at T_{pM} exists both for disordered and ordered films (Table 11.1) indicating it is not directly related to the nanocrystal ordering. Moreover, for any nanocrystal size, T_{pM} is lower for 3D superlattices than disordered films. From ZFC measurements produced by applying a large magnetic field ($H = 500$ Oe instead of 100 Oe), the T_p and T_{pM} remain exactly the same value. This indicates that the appearance of T_{pM} is independent of the initial state of the magnetic moments.

From these data, it is concluded that the magnetic properties of fcc 3D superlattices of single-domain ε-Co nanocrystals with a low size distribution ($\sigma < 7\%$) reveal an enhancement compared with the corresponding disordered films and the layer-by-layer growth of 2D ordered layers. This could be attributed to the coherent alignment of the atomic lattice planes of nanocrystal with one another within the 3D superlattice. Due to dipolar magnetic interaction, the ZFC curves of films are broader than noninteracting nanocrystals. Furthermore, the ZFC curves include three characteristic temperatures, T_p, T_{irr}, and T_{pM}. T_p presented as the average blocking temperature of the samples. The irreversibility point between the ZFC and FC curves, marked by T_{irr}, results from a distribution of magnetic interaction strengths due to the formation of local aggregates (effect of interparticle distances), which is significantly decreased in ordered samples compared with disordered ones. At T_{pM}, in the superparamagnetic regime, the magnetization presents a peak due to the Zeeman coupling of the super-spins with the probe magnetic field.

11.4 Super-Spin Glass Behavior of fcc Supracrystals

Specific properties such as super spin glass of 3D superlattices are observed [44].

In a conventional atomic spin glass, the lack of long-range magnetic order arising from magnetic frustration or highly dilute magnetic ions leads to interesting behavior such as aging, rejuvenation, and memory effects [45]. In a magnetic nanoparticle system, if the nanoparticles are small enough to have a single magnetic domain, that is, the case with 7-nm Co nanoparticles, each nanoparticle will act like a giant or "super"-spin. In sufficiently concentrated systems, these "super-spins" can interact via long-range dipolar interactions, the random nature of which leads to a highly disordered and frustrated magnetic state analogous to that in atomic spin glasses. Most of the interacting systems investigated were disordered aggregate without any regular array of nanoparticles [40, 44–47]. Therefore, the question arises: What is the influence of the long-range ordering of Co nanocrystals in a 3D fcc array on the super-spin glass behavior? The blocking temperature of the Co fcc 3D superlattices is 130 K, which is significantly larger than that observed for a dilute system of Co nanocrystals of a similar size due to the strong magnetic dipolar interactions between the nanocrystals in the 3D superlattice. The FC magnetization curve is nearly temperature independent below T_b, characteristic of spin glass behavior. In less-interacting systems, where the behavior is superparamagnetic, the FC magnetization curve is not temperature independent below T_b but increases with decreasing temperature. Figure 11.16 shows the in-phase (χ') and out-of-phase (χ'') components of the AC susceptibility versus temperature of a fcc Co 3D superlattice, measured in a range of AC frequencies varying by 2 orders of magnitude. Both the χ' and

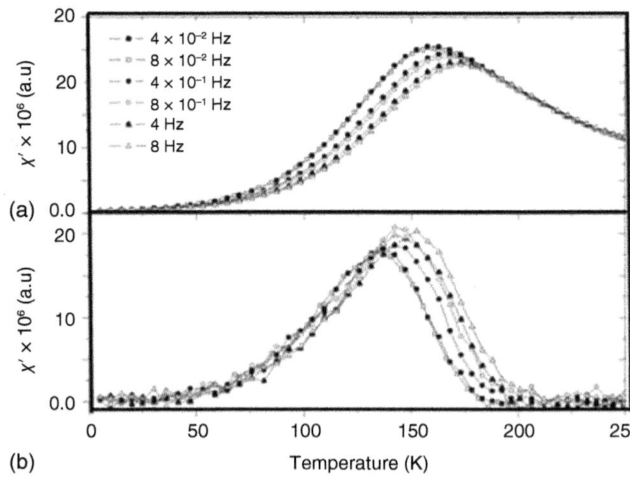

Figure 11.16 (a) In-phase and (b) out-of-phase part of the AC susceptibility versus temperature, measured at frequencies between 0.08 and 8 Hz in an AC driving field of 1.7 Oe for fcc 3D superlattices made of amorphous Co nanoparticles. *Source:* Parker et al. [44], figure 1 (p. 1140)/American Chemical Society.

χ'' curves show a clear frequency dependence where the maximum, T_{peak}, decreases in temperature with decreasing frequency, f. T_{peak} represents the temperature at which the relaxation time of the system, τ, is equal to the observation time, t, which is related to the measurement frequency by $t=1/\omega$, where $\omega = 2\pi f$. Frequency dependence in an AC measurement is observed in both superparamagnetic and super-spin glass materials. To differentiate between these two types of behavior, one can analyze quantitatively the change in T_{peak} with frequency and hence extract a value for τ. For a superparamagnet, where dipolar interactions between the magnetic moments are negligible, the frequency dependence should follow an Arrhenius law $\tau = \tau_0 \exp^{E_a/k_B T}$ (related to the Néel–Brown model) where τ_0 is the angular inverse attempt frequency, E_a is the anisotropy energy, and k_B is the Boltzmann constant. By plotting $\log_{10}\tau$ versus $1/T_{peak}$ and fitting the data to the Arrhenius law, the value of τ_0 can be extracted, and in this case we find $\tau_0 = 10^{-31}$ seconds. This unphysically small value indicates that this system is not best described by simple energy barrier blocking and thermal activation, i.e. the flipping rate of the magnetic moment of a given nanocrystal is no longer dependent solely on its own anisotropy energy but is also influenced by interparticle interaction, analogous to a spin glass. In a spin glass-like system, the dynamic behavior displays a critical slowdown, and hence, the characteristic relaxation time diverges to a finite static glass temperature, T_g, according to a critical power law $\tau = \tau^*(T_{peak}/T_g - 1)^{-z\nu}$, where τ^* is the relaxation time of individual nanocrystal moment and $z\nu$ is the critical exponent. To constrain the analysis, T_g is taken as the maximum in the DC ZFC magnetization curve (Figure 11.16b); the best fit of the data yields $\tau^* = 10^{-8}$ seconds and $z\nu = 13$. The value of τ^* compares quite well with values found for spin glasses, and $z\nu$, although slightly high, is compatible within error to that expected for a canonical spin glass for which $z\nu \approx 10$ [42, 45] and therefore supports the existence of a phase transition from a superparamagnetic to super-spin glass state in a fcc Co 3D superlattice.

Due to their nonequilibrium dynamics, spin glass and super-spin glass materials are known to show aging and memory effects [40, 45, 46, 48–52], that can be demonstrated by a simple DC magnetization experiment. The sample is ZFC from above T_g to a temperature T_s (typically 0.7 T_g) where a wait (t_w) of several thousand seconds is imposed before continuing cooling down to low temperature. A small field is then applied and the magnetization is measured on heating. A deviation from a reference ZFC curve (with no stop during cooling) is observed at T_s, which is known as a "memory dip," so-called as the system has "remembered" the spontaneous relaxation toward a zero value (aging) that occurred when the system was left unperturbed at constant temperature T_S. This measurement protocol has been applied to our Co 3D superlattice, and T_s and t_w are 70 K and 10^4 seconds, respectively. At T_S, the memory dip can be observed and is clear from the difference plot (inset). This effect is less pronounced than that found in atomic spin glasses [45], as has been observed for other super-spin glass systems [44]. These preliminary AC and DC susceptibility investigations provide strong evidence for super-spin glass behavior in Co 3D superlattices, analogous to the disordered and frustrated magnetic state observed in canonical spin glass materials. This work adds another example to the growing list of the correlations between the physical properties of atoms in a bulk phase (or in a nanocrystal) and nanocrystals in a 3D superlattice. In the case of Co 3D superlattices, we have shown that a higher nanoparticle anisotropy leads to a slower approach to saturation in the magnetization versus field curve, similar to the behavior observed in an atomic material [27, 29]. We have also shown evidence of vibrational coherence in Co [53] and Ag [54] 3D superlattices by time-resolved pump–probe spectroscopy and low-frequency Raman and Brillouin scattering, respectively. Now, from a magnetic point of view and based on the magnetic phase diagram of a granular system with an insulating matrix proposed by Kleemann et al. [55], it is reasonable to believe that a variation of the 3D superlattice anisotropy could tune the system to achieve superferromagnetic properties. To further our understanding of the correlation between canonical spin glass and super-spin glass behaviors, we plan to carry out more detailed investigations into the dynamic magnetic responses of these 3D superlattice systems and to investigate the super-spin/superferromagnetic transition by using different annealed 3D superlattice samples characterized by various magnetocrystalline anisotropies [27, 29].

11.5 Alignment of Magnetic Nanocrystals

11.5.1 Do the Mesoscopic Structures Play the Major Role on the Magnetic Properties?

Mesoscopic alignment of magnetic nanocrystals induces marked change in their magnetic properties compared to films [56–60].

In Chapter 6, we know that the evaporation process of colloidal solution having 10-nm spherical γ-Fe_2O_3 nanocrystals coated with either citrate ions or dodecanoic acid produces specific mesoscopic structures (Figures 11.17a,c and 11.18a,c).

With citrate ions as coating agent, in the absence of applied magnetic field, films of highly spherical compact aggregates have an average diameter of 1.5 ± 1 μm (Figure 11.17a). By applying a magnetic field (0.59T) during the deposition process, the nanocrystals are assembled in rather long, chain-like structures with a maximum length of 40 μm and 400 nm in diameter (Figure 11.17c) and with an average distance between chains of around 3 μm.

By increasing the strength of the applied field, the width of the wire-like increases (Figure 11.18).

With dodecanoic acid as coating agent, in the absence (Figure 11.19a) and presence of applied field (Figure 11.19c), homogeneous films are produced.

These changes in the mesoscopic structures are due to the fact that the van der Waals interaction depends drastically on the distance between nanocrystals [59]. With citrate ions as coating agent, the distance between the nanocrystals is determined by the overlap of the repulsive ionic double layers surrounding the nanoparticles. With dodecanoic-coated nanocrystals, the distance between the nanocrystals is correlated to the length of the alkyl chain due to steric repulsion. This large

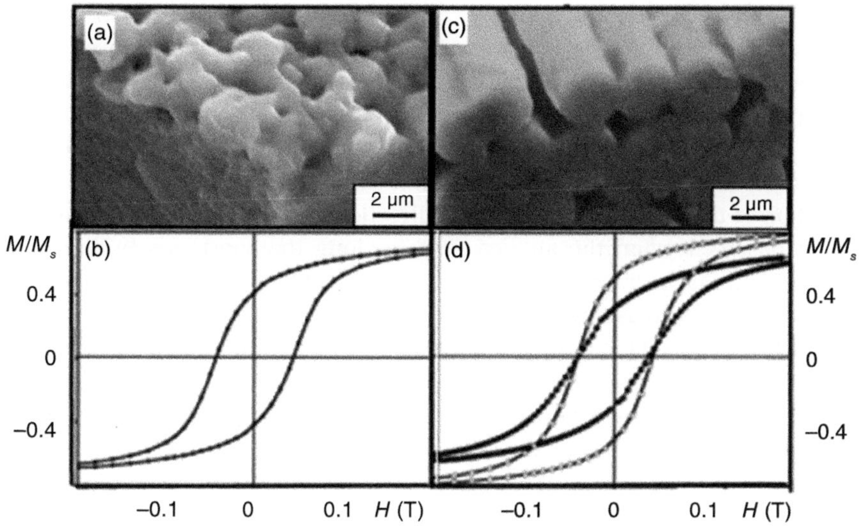

Figure 11.17 (a, c) SEM images of γ-Fe_2O_3 nanocrystals coated with citrate ions deposited on HOPG substrate without applied field (a) and with an applied field (0.59 T) (c). (b, d) Magnetization curves at 3 K corresponding the mesostructures shown on (a, c). (b) Magnetization corresponding to sample shown in (a) and (d) that of (b). The magnetization is recorded with an applied field either parallel (open circles) or perpendicular (solid circles) to the ribbons direction. *Source:* Lalatonne et al. [58, 59]/ American Chemical Society.

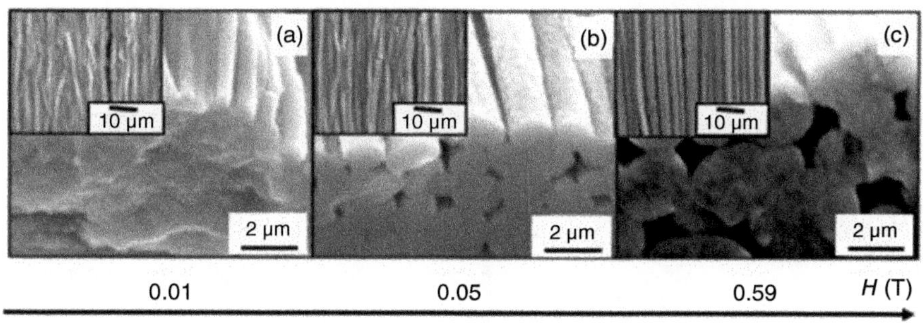

Figure 11.18 SEM images of shaped γ-Fe_2O_3 nanocrystals coated with citrate ions produced at various applied fields during the evaporation process. *Source:* Lalatonne et al. [58, 59]/American Chemical Society.

distance between nanocrystals prevents spherical aggregate formation, and the weak dipolar interaction does not allow the formation of chain-like structures.

Whatever the mesoscopic structures, the magnetic properties, recorded at 3 K, of these films show same value of the saturated magnetization value (77 emu g^{-1}).

With citrate ions used as coating agent, the hysteresis loops, obtained with the rough film shown on Figure 11.17a, are the same when the field is applied along the

x and y directions (Figure 11.17b). They are larger than that obtained for isolated nanocrystals dispersed in a nonmagnetic matrix. In thick films, the reduced remanence and coercive field are 0.41 and 420 Oe, whereas they are 0.30 and 360 Oe, respectively, in diluted solutions. This is due to the increase in the magnetic interactions when nanocrystals form thick films. With wires shown on Figure 11.17c, the hysteresis loops drastically change (Figure 11.17d). When the field is applied along the direction of the long axis of the rods (x direction), the hysteresis loop is squarer than that observed when the applied field is along the short axis (y direction) (Figure 11.17d). Furthermore, in the x direction, the reduced remanence increases, and the coercivity decreases (0.52 and 480 Oe) compared with those observed with films shown in Figure 11.17b (0.41 and 420 Oe). The reduced remanence increases with the wire-like width (Table 11.2).

With dodecanoic acid as coating agents, the hysteresis loops (Figure 11.19b,d) remain unchanged with the field direction (x or y) when the nanocrystals are deposited in the absence or presence of applied field.

Table 11.2 Variation of the reduced remanence with the width of the wire-like.

Width (nm)	1	2	3
M_r/M_s	0.25	0.50	0.60

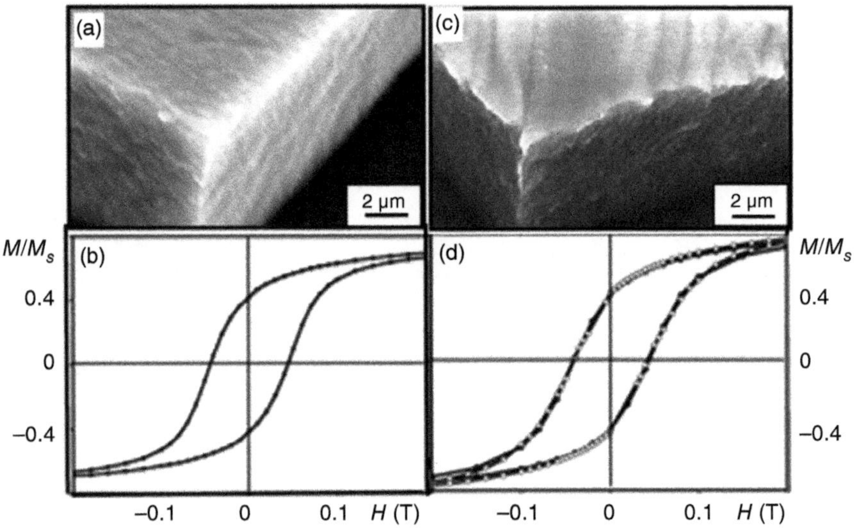

Figure 11.19 (a, c) SEM images of γ-Fe_2O_3 nanocrystals coated with dodecanoic acid deposited on HOPG substrate without applied field (a) and with an applied field (0.59 T) (c). (b, d) Magnetization curves at 3 K corresponding the mesostructures shown on (a, c). (b) Magnetization corresponding to sample shown in (a) and (d) that of (b). The magnetization is recorded with an applied field either parallel (open circles) or perpendicular (solid circles) to the ribbons' direction. *Source:* (a, c) Lalatonne et al [58, 59]/American Chemical Society.

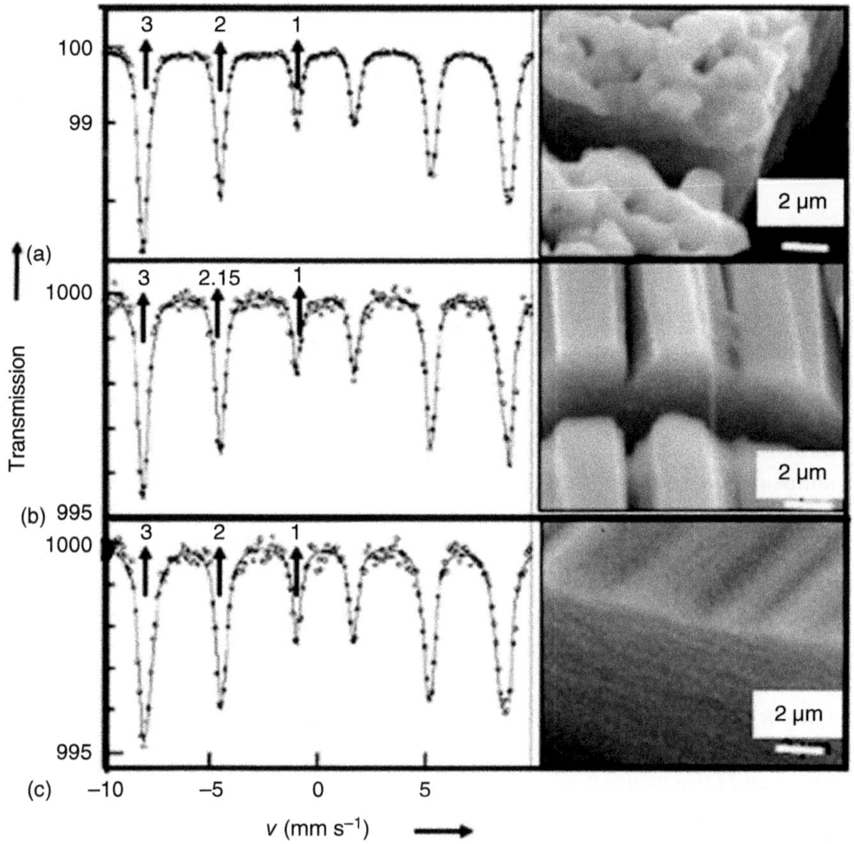

Figure 11.20 Mössbauer spectra of disordered aggregates (a) and wire-like (b) and homogeneous films (c) produced with nanocrystals coated with either citrate ions (a, b) or dodecanoic acid (c). *Source:* Lalatonne et al. [58, 59]/American Chemical Society.

These increases in the reduced remanence with the strength of the applied film and with the formation of tubes are not due to the orientation of the easy axis of nanocrystals [58] even if it is energetically possible [56]. This has been well demonstrated by Mössbauer spectroscopy. From an energetic viewpoint, we could explain this by the orientation of the easy axes of nanocrystals when they are subjected to an applied field. In such a case, this has to be confirmed by Mössbauer spectroscopy. For a total orientation of the easy axes, the relative intensities of the Mössbauer spectra are expected, for γ-Fe_2O_3 nanocrystals, to be 1, 4, and 3, whereas for a total disorder, the Mössbauer spectra are expected to be 1, 2, and 3. Experimentally, for disordered aggregates (Figure 11.20a) and the film (Figure 11.20c), the Mössbauer relative peak intensity is, as expected, 1, 2, and 3, whereas for wires, it is 1, 2.15, and 3 (Figure 11.20b). Such a slight change in the relative Mössbauer intensity (from 2 to 2.15) indicates that very few nanocrystals have their easy axis oriented. Therefore, a slight change in the relative Mössbauer intensity indicates that very few nanocrystals have their easy axis oriented.

From the change in the magnetic properties of the assembly forming wires compared with films, both made with the same batch of nanocrystals and from the increase in the reduced remanence with increasing the width of the wires as with bulk wires, we concluded that the change is due to the shape of the mesoscopic structure [56, 58–60].

11.5.2 Comparison of the Influence or the Easy Axis' Orientation and Dipolar Interactions

The dipolar interactions between magnetic nanocrystals play the major role on the properties compared to that of the easy axes [61, 62].

From data shown above, we know that alignment of magnetic nanocrystals induces a squarer hysteresis loop than the same nanocrystals deposited on a substrate in 2D/3D superlattices. Above we find out that this process can be attributed to the shape of the mesoscopic structure. With 2D assemblies, other factors such as the orientation of the easy axis or dipolar interactions or both can be considered [59]. To enable the distinction between the easy axis orientation and dipolar interaction effect, we study aligned iron oxide nanoparticles called magnetosomes produced from AMB-1 magnetotactic bacteria. The magnetosomes are arranged in chains inside the bacteria, creating a strong magnetic dipole, which is being used by the bacteria to align and swim along the earth's magnetic field. They are ferrite nanocrystals in either the reduced (Fe_3O_4) or oxidized (γ-Fe_2O_3) forms. The size distribution is bimodal with the small magnetosomes having sizes ranging from 4 up to 30 nm and larger ones having sizes lying between 30 and 55 nm. Furthermore, the biological filaments, which surround the magnetosomes, maintain the alignment of the easy axes of the individual magnetosomes [63, 64]. These filaments might remain functional after the extraction of the magnetosome from bacteria cells [61]. Magnetosomes are extracted from AMB-1 magnetotactic bacteria. During the deposition by evaporation process, a magnetic field is applied. Figure 11.21a,b shows the TEM and SEM images of the magnetosome chains organized within wide bands and orientated along the direction of the deposition field. Part of the biogenic material, which binds the chains of magnetosomes to the bacteria, is removed, and the chains attract each other forming a long string of nanoparticles. However, because of the remaining biogenic material, the chains are still bent in several regions.

To estimate the orientation of the easy axes, the crystallographic planes of the magnetosomes with a truncated octahedral structure are identified by HRTEM.

Figure 11.22a shows an HRTEM image of four magnetosomes bound via biogenic material. The enlarged HRTEM images from areas b and c are displayed in Figure 11.22b,c. From the presence of clear {111} planes, the horizontal ⟨111⟩ directions can be deduced. Figure 11.22d also shows a magnetosome containing a twin plane, projected right along the ⟨110⟩ direction. By identifying the twin plane to a (111) crystallographic plane (Figure 11.22d), the [111] crystallographic direction is identified. Figure 11.22a also shows that this direction follows the orientation of the chain, confirming the results obtained by electron holography, which showed that

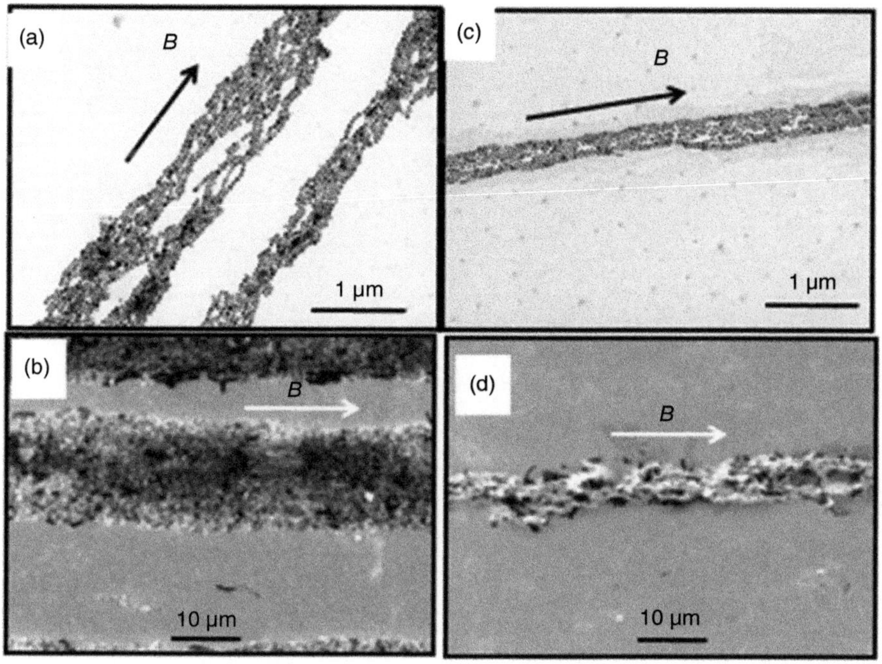

Figure 11.21 TEM (a, c) and SEM (b, d) images of extracted unheated magnetosomes (a, b) and extracted SDS-treated and heated magnetosomes (c, d). The samples are deposited on top of a carbon grid in the presence of a magnetic field *B*, and the arrow below *B* indicates its orientation. *Source:* E. Alphandery, 2009/American Chemical Society.

this direction is [111]. Hence, it can be concluded that the ⟨111⟩ crystallographic directions can be determined either by identifying the projected geometries or by using HRTEM measurements. In what follows, the first method is used since it is easier. Figure 11.23 shows that the [111] crystallographic directions (white arrows) follow the orientation of the chain.

To estimate the influence of the orientation of the magnetic particle easy axes on the magnetic response, the extracted magnetosomes are treated to remove the remaining biological material that binds the magnetosomes together and consequently disorientates the magnetosome easy axes. To do so, the extracted magnetosomes are heated for one hour at 90 °C with 1% sodium dodecyl sulfate (SDS). The treated and untreated magnetosomes are then deposited on the TEM grid. From the TEM and SEM images (Figure 11.21c,d), it is shown that the treated magnetosomes are also aligned forming a slightly more compact assembly than that observed with the untreated magnetosomes (Figure 11.21a,b). Using the same approach as described above, the {111} crystallographic directions (black arrows) are randomly orientated, and the magnetosomes behave like individual nanoparticles with a cubic symmetry (Figure 11.22). This clearly shows that the majority of the magnetosomes do not have one of their ⟨111⟩ crystallographic directions orientated in the direction of the deposition field and therefore possess randomly orientated easy axes.

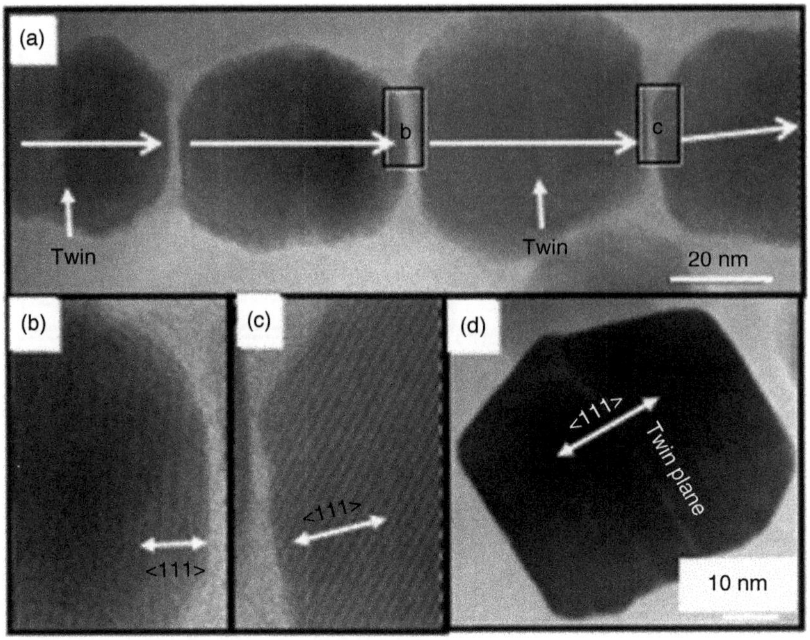

Figure 11.22 (a) HRTEM image of four magnetosomes bound in a chain showing two twin crystals and two magnetosomes of type I. (b) and (c) Enlarged regions showing the {111} diffraction plane and ⟨111⟩ crystallographic directions. (d) HRTEM image of a twin crystal, showing the twin plane and the ⟨111⟩ crystallographic orientation. *Source:* E. Alphandery, 2009/American Chemical Society.

Let us compare the magnetic properties of aligned Fe_3O_4 nanocrystals having the easy axes either aligned (Figure 11.23) or not (Figure 11.24). This permits to differentiate between the influence on the magnetic properties of dipolar interactions and easy axis orientation. The hysteresis loop is squarer when the applied field is parallel noted para (green) to the aligned particles than that recorded when it is perpendicular noted perp (red) (Figure 11.25a). The values of $D_{Mr/Ms} = [M_r/M_{s(para)} - M_r/M_{s(perp)} /M_r/M_{s(para)}]$ and $D_{Hc} = [H_{c(para)} - H_{c(perp)}]$ are 80% and 50%, respectively. Given a mean distance between the magnetosomes of 6 nm, the ratio E_{ms}/E_K is estimated around 9 where E_{ms} and E_K are the magnetostatic and anisotropy energies, respectively. This value is larger than 1 indicating a strong dipolar coupling between the magnetosomes. As shown on Figure 11.23, the [111] crystallographic directions (white arrows) follow the orientation of the chain, whereas some are randomly orientated (dark arrows). This clearly shows that most of the easy axes are orientated along the applied field. Figure 11.25b shows a squarer hysteresis loop when the magnetic field is applied parallel to the oriented magnetosomes than when it is applied perpendicular to this direction. However, compared with the extracted unheated magnetosomes (Figure 11.25a), the hysteresis loop is less square in the parallel field configuration (green line), while it remains identical in the perpendicular field configuration (red line). The values of $D_{Mr/Ms}$ and D_{Hc} decrease from 80% and 50% and from 65% and 30% for the extracted untreated magnetosomes down to treated ones. This decrease cannot be attributed to a loss of dipolar interactions. Given a mean

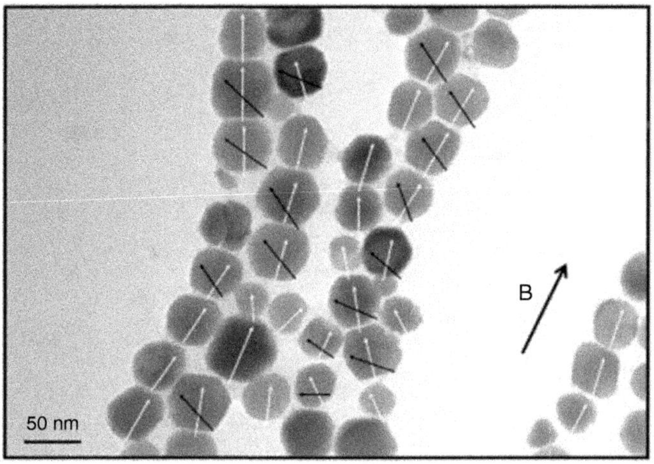

Figure 11.23 Enlarged regions of Figure 11.21a,b. The arrows indicate the ⟨111⟩ crystallographic orientations. The deposition field is designated by *B*, and the arrow below *B* indicates its orientation as in Figure 11.21. White arrows are used when the ⟨111⟩ crystallographic orientations follow the direction of deposition field, while black arrows are used when these orientations do not follow this direction. *Source:* E. Alphandery, 2009/American Chemical Society.

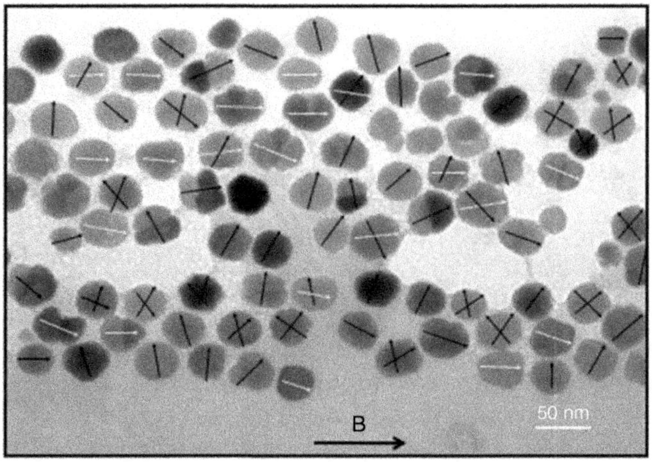

Figure 11.24 Enlarged regions of Figure 11.21c,d. The arrows indicate the ⟨111⟩ crystallographic orientations. The deposition field is designated by *B*, and the arrow below *B* indicates its orientation as in Figure 11.21. White arrows are used when the ⟨111⟩ crystallographic orientations follow the direction of deposition field, while black arrows are used when these orientations do not follow this direction. *Source:* E. Alphandery, 2009/American Chemical Society.

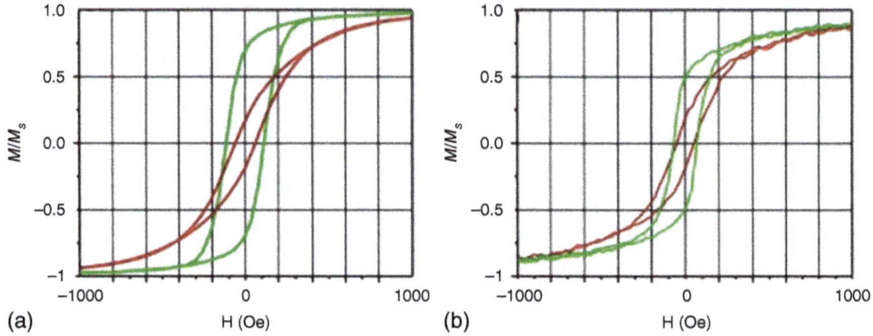

Figure 11.25 Hysteresis loops of the extracted, untreated (a), and SDS-treated (b) magnetosomes. The magnetic field applied during the measurements of the hysteresis loops is either parallel (green line) or perpendicular (red line) to the direction of the alignment. The measurements are carried out at 300 K. *Source:* E. Alphandry, Y. Ding et al. (2009)/American Chemical Society.

Figure 11.26 Zero-field-cooled (ZFC) and field-cooled (FC) magnetization versus temperature curves measured at 10 G for the 7.2-nm Co 3D superlattices (purple) and 4.6-nm Co 3D superlattices (green). *Source:* Yang et al. [65], figure 4a (p. 4911)/John Wiley & Sons.

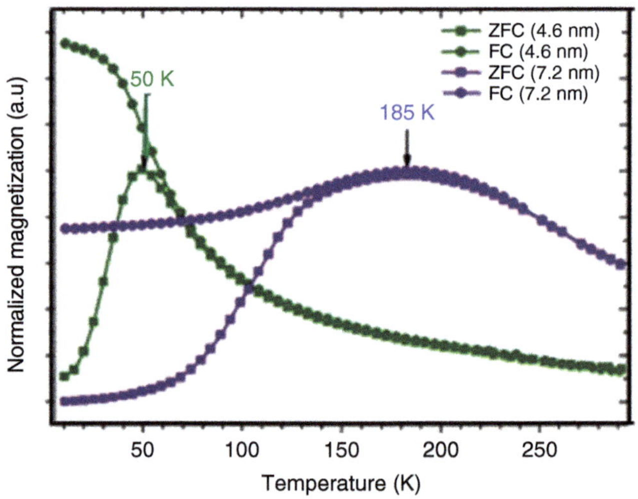

distance between the magnetosomes of around 3 nm, the ratio E_{ms}/E_K is estimated around 12 of the same order of magnitude as that obtained with the extracted, untreated magnetosomes. Therefore, the loss of magnetic anisotropy has to arise from the disorientation of the magnetosome easy axes. Then, it is concluded that the magnetic anisotropy of an assembly of aligned magnetosomes is mainly governed by the dipolar interactions between them and includes a smaller contribution due to the alignment of the easy axes [62]. From the percentages given above, the dipolar interactions represent 70 (10% of the total magnetic anisotropy), whereas the contribution of the easy axes is 30 (10%). This clearly shows that it is possible to establish the origin of the magnetic anisotropy in a system of aligned magnetic nanoparticles and to differentiate between the two contributions, i.e. the nanoparticle dipolar interactions and the orientation of the nanoparticle easy axes. The dominant contribution of the dipolar interactions might arise from the large sizes of the magnetosomes that, for the majority of these, are 20–50 nm.

11.6 Co 3D Superlattice Collective Properties of Amorphous Nanoparticles

Collective properties of amorphous C nanoparticles self assembled in 3D superlattices are obtained [65].

Let us consider the magnetic properties of 7.2- and 4.6-nm amorphous Co nanoparticles. The ZFC and FC magnetization curves were measured in the presence of applied magnetic field of 10 Oe. Figure 11.26 shows the ZFC/FC magnetization curves of both 4.6- and 7.2-nm fcc Co 3D superlattices characterized by an interparticle distance, deduced from grazing-incidence small-angle X-ray scattering (GISAXS), of 3.2 and 3.4 nm. The blocking temperature T_b is 185 and 50 K, respectively, for the 7.2- and 4.6-nm 3D superlattices. An approximate expression for the blocking temperature is

$$k_B T_b = E_a / \ln\left(\tau_\chi / \tau_0\right)$$

where $E_a = K_{eff}V$, K_{eff} being the effective anisotropy energy density and V the nanoparticle volume, and τ_0 is the microscopic time of the order of 10^{-10} seconds. Hence, the higher T_b for the larger-sized Co nanoparticle 3D superlattices is due to the larger nanoparticle volume implying a larger magnetocrystalline anisotropy energy. The FC curve of 7.2-nm Co 3D superlattices shows a slight decrease of below 185 K (T_b), indicating, as observed above, a spin glass-like behavior, while for 4.6-nm Co 3D superlattices, the FC curves show a monotonic increase below the corresponding T_b with a saturating trend at sufficiently low temperatures, indicating relatively weaker interparticle coupling interactions between the nanoparticles.

11.7 Conclusion

In this chapter, we present a large variety of intrinsic properties related due to the self-assemblies of nanocrystals. These properties are related to crystal growth and magnetic properties. The shape of the nanocrystal assemblies plays a major role in the physical properties.

Hence, triangular single crystals with fcc structures are obtained by annealing the ordered nanocrystals under atmospheric pressure at 50 °C. The size of these triangular particles depends on the initial nanocrystal ordering range on the substrate that is local on amorphous carbon and highly extended on HOPG. Hence, the single-crystal size is larger on HOPG than on amorphous carbon. Furthermore, on HOPG, an epitaxial orientation of the triangles is observed.

Co nanocrystals with very low size distribution self-organize into long-range ordered 3D superlattices. We stress that this transition takes place at the atomic scale and that the fcc suprastructure of the superlattice remains unchanged. Magnetic properties of either *amorphous* or hcp Co nanocrystals deposited on a substrate and self-assembled in either *amorphous* or fcc film. It is shown specific magnetic properties due to the ordering of both atoms in the nanocrystals or nanocrystals in the film.

Magnetic properties of Co (ε-phase) of nanocrystals are compared in either disordered or ordered (fcc) assemblies. Multipeak ZFC curves are intrinsic characteristic of these assemblies. A comparison between disordered films and fcc 3D superlattices demonstrates significant differences arising from magnetic interactions whose origin is different from size effect and randomly oriented dipolar interactions in the two systems. This is attributed to the orientational and spatial orderings of single-domain ε-Co nanocrystals markedly changing between ordered and disordered assemblies.

Ferrite nanoparticle organization is controlled by deposition on the substrate. The magnetic properties are directly related to the shape of the mesoscopic structures involving dipolar interactions and not to the orientation of the easy axis.

Magnetosomes are produced by magnetotactic bacteria. The origin of the magnetic anisotropy is explained in an assembly of aligned magnetic nanoparticles. The use of such magnetosome makes it possible to differentiate between the two contributions arising from the alignment of the magnetosome easy axes and the strength of the magnetosome dipolar interactions. The magnetic anisotropy is shown to arise mainly from the dipolar interactions between the magnetosomes.

References

1 Courty, A., Henry, A.I., Goubet, N., and Pileni, M.P. (2007). Large triangular single crystals formed by mild annealing of self-organized silver nanocrystals. *Nat. Mater.* 6: 900–907.

2 Henry, A.I., Courty, A., Goubet, N., and Pileni, M.P. (2008). How do self-ordered silver nanocrystals influence their growth into triangular single crystals? *J. Phys. Chem. C* 112: 48–52.

3 Morriss, R.H., Bottoms, W.R., and Peacock, R.G. (1967). Growth and defect structure of lamellar gold microcrystals. *J. Appl. Phys.* 39: 3016–3021.

4 Cherns, D. (1974). Direct resolution of surface atomic steps by transmission electron microscopy. *Philos. Mag.* 30: 549–556.

5 Germain, V., Li, J., Ingert, D. et al. (2003). Stacking faults in formation of silver nanodisks. *J. Phys. Chem. B* 107: 8717–8720.

6 Wang, Z.L. (1998). Structural analysis of self-assembling nanocrystal superlattices. *Adv. Mater.* 1: 13–30.

7 Lofton, C. and Sigmund, W. (2005). Mechanisms controlling crystal habits of gold and silver colloids. *Adv. Funct. Mater.* 15: 1197–1208.

8 Dasog, M. and Scott, R.W.J. (2007). Understanding the oxidative stability of gold monolayer-protected clusters in the presence of halide ions under ambient conditions. *Langmuir* 23: 3381–3387.

9 Li, L., Yang, J.J., and Minton, T.K. (2007). Morphological changes at a silver surface resulting from exposure to hyperthermal atomic oxygen. *J. Phys. Chem. C* 111: 6763–6771.

10 Chapon, C., Granjeaud, S., Humbert, A., and Henry, C.R. (2001). Structure and morphology of nanometer-sized Pd clusters grown at high temperature on natural graphite single crystals. *Eur. Phys. J.* 13: 23–30.

11 Barkai, M., Grünbaum, E., and Deutscher, G. (1982). The influence of previous substrate heat treatment on the epitaxial growth of silver on mica. *Thin Solid Films* 90: 85–90.

12 Taleb, A., Silly, F., Gusev, A.O. et al. (2000). Electron transport properties of nanocrystals: isolated, and "supra"-crystalline phases. *Adv. Mater.* 12: 633–637.

13 Yang, P., Arfaoui, I., Cren, T. et al. (2012). Unexpected electronic properties of micrometer-thick supracrystals of Au nanocrystals. *Nano Lett.* 12: 2051–2055.

14 Zheng, N., Fan, J., and Stucky, G.D. (2006). One-step one-phase synthesis of monodisperse noble-metallic nanoparticles and their colloidal crystals. *J. Am. Chem. Soc.* 128: 6550–6551.

15 Goubet, N., Portalès, H., Yan, C. et al. (2012). Simultaneous growths of gold colloidal crystals. *J. Am. Chem. Soc.* 134: 3714–3719.

16 Horcas, I., Fernandez, R., Gomez-Rodriguez, J.M. et al. (2007). WSXM: a software for scanning probe microscopy and a tool for nanotechnology. *Rev. Sci. Instrum.* 78: 013705.

17 Jeffrey, L.D. and Al-Amin, D. (2008). Conductance oscillations in molecularly linked Au nanoparticle film-superconductor systems. *Nanotechnology* 19: 025202.

18 Zabet-Khosousi, A., Suganuma, Y., Lopata, K. et al. (2005). Influence of linker molecules on charge transport through self-assembled single-nanoparticle devices. *Phys. Rev. Lett.* 94: 096801.

19 Hamers, R.J., Tromp, R.M., and Demuth, J.E. (1986). Surface electronic structure of Si (111)-(7×7) resolved in real space. *Phys. Rev. Lett.* 56: 1972.

20 Jdira, L., Overgaag, K., Gerritsen, J. et al. (2008). Scanning tunnelling spectroscopy on arrays of CdSe quantum dots: response of wave functions to local electric fields. *Nano Lett.* 8: 4014–4019.

21 Black, C.T., Murray, C.B., Sandstrom, R.L., and Sun, S.H. (2000). Spin-dependent tunneling in self-assembled cobalt-nanocrystal superlattices. *Science* 290: 1131–1134.

22 Middleton, A.A. and Wingreen, N.S. (1993). Collective transport in arrays of small metallic dots. *Phys. Rev. Lett.* 71: 3198.

23 Jeong, U., Teng, X., Wang, Y. et al. (2007). Superparamagnetic colloids: controlled synthesis and niche applications. *Adv. Mater.* 19: 33–60.

24 Lu, A.H., Salabas, E.L., and Schüth, F. (2007). Magnetic nanoparticles: synthesis, protection, functionalization, and application. *Angew. Chem. Int. Ed.* 46: 1222–1244.

25 Lisiecki, I. and Pileni, M.P. (2012). Emergence of new collective properties of cobalt nanocrystals ordered in fcc supracrystals: I, structural investigation. *J. Phys. Chem. C* 116: 3–14.

26 Lisiecki, I., Albouy, P.A., and Pileni, M.P. (2004). Supra-crystal: control of the ordering of self-organization of cobalt nanocrystals at the mesoscopic scale. *J. Phys. Chem. B* 108: 20050–20055.

27 Lisiecki, I., Parker, D., Salzemann, C., and Pileni, M.P. (2007). Face-centered cubic supra-crystals and disordered three-dimensional assemblies of 7.5 nm cobalt nanocrystals: influence of the mesoscopic ordering on the magnetic properties. *Chem. Mater.* 19: 4030–4036.

28 Dormann, J.L., Spinu, L., Tronc, E. et al. (1998). Effect of interparticle interactions on the dynamical properties of γ-Fe_2O_3 nanoparticles. *J. Magn. Magn. Mater.* 183: L255–L260.

29 Parker, D., Lisiecki, I., Salzemann, C., and Pileni, M.P. (2007). Emergence of new collective properties of cobalt nanocrystals ordered in fcc supracrystals: II, magnetic investigation. *J. Phys. Chem. C* 111: 12632–12638.

30 de la Peña O'Shea, V.A., Moreira, I.d.P.R., Roldán, A., and Illas, F. (2010). Electronic and magnetic structure of bulk cobalt: the α, β, and ε-phases from density functional theory calculations. *J. Chem. Phys.* 133: 024701.

31 Puntes, V.F., Zanchet, D., Erdonmez, C.K., and Alivisatos, A.P. (2002). Synthesis of hcp-Co nanodisks. *J. Am. Chem. Soc.* 124: 12874–12880.

32 Held, G.A., Grinstein, G., Doyle, H. et al. (2001). Competing interactions in dispersions of superparamagnetic nanoparticles. *Phys. Rev. B* 64: 012408.

33 Woods, S.I., Kirtley, J.R., Sun, S., and Koch, R.H. (2001). Direct investigation of superparamagnetism in co nanoparticle films. *Phys. Rev. Lett.* 87: 137205.

34 Majetich, S.A. and Sachan, M. (2006). Magnetostatic interactions in magnetic nanoparticle assemblies: energy, time and length scales. *J. Phys. D: Appl. Phys.* 39: R407–R422.

35 Yang, J., Khazen, K., and Pileni, M.P. (2014). How nanocrystallinity and order define the magnetic properties of ε-Co supracrystal. *J. Phys.: Condens. Matter* 26 (29): 295303.

36 De Toro, J.A., Normile, P.S., Lee, S.S. et al. (2013). Controlled close-packing of ferrimagnetic nanoparticles: an assessment of the role of interparticle superexchange versus dipolar interactions. *J. Phys. Chem. C* 117: 10213–10219.

37 Martínez, B., Obradors, X., Balcells, L. et al. (1998). Low temperature surface spin-glass transition in γ-Fe_2O_3 nanoparticles. *Phys. Rev. Lett.* 80: 181.

38 Nikolic, V., Perovic, M., Kusigerski, V. et al. (2015). Experimental evidence for simultaneous relaxation processes in super spin glass γ-Fe_2O_3 nanoparticle system. *J. Nanopart. Res.* 17: 139.

39 Peddis, D., Orrù, F., Ardu, A. et al. (2012). Interparticle interactions and magnetic anisotropy in cobalt ferrite nanoparticles: influence of molecular coating. *Chem. Mater.* 24: 1062–1071.

40 Parker, D., Dupuis, V., Ladieu, F. et al. (2008). Spin-glass behavior in an interacting γ-Fe_2O_3 nanoparticle system. *Phys. Rev. B* 77: 104428.

41 Sun, S. and Murray, C.M. (1999). Synthesis of monodisperse cobalt nanocrystals and their assembly into magnetic superlattices. *J. Appl. Phys.* 85: 4325–4330.

42 Hansen, M.F. and Morup, S. (1999). Estimation of blocking temperatures from ZFC/FC curves. *J. Magn. Magn. Mater.* 203: 214–216.

43 Samia, A.C.S., Hyzer, K., Schlueter, J.A. et al. (2005). Ligand effect on the growth and the digestion of co nanocrystals. *J. Am. Chem. Soc.* 127: 4126–4127.

44 Parker, D., Lisiecki, I., and Pileni, M.P. (2010). Do 8 nm Co nanocrystals in long-range-ordered face-centered cubic (fcc) supracrystals show superspin glass behavior? *J. Phys. Chem. Lett.* 1: 1139–1142.

45 Mydosch, J.A. (1993). *Spin Glasses: An Experimental Introduction*. London: Taylor & Francis.

46 Djurberg, C., Svedlindh, P., Nordblad, P. et al. (1997). Dynamics of an interacting particle system: evidence of critical slowing down. *Phys. Rev. Lett.* 79: 5154–5157.

47 Sahoo, S., Petracic, O., Binek, C. et al. (2002). Superspin-glass nature of discontinuous $Co_{80}Fe_{20}/Al_2O_3$ multilayers. *Phys. Rev. B* 65: 134406.

48 Jonsson, T., Mattsson, J., Djurberg, C. et al. (1995). Aging in a magnetic particle system. *Phys. Rev. Lett.* 75: 4138–4141.

49 Andersson, O., Djurberg, C., Jonsson, T. et al. (1997). Monte Carlo studies of the dynamics of an interacting monodispersive magnetic-particle system. *Phys. Rev. B* 56: 13983.

50 Mamiya, H., Nakatani, I., and Furubayashi, T. (1999). Slow dynamics for spin-glass-like phase of a ferromagnetic fine particle system. *Phys. Rev. Lett.* 82: 4332.

51 Jonsson, P., Hansen, M.F., Svedlindh, P., and Nordblad, P. (2000). Memory effects in an interacting magnetic nano-particle sample. *Physica B* 284: 1754.

52 Bedanta, S., Rhensius, J., Kleemann, W. et al. (2009). Dynamic magnetization properties of a superferromagnetic metal-insulator multilayer observed by magneto-optic Kerr microscopy. *Appl. Phys. Lett.* 105: 07C306.

53 Lisiecki, I., Polli, D., Yan, C. et al. (2013). Coherent longitudinal acoustic phonons in three-dimensional supracrystals of cobalt nanocrystals. *Nano Lett.* 13: 4914–4919.

54 Courty, A., Mermet, A., Albouy, P.A. et al. (2005). Self-organized ag-nanocrystals in fcc "supra" crystals: vibrational coherence. *Nat. Mater.* 4: 395–398.

55 Kleemann, W., Petracic, O., Binek, C. et al. (2001). Interacting ferromagnetic nanoparticles in discontinuous $Co_{80}Fe_{20}/Al_2O_3$ multilayers: from superspin glass to reentrant superferromagnetism. *Phys. Rev. B* 63: 134423.

56 Ngo, A.T. and Pileni, M.P. (2001). Assemblies of ferrite nanocrystals: partial orientation of the easy magnetic axes. *J. Phys. Chem.* 105: 53–58.

57 Ngo, A.T., Richardi, J., and Pileni, M.P. (2005). Mesoscopic solid structures of 11-nm maghemite, γ-Fe_2O_3, nanocrystals: experiment and theory. *Langmuir* 21: 10234–10239.

58 Lalatonne, Y., Motte, L., Russier, V. et al. (2004). Mesoscopic structures of maghemite nanocrystals: collective magnetic properties due to the alignment of nanocrystals. *J. Phys. Chem. B* 108: 1848–1854.

59 Lalatonne, Y., Richardi, J., and Pileni, M.P. (2004). Van der Waals versus dipolar forces controlling mesoscopic organizations of magnetic nanocrystals. *Nat. Mater.* 3: 121–125.

60 Lalatonne, Y., Motte, L., Richardi, J., and Pileni, M.P. (2005). Influence of short-range interactions on mesoscopic organization of magnetic nanocrystals. *Phys. Rev. E* 71: 011404.

61 Alphandery, E., Ngo, A.T., Lefèvre, C. et al. (2008). Difference between the magnetic properties of the magnetotactic bacteria and those of the extracted magnetosomes: influence of the distance between the chains of magnetosomes. *J. Phys. Chem. C* 112: 12304–12309.

62 Alphandry, E., Ding, Y., Ngo, A.T. et al. (2009). Assemblies of aligned magnetotactic bacteria and extracted magnetosomes: what is the main factor responsible for the magnetic anisotropy? *ACS Nano* 3: 1539–1547.

63 Bazylinski, D.A. and Frankel, R.B. (2004). Magnetosome formation in prokaryotes. *Nat. Rev. Micro.* 2: 217–230.

64 Komeili, A., Li, Z., Newman, D.K., and Jensen, G.J. (2006). Magnetosomes are cell membrane invaginations organized by the actin-like protein MamK. *Science* 311: 242–245.

65 Yang, Z., Wei, J., Bonville, P., and Pileni, M.P. (2015). Engineering the magnetic dipolar interactions in three-dimensional binary supracrystals via mesoscale alloying. *Adv. Funct. Mater.* 25: 4908.

12

Mechanical Properties of 3D Superlattices

Here, we investigate the role of nanocrystal size (*d*), number of carbon atoms in the alkyl chain (*n*), interparticle distance (δ_{pp}), solvent vapor pressure during the evaporation, and the 3D superlattice growth processes on the mechanical properties using nanoindentation measurements performed with an Atomic Force Microscope (AFM).

12.1 Measurements of Mechanical Properties Using Atomic Force Microscope (AFM)

Here we use AFM for measurements of 3D superlattices mechanical properties [1–6].

In nanoindentation measurements, the thickness of the 3D superlattices should be at least 10 times larger than the indentation depth (50 nm) in order to avoid the affects from the substrate [7]. Thus, the 3D superlattice thickness should be up to 500 nm and Oliver and Pharr model can be used, but not for thinner 3D superlattices. In such cases, the plate model should be used to estimate the Young modulus.

12.1.1 Oliver and Pharr

The elastic modulus of thick 3D superlattice is determined from the load–displacement curves by applying the Oliver and Pharr model [8, 9] (Figure 12.1a). To determine the Young modulus, it is assumed that the uploading data are obtained from a pure elastic contact and it follows the equations below:

$$E_r = \frac{\sqrt{\pi}}{2} \frac{S}{\sqrt{A}} \tag{12.1}$$

Metal Nano 3D Superlattices: Synthesis, Properties, and Applications, First Edition.

where E_r is the "reduced modulus," A is the contact area, and S is the experimentally measured contact stiffness, defined as

$$S = \frac{dP}{dh} \tag{12.2}$$

in which P and h are the loading force and the elastic displacement, respectively.

The reduced modulus is related with the Young's modulus of the indenter and the sample:

$$\frac{1}{E_r} = \frac{1 - v_i^2}{E_i} + \frac{1 - v_m^2}{E_m} \tag{12.3}$$

E_i, v_i and E_m, v_m refer to the Young modulus and Poisson ratio for the indenter and sample, respectively. The value of P, h, and A can be obtained from the force measurement data and the reduced Young's modulus E_r is calculated from Eqs. (12.1) and (12.2). In practice, the Young's Modulus of the indenter is much bigger than that of sample ($E_i \gg E_m$), hence Eq. (12.3) is usually expressed as

$$\frac{1}{E_r} = \frac{1 - v_m^2}{E_m} \tag{12.4}$$

or the E_m can be calculated by

$$E_m = \left(1 - v_m^2\right) E_r \tag{12.5}$$

Here, the value of E for each sample is obtained by using the commercial software "Punias": The load–displacement curves and calculates the Young's modulus by extracting the force versus penetration (P–h) curve. Each Young's modulus value is obtained with increasing loading force to record the variation of Young's modulus according to the indentation depth with a maximum depth around 50 nm. After each sets of measurements, the cantilever is calibrated with mica substrate in order to obtain the up-to-date parameters of the cantilever for the elastic modulus calculation. For the contact area between the tip and the sample, (i) the manufacture's specifications of the tip are taken into account; (ii) the shape of the AFM probe is checked by SEM; and (iii) the value of the contact area is verified by measuring the Young's modulus of a reference sample as a Teflon foil.

12.1.2 Plate Model

For exploring the mechanical properties of thin 3D superlattices on the transmission electron microscopy (TEM) grid, the Oliver and Pharr model may not be ideal as the films have limited thickness and they may bend over as a whole under the pressure of the AFM tip. Thus, a plate model [10] is used to calculate the Young's modulus of interfacial 3D superlattices and compared to Oliver and Pharr model. The indentation of AFM tip is modeled as a uniform load over a very small central

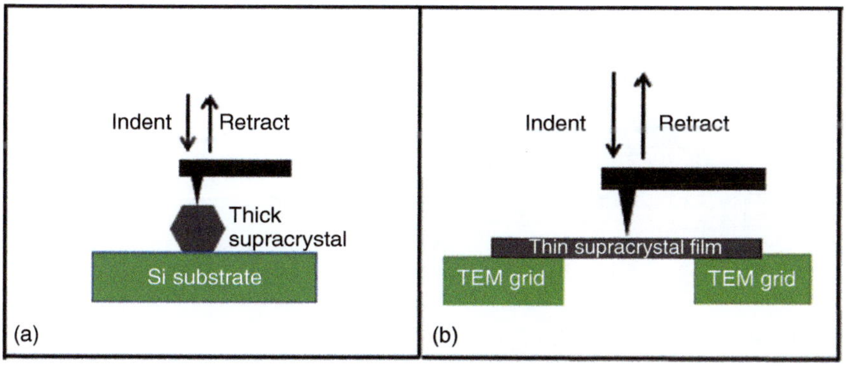

Figure 12.1 Schematic illustration of nanoindentation measurements performed on thick (a) and thin film (b) 3D superlattices.

circular area and the edge of the plate (the thin films) is simply supported (Figure 12.1b).

The displacement, y, of the central point is expressed as

$$y = \frac{Wa^2}{16\pi D}\frac{3+\nu}{1+\nu} \tag{12.6}$$

and

$$D = \frac{Et^3}{12\left(1-\nu^2\right)} \tag{12.7}$$

where E, W, a, t, and ν are the Young's modulus, the total applied load (force), the radius of the plate, the thickness of the plate and the Poisson's ratio, respectively.

The loading force is obtained by the AFM measurements directly and displacement of the central point was calculated by

$$y = h_\text{p} - h_\text{i} - h_\text{r} \tag{12.8}$$

where h_p, h_i, and h_r are the deflection of the cantilever is the displacement of the piezomotor, the indentation depth which measured by AFM profile, and the deflection of the cantilever, respectively. When y is obtained, the Young's modulus E could be calculated from the equation below:

$$E = \frac{3wa^2}{4\pi y t^3}\left(1-\nu\right)\left(3+\nu\right) \tag{12.9}$$

12.1.3 Validity of Mechanical Properties Deduced from AFM

In order to verify that the measured elastic properties of the 3D superlattices are independent of the indenter size, the elastic deformation range is determined using a 5 μm-diameter glass sphere on the film 3D superlattices of 8 nm ε-phase Co nanocrystals. The F–δ curve is recorded by using either a standard AFM tip, an AFM

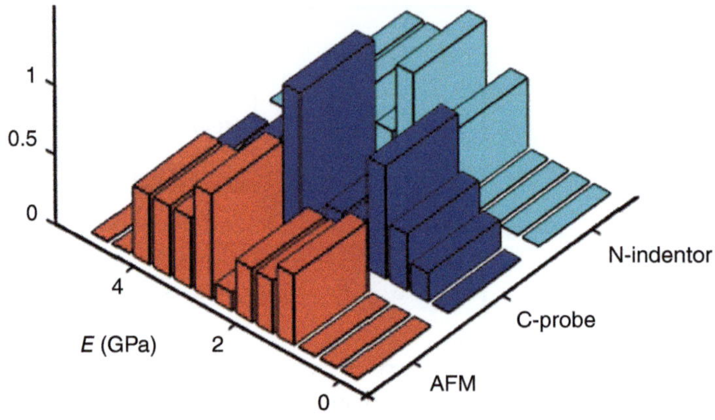

Figure 12.2 Normalized distribution of E^* values of 3D superlattice films of 8 nm ε phase Co nanocrystals deduced from different indenter geometries and testing methods. Red, blue, and cyan colors indicate distribution of E^* values obtained using AFM-based indentation with standard AFM tip, Colloidal probe (C-probe) and nano-indentor (N-indentor), respectively. *Source:* Based on Gauvin et al. [5].

with a colloidal probe (C-probe) mounted on a cantilever of calibrated stiffness or a 5 μm-diameter sphere as an indenter. The Derjaguin–Muller–Toporov (DMT) model is then used to fit the experimental F–δ curves. Figure 12.2 shows a rather good agreement between the various techniques [5].

12.2 3D Superlattices Produced Under Thermodynamic Processes

12.2.1 Influence of the 3D Superlattice Growth Mechanism

Chapter 4 presented that two simultaneous growth processes is obtained by keeping the colloidal solution for at least seven days in saturated regime [1]. A film appears at the air–toluene interface and precipitates are simultaneously assembled at the bottom of the beaker. Both film and precipitates are characterized by an fcc crystalline structure. The film is produced by formation of 3D superlattices, which are floating at the air–toluene interface (Figure 12.3a). The precipitated 3D superlattices result from a homogeneous 3D superlattice growth, which lie on the bottom of the beaker (Figure 12.3d). The size of the precipitated 3D superlattices characterized by well-defined shapes is 1–10 μm. The precipitate surfaces show 2D nucleation without any dislocations. This simultaneous 3D superlattice growth processes are observed for Au nanocrystals in the size range from 5 to 8 nm. The SEM image shows that the interfacial 3D superlattices are of constant thickness (around 300 nm) with highly ordered nanocrystals on a long coherence length (more than a few micrometers). The high-resolution SEM (HRSEM) image (Figure 12.3b) and AFM image (Figure 12.3c) present similar features, as well as straight edges and dislocations.

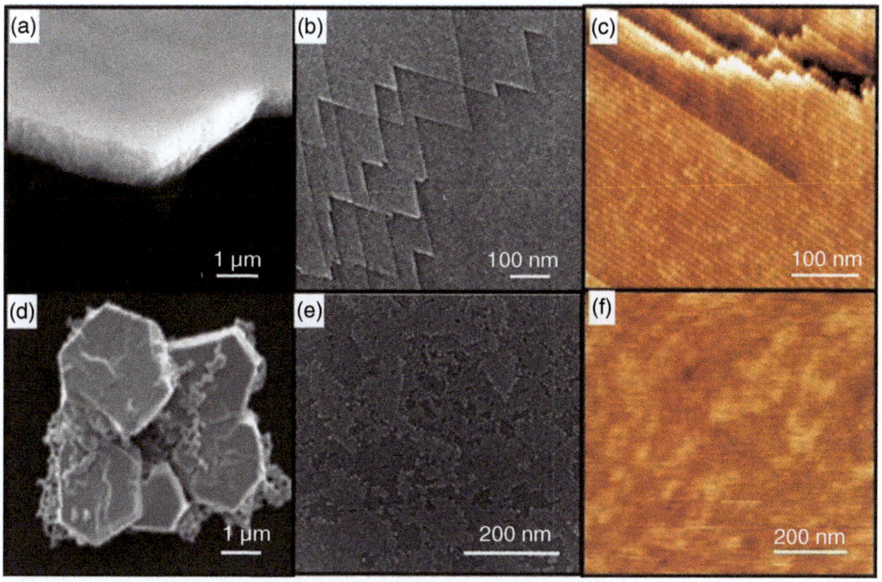

Figure 12.3 (a) Low magnification of SEM image of interfacial 3D superlattices showing constant thickness, (b) high magnification of SEM image showing dislocations, (c) the corresponding AFM image, (d) SEM image of shaped precipitated 3D superlattice solids, (e) high magnification SEM image showing different facets of precipitated 3D superlattices with highly ordered Au nanocrystals, and (f) corresponding AFM image. *Source:* Yan et al. [1]/JOHN WILEY & SONS, INC.

The interfacial 3D superlattices are deposited onto a TEM grid with 2.6 μm diameter circular holes and center-to-center distance of 4 μm. The indentation is performed in the center of the freestanding films (Figure 12.4a). The cracks observed on the films by SEM are avoided during the measurements and a smooth freestanding film is always chosen for the nanoindentation measurements. AFM images are recorded before (Figure 12.4b) and after (Figure 12.4c) indentations. The residual

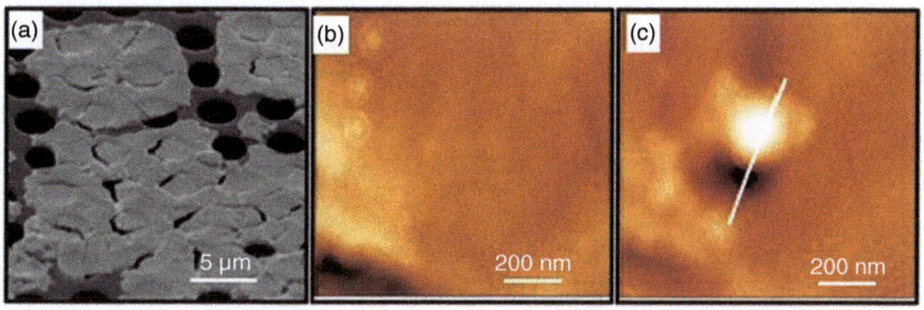

Figure 12.4 (a) SEM image of interfacial 3D superlattices of 5 nm Au nanocrystals on TEM grid. AFM images recorded before (b) and after (c) the indentation. The residual mark clearly shows the pile of nanocrystals. *Source:* Yan et al. [1]/JOHN WILEY & SONS, INC.

mark is observed clearly (Figure 12.4c) and nanocrystals are piled up adjacently. A number of measurements are performed on different spots for each sample and for each nanocrystal size from 5 to 8 nm. Table 12.1 shows that the Young's modulus of interfacial 3D superlattices decreases from 1000 to 110 MPa when increasing the nanocrystal sizes from 5 to 8 nm.

Precipitated 3D superlattices are characterized by well-defined shapes ranging in size from 1 to 10 μm (Figure 12.3d) and their surfaces are characterized by 2D nucleation without any dislocations (Figure 12.3e). Moreover, AFM image (Figure 12.3f) reveals the well-defined surface but much rougher than that of the interfacial films (Figure 12.3c). The mechanical properties of precipitated 3D superlattices are investigated by nanoindentation measurements (Figure 12.1a) using the Oliver and Pharr model. In this case, this model is suitable because the thickness of precipitated 3D superlattices (larger than 1 μm) is more than 10 times larger than the indentation depth (about 80 nm).

Thus, it is reasonable to consider that the substrate does not affect the measurements. Same steps as for interfacial 3D superlattices are repeated to extract the Young's modulus for the precipitated 3D superlattices. AFM images are taken before (Figure 12.5a) and after (Figure 12.5b) the indentation as stated above to check the signature of an indentation. The profile of the residual mark (line in Figure 12.5b), which shows a completely different shape of plastic deformation compared to the interfacial one. Several nanoindentation measurements at different locations are recorded for each sample and 3D superlattices with various nanocrystal sizes are investigated to compare with the counterpart interfacial ones. The Young's modulus E ranges from 5.1 to 0.71 GPa (Table 12.1) for different size of Au nanocrystals.

The E values obtained for precipitated 3D superlattices are consistent with previous reports on shaped 3D superlattices [10, 12]. The origin of the interfacial 3D superlattice softness could be attributed to the layer-by-layer growing process. The interfacial 3D superlattices are formed at the air–solvent interface and the films are floating during the formation, which results in a soft layer-by-layer 3D superlattices. It could be these different growing processes that lead to different mechanical properties. A change in the Young's modulus of 3D superlattices could be due to the degree of nanocrystal ordering [11] the core–ligand interaction [13] or ligand–ligand interaction [14]. As described in Chapter 4, both interfacial and precipitated 3D superlattices are composed of close-packed with fcc structure and same interparticle distances.

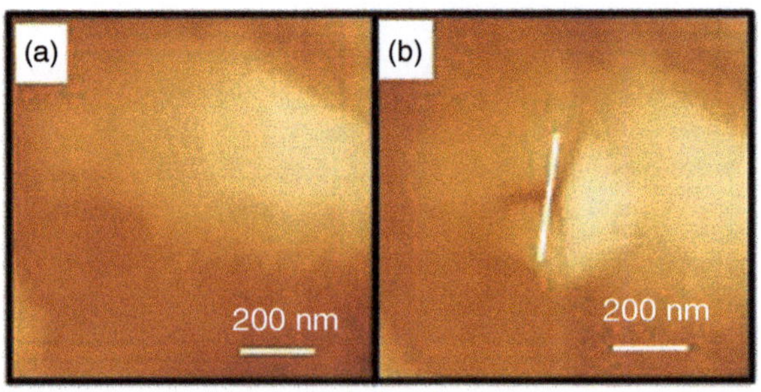

Figure 12.5 AFM image before (a) and after (b) indentation of the shaped 3D superlattices of 5 nm Au nanocrystals with residual mark. *Source:* Yan et al. [1]/ JOHN WILEY & SONS, INC.

Table 12.1 Young's modulus of interfacial/precipitated Au 3D superlattices with various nanocrystal sizes.

E **(GPa)**	**Nanocrystal size**				
3D superlattice	**Models**	**5**	**6**	**7**	**8**
Film	Plate	0.24 ± 0.18	0.10 ± 0.06	0.09 ± 0.05	0.08 ± 0.04
Shaped	Oliver and Pharr	5.1 ± 1.4	3.2 ± 0.9	1.8 ± 0.6	0.7 ± 0.4

Source: Based on Yan et al. [1].

Consequently, it rules out the suggestion that the drop in the elastic moduli could be related to the degree of nanocrystal ordering. Considering that the interfacial and precipitated 3D superlattices are produced from the same batch of colloidal solution, the difference in the elastic Young's moduli cannot be due to the ligand–core interactions. Therefore, such change in elastic moduli could only be attributed to the change of ligand–ligand interactions resulting from the 3D superlattice growth processes. To support such claim, the 5 nm Au precipitated 3D superlattices, characterized by an elastic modulus of 5.1 GPa, are re-dissolved in hexane. The colloidal solution shows a well-defined plasmon resonance peak indicating that the nanocrystals keep their integrity without coalescence. This solution is then slowly evaporated, forming layer-by-layer 3D superlattice films and the Young's modulus is then measured. The average values of the elastic Young's modulus, deduced from Oliver and Pharr model and plate model are 1100 ± 400 and 550 ± 160 MPa, respectively. These values are close to those of interfacial 3D superlattices (1100 ± 300 and 240 ± 180 MPa, respectively), which verify the difference between the layer-by-layer films and the shaped 3D superlattices. The higher value of evaporated 3D superlattices could be due to the roughness and defects in the films that resulting from the evaporation mechanism. In general, the consistency from both models is strong evidence that the mechanical properties of the 3D superlattices could be tuned by changing the 3D superlattice growing process. Therefore, 3D superlattice growing processes tune their mechanical properties.

12.2.2 Tuning of the Stiffness of Au 3D Superlattices

Chapter 1 describes synthesis of Au nanocrystals differing by their average diameter in the range from 4 to 8 nm through inorganic chemical reaction [2, 15]. At the end of the syntheses, Au nanocrystals are composed by a mixture of single domain and MTP crystalline phases. As described previously, simultaneous Au fcc interfacial film and precipitated 3D superlattices are produced, whereas in colloidal solution, nanocrystals remain dispersed in solution. As shown on Table 12.1 and Figure 12.6, whatever the nanocrystal size, the Young's moduli decrease with increasing nanocrystal size. The higher Young moduli obtained from the precipitated 3D superlattices compared to interfacial ones are attributed to the crystal growth mechanism [1]. The decrease of the Young moduli with increasing the nanocrystal size could be attributed to a decrease of the ligand surface coverage. This assumption is supported by the fact that the

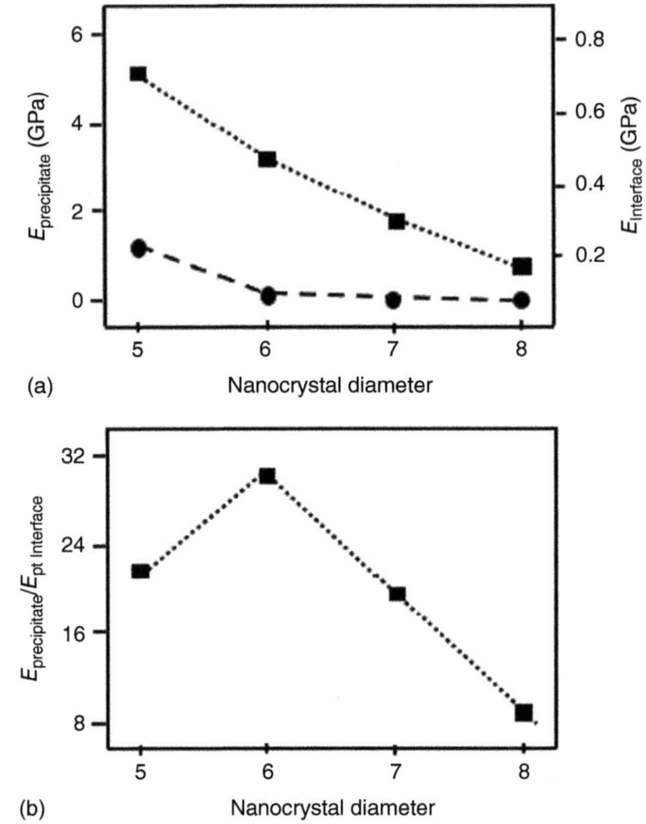

Figure 12.6 (a) Young's modulus of interfacial and precipitated 3D superlattices composed of 5, 6, 7, and 8 nm Au nanocrystals coated with dodecanethiol. (b) Ratio of the Young's moduli of precipitated to interfacial 3D superlattices for various Au nanocrystal sizes. *Source:* Yan et al. [2], figure 2/Royal Society of Chemistry.

interactions between thiol ligand and Au atoms reduce the number of ligands attachment sites at the Au nanocrystal surface [16]. Consequently, the drop of the coverage could induce a drop in the *E* value on increasing the Au nanocrystal size. From such assumption, a homothetic decrease of the Young's modulus is expected as nanocrystal size increases in interfacial and precipitated 3D superlattices. However, this behavior is not observed. Figure 12.6a shows, for the precipitated 3D superlattices, a smooth decrease of the Young's modulus with increasing the nanocrystal size. At variance, in interfacial 3D superlattices, the *E* value dramatically drops from 5 to 6 nm Au nanocrystal size, whereas a smooth decrease is observed from 6 to 8 nm. This behavior is clearly illustrated by plotting the ratio of the Young's moduli of precipitated to interfacial 3D superlattices versus the nanocrystal size (Figure 12.6b).

To understand this phenomenon, the crystalline structure of nanocrystals (single domain or polycrystalline phase) should be considered. As discussed in Chapter 2, the low-frequency Raman scattering (LFRS) spectrum of the polycrystalline nanocrystals is characterized by one band, while two Raman bands corresponding to E_g and T_{2g} modes emerge in the spectrum of single-nanocrystals. In Chapter 4, with 5 nm Au nanocrystals, nanocrystallinity segregation occurs when polycrystal

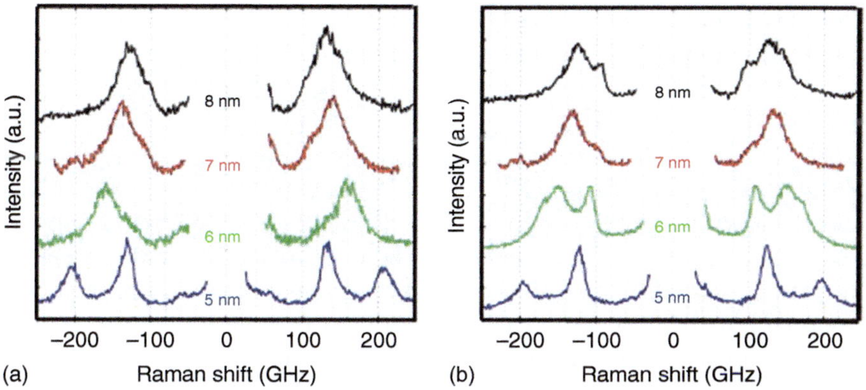

Figure 12.7 Stokes/anti-Stokes LFRS spectra of Au nanocrystals issued from (a) interfacial and (b) precipitated 3D superlattices. The NC size increases from bottom to top in each graph. The spectra were recorded using an excitation wavelength of 532 nm. These spectra are independently scaled and vertically shifted for clarity. *Source:* Yan et al. [2], figure 3/Royal Society of Chemistry.

Au nanocrystals remain in solution, whereas both interfacial and precipitated 3D superlattices are made of single domain nanocrystals, which is confirmed by the two narrow bands, attributed to single domains nanocrystals (Figure 12.7a,b). On the contrary, the spectra of 6, 7, and 8 nm Au nanocrystals involved in interfacial 3D superlattices (Figure 12.7a) exhibit a unique broad Raman band, indicating their main composition of polycrystal nanocrystals, whereas the precipitated 3D superlattices are a mixture of single domain and polycrystalline nanocrystals (Figure 12.7b).

The difference in the relative amounts of single domain and polycrystalline nanocrystals in interfacial and precipitated 3D superlattices could explain the non-homothetic behavior as shown in Figure 12.6b. The nanocrystallinity segregation, observed with 5-nm Au nanocrystals, allows the preparation of 3D superlattices by mixing single-nanocrystals and poly-nanocrystals in controlled amounts. 3D superlattices of the same amount of 5-nm single domain and polycrystal nanocrystals are produced at the end of solvent evaporation process, and the corresponding Young modulus is 126 ± 30 MPa. This value is intermediate between the average Young's moduli determined from the reference 3D superlattices of 5-nm single domain and polycrystals nanocrystals, which are 550 ± 160 and 25 ± 9 MPa, respectively. Such result shows the possibility of synthesizing 3D superlattices with controlled amount of nanocrystallinity and positively demonstrates how the artificially mixed nanocrystallinity changes the Young's moduli of 3D superlattices. Hence, the non-homothetic evolution of the interfacial and precipitated 3D superlattice elastic moduli with increasing the nanocrystals is attributed to change in the relative proportion of single domain nanocrystals compared to polycrystals counterparts involved in the 3D superlattices.

12.3 3D Superlattice Produced Through Heterogeneous 3D Superlattice Growth Process

Here let us consider the varied Young moduli of 3D superlattices by the material used (Au, Co, Ag), their strength of ligand-metallic atoms at the nanocrystal surfaces and ligand–ligand interactions play a key role in the elastic properties of the 3D superlattices. The groups currently used are thiol (SH), amine (NH_2), and carboxylic (COOH) derivatives. The interactions between *thiol ligand* and *noble metal* are very strong (covalent) [17–19]. The carboxylic group is covalently bound to the Co atoms [20]. With amine-terminated ligands (NH_2), their binding on the metallic surfaces are very weak compared to SH and COOH groups [21].

12.3.1 Au 3D Superlattices

The Au 3D superlattices are produced by solvent evaporation of a colloidal solution of single domain and polycrystalline nanocrystals dispersed in toluene.

12.3.1.1 Size and Coating Agent Effect on the Mechanical Properties

Here we compare the mechanical properties deduced from AFM nanoindentation measurements, of 3D superlattices having as building blocks 5 and 7 nm Au nanocrystals coated with alkane thiols differing by their chain length (C_{12}, C_{14}, and C_{16}) [3]. Note that, here, the 3D superlattices are composed by a mixture of single domain and polycrystal nanocrystals. By solvent evaporation of the colloidal solution, under control of the solvent vapor pressure, films and shaped fcc 3D superlattices are produced. As presented in Chapter 3, the solvent vapor pressure plays a major role on the nanocrystal interparticle distance: At zero vapor pressure, the interparticle distance increases linearly with the number of carbon atoms of the coating agent alkyl chains, whereas at larger vapor pressure, the interparticle distance remains unchanged because residual thiol molecules entangle between ligands bound to the nanocrystals.

At zero vapor pressure, fcc film and shaped 3D superlattices of 5 and 7 nm nanocrystals both coated with C_{12} are respectively produced. Their mechanical properties are deduced from the typical force–displacement curve of these fcc 3D superlattices (Figure 12.8a). With 3D superlattices of 7 nm nanocrystals, the maximum indentation depth increases by more than four times, for the same applied force, compared to 3D superlattices of 5 nm nanocrystals. This suggests that 3D superlattices of 7 nm are qualitatively softer than their counter parts of 5 nm. Such differences could be attributed to artifacts due the size difference between 7 and 5 nm nanocrystals under the same AFM tip. In other words, a larger number of 5 nm nanocrystals would be involved in the contact area under the AFM tip when compared to 7 nm nanocrystals and could possibly lead to apparent higher mechanical properties. To make sure that these differences in the stiffness is not due an artifact, it is used a colloidal probe of 5 μm in diameter for which the radius is much larger than the nanocrystal size and the interparticle distance, δ_{pp}. Figure 12.8c,d show SEM images of 5 and 7 nm 3D superlattices, respectively, after indentation

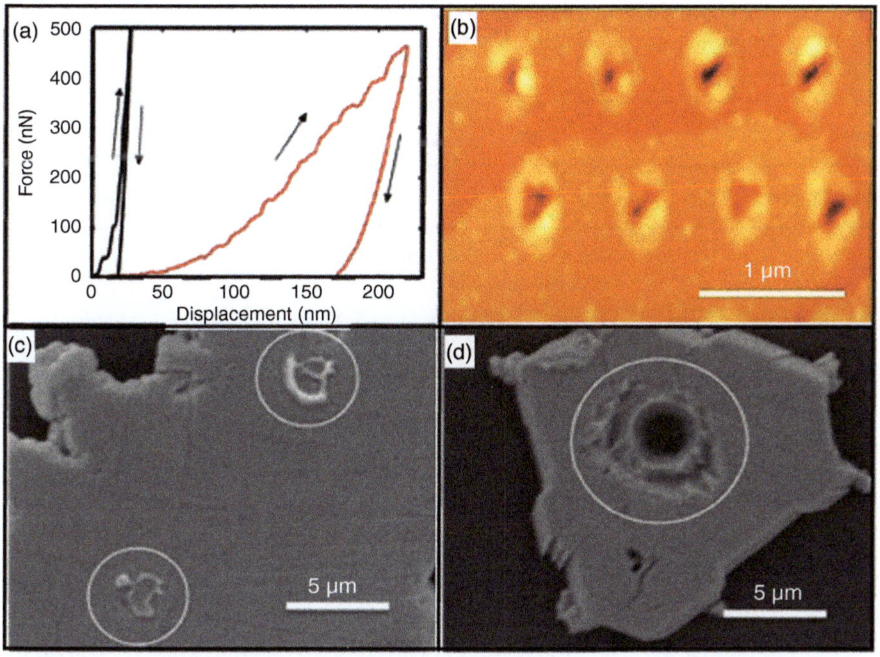

Figure 12.8 (a) Typical force–displacement curves obtained by nanoindentation measurements of fcc 3D superlattices having either 5 or 7 nm Au nanocrystals coated with C_{12} as building blocks. (b) Array of tip prints obtained by AFM imaging in acoustic mode after indenting 3D superlattice surface of 7 nm C_{12} coated nanocrystals. SEM images of (c) 5 nm and (d) after indentation under similar loading conditions using a 5 μm in diameter colloidal probe, 7 nm nanocrystals coated with C_{12} self-assembled, at $P_t = 0\%$, in fcc 3D superlattices produced. Circled zones indicate indented areas.

measurements under similar loading conditions. With 3D superlattice of 5 nm the surface is almost intact after indentation (Figure 12.8c), while the surface 3D superlattice of 7 nm nanocrystals exhibits a deep circular hole revealing the silicon substrate at the center (Figure 12.8d). Qualitatively, 3D superlattice of 5 nm is mechanically stronger than its counterpart independently of the indenter size.

The same experiments as described above are performed with nanocrystals having same average size (5 and 7 nm) and differing by chain length of the coating agents such as tetradecyl (C_{14}) and hexadecyl (C_{16}). Table 12.2 shows that, whatever the chain length when 3D superlattices are formed, the Young modulus drops with 3D superlattices of 7 nm Au nanocrystals compared to the same material of 5 nm. Furthermore, such behavior does not depend on the 3D superlattice morphology (film or shaped). The alkyl chain coverage decreases with decreasing curvature (i.e. with increasing the core size) and tends toward a flat surface. Consequently, we attribute such decrease in the Young moduli to a decrease in the coverage density.

Figure 12.8b shows an array of well-defined and homogeneous Berkovich-like marks after AFM tip indentation of the surface of fcc 3D superlattice of 7 nm Au nanocrystals coated with C_{14}. Moreover, it shows that no sliding occurs in the

Table 12.2 Interparticle distance (δ_{pp}), average elastic modulus (E) of 3D superlattices of 5-nm and 7-nm Au_7 nanocrystals coated by saturated alkyl thiols differing by their chain lengths.

Nanocrystal size		**5 nm**			**7 nm**	
Chain length	C_{12}	C_{14}	C_{16}	C_{12}	C_{14}	C_{16}
Crystalline structure	fcc	fcc amorphous	Amorphous	fcc	fcc	fcc
3D superlattice morphology	Film	Film	Film	Shape	Shape	Shape
δ_{PP}	1.8	2.1	—	1.9	2.0	2.3
E (MPa)	465 ± 125	1538 ± 325	—	63 ± 7	1389 ± 725	$1980 \pm$t373

3D superlattice are produced under zero vapor pressure. *Source:* Based on Gauvin et al. [3].

contact area between the AFM tip and the 3D superlattice surface. This permits to estimate the uncertainty on the absolute value of the contact area.

As in Chapter 3, at zero vapor pressure, the interparticle distance increases with the number of carbon atoms, n, of the alkyl chains used to coat Au nanocrystals [22]. Whatever the 3D superlattice morphology (Table 12.2), at fixed nanocrystal size, the elastic modulus increases continuously with increasing the number of carbon atoms of the alkyl chains (i.e. with increasing the interparticle distance). In such case, the change in the 3D superlattice morphology does not seem to play a role on the elastic moduli. Such trend is similar to what one would expect for simple nanocomposites materials composed of inorganic nanoparticles embedded in an organic host matrix [11, 12]. For a given nanocrystal size, the increase of the alkyl chain length induces an increase in the interparticle distance and consequently a decrease of the alkyl chain density allowing end-groups of the alkyl chains to occupy more space due to excluded-volume effect. This resulting flexibility of the outermost chain region increases with increasing the chain length of the coating agent. The change in chain flexibility of coating alkyl molecules with increasing the number of carbon atoms modulates ligand–ligand interactions between nanocrystals and modifies self-assembly process of nanocrystals during 3D superlattice formation. Hence, the increase in the coating chain flexibility favors alkyl chains entanglement and thereby enhances mechanical cohesion between adjacent nanocrystals within the 3D superlattice and consequently increase the 3D superlattice stiffness. This qualitative picture has appealing similarities with the model of Ohno et al. [16] applied to polymer chains grafted on spherical nanoparticles. Note that, as mentioned in Chapter 3, a rather large amount of alkanethiol molecules remains associated with the coating agent. The scheme proposed above, concerning the change in the conformation of flexible ligands with the number of carbon atoms, does not take into account the presence of free organic molecules in the superlattices, which is expected at zero solvent vapor pressure where the interparticle distance varies linearly with number of carbon atoms.

This could be explained by assuming that the residual thiol containing molecules physisorbed on the alkyl chains used as coating agents is expelled with the solvent.

At vapor pressure up to zero, the morphologies of the aggregates remain similar whereas their mechanical properties slightly differ. The beaker containing the colloidal solution is now placed in confined media instead of the glove box, the partial vapor pressure is P_t =39% and 75% when the nitrogen flow through the sample is saturated or not by toluene, respectively. In such experimental conditions, under P =39%, very few changes in the morphologies and crystalline structures are observed compared to P =0% (Tables 12.2 and 12.3). At P_t =39% and 75%, the fcc superlattice parameter expends resulting in 50% increase of interparticle distances as compared those of fcc 3D superlattices deposited under P =0%. In addition, the 3D superlattice interparticle distance of 5 and 7 nm as building blocks remain a "quasi" constant value of ~3 nm (Table 12.3) [22].

On increasing solvent vapor pressure, the cohesive forces due to the increase of the alkyl chain length decrease by the presence of solvent and thiol containing molecules that quench interactions between flexible ligands inducing a drop of Young modulus. Even though residual molecules may screen out entanglement between flexible ligands bound to nanocrystals, additional cohesive forces between nanocrystals such as van der Waals forces must play a role to maintain the morphology and long-range order in swollen superlattices. Tables 12.2 and 12.3 clearly show no drastic change in the mechanical parameters with solvent vapor pressure, demonstrating that 3D superlattice mechanical properties are not related to their morphologies and consequently to their heterogeneous growth process. Thus, the mechanical properties of Au 3D superlattices, *produced by solvent evaporation (i.e. out of the equilibrium) do not change with their growth mechanism processes*. In conclusion, the Young moduli decrease with decreasing the surface coverage and increase with the coating agent flexibility. Such processes are screened when the 3D superlattices are produced under solvent vapor pressure inducing accumulation of the free organic molecules and solvent in the 3D superlattice interstice.

12.3.1.2 Nanocrystallinity

Thin films of single domain and polycrystal 5-nm Au nanocrystals are produced by slowly evaporating hexane used to disperse them. Whatever the nanocrystallinity of 5-nm Au nanocrystals, the morphology and crystalline structure (fcc) of thin films remains unchanged. Low-angle electron diffraction patterns, recorded from a thin film of 5-nm Au nanocrystals differing by their nanocrystallinity (Figure 12.9a,c), show hexagonal symmetry of the pattern confirming the compact structure of the nanocrystal superlattice. The high-angle diffraction pattern of single domain nanocrystal film exhibits several rings corresponding to electron diffraction by specific planes of atomic lattices with fcc structure with small arcs distinctly observed for most diffraction rings (Figure 12.9b). Similar arcs in the high-angle diffraction pattern were previously observed in superlattices of self-assembled non-spherical building blocks [23], and especially consisting of Ag [24] and Au [25, 26] nanocrystals. Such a feature provides evidence for an average coherent alignment of nanocrystal atomic lattice planes to one another within the

Table 12.3 Interparticle distance (δ_{pp}), average elastic modulus (E) of 3D superlattices of 5 and 7 nm Au nanocrystals coated by saturated alkyl thiol differing by their chain lengths.

	Nanocrystal size	**5 nm**			**7 nm**		
P **(%)**	**Chain length**	C_{12}	C_{14}	C_{16}	C_{12}	C_{14}	C_{16}
39	Crystalline structure	fcc	fcc	Amorphous	fcc	fcc amorphous	fcc amorphous
	3D superlattice morphology	Film	Film	Film	Shape	Film	Film
	δ_{pp}	3.1	3.0	—	2.9	2.9	—
	E (MPa)	385 ± 114	546 ± 192	—	130 ± 60	235 ± 130	—
75	Crystalline structure	fcc	fcc	Amorphous	fcc	fcc	Amorphous
	3D superlattice morphology	Shape Film	Shape Film	Film	Shape	Shape	Film
	δ_{PP}	3.0	3.1	—	2.8	3.0	—
	E (MPa)	658 ± 243	666 ± 202	—	80 ± 33	216 ± 105	—

3D superlattices are produced under 39% and 75% vapor pressure. *Source:* Based on Wan et al. [22].

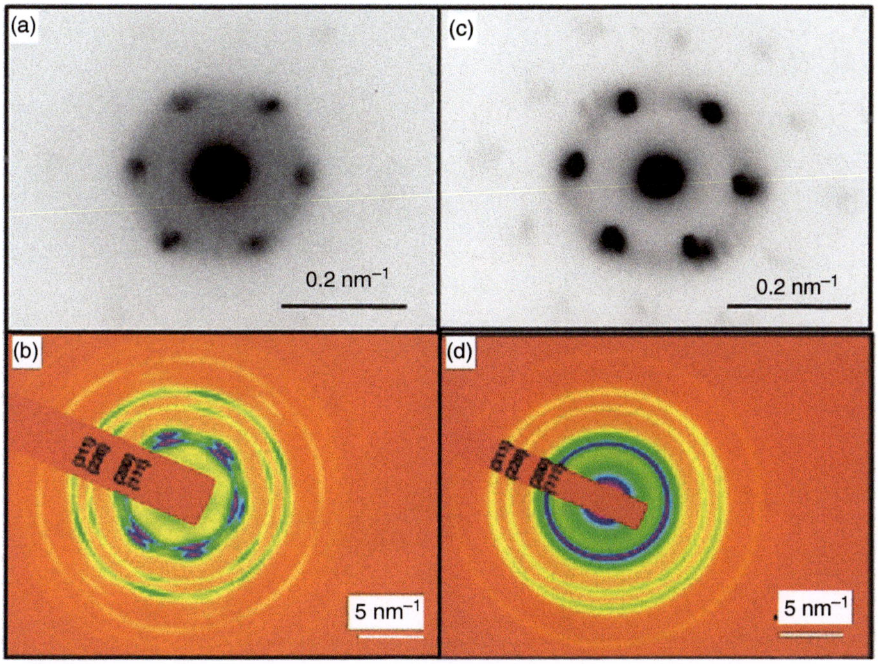

Figure 12.9 Low-angle electron diffraction patterns recorded from a thin film of close-packed of single crystalline (a) and polycrystalline (c) Au nanocrystals coated with dodecanethiol. The high-angle electron diffraction patterns recorded from the same thin film of single crystalline (b) and polycrystalline (d) Au nanocrystals.

superlattice. Such alignment is due to nanocrystal facets large enough to induce face-to-face orientation between neighboring nanocrystals despite the presence of dodecanethiol molecules tethered to their surface. In contrast, when looking no orientational ordering at the high-angle diffraction pattern obtained from polycrystalline nanocrystals (Figure 12.9d), diffraction rings exhibit no similar arcs as those previously mentioned although these nanocrystals are also close-packed in compact superlattice, as shown in Figure 12.9b. This result means that polycrystalline nanocrystals self-assemble into arrays with long-range translational but no orientational ordering [27, 28]. In addition to the crystallinity of nanocrystals and their inherent shape anisotropy, the van der Waals interactions between the surfactant chains and their possible directional distribution should also play an important role in the orientational ordering of nanocrystals, as suggested [29, 30].

The AFM images for single domain Au nanocrystals before (Figure 12.10a,c) and after (Figure 12.10b,d) indentation, respectively. The corresponding 3D superlattice film displaces residual marks with a shape similar to the AFM tip (Figure 12.10b), while the 3D superlattice films composed exclusively of polycrystalline nanoparticles break under the indentation (Figure 12.10d). This change indicates a huge drop of strength of particle–particle interactions in 3D superlattice films when moving from single domain nanocrystals to polycrystalline nanoparticles. A decrease in the Young's modulus by more than 1 order of magnitude between 3D superlattices made up single

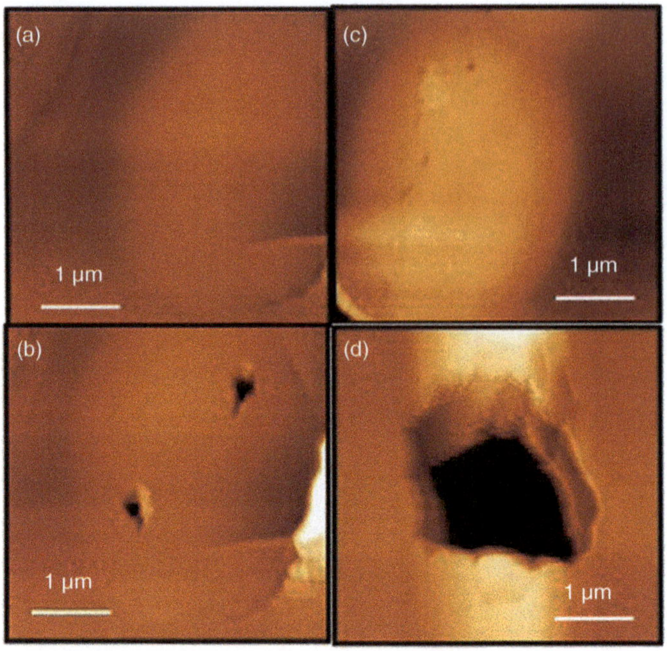

Figure 12.10 AFM images (a, c) before and (b, d) after nanoindentation of 3D superlattice film of (a, b) single and (c, d) polycrystalline 5 nm Au nanocrystals coated with dodecanethiol. *Source:* Goubet et al. [30]/American Chemical Society.

domain and polycrystalline is observed, from 550 ± 160 to 25 ± 9 MPa, respectively. Such change could be explained by the different degree of ordering within the 3D superlattices. As revealed above, single domain nanocrystals self-assemble into 3D superlattices with orientational ordering, which results in strong face-to-face interactions between individual nanocrystals. While for the 3D superlattices made up polycrystalline nanoparticle, where the building blocks are randomly oriented, leading to relatively weak inter-crystal forces and, then, lower Young's modulus.

Hence, Au 3D superlattices differing by the nanocrystallinity of their building blocks, show a strong change in the nanocrystals parking and consequently a marked enhancement of elastic modulus in 3D superlattices made up of single domain nanocrystals respect to those built using polycrystalline domain ones. This drastic change is related to different packing behaviors: while single crystalline nanocrystals exhibit both translational and orientational ordering in the superlattices, their polycrystalline counterparts behave like spheres with no orientational ordering.

12.3.2 Co 3D Superlattices

12.3.2.1 Epsilon Phase Co 3D Superlattice: Influence of Nanocrystal Size and 3D Superlattice Morphology on the Mechanical Properties

Colloidal solutions of 7.1 ± 0.6, 8.2 ± 0.6, and 9.0 ± 0.6 nm in diameter ε-Co nanocrystals coated by oleic acid (OA) with 8% as size distribution are dispersed in toluene.

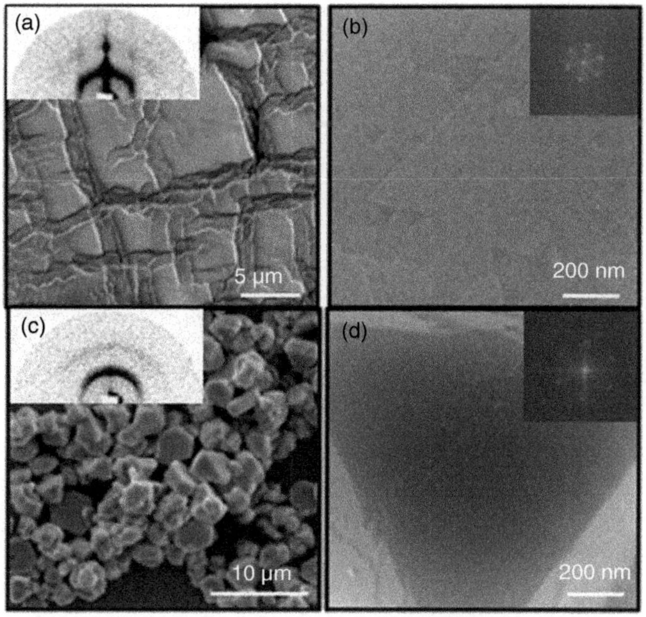

Figure 12.11 SEM (a, c) and high-resolution SEM (b, d) images of film and (a, b) shaped (c, d) 3D superlattices of 8-nm ε-Co nanocrystals. SAXS (inset a, c) and Fourier transform (inset b, d) patterns of the corresponding 3D superlattices.

By slow evaporation of the colloidal solutions, thick films are produced. Similar behavior is observed for the various ε-Co nanocrystals differing by their diameters. Figure 12.11a shows thick film of ε-Co nanocrystals differing by their diameter deposited on a silicon substrate are characterized by fcc structure. The (111) denser planes are lying flat to the substrate. From HRSEM imaging (Figure 12.11b), long-range assembly of ε-Co nanocrystals ordered in a compact hexagonal network is observed, as confirmed by the fast Fourier transform (FFT) pattern. The sixfold symmetry of FFT patterns (inset Figure 12.11b) confirms the orientation of (111) planes parallel to the substrate. The HRSEM shows structural defects such as screw dislocations, terraces, and grain boundaries. The average interparticle distances, δ_{pp}, calculated from the d_{111} spacing, are reported in Table 12.4. On first approximation, it is assumed that the nanocrystal surface is covered by coating agent and most of the extra organic molecules are released. Furthermore, the amount of coating agent decreases with increasing the nanocrystal size.

Figure 12.12a shows normalized histograms of Young modulus values obtained from AFM-based elastic F–δ curves of ε-Co nanocrystal 3D superlattice differing by their nanocrystal size. Note a large distribution of elastic modulus values. This is attributed to the heterogeneity of the 3D superlattices due to large number of structural defects in the superlattices (dislocations and grain boundaries). For any nanocrystal size, histograms of elastic modulus values are centered on 2 GPa and the average interparticle distance $\delta_{pp} \sim 3$ nm is consistent with partial interdigitation of oleic acid ligands between neighboring nanocrystals. At a fixed interparticle distance, an increase of the nanocrystal size from 7 to 9 nm yields a 5% difference in the organic content that has no significant effect on Young moduli of ε-Co 3D superlattices. This result is clearly contrasting with the Young modulus of Au 3D

Table 12.4 Edge-to-edge interparticle distance, δ_{pp}, and Young modulus (E^*) of 3D superlattice films and shaped of 6, 7, and 8 nm ε-Co nanocrystals.

Morphology	Nanocrystal size (nm)	6	7	8
Film 3D superlattice	δ_{pp} (nm)	3.3	3.2	3.3
	E (GPa)	1.7 ± 0.5	2.9 ± 0.9	2.3 ± 0.9
Shaped 3D superlattice	δ_{pp}	2.5	2.3	2.7
	E (GPa)	1.8 ± 0.5	2.1 ± 0.5	2.1 ± 0.4

Source: Based on Gauvin et al. [5].

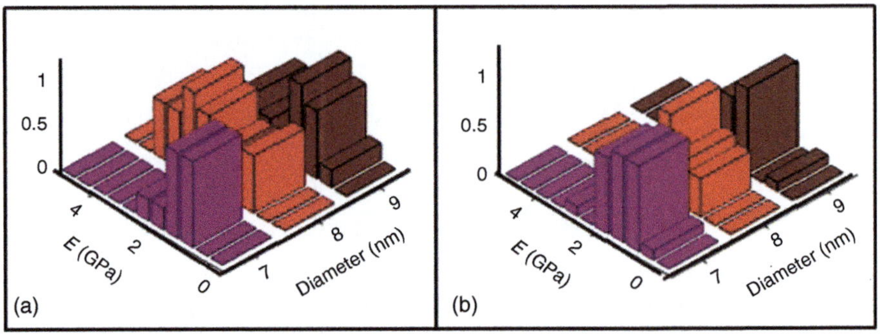

Figure 12.12 Normalized histograms of E^* values of films (a) and shaped 3D superlattices made of ε-Co nanocrystals differing by their nanocrystal size. Magenta, red, and brown colors indicate 3D superlattices of 7 nm-, 8 nm-, and 9 nm-ε-Co nanocrystals, respectively. *Source:* Based on Gauvin et al. [5].

superlattices, described above, that can decrease by 1 order of magnitude as the nanocrystal size increases from 5 to 7 nm at fixed interparticle distance (δ_{pp} = 2.2 nm) [3, 15].

By slow ethanol diffusion, during the evaporation process, in the colloidal solution containing the nanocrystals, shaped 3D superlattices are produced. Figure 12.11c shows an SEM image of several micrometer-thick faceted 3D superlattices of ε-Co nanocrystals. Similar behavior is obtained with the various nanocrystal sizes. The corresponding SAXS pattern (inset Figure 12.11c) shows homogeneous diffraction rings that indicate a random orientation of diffracting entities. Their location is in agreement with a fcc structure. At higher magnification, long-range assembly of ε-Co nanocrystals ordered in a compact hexagonal network is observed, as confirmed by the FFT pattern (inset Figure 12.11d). By comparing the HRTEM images in Figure 12.11b,d, the 3D superlattice films grown layer-by-layer show more structural defects such as screw dislocations, slip bands, and grain boundaries compared to faceted-shape 3D superlattices. The average interparticle distance decreases with faced-shape compared to film 3D superlattices produced from the same batch of nanocrystals (Table 12.4). The reduction in the interparticle distances for shaped 3D superlattices is attributed to a change in the conformation ligands attached to the surface of nanocrystals with formation of bundles due to presence of adsorbed ethanol in the alkyl chains. The elastic properties of shaped 3D superlattices of ε-Co nanocrystals differing by their size are similar to those obtained with films and have a central value on 2 GPa (Table 12.4). However, the width of the histograms which is related to the uncertainty on the elastic modulus values, is narrower in case of shaped 3D superlattices (Figure 12.12b) compared to 3D superlattice films (Figure 12.12a). This could be attributed to a relatively low amount of surface defects of shaped 3D superlattices compared to 3D superlattice films.

Hence, for any nanocrystal sizes studied (from 7 to 9 nm) and for any 3D superlattice morphologies (shaped or film structures), the Young moduli of single domain ε-Co 3D superlattices remain in the same order of magnitude (2 GPa). Furthermore, the Young moduli remain unchanged by decreasing the average distance between nanocrystals within the 3D superlattices from ~3 to 2.5 nm. These data, highly reproducible through various techniques, need to be understood more deeply [5]. Such data are rather surprising if we take into account these data to those obtained with Au nanocrystals produced by solvent evaporation. Here, *for 3D superlattices produced by solvent evaporation*, Tables 12.2–12.4 show that the Young moduli remain unchanged with the Co (ε phase) and Au 3D superlattice morphologies and with the interparticle distance while they markedly differ for the nanocrystal size used as building blocks with unchanged value for Co (ε phase) and a decrease with Au 3D superlattice attributed to a decrease of the alkyl chain coverage due to the decrease of curvature. In first approximation we can assume that effect remains valid. Actually, the change of the curvature between 7 and 9 nm is probably less than from 5 to 7 nm. The excess free molecules physisorbed to the alkyl chain preventing the change in the surface covering and crystal growth process such as to trimethylphosphine (TOPO) or dodecanethiol ($C_{12}H_{25}SH$) for Co (ε phase) and Au nanocrystals, respectively, cannot be retained because they are both present during the chemical reaction.

12.3.2.2 Hierarchical Mechanical Behavior of Co 3D Superlattices Related to Nanocrystallinity

Let us consider fcc 3D superlattices of ≠7-nm Co nanocrystals differing by their nanocrystallinity such as amorphous-Co, ε-Co, and hcp-Co nanocrystals [5]. Table 12.5 shows an average diameter and size distribution of nanocrystals coated with oleic acid. Figure 12.14a–c shows that amorphous-Co nanoparticles exhibit an isotropic spherical shape, while the single domain ε-Co and hcp-Co nanocrystals show polyhedral shapes with quasi-spherical morphology in case of ε-Co nanocrystals and well-developed facets in case of hcp-Co nanocrystals.

By solvent evaporation thick fcc film 3D superlattices of amorphous-Co, ε-Co, or hcp-Co nanocrystals are produced with an unchanged interparticle distance δ_{pp} between adjacent nanocrystals embedded in the 3D superlattices (Table 12.5). The electron diffraction pattern of thin (four to five layers) 3D superlattice films of ≠7-nm amorphous-Co shows a broad and diffused signal characteristic for random atomic arrangement without specific orientational alignment of the isotropic nanoparticles within the 3D superlattices. That of single domain hcp-Co (Figure 12.13d) shows small arc features for the most intense diffraction rings, whereas these arcs

Table 12.5 Average diameter, size distribution (σ), the interparticle distance δ_{pp} of the corresponding 3D superlattices, and Young's modulus of 7 nm Co 3D superlattices with different nanocrystallinities and ligand chain lengths.

Crystalline structure	Amorphous	Amorphous	hcp	ε-phase
Nanocrystal size (nm)	6.9	6.9	7.2	7.1
σ (%)	8.7	8.7	9.7	7.9
Coating agents	$C_{12}H_{25}COOH$	$C_{18}H_{35}COOH$	$C_{18}H_{35}COOH$	$C_{18}H_{35}COOH$
δ_{pp} (nm)	3.0	3.4	3.2	3.4
E (MPa)	80 ± 30	700 ± 400	6600 ± 1500	1700 ± 500

Source: Based on Gauvin et al. [5].

are absent when the 3D superlattice film is made of single domain ε-Co nanocrystals (Figure 12.13e). This is well evidenced through the first and second diffraction ring profiles of hcp- (Figure 12.14a) and ε-Co nanocrystals (Figure 12.14b), indicating that the atomic lattice planes of hcp-Co nanocrystals are coherently aligned with those of neighboring nanocrystals. The presence of arc features is related to single domain hcp-Co nanocrystals being shaped by large facets favoring orientational alignment through face-to-face interactions whereas the facets are hardly observable with the quasi-spherical shape of single domain ε-Co nanocrystals.

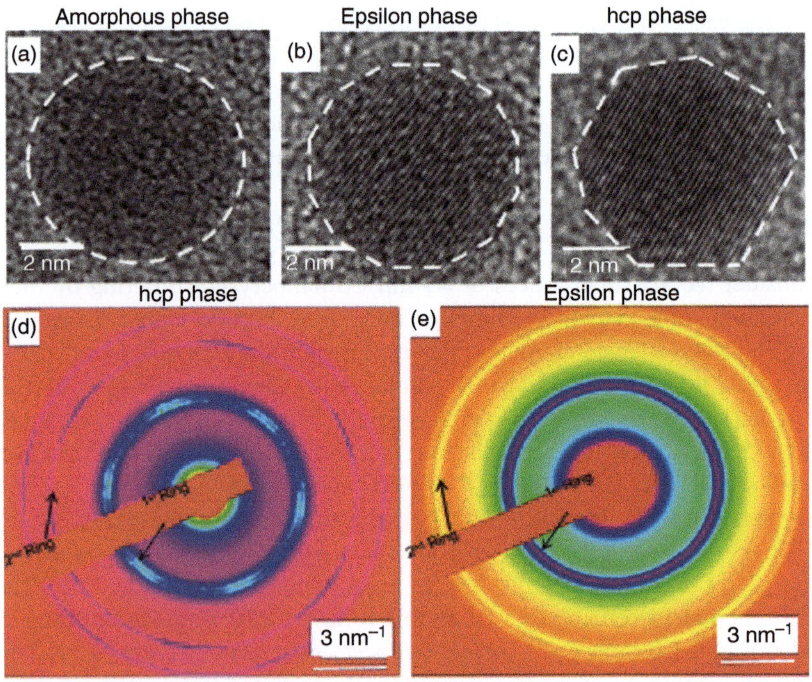

Figure 12.13 High-resolution TEM images of the oleic acid coated 7 nm Co nanoparticles differing by their nanocrystallinity: amorphous-Co (a), ε-Co (b), and hcp-Co (c). Electron diffraction patterns of four to five layers thick superlattice of hcp (d) and ε Co (e) phases. *Source:* Gauvin et al. [5], figure 2ab (p. 3483)/Springer Nature.

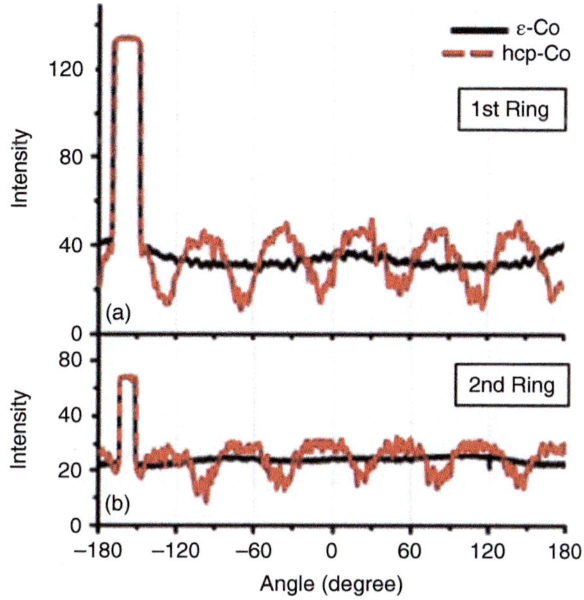

Figure 12.14 Profiles of the electron diffraction patterns shown on Figure 12.13d and 12.13e for superlattices of superlattices having hcp and ε-Co phases respectively. *Source:* Gauvin et al. [5]/ SpringerNature."

Hence 3D superlattices of hcp-Co nanocrystals are characterized by both translational and orientational ordering as already observed [23, 25, 28, 29] whereas those of ε-Co nanocrystals are limited to translational ordering.

The 3D superlattice force–displacement (F–δ) curves of ≠7 nm amorphous-Co, ε-Co, and hcp-Co nanocrystals show a transition in mechanical properties from hcp-Co 3D superlattices being the stiffest material, to a less stiff ε-Co 3D superlattice, both showing elastic deformation, to a soft amorphous-Co 3D superlattice material that is plastically deformed (Figure 12.15). The elastic modulus of ≠7 nm Co 3D superlattices differing by their nanocrystallinity increases by 1 order of

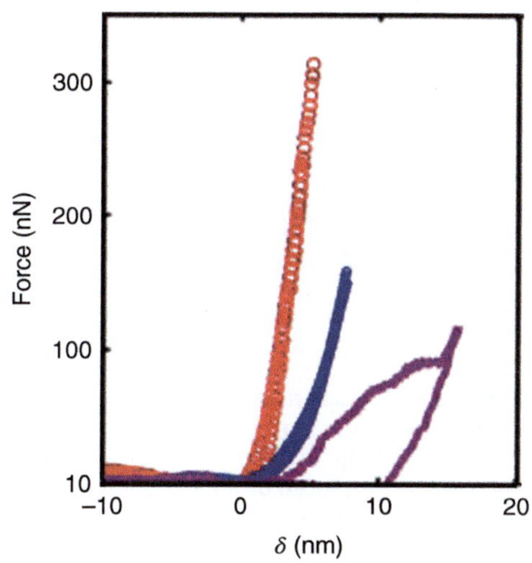

Figure 12.15 Force–distance curves of amorphous-Co, ε-Co, and hcp-Co nanoparticles as indicated by purple solid line, blue circles, and red circles, respectively. *Source:* Gauvin et al. [5], figure 3 (p. 3483)/Springer Nature.

magnitude with improvement of the crystalline symmetry of Co nanocrystals in the following order (Table 12.5): $E_{\text{Amorphous-Co}} < E_{\varepsilon\text{-Co}} < E_{\text{hcp-Co}}$.

With oleic acid as the coating agent, the elastic modulus of ε-Co 3D superlattices, characterized by single domain nanocrystals, is intermediate between the elastic moduli of amorphous-Co and hcp-Co 3D superlattice whereas the interparticle distances (Table 12.5) remain similar. The major change between these various samples differs from their nanocrystal crystalline structures with same nanocrystal size, coating agent and interparticle distance. The correlation between the hierarchical mechanical behavior of 3D superlattices and the nanocrystallinity of their building blocks could be attributed to specific facet–facet interactions tuned by the orientational and translational alignment of nanocrystals within the superlattice. This result agrees with previous data [1, 30, 31, 32]. However, we also have to take into account the fact that such alignment of nanocrystals could be due to dipolar magnetic attractive interactions between facets as observed experimentally [33, 34] and by simulation [35]. On applying 0.3 T to the superparamagnetic Co nanocrystals used here, no changes in the elastic moduli are observed. So, in first approximation, we conclude that magnetic interactions are too weak to significantly alter cohesive interactions between nanocrystals within 3D superlattices even though the ε-Co nanocrystals are soft magnetic material compared to the hcp counterparts.

To analyze such hierarchical mechanical behavior, let us consider a model differing from classical composite material model with inorganic nanoparticles are embedded in an organic matrix. It is based on dimensional analysis of the stress–deformation relationship in an array of rigid spheres connected to their neighbors via compliant junctions. Let us consider an array of rigid spheres of diameter, d, separated by compliant junctions of thickness, h, and of unit cell parameter $a = \sqrt{2}(d + h)$ (Figure 12.16).

We assume that junction stiffness is much smaller than the elastic modulus of the particle, which will subsequently be treated as rigid. Under an applied force F, the displacement δ will localize the strain in the junctions only. As F is applied normally to the (111) plane, the macroscopic strain is $\varepsilon_{\text{macro}} = \delta/d_{111}$, where d_{111} is the (111) plane inter-reticular distance, and the average junction strain is $\varepsilon_{\text{junction}} = \delta/h$. Now, we should consider that the macroscopic stress $\sigma_{\text{macro}} = F/d_{111}{}^2$ is then locally

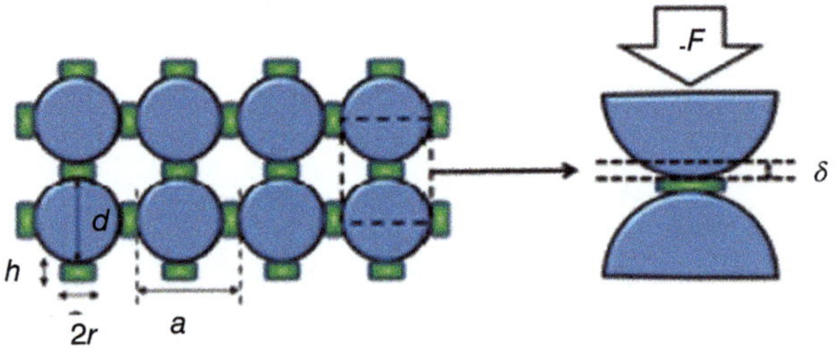

Figure 12.16 Schematics of a periodic array of rigid spheres of diameter d connected via compliant junctions of thickness h and half-width r (left). The periodicity of the array is $a = d + h$. Under applied load F, the compliant junctions are squeezed and their thickness h is reduced by an amount δ while the rigid spheres retain their shape (right). *Source:* Based on Gauvin et al. [5].

distributed in the three junctions of the tetrahedral sub-unit of the fcc array giving an average junction stress $\sigma_{junction} = F\cos\theta/3(\pi r^2)$ where r is the junction radius and $\theta = 45°$. From this scaling analysis we deduce that the effective macroscopic Young's modulus of the fcc array $E_{macro} = \sigma_{macro}/\varepsilon_{macro}$ scales like $E_{macro} = 3\sqrt{3}B\pi r^2/(h(D + h))$ where B is the average bulk modulus of the junction material. The first conclusion of this model is that the Young's modulus of 3D superlattices is directly proportional to the junction area between adjacent nanocrystals.

For oleic acid, nanocrystal size and interparticle distance, we have $E_{hcp\text{-}Co}/E_{\varepsilon\text{-}Co} = (r_{hcp\text{-}Co}/r_{\varepsilon\text{-}Co})^2$ and $E_{hcp\text{-}Co}/E_{Amorphous\text{-}Co} = (r_{hcp\text{-}Co}/r_{Amorphous\text{-}Co})^2$. Herein, from the variation of the Young modulus with nanocrystallinity at fixed nanocrystal size and interparticle distance, $E_{Amorphous\text{-}Co} < E_{\varepsilon\text{-}Co} < E_{hcp\text{-}Co}$, and the radius of the compliant junction between nanocrystals increases with nanocrystallinity as $r_{Amorphous\text{-}Co} < r_{\varepsilon\text{-}Co} < r_{hcp\text{-}Co}$. The increasing trend in the junction radius agrees with the increasing size of nanocrystal facets as observed by HRTEM (Figure 12.13c) and the progressive orientational alignment of nanocrystals facets as observed by electron diffraction (Figure 12.14). This suggests that the coating agent is important for the cohesion between nanocrystals.

To determine the role of the metal core and that of the ligand in the 3D superlattice's thickness, amorphous-Co nanoparticles are coated by lauric acid ($C_{12}H_{24}CO_2$) instead of its oleic ($C_{17}H_{32}CO_2$) counter parts. In other 3D superlattices, the crystalline structure remains unchanged and the interparticle distance is 3 nm instead of 3.4 nm. The Young's modulus decreases by an order of magnitude from 700 ± 400 to 80 ± 30 MPa by changing oleic acid C_{18} to lauric acid C_{12} (Table 12.5). Here, the Young's modulus of the 3D superlattice should be proportional to the bulk modulus B of the protecting ligand layer and that the ratio $E_{C18\text{-}amorphous\text{-}Co}/E_{C12\text{-}amorphous\text{-}Co} = B_{C18}/B_{C12}$ for given junction geometry, nanocrystal size, and interparticle distance. We estimate the bulk modulus $B_{C12} = 460$ MPa using the relationship $B = k/h$, with the junction stiffness $k = 1.4\,\text{N m}^{-1}$ and $h = 3$ nm for lauric acid (C_{12}) coated 6.3 nm Co 3D superlattice films [36]. Then, we estimate $B_{C18} = 4$ GPa using the ratio $E_{C18\text{-}amorphous\text{-}Co}/E_{C12\text{-}amorphous\text{-}Co} = B_{C18}/B_{C12}$. The B_{C12} and B_{C18} values compare well with bulk modulus values of thin surfactant layers reported in the literature [37–40]. Moreover, a similar trend of increasing elastic properties of surfactant layers with their chain length was reported [37]. Hence, the bulk modulus of the junction materials related the nature and the density of the ligand coating the surface of nanocrystals is a key parameter in tuning cohesive interactions within the 3D superlattice. From the data presented here the hierarchichal mechanical behavior is closely related to nanocrystallinity of Co 3D superlattices through orientational and translational ordering of nanoparticles and through facet–facet interactions controlled by nanoparticle shape anisotropy: the Young's modulus increased by 1 order of magnitude when the crystalline structure of Co nanoparticles changed from amorphous to single domain ε-Co and to hcp-Co.

12.3.3 Ag 3D Superlattices: Highly Weak Young Moduli

Let us consider 5.4-nm Ag nanocrystals coated with dodecylamine ($C_{12}H_{25}NH_2$) characterized by icosahedral structure with 20 facets and low size distribution

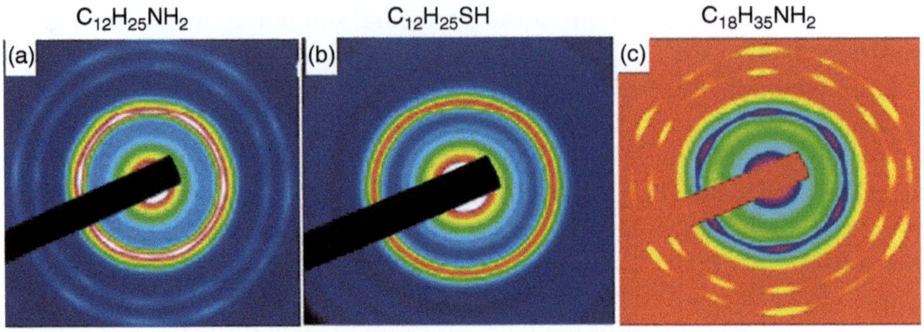

Figure 12.17 High-angle electron diffraction (HAED) patterns of 5.4-nm Ag nanocrystals coated with dodecylamine (a), dodecanethiol (b) and oleylamine (c). *Source:* Çolak et al. [6], figure 1 & 7/Royal Society of Chemistry.

(<10%) dispersed in toluene [6]. Deposition of a drop of the colloidal solution on TEM grid induces according the nanocrystals concentration used either 2D or thin 3D superlattices. The high-angle electron diffraction (HAED) pattern shows several rings assigned to planes of atomic lattices with fcc crystal structures (Figure 12.17a). Actually, small arcs are clearly spotted for some of the diffraction rings. Such behavior is attributed to facet–facet interactions with an average coherent alignment of the atomic lattice planes of nanocrystals within the superlattice [25–27]. This is due to orientational ordering of the atomic lattices planes of icosahedral Ag nanocrystals coated with dodecylamine and indicates a coherent alignment with adjacent isotropic nanocrystals through their faceted interactions. By increasing the colloidal concentration, the nanocrystals deposited on a substrate self-assembled forming a flat thick fcc film 3D superlattices with 1.8 ± 0.1 nm as the average interparticle distance, δ_{pp}, calculated from the d_{111} spacing (Table 12.6).

Figure 12.18 shows a decrease of the Young's modulus with increasing the indentation depth. At the nanoscale, nanoindentation experiments revealed that the Young's modulus of soft materials depends on the indentation depth [41–43]. Near the surface, the Young's modulus is usually higher than that measured at deeper locations; this is known as the indentation size effect and common in rigid materials

Table 12.6 The Young's modulus of (⟨E⟩) of different 3D superlattices of Ag nanocrystals with different average nanocrystal diameter (D), size distribution (σ), interparticle distance (δ_{pp}) and coating agents.

Nanocrystal size (nm)	**5.4**		**4.0**	**5.5**	**8**
σ (%)	**7**	**7**	**7**	**8**	**8**
Coating agents	$C_{12}H_{25}NH_2$	$C_{12}H_{25}SH$	$C_{18}H_{35}NH_2$	$C_{18}H_{35}NH_2$	$C_{18}H_{35}NH_2$
δ_{pp} (nm)	1.8	1.9	3.1	3.1	3.1
E (MPa)	84.6 ± 18.5	23.0 ± 6.9	3.1 ± 0.5	8.8 ± 2.4	125.0 ± 28.7

Source: Çolak et al. [6], table 1/Royal Society of Chemistry.

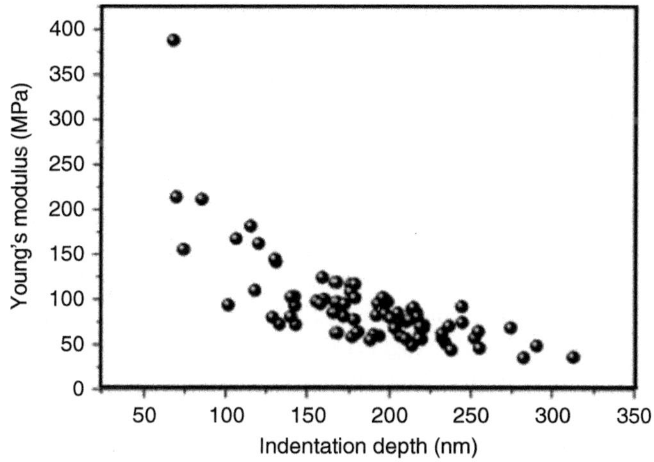

Figure 12.18 Plot of Young's modulus as a function of indentation depth measured for 3D superlattices of 5.4-nm Ag nanocrystals coated with dodecylamine ($C_{12}NH_2$). *Source:* Çolak et al. [6], figure 5/Royal Society of Chemistry.

such as many metals [44]. In soft materials, this is known as "skin effect" and occurs due to both the nonlinearity of stress–strain relation when using sharp probes and ignoring the adhesion forces between the AFM probe and the substrate surface [45]. The Young's moduli become relatively constant when the depth of indentation is greater than ~100 nm, although it is still slightly decreasing. To calculate the Young's modulus, the relatively constant values are averaged while ignoring the increasing values of the Young's modulus at small indentation depths. As a result, for all analyzed 3D superlattice thick films in this study, the plateau averaged value of the Young's modulus, $\langle E \rangle$, is 84.6 ± 18.5 MPa.

When dodecylamine ($C_{12}NH_2$) is replaced by dodecanethiol ($C_{12}SH$) as coating agent, as previously, the 5.4-nm nanocrystals self-assemble in 2D and thin 3D superlattices. The HAED pattern of 5.4-nm Ag nanocrystals coated with dodecanethiol also shows several rings assigned to fcc crystal structures (Figure 12.17b). In contrast to 3D superlattices of Ag nanocrystals coated with dodecylamine (Figure 12.17a), with dodecanethiol as coating agent, no arcs are observed (Figure 12.17b). This clearly indicates that the replacement of dodecylamine with dodecanethiol, i.e. by changing the head group of the coating agent while keeping the alkyl chain the same C_{12}, the coherent alignment of nanocrystals with adjacent nanocrystals through their faceted interactions disappears. This indicates a loss of the orientational ordering while keeping the tendency of being self-assembled with long-range translational order (Figure 12.17b). That is rather surprising, because Ag nanocrystals coated with dodecanethiol are produced by ligand exchange process of their counter parts coated with dodecylamine, characterized by an icosahedral structure. Hence, the ligand exchange process would have the capacity to reduce the facet–facet interactions and consequently would induce the loss of the arcs indicating the formation of more spherical nanocrystals. This unexpected process could be explained by the fact that thiol ligands strongly react with Ag nanocrystal surface inducing formation of a skin of Ag–S and/or Ag_2S. This is supported by the red shift of the localized surface plasmon resonance (LSPR) spectra of these 5.4-nm Ag nanocrystals by replacing dodecylamine with dodecanethiol as coating agent [46]. Such LSPR changes between thiol and amine-coated nanoparticles are attributed to stronger bonding of thiols compared to amines, on Ag atoms at the interface [47]. Hence, the Ag—S bonding would imply specific adsorption sites on Ag nanocrystal surface as it has been demonstrated for Au nanocrystals [48]. Here, it is reasonable to conclude that such specific adsorption sites are mainly located at the edges of the icosahedral edges by transforming the shaped nanocrystals to spherical nanocrystals with the formation of Ag–S/Ag_2S "skin" as proposed previously [45]. No markedly change on the thick film fcc 3D superlattices are observed. The Young's modulus drops from 84.6 ± 18.5 to 23.5 ± 6.9 MPa by ligand exchange (from dodecylamine, $C_{12}NH_2$, to dodecanethiol, $C_{12}SH$) keeping similar average diameter (5.4-nm) and interparticle distance ($\delta_{pp} = 1.9$ instead of 1.8 nm) (Table 12.6). These differences in the Young modulus for 3D superlattices can be correlated to the loss of orientational ordering observed with Ag nanocrystals coated with the change in the 3D superlattice growth on Figure 12.10 with 5-nm Au nanocrystals [1, 30].

By keeping the same Ag nanocrystals (5.4-nm) and replacing the coating agent from dodecylamine ($C_{12}H_{25}NH_2$) to oleylamine ($C_{18}H_{35}NH_2$), as expected, the films of 5.4-nm Ag nanocrystals coated with oleylamine are characterized by a fcc crystalline structure and the δ_{pp} is 3.1 nm. Hence, by changing the coating agent from dodecylamine to oleylamine the δ_{pp} increases from 1.8 to 3.1-nm (Table 12.6). The electron diffraction pattern of thin 3D superlattice film (four to five layers) of Ag nanocrystals coated with oleylamine (Figure 12.17c) shows several rings indicating a long coherence length of fcc arrangement of nanocrystals. Presence of well-defined arcs with intense spots on the first, second, and third rings indicates the atomic lattice planes of Ag nanocrystals coated with oleylamine have long-range coherent alignment with those of neighboring nanocrystals. This further confirms that the rather weak bond between oleylamine and Ag nanocrystal surface keeps the icosahedral shape of nanocrystals and therefore keeps the nanocrystals facet-to-facet interaction. This is supported by the fact that the arcs are observed in the three rings compared to what observed in Figure 12.17a. Hence the weak interactions between the surface atoms and the amine group favor the face-to-face orientation of icosahedral Ag nanocrystals. The Young modulus, determined in the same conditions as above, drops by 1 order of magnitude from 84.6 ± 18.5 to 8.8 ± 2.4 MPa. Such drop in the Young modulus can be explained as follows: As expected the interparticle distance between nanocrystals increases from 1.8 to 3.1 nm, thus inducing better alkyl chain flexibility. Because the amine head group is weakly associated with the surface atoms, the alkyl chain interactions are reduced and the coating agent move freely in the 3D superlattice lattices. This significant change compared to what obtained with Au and Co nanocrystals with coating agents strongly associated with the nanocrystal surface atoms. Keeping the same coating agent (oleylamine), the increase in the nanocrystal size induces an increase in the Young moduli (Table 12.6). However, it remains very low compared to the values obtained with other metallic nanocrystals (Au and Co). Such increase in the elastic properties with nanocrystal size is attributed to an increase in the van der Waals interactions of the nanocrystals.

The Young moduli measured with Ag nanocrystals markedly drop compared to those measured with Co and Au 3D superlattices. It is rather difficult to give a real explanation. Of course, we observed that one process is a chemical reaction between the coating agent and the atom at the surface. However, that is far from an explanation. Part of such behavior could be attributed to the weak interactions between the surface atoms and the coating agent head polar group. However, other causes have to be found. No convincing explanations can be given and such drop of the Young moduli remain an open question.

12.4 Do the Apparent Discrepancies of the Young Moduli Produced with a Large Variety of Metallic Nanocrystals Self-Assembled in fcc Structures Remain Valid or Not?

From the data presented above, it seems that the mechanical properties evolve differently from one system to another. Let list the procedures used to produce 3D superlattices:

(i) *3D superlattice growth process under equilibrium*: Interfacial and precipitated 3D superlattices are produced simultaneously when the colloidal solution is subjected to saturated solvent vapor pressure. The first ones are due to layer-by-layer growth process whereas the second ones grow in solution. This process is obtained with Au nanocrystals coated with dodecanethiol and dispersed in toluene considered as a bad solvent for the alkyl chains and does not take place in hexane as solvent. This is related to the fact that with toluene as solvent marked attractive van der Waals interactions between nanocrystals are shown whereas with hexane, keeping the same nanocrystal size, the attractive interactions between nanocrystals markedly drop [49, 50]. The drop of the elastic modulus of films compared to shaped 3D superlattices is attributed to the change in the 3D superlattice growth process [1, 30]. Both cases (film and shaped 3D superlattices), the Young moduli decrease on increasing the nanocrystal size used as building blocks. This procedure is highly reproducible but cannot be produced with other metallic nanocrystals (Co and Ag nanocrystals). No obvious explanation can be given.

(ii) *Presence of non-solvent molecule for the alkyl chains*: Non-solvent molecule vapor pressure for the ligand chains subjected to a colloidal solution induces a change in the conformation of the coating agents in solution and favors an increase in the nanocrystal interactions and consequently growth process in solution [51]. Very few studies compare the mechanical properties of film 3D superlattices and those shaped counterparts produced as the procedure presented here. With ε-Co 3D superlattices no change in the average value of the Young moduli is observed between films and shaped 3D superlattices [5].

(iii) *Solvent evaporation of the colloidal solution*: By slow solvent evaporation process both film and shaped 3D superlattices are produced. Here solvent, adsorbed reactive and ligand molecules on the nanocrystal surfaces are trapped in the 3D superlattices, even though it is dried. The solvent vapor pressure during the 3D superlattice growth process has a major impact on its

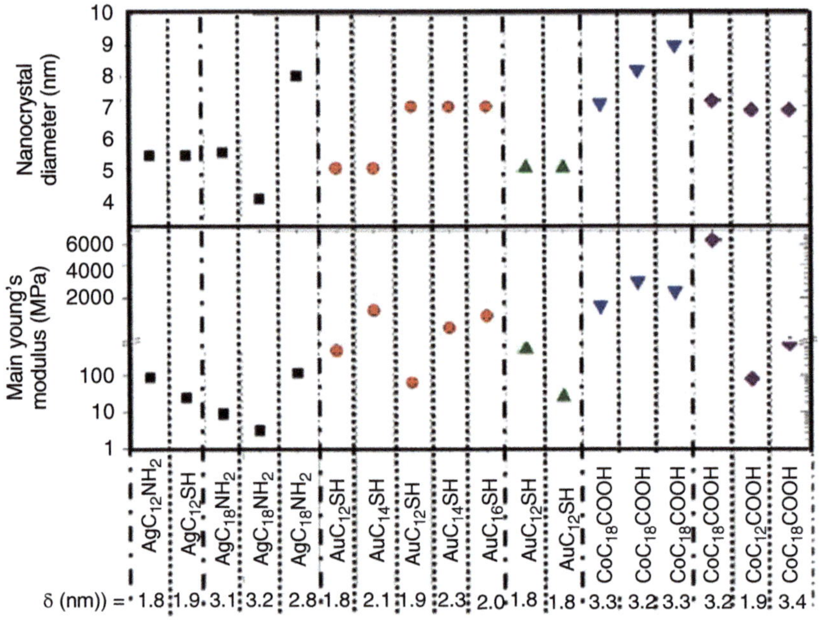

Figure 12.19 (up) The average nanocrystal diameter (*D*), (bottom) the Young's modulus (*E*), of various materials with different coating agents from different studies. The interparticle distances (δ_{pp}) of each materials are also shown in the bottom of the figure.

internal structure and on the interparticle distances. In spite of the purification process after the nanocrystal synthesis, the presence of free ligand molecules within the colloidal solutions, and hence during the evaporation process, has to be considered. During the evaporation process and under zero vapor pressure a part of the residual "unreacted" ligand-containing molecules physisorbs on the alkyl chains used as coating agents is likely expelled with the solvent during evaporation. This process markedly drops on increasing the solvent vapor pressure. Then, the excess ligand and reactive derivatives cannot be excluded with the solvent molecules by depletion forces and consequently the solvent and ligand-containing molecules remain in the superlattices.

Let us consider the various experimental data obtained with 3D superlattices produced by solvent evaporation. Figure 12.19 summarizes the Young moduli from one system to another. These 3D superlattices are made of various metallic nanocrystals differing by their coating agents and the interparticle distance of the building blocks. Obviously, such comparison permits to face most of the apparent of discrepancies obtained from one to other systems.

Let us consider the various parameters involved to control the elastic modulus of 3D superlattices produced by solvent evaporation.

(i) *Ligand head polar group-surface atoms interactions*: The ligand head polar groups, currently used, are thiol (SH), amine (NH_2), and carboxylic (COOH) derivatives. The interactions between the head polar group and the surface atoms markedly change:
The thiol ligands are "quasi" covalently bound with Au atoms as demonstrated experimentally [17–19] and confirmed by atomistic simulation [52]. At the Au surface, significant damages and/or amorphous-like structures are observed [16]. With Ag atoms, the strength of the bonding remains strong but is less and not very well known. Furthermore, a chemical reaction between thiol ligand and Ag atoms takes place at the interface and Ag_2S/AgO layer grows on the Ag surface. Such chemical reaction markedly modifies the nanocrystals surface as well demonstrated above. Hence, contrary to Au nanocrystals, the chemisorption with "quasi-covalent" bond is followed by a chemical reaction between Ag atoms and sulfur inducing formation of a shell of Ag_2S at the nanocrystal interface.
The carboxylic group is covalently bound with Co atoms [20].
The strong interactions of thiol and carboxylic head groups could be one of the key parameter to produce 3D superlattice with rather large Young moduli.
With amine-terminated ligands (NH_2), the bindings on the metallic surface are very weak compared to SH and COOH groups [21]. NH_2 terminated-ligand do not produce almost any distortion in the crystalline structure of the metallic nanoparticles This explains the efficiency in the ligand exchange process from amines to thiols, and not the other way around [53]. The drop of the Young moduli obtained with Ag 3D superlattices is probably due to these weak interactions.

(ii) *Ligand–ligand interactions*: Let first consider the configuration of the alkyl chains.
The saturated alkyl chains as coating agent (C_nH_{2n+1}) are well known to be characterized by a zig-zag molecular structure. At low solvent vapor pressure, by evaporating solvent van der Waals interactions between the ligand–ligand chains tend a "perfect" interdigitation (Figure 12.20a). The calculated length of a zig-zag chain [54, 55] is similar to the measured interparticle distance, δ_{pp}, between nanocrystals in 3D superlattices. This clearly indicates a "perfect" interdigitation between the alkyl chains.
Oleyl alkyl chains ($C_{18}H_{35}$) consist of two alkyl chains (C_7H_{16} and C_9H_{18}) bonded by a double bond. This provides a larger flexibility of the chains than that of the zig-zag configuration of saturated alkyl chains. The interactions between oleyl chains can be schematized as shown on Figure 12.20b. This most likely structure is supported by the fact that polymerization process

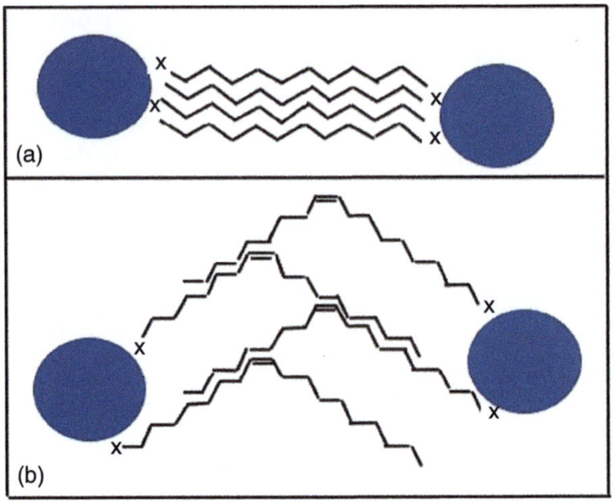

Figure 12.20 Molecular structures of the coating agents that are used in this study. X is either SH or NH_2.

through the double bonds is produced by thermal treatment [56]. The dynamic processes of oleic acid are enhanced compared to saturated alkyl chains.

(iii) *Influence of the interparticle distance, δ_{pp}*: The change in δ_{pp} is obtained by either increasing the number of carbon atoms of the saturated alkyl chain (C_nH_{2n+1}) or changing the molecular structure of the coating agent such as oleyl.

At zero solvent vapor pressure, by using similar procedure to produce 3D superlattices, the δ_{pp} does not change from one metal to another (Co, Au, Ag) (Table 12.7 and Figure 12.19). With C_nH_{2n+1} alkyl chain, δ_{pp} increases with n. This confirmed and supported by atomistic molecular dynamics simulations [52, 57]. The Young moduli increases with increasing the interparticle distance. This is attributed by the fact that the increase in the interparticle distance induces more flexibility of the alkyl chains and consequently favor their interactions.

With oleyl group (C_{18}) δ_{pp} is around 3 nm. On increasing the solvent vapor pressure during the evaporation process, because the residual chemical

Table 12.7 Range of the elastic moduli deduced with various 3D superlattices differing by size in the range of 1.5-nm to 13-nm, by coating agent and by procedure to produce such assemblies.

Materials	Molecular structure	3D superlattices Young's moduli
Ag	Zig-zag form oleyl group	$3\,MPa < E < 120\,MPa$
Au	Zig-zag form	$23\,MPa < E < 2000\,MPa$
Co	Zig-zag form oleyl group	$700\,MPa < E < 7000\,MPa$
PbS	Oleyl group	$200\,MPa < E < 6000\,MPa$
Pt	Foam	$900\,MPa < E < 2000\,MPa$

Molecular structure (of the ligand either in zig-zag form or as oleyl group).

remains absorbed in the 3D superlattice interstice, the distance between nanocrystals remains unchanged.

(iv) *Orientational ordering along the atomic lattices of nanocrystals*: Nanocrystals with fcc structures are characterized by decahedral, icosahedral MTPs with uniformly sized triangular facets corresponding to {111} crystallographic surfaces whereas cuboctahedron and the truncated octahedron both {111} and {100} facets are present [58]. The two latter correspond to single domain nanocrystals. The presence of facets in typical morphologies play a key role in the crystallinity segregation by strengthening the interactions between nanocrystals, thus favoring their self-assembling. This orientational ordering of the atomic lattices of nanocrystals is indicative of their faceted morphology. Translational ordering exists with the formation of the crystalline structure of the 3D superlattices (fcc). When the atomic lattice planes are aligned within the superlattice due to facets interaction, the Young's modulus of the 3D superlattice markedly increases compared to the lattice plane are randomly oriented. This claim is valid for Au and Co 3D superlattices, but not with Ag ones. No obvious explanation can be given.

(v) *Nanocrystal size*: The major discrepancies concern the change of the Young's moduli with the nanocrystal size involved in the building blocks. In some cases, the Young's moduli increase with increasing the nanocrystal size [6, 11] and was attributed to an increase in the van der Waals interactions, whereas in other cases they decrease in relation to ligand coverage at the interface [1, 2, 30, 52, 57]. Another behavior is related to the same elastic properties on increasing the nanocrystal size [1, 3]. Obviously, none of these explanations remain valid. In the bulk phase the Young's moduli of such materials are similar. Thus, in first approximation, it does not seem to be reasonable to attribute such changes to the material itself. The major differences between these various materials remain to procedure developed to produce the building blocks. Excess of reactive molecules adsorbed on the ligands could be one of the parameter. However, it is rather difficult to bring proof for assigning such changes to this parameter.

12.5 Mesoscopic Assemblies of Co Nanocrystals Differing by Their Size Distribution: Mechanical Intrinsic Properties

Let us consider a slow solvent evaporation of a colloidal solution subjected to a magnetic field perpendicular to the substrate (see Chapter 6 for more information) [59–61]. The Co nanoparticles are characterized by same average size and differ from their size distribution. With low size distribution, mainly dots (upright or fallen columns) are produced at the end of the evaporation (Figure 12.21a). Conversely, when the size distribution increases, a large number of flower-like

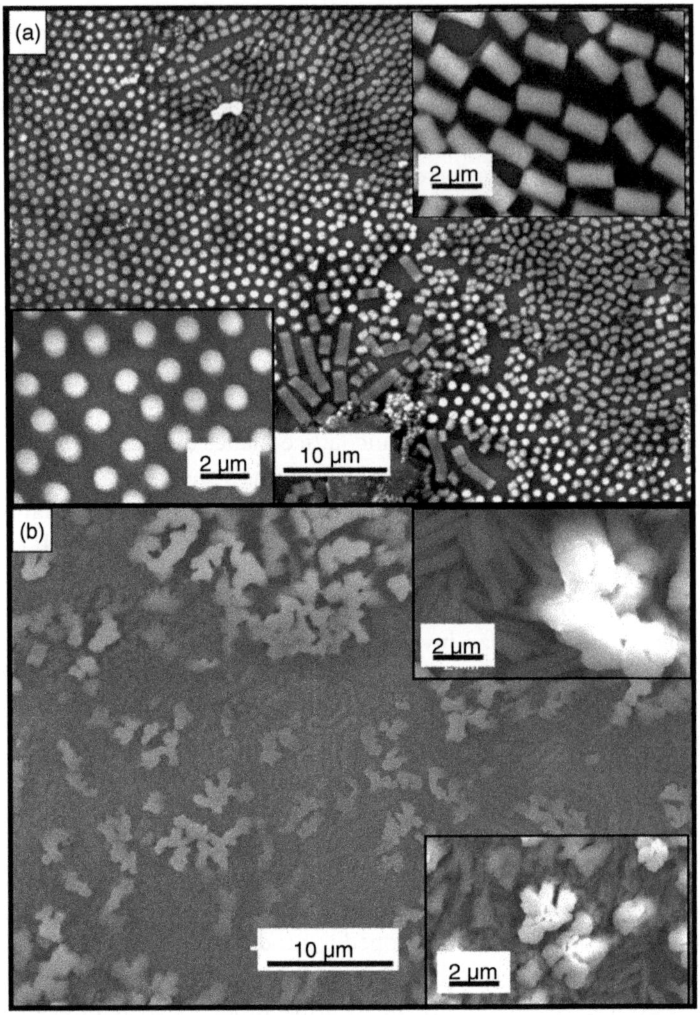

Figure 12.21 SEM patterns of mesostructures of 5.7 nm cobalt nanocrystals having 13% (a) and 18% (b) size distributions. *Source:* Isabelle Lisiecki (2009)/Elsevier.

structures as a result of coalescence of either upright or fallen columns forming worm-like and labyrinthine shapes are produced (Figure 12.21b). The proposed mechanism of pattern formation is the following: during the evaporation process, a liquid–gas phase transition occurs with the formation of a concentrated solution of nanocrystals in equilibrium with a diluted one. In the concentrated phase, columns (dots) are progressively formed and tend to migrate to self-organize in hexagonal patterns.

When the size distribution is low enough, the nanocrystals dispersed in solution tend to self-assemble in fcc 3D superlattices with the formation of well-defined and compact columns, whereas at higher size distribution values, the interactions between particles markedly decrease and the columns are formed with disordered entities. This creates defects. The cohesive forces between columns are not large enough to keep them ordered, and columns tend to fuse to form labyrinths. Hence, the mesoscopic structure of cobalt nanocrystals is tuned from well-defined dots to labyrinths with an increasing nanocrystal size distribution while keeping the same average nanocrystal size. From that it is claimed that the mechanical properties of the column formation control the final patterns.

12.6 Conclusions

Elastic properties of highly ordered 3D superlattices depend on (i) the 3D superlattice growth process, (ii) the procedure, (iii) the nanocrystallinity, and (iv) coating agent used to maintain the integrity of nanocrystals used as the building blocks. The Young's modulus evolves from few GPa to a few MPa.

References

1 Yan, C., Arfaoui, I., Goubet, N., and Pileni, M.P. (2013). Soft supracrystals of Au nanocrystals with tunable mechanical properties. *Adv. Funct. Mater.* 23: 2315–2321.

2 Yan, C., Portalès, H., Goubet, N. et al. (2013). Assessing the relevance of building block crystallinity for tuning the stiffness of gold nanocrystal superlattices. *Nanoscale* 5: 9523–9527.

3 Gauvin, M., Wan, Y., Arfaoui, I., and Pileni, M.P. (2014). Mechanical properties of Au supracrystals tuned by flexible ligand interaction. *J. Phys. Chem. C* 118: 5005–5012.

4 Gauvin, M., Yang, N., Barthel, E. et al. (2015). Morphology, nanocrystallinity and elastic properties of single domain ε-Co supracrystals. *J. Phys. Chem. C* 119: 7483–7490.

5 Gauvin, M., Yang, N., Yang, Z. et al. (2015). Hierarchical mechanical behavior of cobalt supracrystals related to nanocrystallinity. *Nanoresearch* 8: 3480–3487.

6 Çolak, A., Wei, J., Arfaoui, I., and Pileni, M.P. (2017). Coating agent-induced mechanical behavior of 3D self-assembled nanocrystals. *Phys. Chem. Chem. Phys.* 19: 23887.

7 Chudoba, T., Schwarzer, N., and Richter, F. (2000). Determination of elastic properties of thin films by indentation measurements with a spherical indenter. *Surf. Coat. Technol.* 127: 9.

8 Oliver, W.C. and Pharr, G.M. (2004). Measurement of hardness and elastic modulus by instrumented indentation: advances in understanding and refinements to methodology. *J. Mater. Res.* 19: 3–20.

9 Oliver, W.C. and Pharr, G.M. (1992). An improved technique for determining hardness and elastic-modulus using load and displacement sensing indentation experiments. *J. Mater. Res.* 7: 1564.

10 Young, W.C. and Budynas, R.G. (2002). *Roark's Formulas for Stress and Strain*. McGraw-Hill.

11 Podsiadlo, P., Krylova, G., Lee, B. et al. (2010). The role of order, nanocrystal size, and capping ligands in the collective mechanical response of three-dimensional nanocrystal solids. *J. Am. Chem. Soc.* 132: 8953.

12 Tam, E., Podsiadlo, P., Shevchenko, E. et al. (2010). Mechanical properties of face-centered cubic supercrystals of nanocrystals. *Nano Lett.* 10: 2363.

13 Lee, B., Podsiadlo, P., Rupich, S. et al. (2009). Comparison of structural behavior of nanocrystals in randomly packed films and long-range ordered superlattices by time-resolved small angle X-ray scattering. *J. Am. Chem. Soc.* 131: 16386–16388.

14 He, J., Lin, X.M., Divan, R., and Jaeger, H.M. (2011). In-situ partial sintering of gold-nanoparticle sheets for SERS applications. *Small* 7: 3487–3492.

15 Goubet, N., Portalès, H., Yan, C. et al. (2012). Simultaneous growths of gold colloidal crystals. *J. Am. Chem. Soc.* 134: 3714–3719.

16 Olmos-Asar, J.A., Luduena, M., and Marisca, M.M. (2014). Monolayer protected gold nanoparticles: the effect of the headgroup-Au interaction. *Phys. Chem. Chem. Phys.* 16: 15979–15987.

17 Xue, Y., Li, X., Li, H., and Zhang, W. (2014). Quantifying thiol–gold interactions towards the efficient strength control. *Nat. Commun.* 5: 4348.

18 Gronbeck, H., Curioni, A., and Andreoni, W. (2000). Thiols and disulfides on the Au(111) surface: the headgroup-gold interaction. *J. Am. Chem. Soc.* 122: 3839–3842.

19 Jin, R., Zeng, C., Zhou, M., and Chen, Y. (2016). Atomically precise colloidal metal nanoclusters and nanoparticles: fundamentals and opportunities. *Chem. Rev.* 116: 10346–10413.

20 Legrand, J., Petit, C., Bazin, D., and Pileni, M.P. (2000). Collective effect on magnetic properties of 2D superlattices of nanosized cobalt particles. *Appl. Surf. Sci.* 164: 186–192.

21 Hiramatsu, H. and Osterloh, F.E. (2004). A simple large-scale synthesis of nearly monodisperse gold and silver nanoparticles with adjustable sizes and with exchangeable surfactants. *Chem. Mater.* 16: 2509–2511.

22 Wan, Y., Goubet, N., Albouy, P.A. et al. (2013). Hierarchy in Au nanocrystal ordering in a supracrystal: II. Control of interparticle distances. *Langmuir* 29: 13576–13581.

23 Quan, Z. and Fang, J. (2010). Superlattices with non-spherical building blocks. *J. Nano Today* 5: 390–411.

24 Harfenist, S.A., Wang, Z.L., Alvarez, M.M. et al. (1996). Highly oriented molecular Ag nanocrystal arrays. *J. Phys. Chem.* 100: 13904–13910.

25 Brown, L.O. and Hutchison, J.E. (2001). Formation and electron diffraction studies of ordered 2D and 3D superlattices of amine-stabilized gold nanocrystals. *J. Phys. Chem. B* 105: 8911–8916.

26 Stoeva, S.I., Prasad, B.L.V., Uma, S., et al. Face centered cubic and hexagonal closed-packed nanocrystal.

27 Korgel, B.A., Fullam, B.A., Connolly, S., and Fitzmaurice, D.J. (1998). Assembly and self-organization of silver nanocrystal superlattices: ordered "soft spheres". *J. Phys. Chem. B* 102: 8379–8388.

28 Li, R., Bian, K., Hanrath, T. et al. (2014). Decoding the superlattice and interface structure of truncate PbS nanocrystal-assembled supercrystal and associated interaction forces. *J. Am. Chem. Soc.* 136: 12047–12055.

29 Portalès, H., Goubet, N., Sirotki, S. et al. (2012). Crystallinity segregation upon selective self-assembling of gold colloidal single nanocrystals. *Nano Lett.* 12: 5292–5298.

30 Goubet, N., Yan, C., Polli, D. et al. (2013). Modulating physical properties of isolated and self-assembled nanocrystals through change in nanocrystallinity. *Nano Lett.* 13: 504–508.

31 Bian, K., Wang, Z., and Hanrath, T. (2012). Comparing the structural stability of PbS nanocrystals assembled in fcc and bcc superlattice allotropes. *J. Am. Chem. Soc.* 134: 10787–107890.

32 Wang, Z., Schliehe, C., Wang, T. et al. (2011). Deviatoric stress driven formation of large single-crystal PbS nanosheet from nanoparticles and in situ monitoring of oriented attachment. *J. Am. Chem. Soc.* 133: 14484–14487.

33 Ku, J., Aruguete, D.M., Alivisatos, A.P., and Geissler, P.L. (2011). Self-assembly of magnetic nanoparticles in evaporating solution. *J. Am. Chem. Soc.* 133: 838–848.

34 Vella, D., du Pontavice, E., Hall, C.L., and Goriely, A. (2013). The magneto-elastica: from self-buckling to self-assembly. *Proc. R. Soc. Math. Phys. Eng. Sci.* 470: 20130609–20130609.

35 Widmer-Cooper, A. and Geissler, P. (2014). Orientational ordering of passivating ligands on CdS nanorods in solution generates strong rod–rod interactions. *Nano Lett.* 14: 57–65.

36 Lisiecki, I., Halté, V., Petit, C. et al. (2008). Vibration dynamics of supra-crystals of cobalt nanocrystals studied with femtosecond laser pulses. *Adv. Mater.* 20: 4176–4179.

37 DelRio, F.W., Jaye, C., Fischer, D.A., and Cook, R.F. (2009). Elastic and adhesive properties of alkanethiol self-assembled monolayers on gold. *Appl. Phys. Lett.* 94: 131909.

38 Chen, Y.L., Helm, C.A., and Israelachvili, J.N. (1991). Measurements of the elastic properties of surfactant and lipid monolayers. *Langmuir* 7: 2694–2699.

39 Zanoni, R., Naselli, C., Bell, J. et al. (1986). Elastic properties of Langmuir–Blodgett films. *Phys. Rev. Lett.* 57: 2838–2840.

40 Smaali, K., Desbief, S., Foti, G. et al. (2009). Osmotic stress on concentrated colloidal suspensions: a path towards equilibrium? *Braz. J. Phys.* 39: 156–162.

41 Dimitriadis, E.K., Horkay, F., Maresca, J. et al. (2002). Determination of elastic moduli of thin layers of soft material using the atomic force microscope. *Biophys. J.* 82: 2798–2810.

42 Clifford, C.A. and Seah, M.P. (2005). Quantification issues in the identification of nanoscale regions of homopolymers using modulus measurement via AFM nanoindentation. *Appl. Surf. Sci.* 252: 1915–1933.

43 Hou, H.Y., Chang, N.K., and Chang, S.H. (2006). Dynamic indentation of polymers using the atomic force microscope. *Nanomech. Mater. Struct.* 171–180.

44 Arsenlis, A. and Parks, D.M. (1999). Crystallographic aspects of geometrically-necessary and statistically-stored dislocation density. *Acta Mater.* 47: 1597–1611.

45 Dokukin, M.E. and Sokolov, I. (2012). On the measurements of rigidity modulus of soft materials in nanoindentation experiments at small depth. *Macromolecules* 45: 4277–4288.

46 Wei, J., Schaeffer, N., and Pileni, M.P. (2014). Ag nanocrystals: 1. Effect of ligands on plasmonic properties. *J. Phys. Chem. B* 118: 14070–14075.

47 Romann, J., Wei, J., and Pileni, M.P. (2015). Computational matching of surface plasmon resonance: interactions between silver nanoparticles and ligands. *J. Phys. Chem. C* 119: 11094–11099.

48 Djebaili, T., Richardi, J., Abel, S., and Marchi, M. (2013). Atomistic simulations of the surface coverage of large gold nanocrystals. *J. Phys. Chem. C* 117: 17791–17800.

49 Goubet, N., Richardi, J., Albouy, P.A., and Pileni, M.P. (2011). Which forces do control the supracrystal nucleation in organic media? *Adv. Funct. Mater.* 21: 2693–2704.

50 Goubet, N., Richardi, J., Albouy, P.A., and Pileni, M.P. (2011). How to predict the growth mechanism of supracrystals from gold nanocrystals. *J. Phys. Chem. Lett.* 2: 417–422.

51 Talapin, D.V., Shevchenko, E.V., Kornowski, A. et al. (2001). New approach to crystallization of CdSe nanoparticles into ordered three-dimensional superlattices. *Adv. Mater.* 13: 1868–1871.

52 Liu, X.P., Ni, Y., and He, L.H. (2016). Elastic properties of gold supracrystals: effects of nanocrystal size, ligand length, and nanocrystallinity. *J. Chem. Phys.* 144: 144507.

53 Linnert, T., Mulvaney, P., and Henglein, A. (1993). Surface chemistry of colloidal silver: surface plasmon damping by chemisorbed iodide, hydrosulfide (SH–), and phenylthiolate. *J. Phys. Chem.* 97: 679–682.

54 P. Schapotschnikow, ; Pool, R and T.J.H. Vlugt. Molecular simulations of interacting nanocrystals. *Nano Lett.* 2008, 8, 2930-2934.

55 Schapotschnikow, P. and Vlugt, T.J.H. (2009). Understanding interactions between capped nanocrystals: three-body and chain packing effects. *J. Chem. Phys.* 131: 124705.

56 Dreyer, A., Feld, A., Kornowski A. ; , YilmazI, E.D. ;, Noel H. ; , Meyer, A. ; ,Krekeler, T. ; Jiao, C. ; Stierle, A. ; Abetz, V., Weller, H. ; and Schneider, G. Organically linked iron oxide nanoparticle supercrystals with exceptional isotropic mechanical properties *Nat. Mater.*, 2016, 15, 522-528

57 Liu, X.P., Ni, Y., and He, L.H. (2016). Molecular dynamics simulation of interparticle spacing and many-body effect in gold supracrystals. *Nanotechnology* 27: 135707.

58 Wang, Z.L. (2000). Transmission electron microscopy of shape-controlled nanocrystals and their assemblies. *J. Phys. Chem. B* 104: 1153–1175.

59 Germain, V. and Pileni, M.P. (2005). Dots and labyrinthines of cobalt nanocrystals. *Adv. Mater.* 17: 1424–1230.

60 Germain, V. and Pileni, M.P. (2005). Mesostructures of cobalt nanocrystals: mechanism. *J. Phys. Chem.* 109: 5548–5553.

61 Germain, V., Richardi, J., Ingert, D., and Pileni, M.P. (2005). Mesostructures of cobalt nanocrystals: experiment and theory. *J. Phys. Chem. B* 109: 5541–5547.

13

Cracks in Nanocrystal Film

Crack patterns in various materials have been studied over several decades. The similarities between the geometries of the different patterns are observed over a wide range of scales [1–4] from millimeters to kilometers. This suggests a universal mechanism. Different morphologies of crack patterns such as isotropic, directional, radial, ring, spiral, and others can be observed depending on the drying process of a colloidal suspension [5–9]. Experiments [1, 9] show that the areas and distances between cracks increase with film height. However, different scaling laws were, surprisingly, observed for linear (1D) and isotropic (2D) cracks. A deviation from linear scaling of the average crack distance with height, h, in their directional drying experiments [1] was observed and was in good agreement with a theoretical model [10]. This predicted that the distance between cracks varied as $h^{2/3}$. This is different from the linear relationship between the square root of the area and the height obtained experimentally [9] that was confirmed by theoretic model for the isotropic crack patterns. To solve this puzzle, the scaling behavior of 1D and 2D crack patterns is here directly compared in the same maghemite nanoparticle films over 2 orders of magnitude. Moreover, a new method to align the cracks is employed by applying a magnetic field to the film during the drying process. To compare with the simulation results, 1D and 2D crack formations are theoretically studied using a recently developed model comparing them with those observed in the experiments.

13.1 Cracks of Nanocrystal Films

The γ-Fe_2O_3 nanocrystals are made by a slightly modified synthesis as reported elsewhere [11]. The produced nanocrystals are dispersed in chloroform to give an oily ferrofluid. A drop of this solution is deposited on a carbon grid. The TEM image shows a very large coverage of nanocrystals. A 2D hexagonal network with a large coherence length is observed and confirmed by the Fourier transform (inset) of the

Metal Nano 3D Superlattices: Synthesis, Properties, and Applications, First Edition.

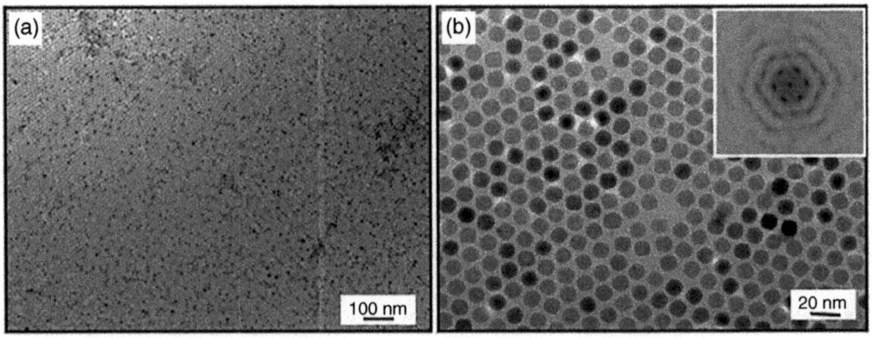

Figure 13.1 TEM image of 11-nm γ-Fe_2O_3 nanocrystals organized in a compact hexagonal network at low (a) and short (b) magnifications. Inset of (b): Fourier transform of (a) selected area. *Source:* Ngo et al. [12]/Royal Society of Chemistry.

image (Figure 13.1). The average nanocrystal diameter and size distribution are 11 nm and 5%, respectively. The electron diffraction spots in the measurements of nanocrystals are indexed to the maghemite structure (γ-Fe_2O_3).

The ferrofluid of the nanocrystal concentration from 6% to 12% by weight is deposited on silicon wafer placed at the bottom of a glass beaker. During solvent evaporation process, the solution is subjected or not to an applied magnetic field (0.2 T) parallel to the substrate. As the chloroform evaporates, the concentration of the solution increases with the formation of a gel that adheres to the silicon wafer. Further evaporation induces shrinkage of the thin-film layer due to large stresses, but the adhesion to the silicon substrate limits the contraction of the film. Isotropic and directional cracks are observed.

13.1.1 Isotropic Cracks

Figure 13.2a shows typical crack arrays formed in a nanocrystal film. A careful analysis of such patterns at higher magnifications clearly shows marked changes in the crack networks. From the periphery to the center of the sample, (i) the crack surface and the height of film progressively decrease. (ii) The cracks progressively lose their anisotropic features with elongated fragments (Figure 13.2b) to become isotropic (Figure 13.2c).

The fragment area and the film height are controlled by changing the percentage of γ-Fe_2O_3 nanocrystals dispersed in chloroform. The analysis of the fragment area, A, bounded by cracks is made by using a public domain image processing and analysis program (NIH Image) using about 100 and 200 fragments on average [13]. The heights, h, of the nanocrystal film are measured over the whole substrate on tilting by 45°. The layer height of the crack is controlled by solutions containing 6–12% by weight of nanoparticles.

Figure 13.3a,b shows that the increase of nanocrystal concentration by a factor of 2 (from 6% to 12% in nanocrystal weight) induces a decrease of the film height by a *factor 10* (from 23.6 ± 0.3 nm to 2.3 ± 0.1 μm). In these concentration limits, the typical isotropic 2D crack patterns remain unchanged.

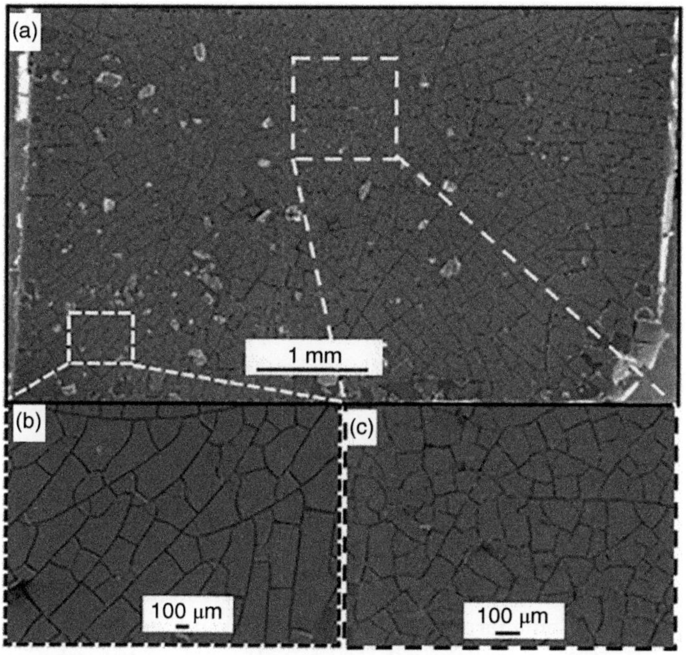

Figure 13.2 SEM images of typical cracks in a film of 3D superlattices made of 11-nm γ-Fe_2O_3 nanocrystals at zero angle tilt and long range (a) low magnification (b) at the periphery, and (c) at the center. *Source:* Ngo et al. [12]/Royal Society of Chemistry.

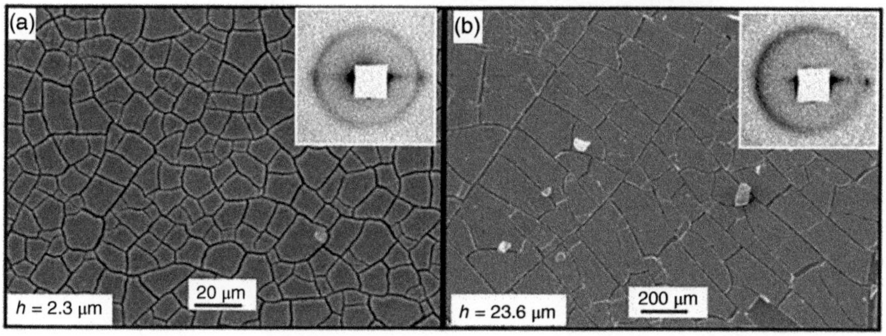

Figure 13.3 SEM (a, b) images of isotropic 2D crack patterns of 3D superlattices made of 11-nm maghemite nanocrystals at zero angle tilt and their corresponding SAXRD characterization inset. The average layer heights are 2.3 ± 0.1 μm (a) and 23.6 ± 0.3 μm (b), respectively. *Source:* Ngo et al. [12]/Royal Society of Chemistry.

At any height, the top surface of the fragments bounded by cracks is covered with isotropic patterns (Figure 13.4a). A recombination of the fragments by elimination of the cracks (Figure 13.4b) clearly shows that these patterns on the surface are continuous even when they cross a crack. This implies that the top surface patterns are formed before the cracking process. From this, it is concluded that these top surface patterns are directly related to the nanocrystal ordering.

Small-angle X-ray diffraction patterns obtained in grazing-angle geometry (Figure 13.3a) show two equatorial well-defined spots with a homogeneous thin diffraction ring. This indicates that part of the nanocrystals forms well-stacked layers

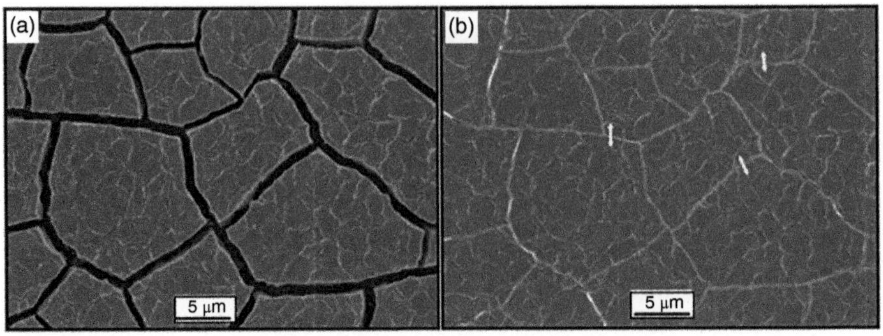

Figure 13.4 SEM image (a) of a magnification of a region in Figure 13.2a and reconstruction of the fragments bounded by cracks (b). *Source:* Ngo et al. [12]/Royal Society of Chemistry.

parallel to the substrate. The homogeneous ring is attributed to similar nanocrystal ordering with a random orientation. The equatorial reflection most probably corresponds to the stacking of such planes with a thickness of 100 nm. A further insight into the structure in Figure 13.5 shows 2D nanocrystal well ordered in hexagonal packing. Some terraces are clearly visible in Figure 13.5, and the underneath layer seems to be in registry with the upper layer. This shows the formation of 3D ordered nanocrystal assemblies, so-called 3D superlattices.

To estimate the average domain size of superlattices made of γ-Fe_2O_3 nanocrystals, the orientation of the nanocrystal assemblies is determined by Fourier transforms at several spots of the film. Let us consider fragments with isotropic surface patterns with the thinnest sample ($h = 2.3\,\mu m$). A top surface pattern magnification corresponding to the square in Figure 13.6a shows the 2D hexagonal packing of nanocrystals, and this is confirmed by the Fourier transforms (Figure 13.6b).

Here, two domains of 3D superlattices with different orientations are clearly seen by the Fourier transforms (areas delimited by various dashed lines). Some defects and

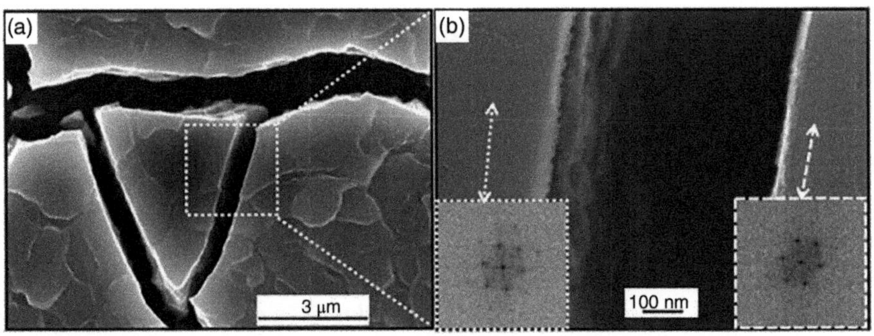

Figure 13.5 Field-emission gun scanning electron microscope (FEG-SEM) images of a selected region. (a) Low magnification. (b) Top surface of two fragments side by side bounded by a crack and their corresponding Fourier transforms. Inset: low magnification. *Source:* Ngo et al. [12]/Royal Society of Chemistry.

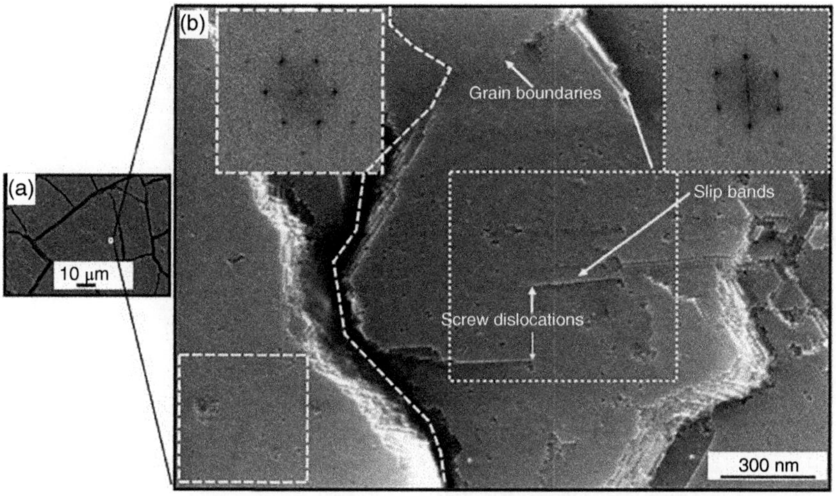

Figure 13.6 FEG-SEM image of the top surface corresponding to the square shown in inset (SEM image). Point defects, grain boundaries, slip bands, and dislocations are observed. Insets: Fourier transforms of areas delimited by solid and dashed line. Low (a) and high (b) magnifications. *Source:* Ngo et al. [12]/Royal Society of Chemistry.

dislocations are observed on the top surface, including point defects, screw dislocations, slip bands, and grain boundaries. The ferrofluid gel produced by chloroform evaporation and adhering to the silicon wafer is made of stacking layers of nanocrystals having some solvent molecules in their surroundings. Further evaporation induces stresses that lead to dislocations of superlattice blocks moving along slip planes. Such dislocations are usually not observed in amorphous films due to the absence of slip planes. At the final stage of evaporation, the thin-film layer starts to crack due to the large stresses caused by the adhesion to the silicon substrate. The crack patterns spanning the whole substrate appear. The cracks go straight from the top to the bottom of the film. From these, it is concluded that 3D superlattice films are formed before the appearance of cracks. To estimate the average domain size of superlattices made of γ-Fe_2O_3 nanocrystals, the orientation of the nanocrystal assemblies is determined by Fourier transforms at several spots of the film. For the film with $h = 2.3\,\mu m$, an average domain size of superlattices, keeping the same orientation, is about 1 µm. On increasing the height of the 3D superlattice film ($h = 23.6\,\mu m$), larger areas of ordered nanocrystals keeping the same orientation with diameters up to 3 µm are observed. The average distance between cracks depends on the film height. As examples for height of 2.3 and 23.6 µm, the average distance evolves from 11 and 135 µm, respectively. Hence, the diameter of the superlattices is less than 10% of the size of fragments bound by the cracks. Consequently, these fragments are composed of randomly oriented 3D superlattices as mentioned above to explain the thin diffraction ring obtained in the SAXRD patterns (inset Figure 13.3). From these data, it is concluded that the height of the film tunes the coherent length with a marked increase from less than 1 to 3 µm when the height of the film increases by a factor of 10 from 2.3 to 23.6 µm.

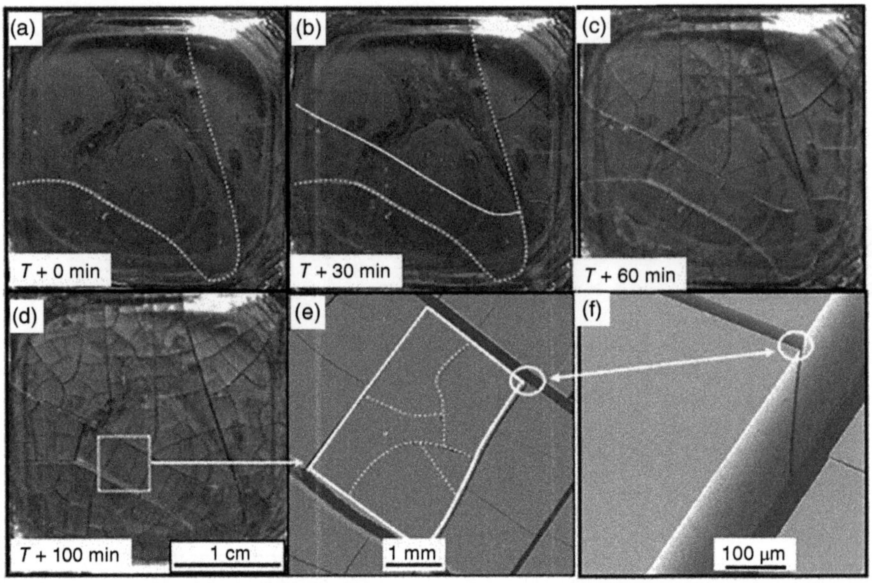

Figure 13.7 Photographs (a–d) and SEM (e, f) images of isotropic 2D crack patterns made of 10-nm γ-Fe_2O_3 nanocrystals produced without a magnetic field during the evaporation process with an average layer height of 415 μm. (a–d) Photographs of the formation of centimeter-size isotropic cracks. Formation of a first crack that splits the magnetic film into three regions (a). Later on, a second crack appears and meets the first orthogonally. New cracks form and meet the two first cracks usually with junction angles of 90° (b, c). Isotropic crack patterns obtained at the end of the drying process (d). (e, f) SEM image at zero angle (e) of a primary isotropic 2D crack (solid lines) with secondary 2D cracks (dashed lines) and by 45° tilt (f). *Source:* Ngo et al. [14, 15]/ American Chemical Society.

Let us follow the formation of crack patterns for the thickest sample, $h = 415\,\mu m$, by taking photographs at different times using a digital camera. The formation of isotropic 2D crack patterns obtained without a magnetic field is shown in Figure 13.7a–d. Figure 13.7a shows the formation of a first crack (solid line) that splits the magnetic film into two regions. After seven minutes, a second crack appears and meets the first orthogonally (dashed lines in Figure 13.7b). Then, new cracks form and meet the two first cracks usually with junction angles of 90° (Figure 13.7b,c). Finally, isotropic fragments bounded by cracks invade all the sample (Figure 13.7d). As already observed for other experimental systems [4, 9], the distribution of crack junction angles is peaked at 90°, and the fragments bounded by cracks are mainly four-sided with six neighbors [16]. A magnification of the region in Figure 13.7d shows that a typical primary 2D crack fragment delimited by solid lines (Figure 13.7e) is broken into smaller ones (dashed lines). Again, by tilting the sample, it is clearly shown that a typical crack goes from the top to the bottom of the film (Figure 13.7f). To analyze the isotropic 2D crack patterns, we use the same criterium as for directional cracks to separate the primary cracks from the secondary ones. Thus, all cracks with an aperture six times smaller than those of the two first cracks (Figure 13.7a,b) are called secondary 2D cracks observed.

13.1.2 Orientational Cracks

Directional cracks are obtained by subjecting the colloidal solution to an applied magnetic field (0.2T) during the evaporation process. This provokes large stresses that are at the origin of the crack formation. The crack patterns spanning the whole substrate appear after drying. The layer height of the crack patterns is controlled by solutions containing 0.35–28% in weight of nanocrystals. Samples with average layer heights, h, varying from 7 ± 0.2 to $107 \pm 3.5\,\mu m$ are obtained using a silicon wafer.

To analyze the experimental data quantitatively, we need to define primary 1D and 2D cracks and secondary 2D cracks. The straight cracks parallel to the direction of the applied field correspond to the first generation of cracks and are called primary 1D cracks. The fragments bounded by these primary 1D cracks and the perpendicular ones (second generation of cracks) are called primary 2D cracks.

Let us consider the variation of crack patterns with height that is produced at various nanocrystal concentrations. The crack distances between regular linear cracks, D, are measured over the whole substrate and their heights, h, on tilting the sample by 45°. The size and height of the crack patterns depend on the initial concentration of nanocrystals deposited with the formation of directional cracks at the micrometer (Figure 13.8a) to millimeter (Figure 13.8b) and centimeter scales (Figure 13.8c). Note that samples having 7 μm (Figure 13.8a) and 14 μm as average layer heights are characterized by only parallel cracks (primary 1D cracks), whereas, at 24 μm and above, perpendicular and secondary cracks are observed within the bands formed by the primary linear cracks. On tilting the samples by 45° tilt, Figure 13.8b–f shows typical cracks that go from the top to the bottom of the film and increase with the amount of nanocrystals dispersed in solution.

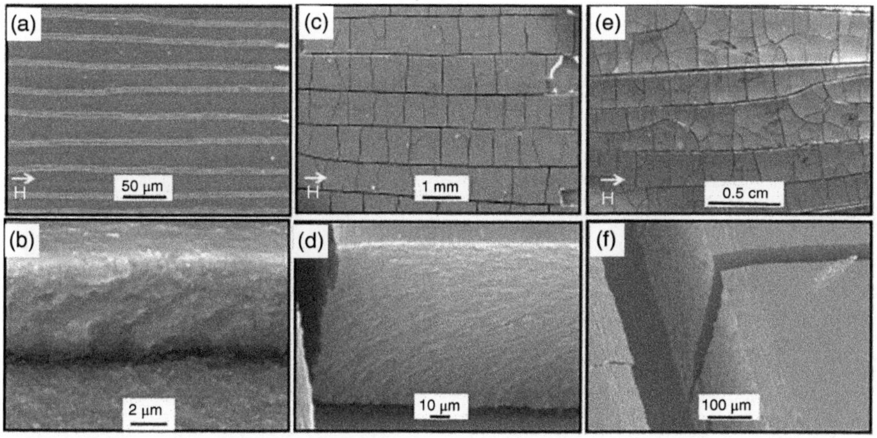

Figure 13.8 Micrometer-size directional crack patterns obtained at the end of the drying process: SEM patterns of directional (1D–2D) crack patterns made of 11-nm γ-Fe_2O_3 nanocrystals produced by applying a magnetic field (0.2T) during the evaporation process. Part figures (b, e, and f) are the corresponding images of (a, c, and e) tilted by 45°. The average layer heights are $7.0 \pm 0.2\,\mu m$ (a, b), $107 \pm 4\,\mu m$ (c, d), and $415 \pm 5\,\mu m$, respectively. *Source:* Ngo et al. [14, 15]/American Chemical Society.

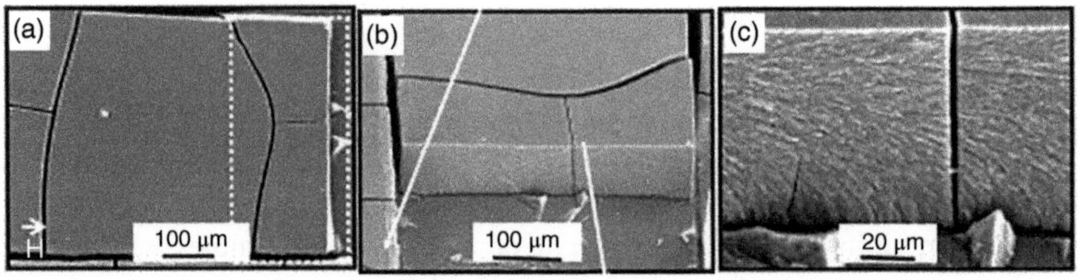

Figure 13.9 SEM images at zero angle (a) and by 45° tilt (b, c) of directional (1D–2D) crack patterns made of 10-nm γ-Fe_2O_3 nanocrystals produced by applying a magnetic field (0.2 T) during the evaporation process. The average height is 107 ± 4 μm. *Source:* Ngo et al. [14, 15]/American Chemical Society.

A careful observation of the patterns shows that all primary and secondary cracks go straight from the top to the bottom of the film when deposition takes place under chloroform vapor pressure. Note that at the bottom of the cracks, layers of nanocrystals can be observed. Figure 13.9 shows how a typical secondary 2D crack goes all the way through the film.

Let us consider an elongated fragment of 3D superlattices (Figure 13.10a) obtained at the border of the thickest sample (h = 23.6 μm). The top surface patterns are mainly in the direction of the long axis of the fragment. As observed for isotropic cracks, the patterns on the surface, shown on Figure 13.11a, are continuous even when crossing a crack (see double arrows in Figure 13.11b). This confirms that, even at the edges of the films, they are formed before the cracking process. Here, the elongated fragments are very large, and the domain size of hexagonal packing of nanocrystals keeping the same orientation covers the entire FEG-SEM image with a display size of 1.9 × 2.5 mm^2 used to have a resolution of the nanoparticles. Therefore, to estimate the 3D superlattice size, several FEG-SEM images are taken along the short and long axes of the fragment (Figure 13.10a), and the orientations of the nanocrystal organization are again deduced from their Fourier transform patterns. Typical images of top surfaces along the short and long axes of the fragment are

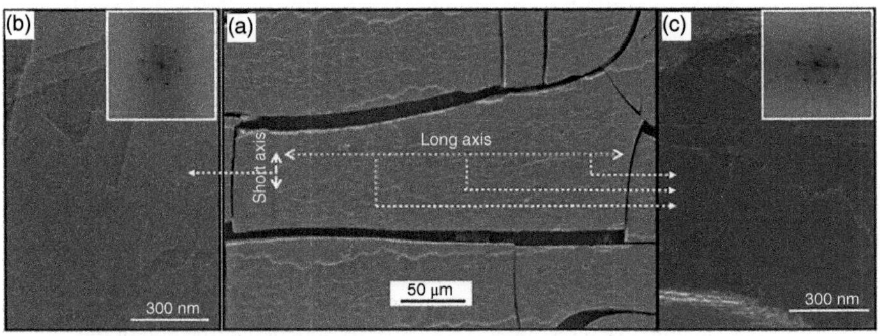

Figure 13.10 (a) SEM image of elongated crack patterns of 3D superlattices made of 11-nm γ-Fe_2O_3 nanocrystals at zero angle tilt and (b, c) FEG-SEM images of the top surface composed of nanocrystals in 2D hexagonal packing. Insets: Fourier transforms of (b) and (c). *Source:* Ngo et al. [12]/Royal Society of Chemistry.

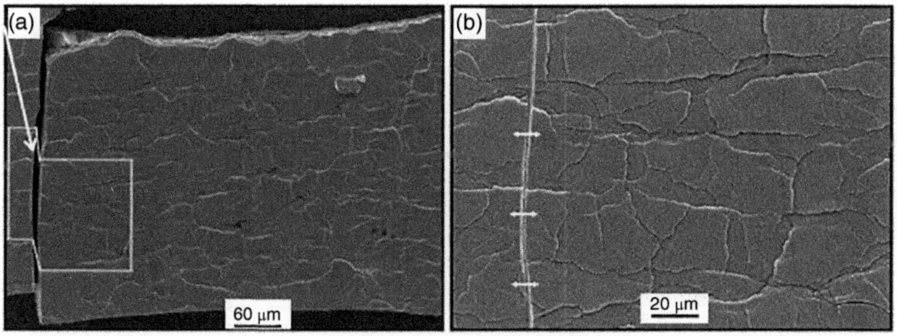

Figure 13.11 (a) SEM image of 3D superlattices elongated fragment (b) the corresponding reconstruction of two fragments bounded by a crack. This corresponds to the area delimited by dashed line shown in inset (magnification of a region in Figure 13.10a). *Source:* Ngo et al. [12]/Royal Society of Chemistry.

shown in Figure 13.10b,c, respectively. From these, it is concluded that the coherent lengths along the long and short axes of the fragment are around 250 and 12 μm, respectively. This shows that along the long axis, the size of 3D superlattices may be close to the separation of cracks. As for isotropic crack patterns (Figure 13.5), with elongated fragments, it is observed that the hexagonal packings of nanocrystals on both sides of a crack usually have the same orientation. This clearly shows that for anisotropic crack fragments observed here, the size of the 3D superlattices may be limited by the formation of cracks.

We first follow the formation of cracks for the thickest sample, h =415 μm, by taking photographs at different times using a digital camera (Figure 13.12a–d). Figure 13.12a shows the formation of the first crack in the direction of the applied field, at the middle of the substrate where the stress in the fragment is the largest [17]. Successive domain divisions occur mainly parallel to the applied field leading to equally spaced linear cracks (Figure 13.12b,c). One hour after the formation of the first crack, cracks perpendicular to the directional ones appear and invade all the sample as the evaporation continues (Figures 13.12d). The formation of directional cracks produced by applying a magnetic field parallel to the substrate is explained as follows. In the liquid state, the nanocrystals are free to rotate, and their magnetic moments are aligned along the applied magnetic field direction [18]. This favors the attractions between nanocrystals in the direction parallel to the applied field, whereas repulsions take place in the perpendicular direction, leading to an anisotropic stress within the film.

A magnification of a region in Figure 13.12e by using SEM shows that a typical primary 2D crack fragment delimited by full lines (Figure 13.12f) is four-sided with 90° angles. The mean crack aperture of parallel cracks (first generation of primary 1D cracks) is about twice as large as that of the perpendicular ones (second generation of cracks) and is clearly seen in Figure 13.12f. At the end of the drying process, the crack patterns (Figure 13.12e) show that the primary 2D crack fragments are broken into smaller ones (from two to eight domains) corresponding to the third and later generations of cracks, as indicated by the dashed lines in Figure 13.12f.

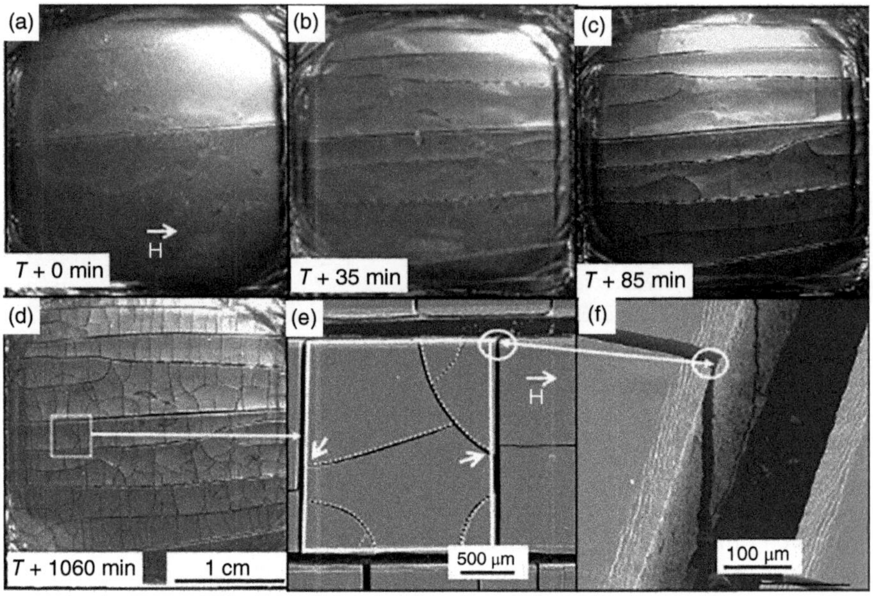

Figure 13.12 Photographs (a–d) and SEM (e, f) images of directional (1D–2D) crack patterns made of 10-nm γ-Fe_2O_3 nanocrystals produced by applying a magnetic field (0.2 T) during the evaporation process with an average layer height of 415 μm. (a–d) Photographs of the formation of centimeter-size directional cracks. Formation of the first crack in the direction of the applied field, at the middle of the substrate (a). Successive domain divisions occur mainly parallel to the applied field, leading to equally spaced linear cracks (b), (c) obtained after 15 and 65 minutes of the formation of the first crack, respectively. Formation of perpendicular cracks (d) and final crack pattern obtained after 105 and 1060 minutes of the formation of the first crack, respectively. (e, f) SEM images of primary 2D crack (solid lines) with secondary 2D cracks (dashed lines). *Source:* Ngo et al. [14, 15]/American Chemical Society.

These new generations of cracks are called secondary 2D cracks. Note that the new generation of cracks tends to appear at the middle of the sides of existing fragments bounded by primary cracks (see arrows in Figure 13.12f), which is consistent with the theoretical model and the simulations [17].

The formation of directional cracks produced by applying a magnetic field parallel to the substrate is explained. In the liquid state, the nanocrystals are free to rotate, and their magnetic moments are aligned along the applied magnetic field direction [18]. This favors the attractions between nanocrystals in the direction parallel to the applied field, whereas repulsions take place in the perpendicular direction, which leads to an anisotropic stress within the film. Therefore, the stress σ_{yy} perpendicular to the applied field is larger than that (σ_{xx}) parallel to the field according to the following equation [19]:

$$\sigma_{yy} - \sigma_{xx} = \mu_0\left(H^2 + M \cdot H\right) > 0$$

where μ_0, M, and H are the magnetic permeability of vacuum, the magnetization of the film and the field, respectively. The higher stress σ_{yy} is relieved first, leading to the parallel cracks shown in Figure 13.12a,b. Later on, the stress parallel to the direction of the field will also be relieved, inducing the formation of perpendicular cracks (Figure 13.12c).

13.1.3 Universal Feature

We now focus our study on the scaling behavior of 1D and 2D primary crack patterns shown in Figures 13.7 and 13.12. The properties of cracks produced by water or chloroform evaporation processes remain unchanged within the experimental errors. The area of the fragments bounded by cracks (A) and the crack distances between regular linear cracks (D) are measured over the whole substrate and their heights (h) on tilting the sample by 45°. The area of directional 2D cracks and isotropic 2D cracks are called A_{lp} and A_{ip} respectively. The area analysis of the individual fragments of primary 2D cracks is made by using a public domain image processing and analysis program (NIH Image) [16] over the whole sample on 100 and 600 fragments in average, respectively.

Let us consider isotropic 2D primary crack patterns. A linear variation, on a logarithmic scale, of the average square root area of crack domains ($\sqrt{A_{ip}}$) with the layer height is observed (full circles in Figure 13.13). Using the average experimental value of the dimensionless ratio $K = \sqrt{A_{ip}}/h$ ($K = 5.42 \pm 0.32$), the linear fit gives a slope of 1.00 in good agreement with a universal scaling law, $A = (KH)^2$. The scaling law for the isotropic crack patterns is shown over 3 orders of magnitude. With orientational crack patterns, a linear variation, on a logarithmic scale, (h) is also observed (squares in Figure 13.13). Using the average experimental value of the dimensionless ratio $K = D/h$ ($K = 5.31 \pm 0.25$), the linear fit gives a slope of 1.00 in good agreement with a universal scaling law, $D = Kh$. From this, within the experimental errors, similar dimensionless K values are obtained for 1D and 2D primary cracks.

The scaling of directional and isotropic crack patterns with height is now theoretically studied using a recently introduced bundle-spring-block model [6]. Simple spring-block models have been widely used to successfully explain experimentally

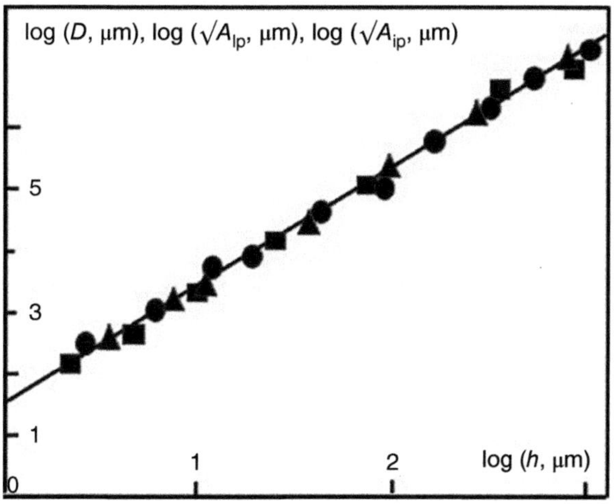

Figure 13.13 Dependence of the average crack distance (D) and the average square root areas of fragments bounded by primary cracks ($\sqrt{A_{lp}}, \sqrt{A_{ip}}$) on the layer height. The average crack distance, D, versus the layer height, h, designated by open squares, fits a linear curve on the logarithmic scale with $\log D\ 0.72 + 1.00 \log h$ (solid lines). Solid triangles and circles correspond to the average square root areas of directional and isotropic primary 2D cracks, $\sqrt{A_{lp}}$ and $\sqrt{A_{ip}}$, respectively.

observed crack patterns [6, 20–22]. Within this approach, the layer is represented by a large two-dimensional hexagonal or square array of blocks where each neighbor is connected by coil springs. To study the variation of crack patterns with height, the single springs between blocks were replaced by a bundle of h springs with the bundle-spring model [22]. h is a dimensionless number that plays the role of the height.

With isotropic crack patterns, the bundle-spring model predicts a linear increase in the square root of the average fragment area bounded by cracks with height, in good agreement with our experiments (full triangles in Figure 13.13). Here, the bundle-spring model is further developed to answer two questions posed by the experiments:

(i) Does the isotropic and directional crack patterns obtained by theory follow the same scaling law as observed in the experiments?
(ii) Does the simulations reproduce the anisotropic crack patterns observed by applying a magnetic field?

To answer the first question, at the start of each simulation, an initial stress is generated by choosing larger side lengths L_x and L_y than the values L_x^0 and L_y^0 of the relaxed block array. The initial extension of the arrays is defined by the formula: $s_x = (L_x - L_x^0)/L_x$ and $s_y = (L_y - L_y^0)/L_y$. For $s_x = s_y = 0.1$, isotropic 2D crack patterns are obtained in agreement with Ref. [23]. The crack pattern obtained for a square array in Figure 13.14a shows that the propagation of cracks is markedly influenced by the arrangements of blocks. Figure 13.14b shows the final crack pattern obtained using a square array. The linear cracks are often not continuous, which is explained by an inefficient propagation of the stress built up at the tips of cracks. Isotropic and directional crack patterns were obtained using hexagonal or square arrays in a range of heights between 5 and 50. To study the scaling laws in the isotropic and directional cases, the average area A of the surface limited by the cracks and the average distance D between the cracks, respectively, are determined. Figure 13.15 shows a log–log plot of D and $\sqrt{A}$ as a function of h for the directional and isotropic cracks, respectively.

For the directional cracks produced by a unilateral stress, the patterns are analyzed by the average distance D between the cracks. Figure 13.15 shows a log–log

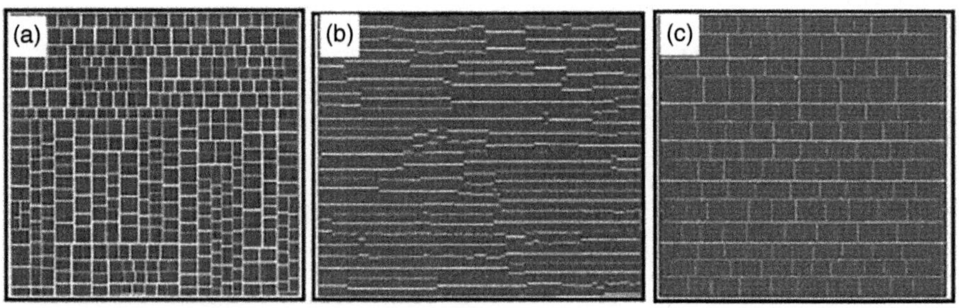

Figure 13.14 Simulation of isotropic (2D) and linear (1D) cracks: (a, b) crack patterns obtained by simulations for isotropic and uniaxial stress using a square arrangement of blocks (h 50). (c) Crack patterns obtained by simulations for biaxial stress (s_x 0.50, s_y 0.25) using a square arrangement of blocks (h 50). *Source:* Ngo et al. [14, 15]/American Chemical Society.

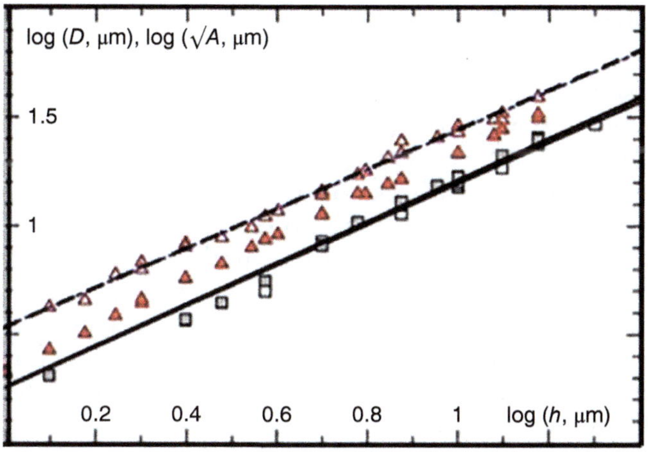

Figure 13.15 Scaling of mean separation D and square root area $\sqrt{A}$ of cracks with height. A linear increase $D \sim h$ well describes the data within the statistical accuracy of simulation. Squares and triangles give square and hexagonal arrays with $\log D$ $0.25+0.94 \log h$ (solid lines) and $\log D$ $0.53+0.91 \log h$ (dashed lines), respectively. The solid symbols show the square roots of the mean area A for isotropic extension (square [solid squares] and hexagonal [solid triangles] arrays).

plot of D and $\sqrt{A}$ as a function of H. All data are well fitted by straight lines with slope 1. For the square array, D and $\sqrt{A}$ are identical within the simulation errors at a given height as observed in the experiments. For the square array, D (open squares) and $\sqrt{A}$ (solid squares) are identical within the simulation errors at a given height. This is in very good agreement with the experimental behavior observed in Figure 13.13. However, for the hexagonal array, $\sqrt{A}$ (filled triangles in Figure 13.15) is significantly smaller than D (open triangles). This difference is explained by the large number of triangles with a smaller surface area than the four-sided surfaces present in the crack patterns for hexagonal arrays. Obviously, simulations using the square array better describe the experimental data, in contrast to those obtained using a hexagonal array. Let us now turn to the second question, whether the simulation can reproduce the anisotropic crack patterns observed by applying a magnetic field in Figure 13.12. To answer this, the dipolar interactions in the film are taken into account by applying a biaxial stress at the beginning of the simulation. A study of the experimental crack patterns obtained at large heights (Figure 13.12d) yields the following range of values for the pre-strain: $s_x = 0.50–0.60$ and $s_y = 0.25–0.26$. The simulations are carried out using a square arrangement of blocks. Figure 13.14 shows the crack patterns obtained for $s_x = 0.50$ and $s_y = 0.25$ within the range of the experimental values (after two and six million simulation steps, respectively). The crack patterns (Figure 13.14c) obtained by simulations using the square array agree well with the structures experimentally observed in Figure 13.12d neglecting the presence of the secondary cracks. The secondary cracks may be explained by the hypothesis that the drying of the film continues after the formation of the primary cracks.

13.2 Cracks in Nature and Current Life

The magnificent Nature surprises all of us. There is no better artist than Mother Nature. The perfection of its creations is true works of art that sometimes makes your jaw drop. Here, we feature a collection of crack patterns found in different parts of Nature such as landscape, trees, and rocks that only it can produce.

Figure 13.16a,b shows cracks forming when wet earth dries up and shrinks. The cracks form over a range of scales and share the same characteristic shapes. These cracks form perfect sites for seeds to sprout. These well-defined isotropic cracks are observed in very high scale as observed in Northern Mexico with the appearance of massive half-mile-long crack (Figure 13.16c,d). From Figure 13.16d, the reader can imagine the size of the cracks taking into account the size of car and people in the photograph. These cracks are mainly due to drying processes taking place during the water evaporation. These are mainly due to change in the ground temperature. Similar isotropic cracks are observed with rocks (Figure 13.17a,b). In such cases, these cracks are due to fast decrease in magma temperature. Figure 13.17a shows subhorizontal fracture segment with a distinct disk shape and convex upward and downward. Figure 13.17b shows that the cracks are characterized by a semi-regular pattern. Very distinctive landscape is due to the erosion process.

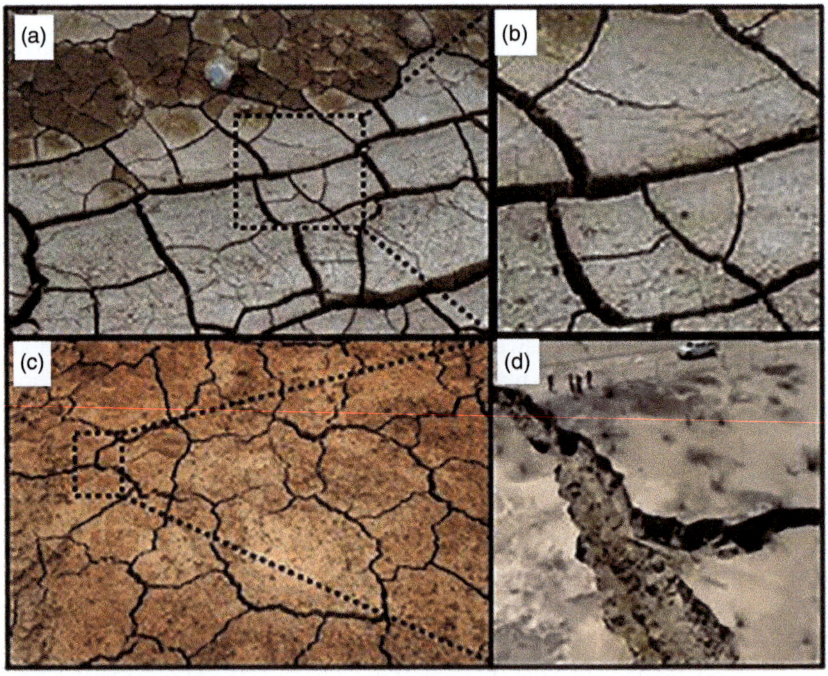

Figure 13.16 Isotropic cracks form when wet earth dries up and shrinks: (a, b) a landscape image. (c, d) Massive half-mile-long crack appears in the ground in Northern Mexico.

Figure 13.17 (a) and (c) Jointed basalts (volcanic flows) from Giant's Causeway (Ireland). (b) Granite cracks from the south and east of Laramie, Wyoming, USA.

Isotropic and orientational cracks are observed in homemade arts. Here it is shown, as examples, isotropic pottery cracks resulting from stress within the ceramic body due to thermal expansion and contraction and shrinkage (Figure 13.18a,b) and orientational painting cracks on Mona Lisa painting (Figure 13.18c,d).

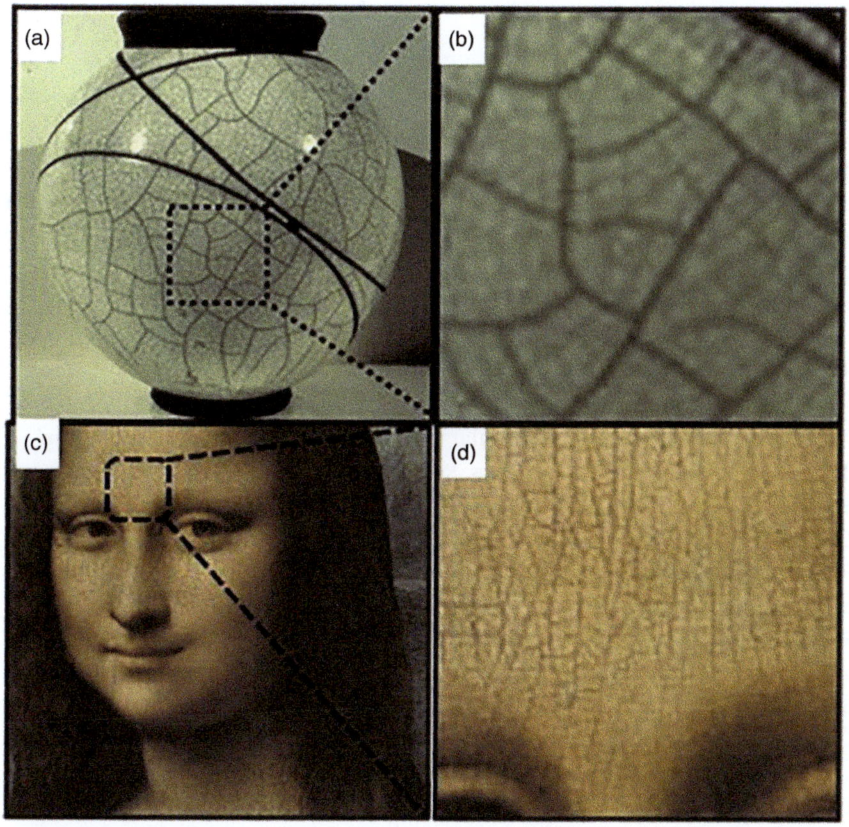

Figure 13.18 Isotropic (a, b) and orientated (c, d) cracks on ceramics and on Mona Lisa painting.

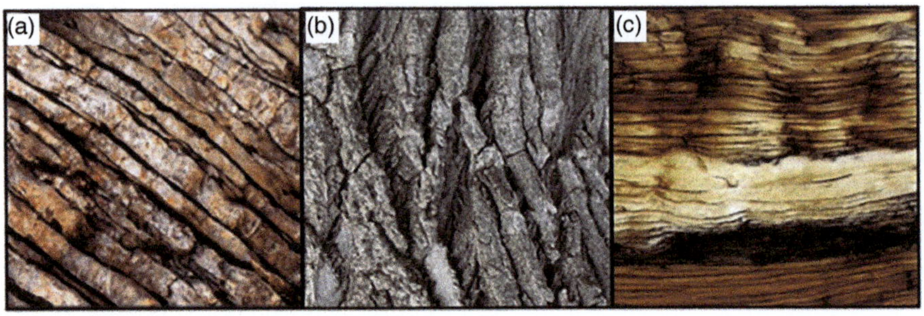

Figure 13.19 Orientated cracks observed (a) in the rock on cliff face (b). Tree bark from an old cottonwood (c) or dried scales cover the trunk of this foxtail pine in California's Sierra Nevada Mountains.

Orientation cracks are observed in a large variety of natural materials and men's work. Such orientation is highly diversified. Hence as shown in Figure 13.19, cliff face and tree bark or seasoned scales cover the trunk of this foxtail pine; we observe orientation cracks. Similarly, basaltic volcanic flows induce column formation with perfectly hexagonal features as shown in Figure 13.17c.

13.3 Conclusions

Cracks are due to a shrinking material restricted by adhesion. They are observed in several surfaces we see every day, for example, in Nature, painting, glass, and films. Here, we used films of nanocrystals to demonstrate that cracks follow a universal law with a clear relation between the average surface and the film height. This has to be extended to what is observed in Nature.

References

1 Allain, C. and Limat, L. (1995). Regular patterns of cracks formed by directional drying of a colloidal suspension. *Phys. Rev. Lett.* 74: 2981–2984.

2 Belousov, V.V. and Gzovsky, M.V. (1965). Experimental tectonics. *Phys. Chem. Earth.* 6: 409–472.

3 Pileni, M.P., Lalatonne, Y., Ingert, D. et al. (2004). Self assemblies of nanocrystals: preparation, collective properties and uses. *Faraday Discuss.* 125: 251–264.

4 Shorlin, K.A., de Bruyn, J.R., Graham, M., and Morris, S.W. (2000). Development and geometry of isotropic and directional shrinkage-crack patterns. *Phys. Rev. E* 61: 6950–6957.

5 Colina, H. and Roux, S. (2000). Experimental model of cracking induced by drying shrinkage. *Eur. Phys. J. E* 1: 189–194.

6 Leung, K.T., Jozsa, L., Ravasz, M., and Neda, Z. (2001). Spiral cracks without twisting. *Nature* 410: 166.

7 Xia, Z.C. and Hutchinson, J.W. (2000). Crack patterns in thin films. *J. Mech. Phys. Solids* 48: 1107–1131.

8 Nakahara, A. and Matsuo, Y. (2006). Transition in the pattern of cracks resulting from memory effects in paste. *Phys. Rev. E* 74: 045102-1–045102-4.

9 Groisman, A. and Kaplan, E. (1994). An experimental study of cracking induced by desiccation. *Europhys. Lett.* 25: 415–420.

10 Komatsu, T.S. and Sasa, S. (1997). Pattern selection of cracks in directionally drying fracture. *J. Appl. Phys.* 36: 391–395.

11 Park, J., An, K., Hwang, Y. et al. (2004). Ultra-large-scale syntheses of monodisperse nanocrystals. *Nat. Mater.* 3: 891–895.

12 Ngo, A.T., Richardi, J., and Pileni, M.P. (2013). Crack patterns in superlattices made of maghemite nanocrystals. *Phys. Chem. Chem. Phys.* 15: 10666–10672.

13 NIH *Image* is a public domain image processing and analysis program for the Macintosh. http://rsb.info.nih.gov/nih-image/about.html.

14 Ngo, A.T., Richardi, J., and Pileni, M.P. (2008). Cracks in magnetic nanocrystal films: do directional and isotropic crack patterns follow the same scaling law? *Nanoletters* 8: 2485–2489.

15 Ngo, A.T., Richardi, J., and Pileni, M.P. (2008). Do directional primary and secondary crack patterns in thin films of maghemite nanocrystals follow a universal scaling law? *J. Phys. Chem. B* 112: 14409–14414.

16 Bohn, S., Douady, S., and Courder, Y. (2005). Four sided domains in hierarchical space dividing patterns. *Phys. Rev. Lett.* 94: 054503-1–054503-4.

17 Hornig, T., Sokolov, I., and Blumen, A. (1996). Patterns and scaling in surface fragmentation processes. *Phys. Rev. E* 54: 4293–4298.

18 Hendriksen, P.V., Bodker, F., Linderoth, S. et al. (1994). Ultrafine maghemite particles. I. Studies of induced magnetic texture. *J. Phys. Condens. Matter* 6: 3081–3090.

19 Rosensweig, R.E. (1997). *Ferrohydrodynamics*, 103–108. Mineola, NY: Dover Publications Inc.

20 Skjeltorp, A.T. and Meakin, P. (1998). Fracture in microsphere monolayers studied by experiment and computer simulation. *Nature* 335: 424–426.

21 Walmann, T., Malthe-Sorensen, A., Feder, J. et al. (1996). Scaling relations for the lengths and widths of fractures. *Phys. Rev. Lett.* 77: 5393–5396.

22 Leung, K.T. and Néda, Z. (2000). Pattern formation and selection in quasistatic fracture. *Phys. Rev. Lett.* 85: 662–665.

23 Richardi, J., Ngo, A.T., and Pileni, M.P. (2010). Simulations of cracks supported by experiments: the influence of the film height and isotropy on the geometry of crack patterns. *J. Phys. Chem. C* 114: 17324–17332.

14

Water-Dispersive Hydrophobic Suprastructures: Specific Properties

Energy is a vital element for our modern society. This explains the tremendous efforts, both in experimental and theoretical researches, which have been made over these last three decades in the area of energy. The topic of thermal energy storage and release has recently attracted increasing interest due to the important potential applications not only in solar energy production and waste heat recycling but also in water heating, cooling, or air-conditioning and in space research and catalysis. Self-assemblies of inorganic nanocrystals in 3D crystalline structures called supercrystals/3D superlattices, as well as nanostructured materials, have attracted great interest in recent years because of the unusual mechanical, electrical, magnetic, and optical properties endowed by confining the dimensions of such materials. The properties of these materials are neither those of the bulk phase nor those of the nanomaterials used as building blocks but are dictated by their ordering in 3D superlattices [1–4]. Several multidisciplinary researches also focus on innovative solutions involving hybrid and inorganic nanomaterials. These approaches need a strong and broad range of competences in synthesis, characterization, and modeling of organic and inorganic materials. The unique chemical and physical properties of these hybrid structures are needed to reach relevant applications. Here, we develop new concepts to disperse hydrophobic 3D superlattices in aqueous solution. The optical scattered spectra at larger wavelength are very broad and redshifted linearly with the wavelength as the assembly size increases. This metamaterial property is added to the fact that 3D superlattices keep the fingerprint of isolated nanocrystals. These water-dispersive hybrid structures, characterized by very low stiffnesses and Young's moduli, subjected to light can operate as very efficient universal nanoheaters [5–8].

Metal Nano 3D Superlattices: Synthesis, Properties, and Applications, First Edition.

14.1 Au and Co "Clustered" Structures

14.1.1 Fabrication and Characterization

To produce Co and Au "clustered" structures [6, 7], we need to build up an "organic molecular parachute" to disperse hydrophobic shaped 3D superlattices in aqueous solution. For this purpose, we use two steps:

(i) *Production of shaped 3D superlattices*: shaped Co or Au 3D superlattices are obtained by controlling their growth processes [4, 9, 10]. The affinity between the solvent and the coating agent could control the final morphology of nanocrystal assemblies to obtain either layer-by-layer films (heterogeneous growth) or well-defined shaped 3D superlattices (homogeneous growth). Subtle changes in the Hildebrand solubility parameter strongly impact the interaction potential and are sufficient to switch from an attractive to a repulsive situation, i.e. from a homogeneous to a heterogeneous growth. The interaction potentials of Co or Au nanocrystals coated with either oleic acid or octadecanethiol are appealing when nanocrystals are dispersed in toluene and repulsive in hexane. The addition of traces of non-solvent, either ethanol or isopropanol, induces a marked increase in the attractive interactions between nanocrystals, consequently favoring a homogeneous growth process. The scanning electron microscopy (SEM) images of the collected precipitates (Figures 14.1a and 14.2a) show the formation of well-defined aggregates with various shapes and a rather large size distribution. Most of these shaped aggregates are characterized by a flat surface with well-defined facets in hexagonal symmetry while others present some defects. The small-angle X-ray scattering (SAXS) pattern-related planes of 3D superlattices and profile (Figures 14.1d, 1 4.2d, 14.1g, 14.2g) show the formation of face-centered cubic (fcc) Co (Figure 14.1) or Au (Figure 14.2) shaped 3D superlattices.

(ii) *Molecular parachute of 3D superlattices through bridges between vesicles*: to disperse such hydrophobic 3D superlattices in aqueous solution, we need to control the interdigitation processes and to produce multiwalled structures. The coating agent of nanocrystals used as building blocks must interact strongly with the biological molecules. For this purpose, we use a concept developed several years ago, to build up thermodynamically stable supra-aggregates [11]. The major parameters are the surfactant volume fraction, the volume of water, and the number of surfactant molecules. The surfactant parameter, σ, is given by $\sigma = v_S/a_S\, l_s$ where v_S, a_S, and l_s are the surfactant volume fraction, the surface per head polar group, and the surfactant tail length, respectively. Finally, the packing of multiwall structures depends on the radius of the vesicular water core, the thickness of the bilayer, and the volume of water in the core. According to the packing parameter, the best interdigitation process is obtained with long-tail surfactant molecules and large head group surface area [12]. Furthermore, we know that the energy gain due to the addition of a $-CH_2-$ group into the alkyl chain is 8×10^{-21} J. The curvature of nanocrystals creates

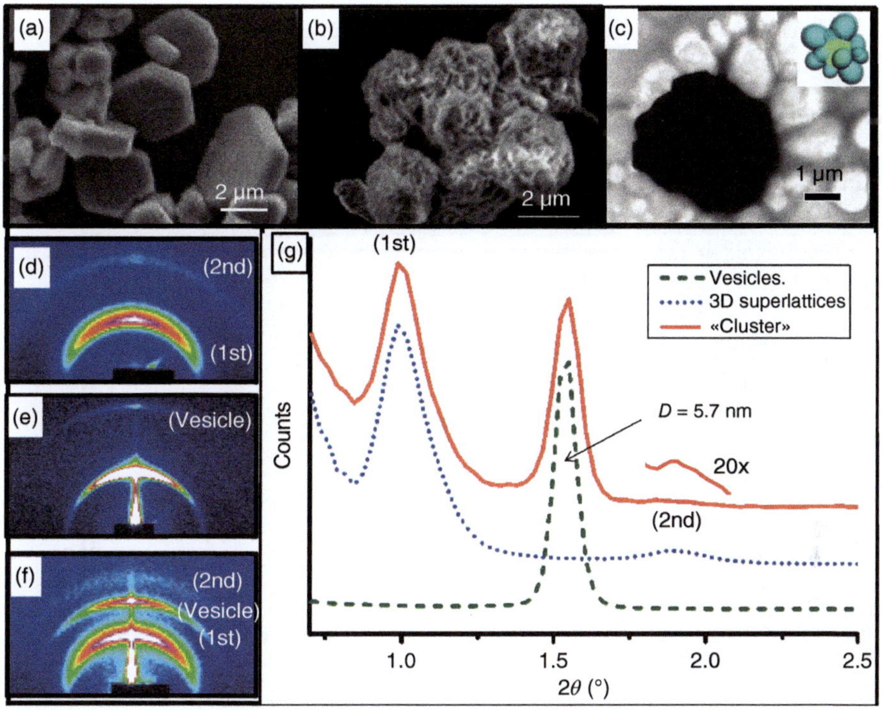

Figure 14.1 (a) and (b) are the SEM images of Co shaped nanocrystals before (a) and after (b) DPPC treatments. (c) TEM image corresponding to image of (b) showing that the vesicles are closely attached to the 3D superlattices forming cap-like structures. (d) SAXS characterization of the 3D superlattices, (e) the vesicles, (f) the clustered structure, and (g) the profile of the different SAXS patterns. The two red peaks in (f) can be attributed to the diffraction of the 3D superlattices and vesicles, respectively. *Source:* (a–c) Yang et al. [5, 6]/American Chemical Society. (d–f) Marie Paule Pileni (2020)/Royal Society of Chemistry.

voids between two consecutive coating agent molecules at the nanocrystal surface. Because voids are energetically unfavorable, the dipalmitoylphosphatidylcholine (DPPC) alkyl chains and the C_{18} chains of PEG-2000-DOPE (1,2-dioleoyl-sn-glycero-3-phosphoethanolamine-*N*-[azido(polyethylene glycol)-2000] ammonium salt) fill them up. This induces strong interactions between DPPC (dipalmitoyphosphatidylcholine), PEG-2000-DOPE, and nanocrystals coated with C_{18} $H_{37}X$, where X is either carboxylic group or thiol for Co and Au nanocrystals, respectively. The PEG2000-DOPE hydrophilic chain (C_{91}) makes bridges from one vesicle to another and consequently plays the role of "parachutes" preventing gravity processes. These assumptions permit to design the following mechanism (Figure 14.3). The nanocrystals dispersed in solution and subjected to a non-solvent induced the nanocrystal aggregate formation (nucleation process). By cooling down the samples, the nanocrystals coated with C_{18} alkyl chains self-assembled (growth process) produced shaped 3D superlattices. DPPC (C_{16}) and PEG2000-DOPE (C_{18}–C_{91})

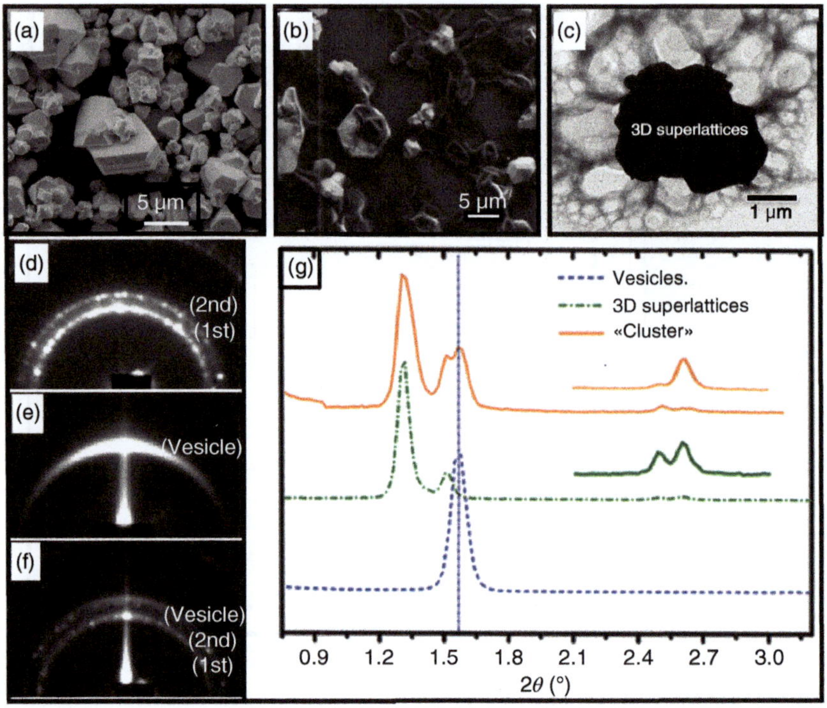

Figure 14.2 (a) and (b) are the SEM images of Au shaped nanocrystals before (a) and after (b) DPPC treatments. (c) TEM image corresponding to image of (b) showing that the vesicles are closely attached to the 3D superlattices forming cap-like structures. (d) SAXS characterization of the 3D superlattices, (e) the vesicles, (f) the clustered structure, and (g) profile of the different SAXS patterns. The two red peaks in (f) can be attributed to the diffraction of the 3D superlattices and vesicles, respectively. *Source:* (a, b, d–f) Yang et al. [7]/American Chemical Society. (c) Marie Paule Pileni (2020)/Royal Society of Chemistry.

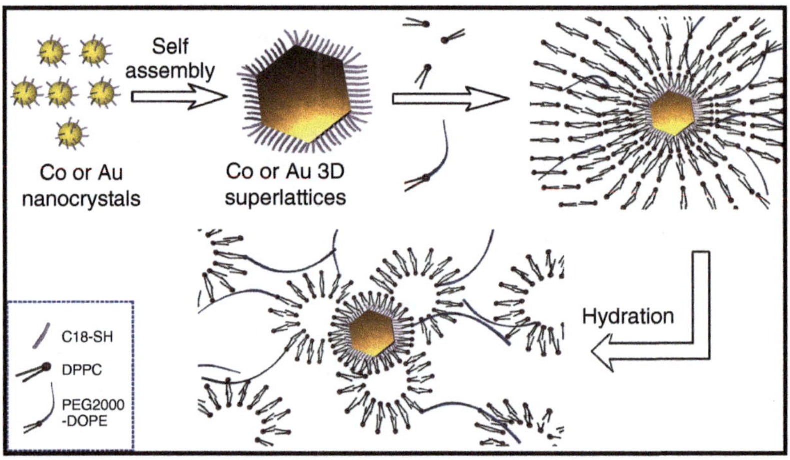

Figure 14.3 Expected mechanism to disperse hydrophobic 3D superlattices in aqueous solution with the assistance of DPPC/PEG2000-DOPE vesicles and PEG2000-DOPE as bridges between vesicles.

both characterized by similar long alkyl chains (C_{16} and C_{18}) interdigitated with Au_{C18} nanocrystals. By adding water, the 3D superlattices were dispersed in solution.

The SEM (Figures 14.1b and 14.2b) and the corresponding transmission electron microscopy (TEM) (Figures 14.1c and 14.2c) images reveal the structure of individual composites, with the presence of large vesicles attached to the 3D superlattice surface and drape-like structures linked to it. These drape-like structures are due to the drying process because during TEM sample preparation, water is released from vesicles and collapsed shells are left.

The SAXS pattern of structural studies acquired on shaped 3D superlattices after DPPC/PEG-DOPE treatment shows well-defined rings (Figures 14.1f and 14.2f). The new ring appearing in the diffraction pattern is attributed to the presence of vesicles. To support such statement, a blank experiment is performed, in the absence of 3D superlattices, with DPPC/PEG-DOPE molecules. After hydration, the solution is dried, and the SAXS pattern (Figures 14.1e and 14.2e) shows a strong signal due to the diffraction of periodic vesicle bilayers. The calculated bilayer distance is 5.7 nm, which agrees with the thickness of the liposome shell [13]. The profile (Figures 14.1g and 14.2g) clearly shows that the diffraction peaks of the composite are a combination of the peaks of isolated Co or Au 3D superlattices and of vesicles indicating that the DPPC/PEG2000-PE vesicles strongly interact with Co or Au 3D superlattices.

From these data, we conclude that "clustered" structures suspended in water retain their 3D crystalline ordering, i.e. nanocrystals (Au/Co) used as building blocks remain in their metal form.

With Co and Au "clustered" structures, nanocrystals are highly ordered at the edge of the 3D superlattices (Figure 14.4b,d). However with Co "clustered" structure, undefined layers probably attributed to CoO nanocrystals are formed at the 3D superlattice edges (Figure 14.4a,b). With Au counterparts, parallel multilayers of lipids are attached to its surface (Figure 14.4c,d).

To further analyze the composite *in situ*, i.e. without drying the Co "clustered" structures, we used laser-scanning confocal fluorescence microscopy taken with 1% head-labeled fluorescent lipid (16 : 0 Liss Rhod PE) introduced into the vesicles. Figure 14.5 shows fluorescence images of 3D superlattice–vesicle composites with and without fluorescent labeling. In the labeled sample, fluorescent signal (Figure 14.5b) corresponds to the locations of 3D superlattices in the bright-field image (Figure 14.5a). The lack of a fluorescence signal in the control sample without fluorescent lipid (Figure 14.5c,d) indicates that the 3D superlattices are coated with one or more layers of lipids. Additionally, we observed several areas of fluorescence is close proximity to 3D superlattices, indicating the attachment between lipid vesicles and 3D superlattices. Moreover, structures similar to the one shown by electron microscopy in Figure 14.4b are also observed (arrows in Figure 14.5b and inset): an empty vesicle is attached to the 3D superlattice.

At this stage, we need to propose a mechanism to explain the dispersion of a few-micrometer Co 3D superlattices in aqueous media. We have to consider the

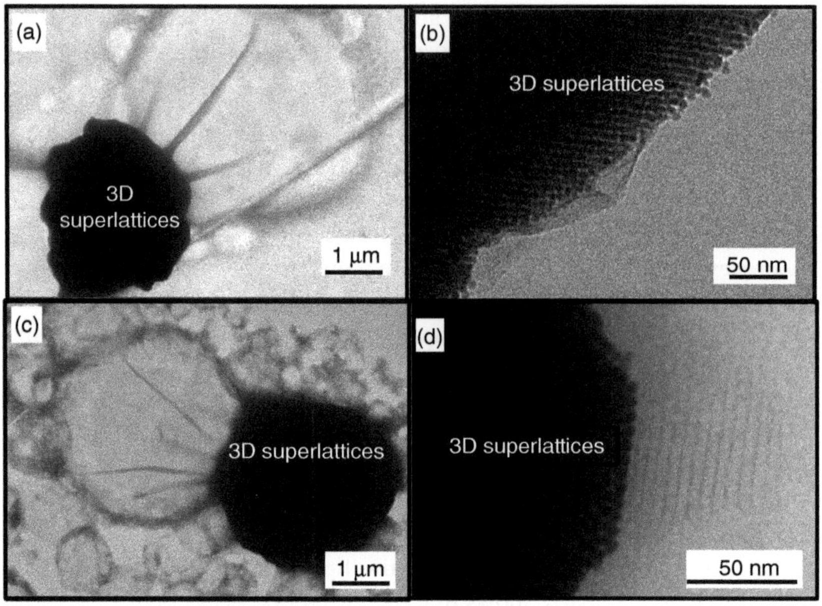

Figure 14.4 TEM images of the Co (a, b) and Au (c, d) "clustered" structures submitted to hydration and then dried. The 3D superlattices are surrounded and joined by the vesicles (a, c). The bilayer vesicle is well joined to the surface of the 3D superlattice, and the ordered structure of the 3D superlattice is still kept in the composite (b, d). *Source:* (a, b) Yang et al. [5, 6]/American Chemical Society. (c, d) Yang et al. [7]/American Chemical Society.

structures of the various molecules involved in coating the 3D superlattices. Here, DPPC has a similar hydrophobic alkyl chain as oleic acid or hexadecanethiol used to coat on either Co or Au nanocrystals, the building blocks of 3D superlattices. PEG2000-PE has both the same length of hydrophobic alkyl chain and the double bond as oleic acid (Figure 14.6).

This indicates that these alkyl chains would have strong mutual affinity, and herein there should be an association between the hydrophobic chains of phospholipid and oleic acid, which is covalently bonded on the surface of 3D superlattices. It could be assuming that, due to the strong interactions between the 3D superlattice ligands and the hydrophobic tails of the phospholipid, lipid–3D superlattice–lipid sandwich structure is formed first (Figure 14.7a).

During hydration, some of these attached amphipathic phospholipids could permit to form vesicles, exposing the hydrophilic heads to water. Some lipids still intertwine with oleic acid, keeping the structure of multilayers. Meanwhile, the long hydrophilic chains (91 carbon and 47 oxygen atoms) of PEG2000-PE favorably extend into water (Figure 14.7b). This long chain is abundant in oxygen, which renders it easy to form hydrogen bond in aqueous solution. The PEG chains of vesicles on a 3D superlattice or the free vesicles diffused in solution interact with each other, permitting to create a network responsible for the formation of a colloidal solution

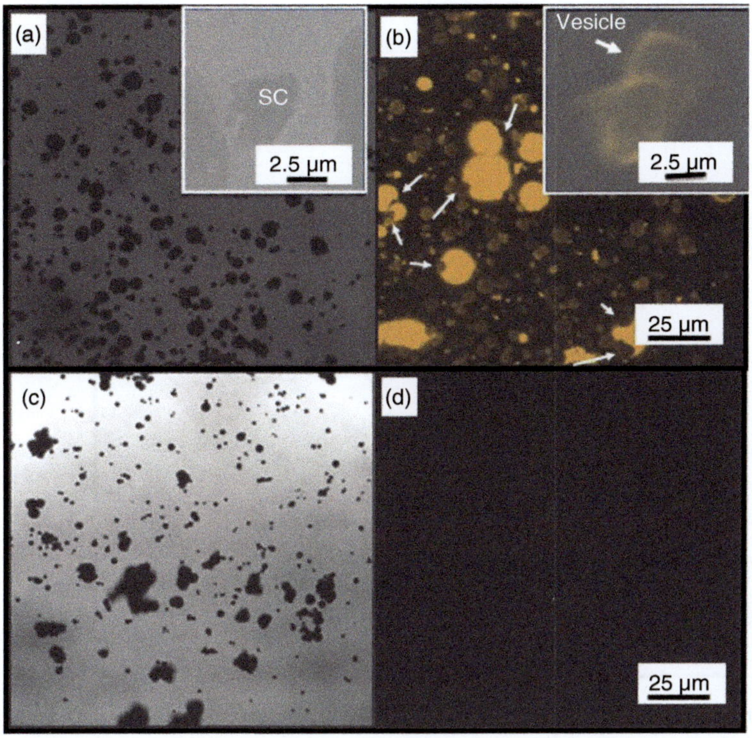

Figure 14.5 Confocal fluorescence microscopy of the 3D superlattice–vesicle composite with and without 1% Liss Rhod PE labeling. (a, b) are the bright-field and fluorescence images for 3D superlattice–vesicle composites with fluorescent labeling. The white arrows and inset image in (b) indicate similar structures observed in the TEM study (Figure 14.4a,b). (c, d) are the bright-field and fluorescence images for 3D superlattice–vesicle composites without fluorescent lipids. *Source:* Yang et al. [5, 6]/ American Chemical Society.

containing large 3D superlattices in aqueous solutions (Figure 14.7). This is also supported by the fact that when the residue (vesicle/Co 3D superlattices) is washed with water, quite a lot of organic molecules are released and a redispersed solution, with a short lifetime (15 minutes versus more than 12 hours, note the nearly same density of the different solutions), is produced. Meanwhile, during the hydration process, the DPPC double layer curls into vesicles, while for the attached parts between the lipid and 3D superlattice, the alkyl chains will intertwine with each other much more tightly due to the hydrophobic effect. In other words, the 3D superlattice remains stable in the hydrophobic environment, owing to the attached vesicles and multilayers. Note that both PEG2000-PE and DPPC molecules are needed to disperse Co 3D superlattices in aqueous media. This is due to the fact that PEG2000-PE has a molecular structure with the contribution of both hydrophobic tails and the long hydrophilic head, which is abundant of oxygen atoms, favoring to form hydrogen bonds in aqueous solution and making this system stable. However, PEG2000-PE will not form the vesicles after hydration without DPPC [14–16]. The

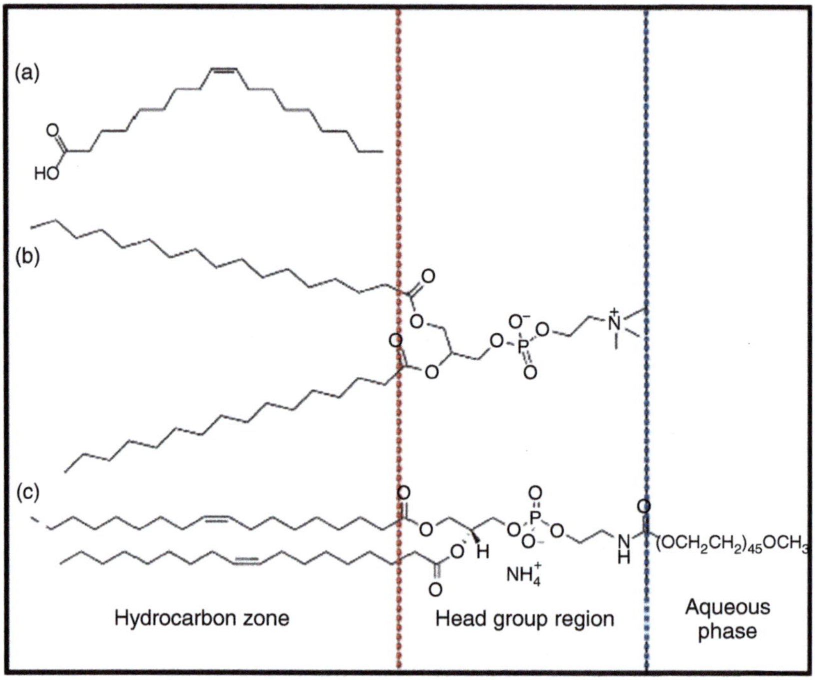

Figure 14.6 Structures of (a) OA, (b) DPPC, and (c) PEG2000-PE.

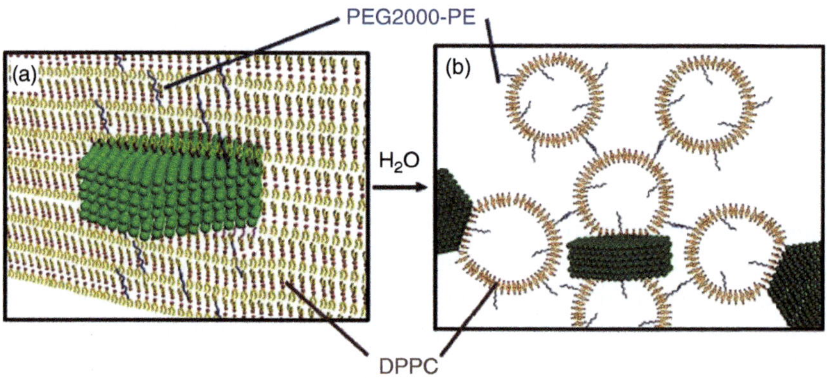

Figure 14.7 Dried (a) and dispersed in water (b) of "Clustered" structures.

control of the DPPC/PEG2000-PE ratio indicates that the hydrophobic/hydrophilic balance is a key parameter in solubilization of hydrophobic Co 3D superlattices in aqueous solution. For 90 : 10 < DPPC/PEG2000-PE < 95 : 5, the solution is stable for more than 12 hours. With time increasing, a precipitate is observed due to the network breaking by gravity. However, by a mild vortex, the Co 3D superlattices are dispersed again. Hence, the stability of the nanocrystals is due to the existence of the ligands. On the one hand, these ligands have a strong interaction with the nanoparticles, and, on the other hand, they have a good solubility in the solvent. Of course, the reader could estimate that such mechanism is rather speculative because not all the 3D superlattices are observed with vesicles coating from the microscopy images. However, the 3D superlattices do not aggregate, and the protecting layer on these 3D superlattices is also observed, confirming the presence of thin lipid bi- or multilayers. Note that the vesicles are not observed in the majority of the 3D superlattices attributed to the washing process (three times) needed for the microscopic studies.

14.1.2 Specific Properties

14.1.2.1 Magnetic Properties

Co "clustered" structures dispersed in water keep their magnetic properties [6].

A small beaker is filled with the colloidal solution described above that is subjected to a small commercial magnet (0.1 T). After few minutes, the black color moves closer to the magnet (Figure 14.8). By removing the magnet, progressively the colloidal solution reaches the initial gray color.

Let us look more carefully by using an optical microscope what happens when the colloidal solution is subjected to a magnet. Figure 14.9a shows the initial state, and the Co "clustered" structures are visible as black dots. Then, a magnet is placed near the beaker. The suspended Co 3D superlattices start to move along the magnetic field lines. With increasing time, the Co 3D superlattices successively migrate along the field direction, inducing a progressive association of 3D superlattices and forming a linear structure (Figure 14.9b–d). After 120 seconds, the magnetic field is removed, and the linear composites separate to small pieces within 30 seconds (Figure 14.9e). By changing the magnetic field orientation, the 3D superlattice alignment follows the field direction (Figure 14.9f), and their movement is reversible. Note that in Figure 14.9e the dots are larger than what we observed in Figure 14.9a. This is due to the fact that the magnetic interactions make the Co 3D superlattices closer, which are difficult to separate from each other by Brownian motion in a short period. However, by shaking the colloidal solution, the dots disappear, and similar feature as shown in Figure 14.9a is observed. This phenomenon indicates that no coalescence between Co 3D superlattices takes place. The TEM pattern obtained after subjecting the Co "clustered" structures to a magnetic field remains similar to that shown in Figure 14.4a. This supports the fact that the integrity of Co "clustered" structures is retained during magnetic actuation.

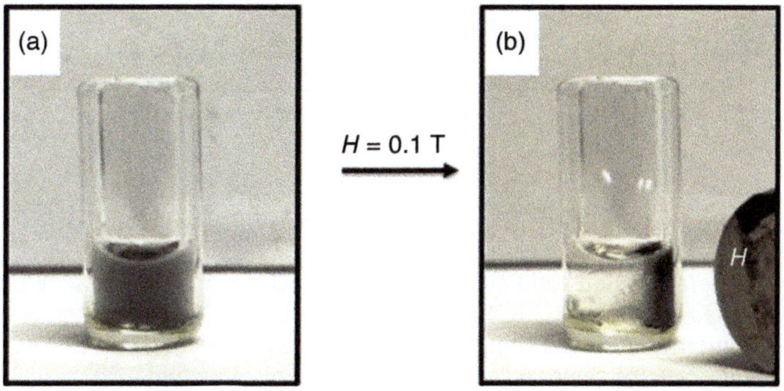

Figure 14.8 Co "Clustered" structures dispersed in aqueous solution in absence (a) and presence (b) of a applied magnetic field. *Source:* Yang et al. [5, 6]/ American Chemical Society.

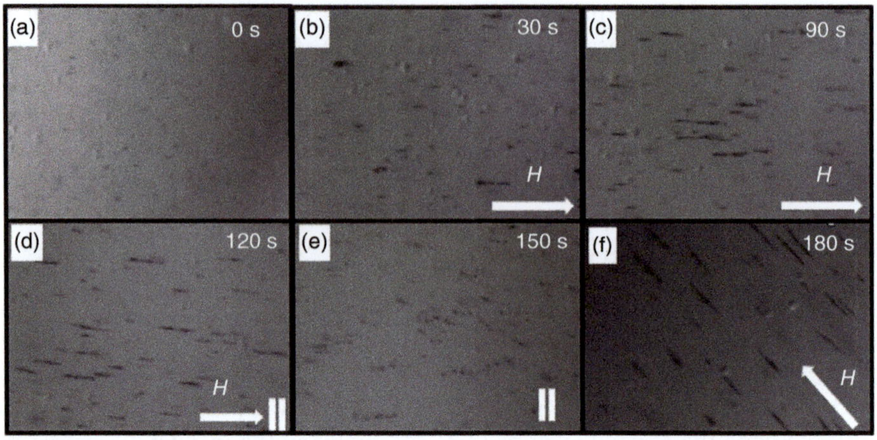

Figure 14.9 Mobility of the 3D superlattices in a magnetic field gradient. (a) No applied field. (b–d) Snapshots of 3D superlattices magnetic induced assembly and movement. (e) After removing the magnetic field for 30 seconds, the lined 3D superlattices separate into smaller ones. (f) Upon a second application of the magnetic field, the isolated 3D superlattices align and move again. *Source:* Yang et al. [5, 6]/ American Chemical Society.

To confirm the fact that Co "clustered" structures retain their magnetic properties, the measurement of the zero-field-cooled (ZFC)/field-cooled (FC) magnetization curves for Co 3D superlattices subjected or not to DPPC/PEG2000-PE is shown in Figure 14.10. They are recorded to determine their blocking temperatures (T_b)

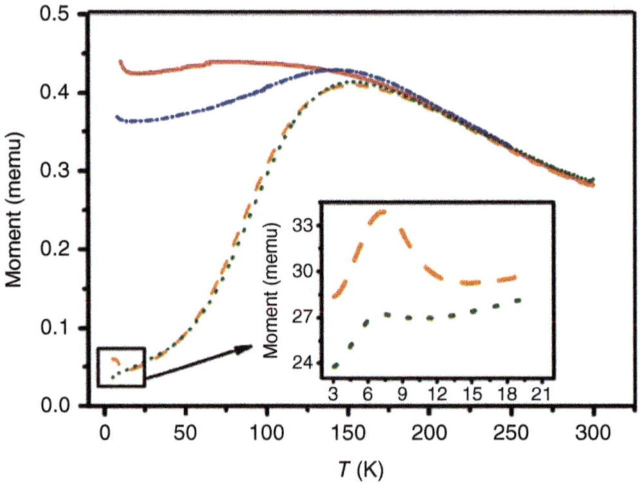

Figure 14.10 ZFC/FC magnetization curves for Co 3D superlattices subjected or not to DPPC/PEG2000-PE coating processes. The inset curves are the precise scan at low temperature. Red solid curve: FC curve of 3D superlattice–vesicle composite. Blue dashed-dotted curve: FC curve of 3D superlattice. Yellow dashed curve: ZFC curve of 3D superlattice–vesicle composite. Olive dotted curve: ZFC curve of 3D superlattice. *Source:* Yang et al. [6], figure 6 (p. F)/American Chemical Society.

before and after phospholipid treatment. The colloidal solution of Co "clustered" structures is deposited on a silica wafer and dried. Simultaneously, the "untreated" Co 3D superlattices are kept in the same aqueous solution as that of the Co "clustered" structures. Because of their hydrophobic character, the Co 3D superlattices keep their integrity either aggregated or pasted on the wall of the flask. The magnetization measurements are performed 10 days after water evaporation and keeping the samples without any precautions at ambient atmosphere. The untreated Co 3D superlattices are kept under vacuum during the same period of time. Figure 14.10 shows that, for both Co 3D superlattices and Co "clustered" structures, T_b is similar at 152 K (−121 °C). This clearly shows that the Co 3D superlattices coated or not coated with DPPC/PEG2000-PE retain their magnetic properties. This confirms that the 3D superlattices retain their integrities after vesicle treatments with unchanged values of nanocrystal diameters and interparticle distances. However, a more careful investigation at low temperature shows that the composite has a peak at 7 K in the ZFC curve (see inset of Figure 14.10). It is characteristic of very small nanoparticles or thin layers of cobalt oxide, CoO. Concerning the unchanged T_b, it is clearly demonstrated that only the surface of Co 3D superlattice is passivated, forming a passivation layer. This explains the change in the contrast at the edge of the Co 3D superlattice–vesicle shown in Figure 14.4b. Importantly, this CoO passivation layer protects the Co 3D superlattice against further oxidation and retains the high stability of Co 3D superlattices.

With Co "clustered" structures, the magnetic properties of shaped 3D superlattices before and after biological treatment remain unchanged indicating that oxygen molecules do not diffuse inside the 3D superlattice. This is due to the oxidation of Co nanocrystals located at the edges of the "clustered" to CoO nanocrystals, forming a barrier to oxygen diffusion inside the crystalline structure.

14.1.2.2 Optical Properties

Au "clustered" structures are characterized by two specific absorption spectra: The first one is due to localized surface plasmon resonance indicating the finger print of the nanocrystals used as building blocks and a spectrum assigned to a photonic (i.e. collective) mode with a maximum red shifted with the cluster size [7].

To control the size of "clustered" structures of 5.5-nm Au nanocrystals (Figure 14.11a), we need to control the nucleation (t_1) and growth (t_2) time processes. Figure 14.11 shows a progressive increase in the Au "clustered" structures. 3D superlattices are obtained for nucleation and growth times up to 6 and 10 hours, respectively (Figure 14.11e,f). The size of the 3D superlattices increases with increasing these times. At 16- and 60-hour nucleation and growth process times, respectively, the 3D superlattices produced are too large to be stabilized for a long time by the "molecular parachute." Below these time values (6 and 10), the assemblies of Au nanocrystals are amorphous aggregates (Figure 14.11b–d).

The absorption spectrum of Au "clustered" is redshifted on increasing its average size (Figure 14.12c–f), and a shoulder around 530 nm is observed.

To simulate the absorption spectra, we first calculate the effective dielectric function of Au assemblies by using Maxwell–Garnett effective medium theory

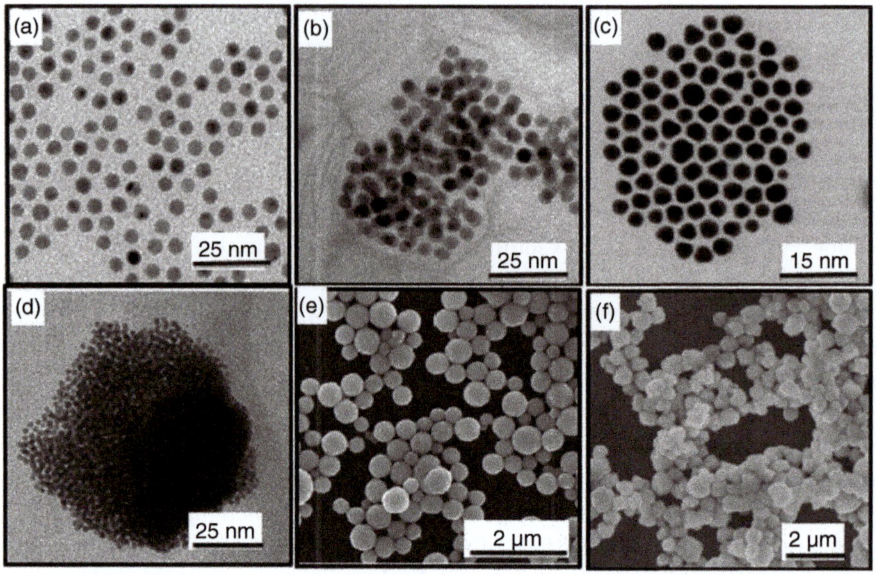

Figure 14.11 (a) TEM image of isolated water-soluble zwitterionic Au nanocrystals dispersed in vesicular solution and with a diameter of 5.5 ± 0.3 nm, (b–f) aggregate of 5.5-nm Au nanocrystals having an average size of 52, 125, 198, 413, and 611 nm respectively. Note that assemblies of 52, 125, and 198 nm reveal a slight change in the interparticle distances indicating that the aggregates are disordered. However, we cannot neglect some nanocrystal ordering in the dark region as shown in (d). For 413 and 611 nm, the nanocrystals used as building blocks are ordered in fcc crystalline structures. *Source:* Yang et al. [7]/American Chemical Society.

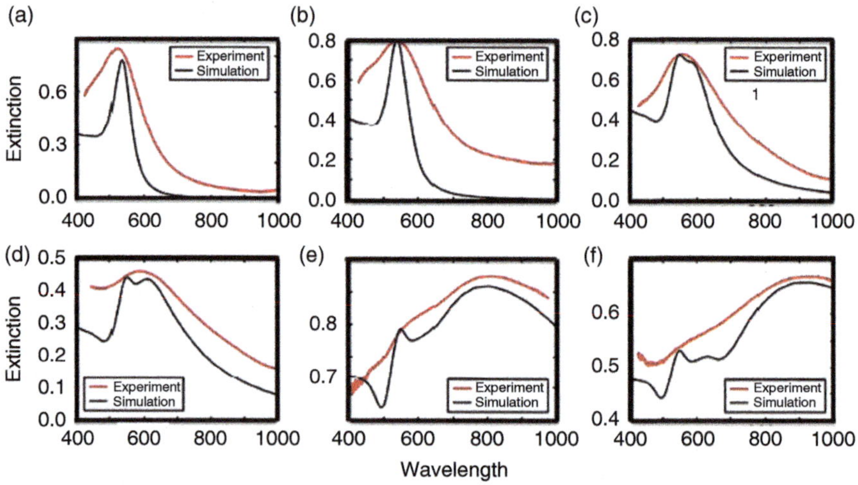

Figure 14.12 Extinction spectra for isolated Au assemblies with various sizes. (a) Dispersed 5.5-nm Au nanocrystals. Au amorphous aggregates having various sizes of (b) 52 nm, (c) 125 nm, and (d) 198 nm; Au 3D superlattices with sizes of (e) 413 nm and (f) 611 nm. The measured size distribution was also indicated. The following sizes were used to fit the measured data: 5.5, 52, 140, 198, 480, and 611 nm. *Source:* Yang et al. [7], figure 5/American Chemical Society.

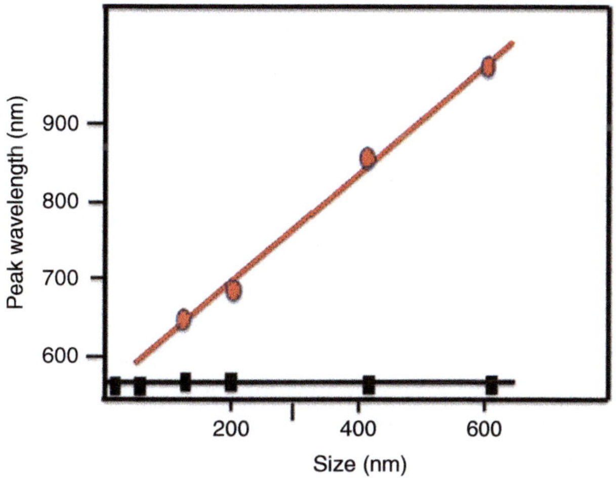

Figure 14.13 Variation of the peak wavelength as a function of the assembly size. The black curve illustrates the absorption peak of dispersed Au nanocrystals used as building blocks. The scattering peak represented in the red curve redshifts as the assembly size increases. *Source:* [7], figure 7/ American Chemical Society.

(EMT) [17]. Then, Mie theory is used to calculate the spectral response of Au assemblies dispersed in water. The distribution over the size of assemblies is deduced from SEM images (Figure 14.11) and accounted for in our calculations. Figure 14.12 shows a good agreement between experiments and calculated data. Note that the absorption spectra of the assemblies are the sum of two spectra. At short wavelength, one is found at fixed wavelength (530 nm), whereas the second one is redshifted with increasing the aggregate size. At shorter wavelength, the peak fulfills the resonance condition of spherical Au nanocrystals in a ligand-based medium. The scattered spectra at larger wavelength are very broad and redshifted linearly with the wavelength as the assembly size increases (Figure 14.13). Such optical response at large wavelength is ascribable to a photonic (i.e. collective) mode of the assembly, since its spectral position scales linearly with the size of the 3D superlattice when the fingerprint of dispersed nanocrystals used as building blocks is retained.

14.2 Colloidosomes and Supraballs

Shell and spherical solid assemblies of nanocrystals are called colloidosomes and supraballs respectively [5, 18].

14.2.1 Colloidosome: Fabrication and Characterization

Pickering originally proposed the concept of colloidosomes, more than century ago, when they discovered that microsized solid particles stand at the interface between two immiscible fluids [19]. In recent two decades, much diversified hollow capsules consisting of coagulated, fused, or close-packed layers of colloidal particles with sizes ranging from micrometer to millimeter were produced and investigated [20–22]. Some of these new superstructures were produced via assembly into shells around water-in-oil emulsion of latex particles, large porous nanoparticles, and layer-by-layer assembly of multilayer shells consisting of alternating cationic and anionic polyelectrolytes. It was also demonstrated that solid microspheres in oil–water emulsion can be produced in the conditions of a weak nanoparticle confinement at the fluid interface affected by the thermal fluctuations [23–25]. When the size of particles decreases, thermal energy fluctuations decrease the adhesion forces of particles at the liquid–liquid interface [26, 27]. *Consequently, it makes it impossible to produce colloidosomes.* This can be achieved via tuning surface properties using various ligand modifications such as chemical cross-linking, specific ligand interactions, and other nanoparticle stabilizers at the fluid interface. Recently, we developed a new concept to lock Fe_3O_4 nanocrystals at the liquid–liquid interface: a "Janus bilayer" shell for the nanocrystals coated with oleic acid (C_{18}–COOH) adsorbed at the liquid–liquid interface [5] inducing the formation of colloidosomes. The inner shell of the bilayer is the tail part of oleic acid covalently bound to Fe_3O_4 nanocrystals, whereas the outer shell is composed of lauryl trimethyl ammonium bromide (DTAB, C_{12}–$NH_3^+Br^-$) in water phase and octadecene (ODE, $C_{18}H_{36}$) in chloroform. Specific interactions between aliphatic ODE and oleic acid, covalently bound at the nanocrystal surface, take place, inducing the formation of a monolayer–nanocrystal–membrane. This favors the adsorption of nanocrystals at the oil–water interface and consequently locks them at liquid–liquid interface [28–30]. This approach helped to avoid complicated procedures, such as those found in nanocrystal polymersomes [31, 32]. Hence, the shell of Fe_3O_4 nanocrystals called colloidosomes are produced (Figure 14.14a,b). This can be composed of a few layers of nanocrystals.

From atomic force microscope (AFM) measurements, 3D topography image of an individual colloidosome, in water, attached to the modified silica surface shows that the shape of such suprastructure is almost spherical with an approximate convoluted diameter of about 70 nm (Figure 14.14c). Its shape remains unchanged for a load ≤1.5 nN. The colloidosome deformation increases with increasing load. At 3 nN loaded, individual nanocrystals are distinguished. This suggests that the nanocrystals are mobile in the colloidosome and therefore shift their positions slightly due to interactions with the tip. At high enough loads, the interactions involved in keeping the nanocrystal shell structure are too weak compared with the applied forces, and a complete colloidosome destruction is observed at 5 nN. Thus, the collapse leads to the formation of a film of nanocrystals.

Figure 14.14d shows steps at the colloidosome edges. The height of the step-edge structure of colloidosome is about 10 nm that is consistent with the size of the nanocrystal (about 9.6 nm) and the associated organic coating agent. Figure 14.15 shows the deformation and adhesion at the edge regions. It is clearly shown that these steps are followed by clear peaks in deformation, suggesting very local nanocrystal movement within the layer. It should be noted that the adhesion force depends on the contact area, and this is likely to contribute to the increased adhesion at the edges. The magnitude of the deformation at the first edge step and further toward the colloidosome center is rather high. Such adhesion and deformation are probably due to the organic molecules involved to produce colloidosomes. As described above, we have to remember that not only oleic acid and DTAB but also ODE molecules are

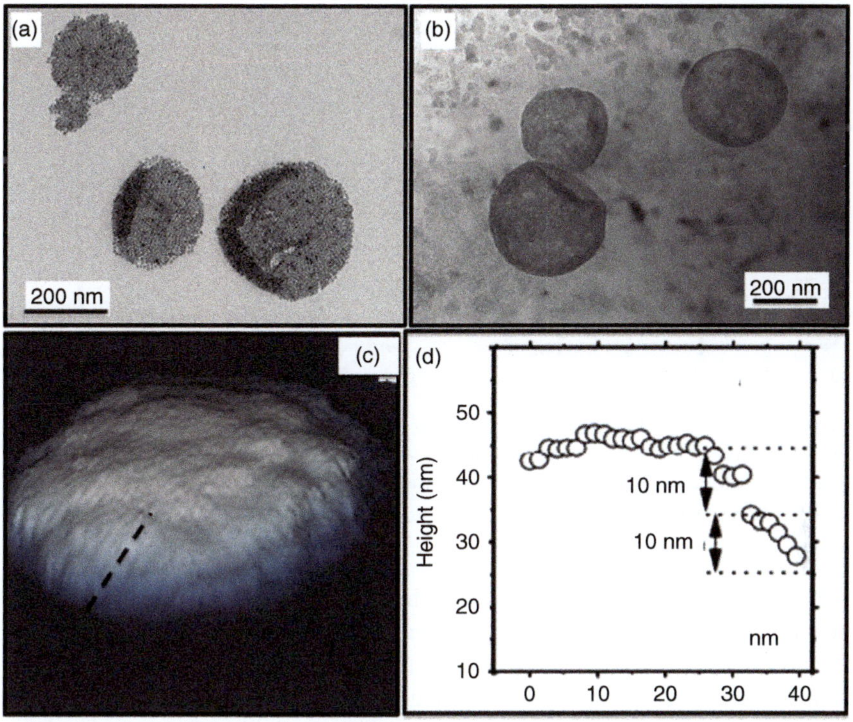

Figure 14.14 (a) TEM, (b) cryo-TEM, (c) topography measured at an applied force of 2 nN, and (d) the height variation along the black dash lines of colloidosomes. *Source:* (a–c) Dobryden et al. [18]/John Wiley & Sons, Inc.

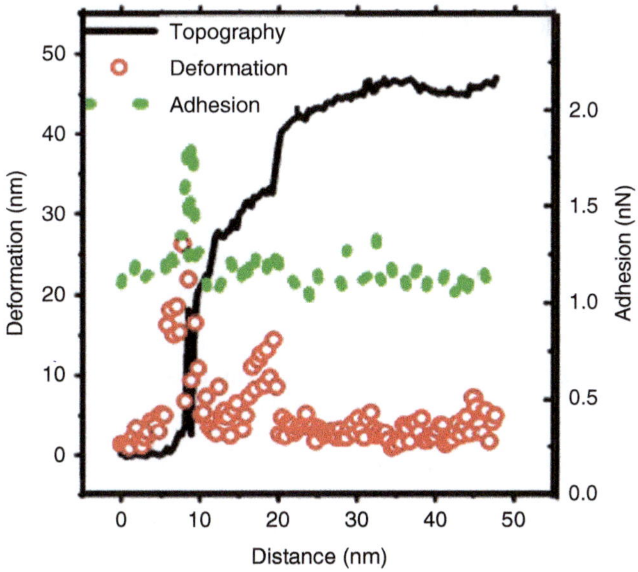

Figure 14.15 The cross-sectional data for high-resolution topography images measured at the edge of a colloidosome using an imaging load of 2 nN. *Source:* Dobryden et al. [18]/John Wiley & Sons.

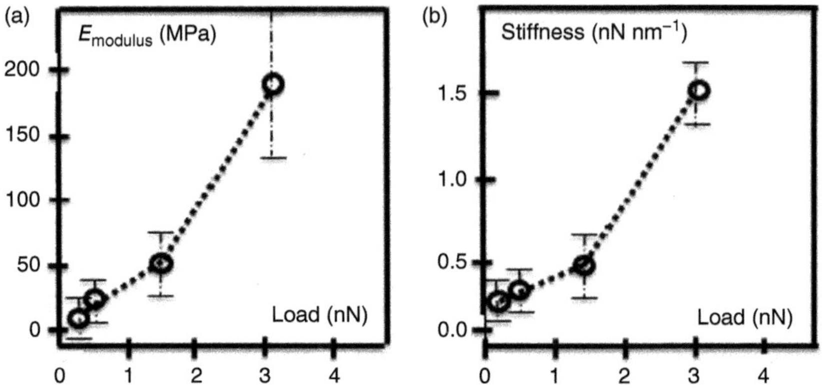

Figure 14.16 The load dependence of (a) the elastic modulus and (b) stiffness of colloidosome. The error bars represent property variations for each studied hybrid. The values are confirmed for some more colloidosomes at the same loads. *Source:* Modified from Dobryden et al. [18].

needed. Colloidosomes cannot be produced in the absence of ODE. The existence of such chemicals on the surface and in between the nanocrystals increase the adhesion forces and the deformation at these locations. This claim will be confirmed below with supraballs, produced by the same produced of colloidosome but in the absence of ODE. In such case, the adhesion and deformation markedly drop. Consequently, it is concluded that ODE molecules in the shell of the colloidosome favor the increase in deformation and adhesion force at the edges. Note that the washing procedure of suprastructures does not prevent adsorption of ODE and DTAB to the oleic acid used to coat the Fe_3O_4 nanocrystals. Interestingly, these steps are followed by clear peaks in the deformation, suggesting very local nanocrystal movement within the layer. This, again, demonstrates that the colloidosomes are soft even at the nanoscale. This is due to the larger volume between the nanocrystals in the colloidosomes.

The apparent elastic modulus and surface stiffness of the colloidosome increase with the load (Figure 14.16). The increase in elastic modulus with increasing load is pronounced, and a significant and irreversible shape changes at a load of 3 nN. At high load, the elastic modulus is rather large. This could be due to the fact that, at higher applied loads, the soft colloidosome becomes sufficiently compressed to induce a high nanocrystal packing that could be similar to layered nanocrystals.

14.2.2 Supraballs: Fabrication and Characterization

Similar procedure used to produce colloidosome is applied for supraballs. All the experimental conditions remain similar except that the suprastructure growth takes place in the absence of octadecene, ODE, used to lock nanocrystals at liquid–liquid interface. Hence, the high mobility of the nanocrystals at the interface induces the formation of large spherical solid dense-packed assemblies of nanocrystals called

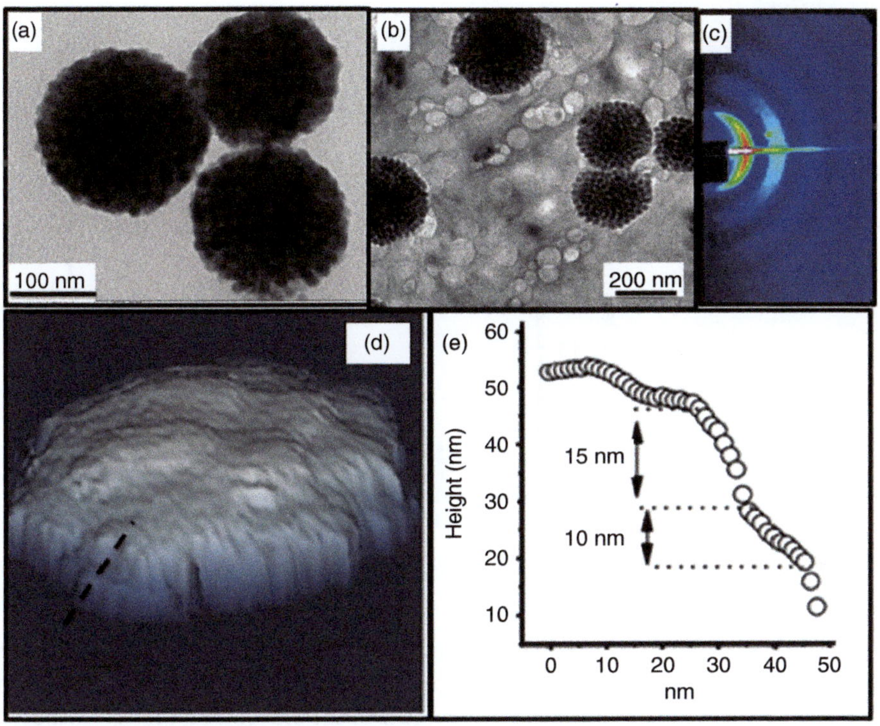

Figure 14.17 (a) TEM, (b) cryo-TEM, (c) SAXS pattern, (d) topography measured at an applied force of 2 nN, and (e) the height variation along the black dash lines of supraballs. *Source:* Dobryden et al. [18]/John Wiley & Sons, Inc.

supraballs (Figure 14.17a,b) instead of colloidosomes (Figure 14.14). From SAXS experiments, Figure 14.17c shows that the nanocrystals in the supraballs are self-assembled in fcc crystalline structure.

A 100-nm single fcc supraball deposited on a modified silica surface is imaged in water (Figure 14.17d). The supraball shape is almost spherical with a possible small deformation due to its attachment to the surface. The supraball shape remains unchanged below and at 5 nN load. The irreversible shape is obtained at the imaging force of 10 nN with a deformation reaching about half the value of its height. As with colloidosomes, the AFM tomography image shows step-edge structure. This structure is observed as with colloidosome (Figure 14.17e). The height of the step edges between the nanocrystal layers is about 10–15 nm that is consistent with the size of the nanocrystal and coating agent. Note that similar fcc spherical particles have been previously obtained with either 6-nm cobalt iron oxide nanocrystals [33] or large colloidal silica particles [34].

Figure 14.18 shows the deformation and adhesion forces of supraball. The magnitude of the deformation at the first edge step and further toward the supraball center

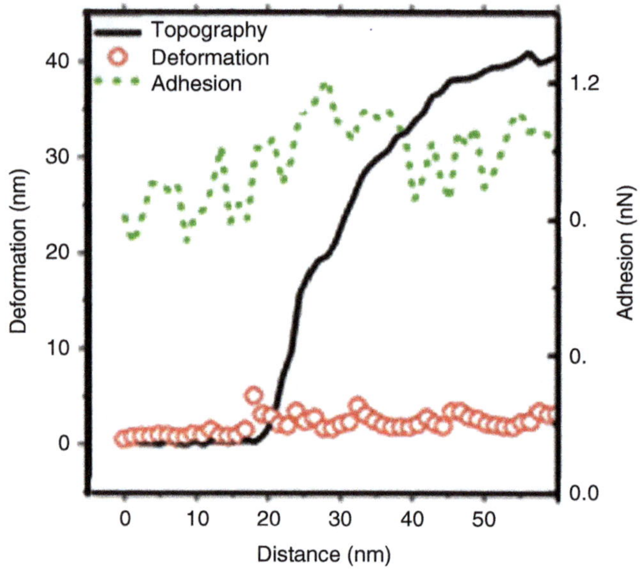

Figure 14.18 The cross-sectional data for high-resolution topography images measured at the edge of a supraball using an imaging load of 2 nN. *Source:* Dobryden et al. [18]/John Wiley & Sons.

is rather small. Similarly, the adhesion forces show a slight increase in the edge region and very little variation at the step edges. Both deformation and adhesion measured with supraballs are smaller than those of colloidosomes. This is attributed to the absence of ODE in the supraball. Actually, the organic molecules are limited to oleic acid and DTAB.

The apparent elastic modulus and stiffness progressively increase with increasing load (Figure 14.19). This confirms that supraballs keep their integrity for a rather large load compared with colloidosomes.

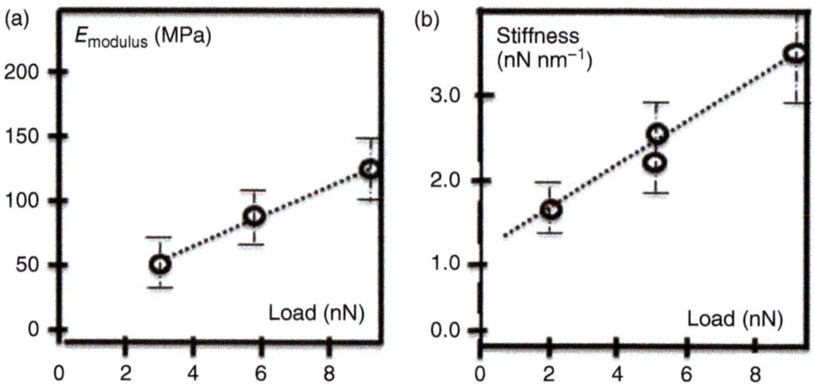

Figure 14.19 The load dependence of (a) the elastic modulus and (b) stiffness of supraball; the error bars represent property variations for each studied hybrid. The values are confirmed for some more colloidosomes at the same loads. *Source:* Based on Mazzanti et al. [8].

14.3 Nanoheaters

Here we demonstrated that the suprastructures described above behave a nanoheaters [8].

Let us consider the Au "clustered" structures of 5.5-nm Au nanocrystals similar to those shown in Figure 14.2. Extinction spectra of Au "clustered" assemblies show two peaks that are assigned to a scattering mode that redshifted with increasing the assembly size and an absorption mode associated with localized surface plasmon resonance, which is independent of their size. Au "clustered" structure having an average size of 198 nm is characterized by a large peak centered around 700 nm assigned to a photonic (i.e. collective) mode of the assembly (Figure 14.12d) and a peak centered around 530 nm corresponding to the absorption of isolated Au nanocrystals.

The energy flow upon light irradiation is investigated by pump–probe experiments. The differential transmission ($\Delta T/T$) spectra are recorded as a function of time and of the probe wavelength, λ. At the initial time scale of 10 ps, the typical $\Delta T/T$ map is due to the transient spectra of isolated nanocrystals at various probe wavelengths. They are dominated by shift and broadening of the plasmonic resonances as observed previously [35]. These spectra exhibit a decay constant of few picoseconds (Figure 14.19), corresponding to the time scale of electron–phonon scattering in noble metals [36]. Both dynamic and static data confirm that Au "clustered" structures at room temperature keep the fingerprint of isolated nanocrystals used as building blocks. This was already observed by scanning tunneling microscopy/scanning tunneling spectroscopy (STM/STS) experiments at low temperature, with dried and thick 3D superlattices [37].

Interestingly, at longer times scales, the $\Delta T/T$ map of Au "clustered" structures (Figure 14.20) exhibit *very distinct features* that, to the best of our knowledge, have not been reported for any nanostructures. Instead of a monotonic decrease over time, we observe the buildup of a positive signal that is redshifted by about 100 nm compared with early spectra.

This novel phenomenon cannot be ascribed to mechanical oscillations that, in Au nanocrystals, are well known to have a strong impact on the ultrafast optical response at this time scale [38]. To explain the origin of such unexpected transient optical response at the 100-ps time scale, remaining constant at the nanosecond time scale, a model to simulate the optical experiments was developed [8]. Here, we have to remember how 3D superlattices are formed. The alkyl chains used to coat nanocrystals interdigitate and play a role in ordering nanocrystals in crystalline structure. The static optical response indicates that such 3D superlattices behave as metaparticles rather than finite arrays of nanocrystals and the collective effects are dominated by the onset of an effective medium. Due to higher penetration depth of light in the 3D superlattice structure (about 60 nm for Au nanocrystals), despite the thickness of the assembly, the power absorbed per unit volume is more uniform compared with that in a nanosphere of the same diameter. The pump pulse is absorbed by the metallic phase of the 3D superlattice and initiates a chain of energy transfer processes. According to this model (see Ref. [8]), the delayed buildup of the $\Delta T/T$ signal on the

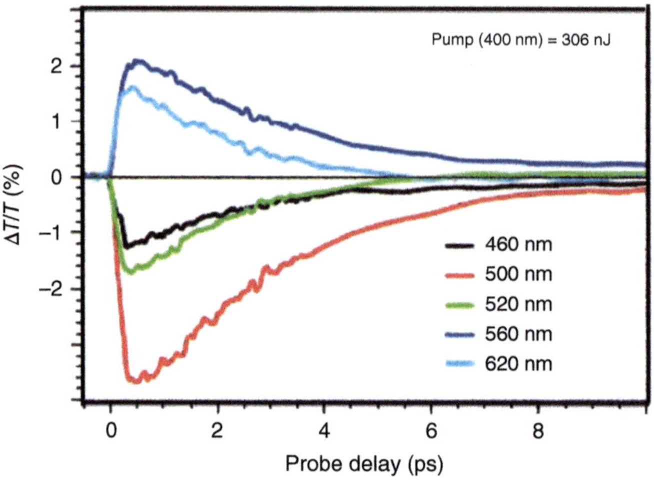

Figure 14.20 Dynamics of the $\Delta T/T$ at selected probe wavelengths, at short time scale of Au "clustered" structure. Light–heat conversion dynamics in highly diversified water-dispersed hydrophobic nanocrystal assemblies. *Source:* Modified from Dobryden et al. [18].

few hundred-picosecond time scale is the signature of the final step in the light–heat conversion process that ultimately results in the heating of the matrix of organic ligands. The very good agreement between experiments and the model confirms a collective regime of photo-temperature generation enabled by the assembling. It is concluded that the coating agent, even though not directly absorbing, acts as an internal reservoir for efficient accumulation of energy within few hundreds of picoseconds. Very surprisingly, without any simple explanation, such long-time-scale transient has *never been observed in dried system*, i.e. when the same untreated 3D superlattices are deposited in a substrate. Hence, water molecules are needed to observe such nanoheater behavior.

At this point, we ask ourself if such nanoheater could be an intrinsic property of any water dispersive hybrid assembly of hydrophobic nanocrystals, regardless of the peculiar nature (i.e. plasmonic or not) of their building blocks. With supraballs dispersed in water, similar behaviors in the transient optical responses are experimentally observed with both Fe_3O_4 superballs (Figure 14.21).

Similar data were obtained with other 3D superlattices. It is so concluded that self-assembled nanocrystals, i.e. 3D superlattices, behave as universal nanoheaters.

At this point, another question arises: Could such "nanoheater" exist with shell of nanocrystals as obtained with colloidosomes? Figure 14.22 shows a very slight increase in the long-time transient. This clearly shows that the nanoheater cannot be efficient when one or few layers of nanocrystals are self-assembled. The amount of chain to favor the energy transfer is not large enough. Nevertheless, it still exists.

Few years ago, we demonstrated that by subjecting 3D superlattices to light, inter-nanocrystal coherence inside fcc 3D superlattices takes place. Even though the forces involved between atoms in an atomic crystal and nanocrystals in a 3D

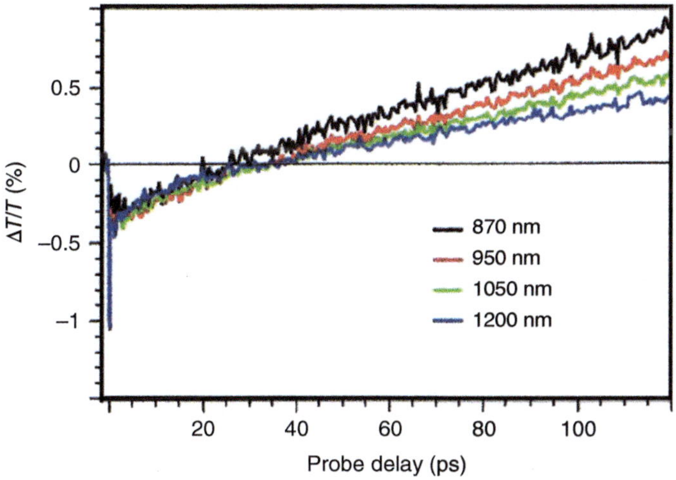

Figure 14.21 Dynamics of the $\Delta T/T$ at selected probe wavelengths, at long time scale of Au "clustered" structure. Light–heat conversion dynamics in highly diversified water-dispersed hydrophobic nanocrystal assemblies. *Source:* Based on Mazzanti et al. [8].

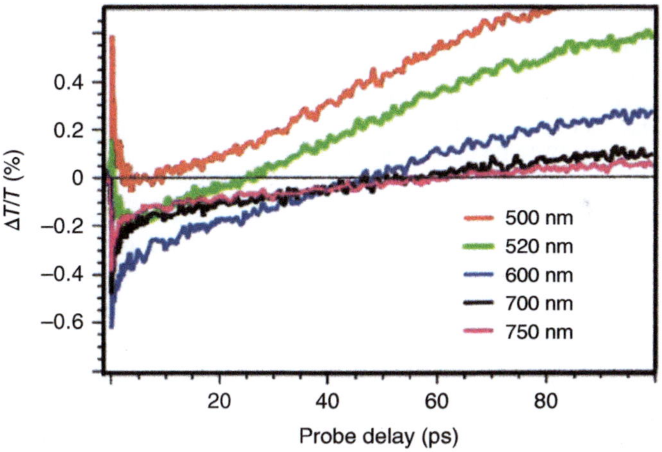

Figure 14.22 Dynamics of the $\Delta T/T$ at selected probe wavelengths, at long time scale of fcc Fe_3O_4 supraballs. Light–heat conversion dynamics in highly diversified water-dispersed hydrophobic nanocrystal assemblies. *Source:* Mazzanti et al. [8], figure 5d (p. 5)/National Academy of Sciences.

superlattice are not the same, nanocrystals in a 3D superlattice behave as atoms in a nanocrystal [39]. With water-dispersive Au "clustered" structures, Figure 14.23 shows an experimental oscillation period of 300 ps. We attribute this oscillation to the period of mechanical oscillation due to a breathing mode of the whole assembly. This value is consistent with that deduced from the calculation. We assume that the calculated period is given by $T = 2D/\nu_s$ where D is the average diameter (198 ± 84 nm) of the assembly and ν_s is the speed of sound inside the 3D superlattice that is

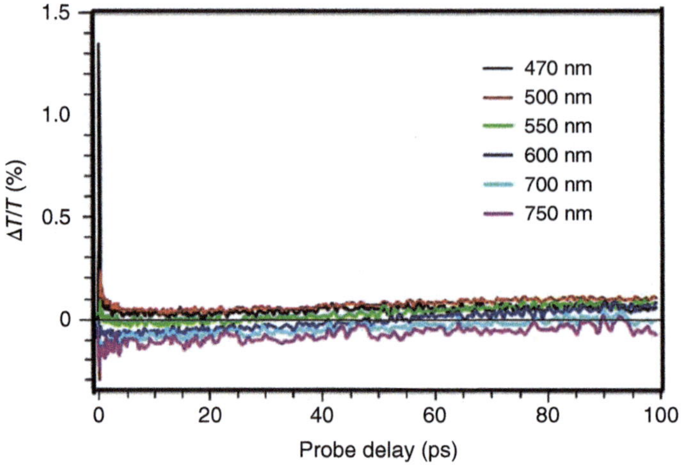

Figure 14.23 Dynamics of the $\Delta T/T$ at selected probe wavelengths, at long time scale of Fe_3O_4 shell called colloidosomes. Light–heat conversion dynamics in highly diversified water-dispersed hydrophobic nanocrystal assemblies. *Source:* Based on Mazzanti et al. [8].

Figure 14.24 Dynamics of the $\Delta T/T$ at selected probe wavelengths, on nanometer time scales of energy flow in the Au clustered structures. Light–heat conversion dynamics in highly diversified water-dispersed hydrophobic nanocrystal assemblies. *Source:* Mazzanti et al. [8], figure 3c (p. 3)/National Academy of Sciences.

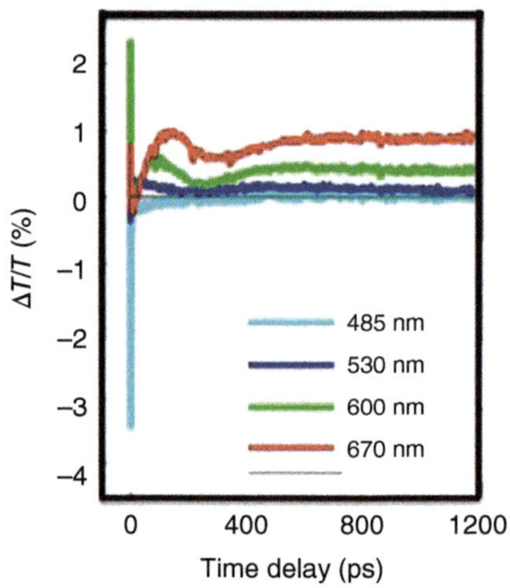

estimated to be $1235 \pm 12\,\mathrm{m\,s^{-1}}$ at room temperature for fcc 3D superlattices of 7.1-nm Co nanocrystals used as building blocks [40]. The resulting oscillation period of the Au "clustered" structure is then $236 \pm 77\,\mathrm{ps}$. This calculated value is consistent with deduced from experiments (Figure 14.24).

14.4 Conclusion

Here, we report three series of new organic/inorganic supracrystalline hybrid structures. The first structure ("clustered") is composed of 3D superlattices made of hydrophobic nanocrystals (Co or Au) with the organic material adapting to the inorganic component and surrounding the 3D superlattices to produce a "parachute." This favors the dispersion of rather large 3D superlattices in aqueous solution. The two others are self-assemblies of densely packed Fe_3O_4 nanocrystals forming either shells or fcc solid spherical particles called colloidosomes or supraballs, respectively.

The physical properties of such hybrids remain at the early stage. Due to the 3D self-assemblies, some collective properties are maintained such as the appearance of optical photonic (i.e. collective) mode due to the formation of metamaterial keeping the fingerprint of isolated nanocrystals. These water-dispersive hybrid structures subjected to light can operate as very efficient universal nanoheaters.

References

1 Pileni, M.P. (2001). Nanocrystals self-assemblies: fabrication and collective properties. *J. Phys. Chem. B* 105: 3358–3372.

2 Pileni, M.P. (2007). Self-assembly of inorganic nanocrystals: fabrication and collective intrinsic properties. *Acc. Chem. Res.* 40: 685–693.

3 Pileni, M.P. (2017). Impact of the metallic crystalline structure on the properties of nanocrystals and their mesoscopic assemblies. *Acc. Chem. Res.* 50: 1946–1955.

4 Boles, M.A., Engel, M., and Talapin, D.V. (2016). Self-assembly of colloidal nanocrystals: from intricate structures to functional materials. *Chem. Rev.* 116: 11220–11289.

5 Yang, Z., Altantzis, T., Zanaga, D. et al. (2016). Supracrystalline colloidal eggs: epitaxial growth and freestanding three-dimensional supracrystals in nanoscaled colloidosome. *J. Am. Chem. Soc.* 138: 3493–3500.

6 Yang, N., Yang, Z., Held, M. et al. (2016). Dispersion of hydrophobic Co supracrystal in aqueous solution. *ACS Nano* 10: 2277–2286.

7 Yang, N., Deeb, C., Pelouard, J.L. et al. (2017). Water-dispersed hydrophobic Au nanocrystal assemblies with a plasmon fingerprint. *ACS Nano* 11: 7797–7806.

8 Mazzanti, A., Yang, Z., Silva, M.G. et al. (2019). Light–heat conversion dynamics in highly diversified water-dispersed hydrophobic nanocrystal assemblies. *Proc. Natl. Acad. Sci. U.S.A.* 116: 8161–8166.

9 Thanh, N.T.K., Maclean, N., and Mahiddine, S. (2014). Mechanisms of nucleation and growth of nanoparticles in solution. *Chem. Rev.* 114: 7610–7630.

10 Goubet, N., Richardi, J., Albouy, P.A., and Pileni, M.P. (2011). Which forces control supracrystal nucleation in organic media? *Adv. Funct. Mater.* 21: 2693–2704.

11 Lisiecki, I., André, P., Filankembo, A. et al. (1999). Mesostructured fluids. 1. $Cu(AOT)_2$–H_2O–isooctane in oil rich regions. *J. Phys. Chem. B* 103: 9168–9175.

12 Kranenburg, M., Veturoli, M., and Smit, B. (2003). Phase behavior and induced interdigitation in bilayers studied with dissipative particle dynamics. *J. Phys. Chem. B* 107: 11491–11501.

13 Nagle, J.F. and Tristram-Nagle, S. (2000). Structure of lipid bilayers. *Biochim. Biophys. Acta, Biomembr.* 1469: 159–195.

14 Allen, C., Dos Santos, N., Gallagher, R. et al. (2002). Controlling the physical behavior and biological performance of liposome formulations through use of surface grafted poly(ethylene glycol). *Biosci. Rep.* 22: 225–250.
15 Bedu-Addo, F.K. and Huang, L. (1995). Interaction of PEG-phospholipid conjugates with phospholipid: implications in liposomal drug delivery. *Adv. Drug Delivery Rev.* 16: 235–247.
16 Edwards, K., Johnsson, M., Karlsson, G., and Silvander, M. (1997). Effect of polyethyleneglycol-phospholipids on aggregate structure in preparations of small unilamellar liposomes. *Biophys. J.* 73: 258–266.
17 Aspnes, D.E. (1982). Local-field effects and effective-medium theory: a microscopic perspective. *Am. J. Phys.* 50: 704–709.
18 Dobryden, I., Yang, Z., Claesson, P.M., and Pileni, M.P. (2020). Water dispersive suprastructures: an organizational impact on nanomechanical properties. *Adv. Mater. Interfaces* 8 (3): 2001687.
19 Pickering, S.U. (1907). Emulsions. *J. Chem. Soc. Trans.* 91: 2001–2021.
20 Velev, O.D. and Nagayama, K. (1997). Assembly of latex particles by using emulsion droplets. 3. Reverse (water in oil) system. *Langmuir* 13: 1856–1859.
21 Velev, O.D., Furusawa, K., and Nagayama, K. (1996). Assembly of latex particles by using emulsion droplets as templates. 2. Ball-like and composite aggregates. *Langmuir* 12: 2385–2391.
22 Velev, O.D., Furusawa, K., and Nagayama, K. (1996). Assembly of latex particles by using emulsion droplets as templates. 1. Microstructured hollow spheres. *Langmuir* 12: 2374–2384.
23 Montanarella, F., Geuchies, J.J., Dasgupta, T. et al. (2018). Crystallization of nanocrystals in spherical confinement probed by *in situ* X-ray scattering. *Nano Lett.* 18: 3675–3681.
24 Yi, G.-R., Manoharan, V.N., Klein, S. et al. (2002). Monodisperse micrometer-scale spherical assemblies of polymer particles. *Adv. Mater.* 14: 1137–1140.
25 Harrison, R.G., Washburn, A.L., Pickett, A.T., and Call, D.M. (2008). Assembly of CdSe nanoparticles into microspheres by a liquid droplet emulsion process. *J. Mater. Chem.* 18: 3718–3722.
26 Lin, Y., Skaff, H., Emrick, T. et al. (2003). Nanoparticle assembly and transport at liquid–liquid interfaces. *Science* 299: 226–229.
27 Duan, H., Wang, D., Sobal, N.S. et al. (2005). Magnetic colloidosomes derived from nanoparticle interfacial self-assembly. *Nano Lett.* 5: 949–952.
28 Lacava, J., Born, P., and Kraus, T. (2012). Nanoparticle clusters with Lennard-Jones geometries. *Nano Lett.* 12: 3279–3282.
29 Glaser, N., Adams, D.J., Böker, A., and Krausch, G. (2006). Janus particles at liquid–liquid interfaces. *Langmuir* 22: 5227–5229.
30 Zhang, X., Zhu, Y., and Granick, S. (2002). Hydrophobicity at a Janus interface. *Science* 295: 663–666.
31 Hickey, R.J., Koski, J., Meng, X. et al. (2014). Size-controlled self-assembly of superparamagnetic polymersomes. *ACS Nano* 8: 495–502.
32 Howse, J.R., Jones, R.A.L., Battaglia, G. et al. (2009). Templated formation of giant polymer vesicles with controlled size distributions. *Nat. Mater.* 8: 507–511.

33 de Nijs, B., Dussi, S., Smallenburg, F. et al. (2014). Entropy-driven formation of large icosahedral colloidal clusters by spherical confinement. *Nat. Mater.* 14: 56.

34 Rossi, L., Soni, V., Ashton, D.J. et al. (2015). Shape-sensitive crystallization in colloidal superball fluids. *Proc. Natl. Acad. Sci. U.S.A.* 112: 5286–5290.

35 Silva, M.G., Teles-Ferreira, D.C., Siman, L. et al. (2018). Universal saturation behavior in the transient optical response of plasmonic structures. *Phys. Rev. B: Condens. Matter* 98: 115407.

36 Sun, C.K., Vallée, F., Acioli, L. et al. (1994). Femtosecond-tunable measurement of electron thermalization in gold. *Phys. Rev. B: Condens. Matter* 50: 15337–15348.

37 Yang, P., Arfaoui, I., Cren, T. et al. (2012). Unexpected electronic properties of micrometer-thick supracrystals of Au nanocrystals. *Nano Lett.* 12: 2051–2055.

38 Hartland, G.V. (2011). Optical studies of dynamics in noble metal nanostructures. *Chem. Rev.* 111: 3858–3887.

39 Courty, A., Mermet, A., Albouy, P.A. et al. (2005). Vibrational coherence of self-organized silver nanocrystals in f.c.c. supra-crystals. *Nat. Mater.* 4: 395–398.

40 Lisiecki, I., Polli, D., Yan, C. et al. (2013). Coherent longitudinal acoustic phonons in three-dimensional supracrystals of cobalt nanocrystals. *Nano Lett.* 13: 4914–4919.

15

Nanocrystal Self-Assembly in Cells

15.1 Ferrite Colloidosomes and Supraballs

As shown in Chapter 14, colloidosomes and supraballs are suprastructures of Fe_3O_4 nanocrystals coated with oleic acid and dispersed in aqueous solution [1]. Whatever the coating agent, the average diameter of Fe_3O_4 nanocrystals is 10 ± 1 nm. Colloidosomes (Figure 15.1a) are shell of self-assembled nanocrystals, whereas supraballs (Figure 15.1b) are spherical fcc 3D superlattices. Their diameters are 170 ± 90 and 69 ± 15 nm, respectively. Below, as a control, we use the Fe_3O_4 nanocrystals coated with dopamine and consequently freely dispersed in solution.

15.2 Intracellular Fate of Hydrophobic Nanocrystal Self-Assemblies in Tumor Cells

15.2.1 Ability of Colloidosomes and Supraballs to be Uptaken into Tumor Cells

We investigate the interactions of dispersed nanocrystals, colloidosomes, and supraballs with A431 epidermoid carcinoma cells [2]. The cells are exposed to the nanomaterials in serum-complemented culture medium for 30 minutes to 24 hours. No sign of nanocrystal aggregation in culture medium is observed. None of the nanocarriers present cytotoxicity at concentration [Fe] from 11 to 112 $\mu g\,ml^{-1}$ as assessed by metabolic activity measurements that are similar to the control nonexposed cells. Consistently, the necrosis level did not exceed 7% in all conditions.

Let us investigate the transfer of magnetic properties to the tumor cells incubated with dispersed nanocrystals, colloidosomes, and supraballs: after 24 hours' incubation, the cells are rinsed, harvested from culture flasks, and then suspended. They are exposed to a calibrated and uniform gradient magnetic field to measure their magnetophoretic mobility. The velocity of single cells toward the magnet directly quantifies each cell magnetization. This reflects the balance between the magnetic

Metal Nano 3D Superlattices: Synthesis, Properties, and Applications, First Edition.

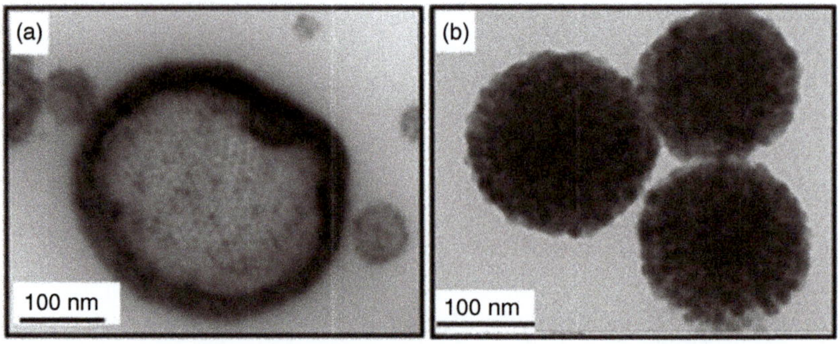

Figure 15.1 TEM images of colloidosomes (a) and supraballs (b) of 10-nm Fe_3O_4 nanocrystals. *Source:* (a) Nicolás-Boluda et al. [2]/JOHN WILEY & SONS, INC and (b) Dobryden et al. [3]/JOHN WILEY & SONS, INC.

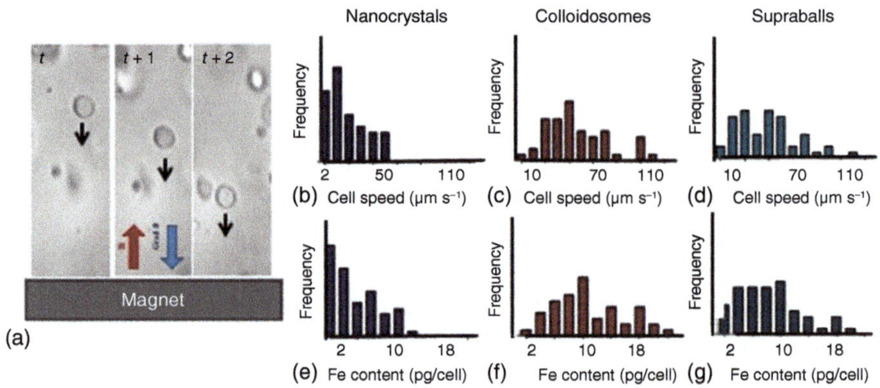

Figure 15.2 Magnetophoretic mobility of A431 cells incubated with dispersed nanocrystals, colloidosomes, and supraballs, respectively, during one day. (a) Illustration of magnetophoresis setup. Loaded cells in suspension are subjected to a magnetic gradient, and their motion toward the magnet is recorded by videomicroscopy. Images are taken every 0.5 seconds ($t = 0.5, t + 1 = 1, t + 2 = 1.5$ seconds). This reflects the balance between magnetic force (proportional to cell magnetic moment) and viscous forces (proportional to cell radius and cell velocity) in a permanent regime of constant velocity (b, c, d). Frequency histogram of A431 cell speed distribution loaded with dispersed nanocrystals (b), colloidosomes (c), and supraballs (d). (e, f, g) Frequency histogram of Fe content (iron mass per cell) was obtained by calculating the average magnetic load per cell. Frequency histogram of Fe content in A431 cells incubated with dispersed NCs (e), colloidosomes (f), and supraballs (g). *Source:* (a) Nicolás-Boluda et al. [2]/JOHN WILEY & SONS, INC.

force (proportional to cell magnetic moment) and the viscous force (proportional to cell radius and cell velocity) in the constant velocity regime (Figure 15.2).

Histograms of cell velocity (Figure 15.2b–d) show markedly enhanced magnetic mobility when the cells are exposed to colloidosomes and supraballs as compared with nanocrystals in their dispersed form (Table 15.1).

To evaluate the iron mass per cell from the measured magnetophoretic mobility, we assume identical saturation magnetization for dispersed nanocrystals and

Table 15.1 Average cell velocity, v, toward the magnet and mean iron mass per cell m_{Fe} measured from the cell magnetophoresis assay.

Materials	Dispersed nanocrystals	Colloidosomes	Supraballs
v	$28 \pm 16\,\mu m\,s^{-1}$	$57 \pm 26\,\mu m\,s^{-1}$	$48 \pm 25\,\mu m\,s^{-1}$
m_{Fe}	5.5 ± 3.5 pg per cell	13 ± 7 pg per cell	10 ± 7 pg per cell

Source: Based on Nicolás-Boluda et al. [2].

suprastructures. The data show that the iron uptake per cell is a factor of 2 larger for colloidosomes and supraballs than for dispersed nanocrystals. Hence, the suprastructures have a better ability to be internalized by tumor cells in comparison with the dispersed nanocrystals at equivalent iron dose.

15.2.2 Internal Distribution of Colloidosomes and Supraballs in Tumor Cells

15.2.2.1 Nanocrystal Dispersions

To elucidate the intracellular distribution and fate of nanocrystals in tumor cells, we follow the particle's uptake by TEM at different points of time for up to one week after exposure to the two suprastructures and to dispersed nanocrystals. Dopamine-coated dispersed nanocrystals early interact as loose clusters with the cell plasma membrane (30 minutes' incubation) and are promptly internalized into membrane-closed endosomes in which their relatively dispersed state is preserved (Figure 15.3).

Figure 15.4 shows that, after two days, the nanocrystals become more concentrated in lysosomes and are seen as aggregates surrounded by an electron-dense lysosomal matrix, which suggests the implication of endogenous lysosomal proteins in the process of nanocrystal transformation. The diameters of the intracellular nanocrystals are significantly reduced at day 5 and day 8 in comparison with the first two days while the spherical shape is conserved. This suggests a gradual dissolution of the nanocrystals, as observed in previous studies for iron oxide nanoparticles with different shapes and coatings that underwent degradation in lysosomes and Fe recycling into ferritin proteins [5–7].

15.2.2.2 Colloidosomes

In contrast to dispersed nanocrystals, colloidosomes show very distinct intracellular distribution. The most striking difference is that the nanocrystal shells of colloidosomes are closely located to the cell membranes, adapting to the shape of intracellular vesicles (Figure 15.5).

This is clearly facilitated by their flexible structure with a low stiffness (see Chapter 14). Colloidosome internalization is slightly delayed compared with that of the smaller dispersed nanocrystals and is barely observed at 30 minutes' and four hours' incubation. After one-day incubation, most of colloidosomes are entirely encapsulated within intracellular membranes in the cell cytoplasm. In comparison

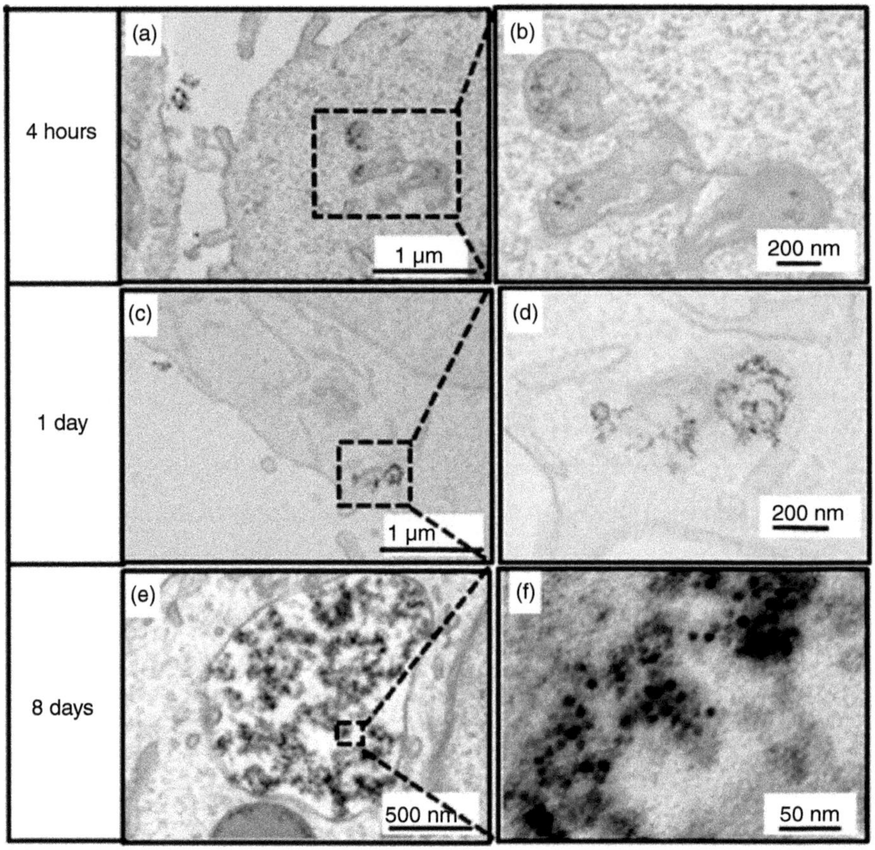

Figure 15.3 2D TEM slices images of A431 cells incubated with Fe_3O_4 nanocrystals ([Fe]=28 μg l^{-1}) for different points of times: 4 hours (a, b) 1 day (c, d) and at 8 (e, f) days. *Source:* Nicolás-Boluda et al. [2]/JOHN WILEY & SONS, INC.

to the size of colloidosomes in water suspension, the domain of self-assembled nanocrystals increases in cells from early time points, suggesting cell-induced deformation, rearrangement, and/or fusion of the self-assemblies (Figure 15.4).

In the 2D TEM slices, some curved fragments of colloidosomes embedded into intracellular membranes without conspicuous lumen are also found (Figure 15.5). This shows that colloidosomes can be flattened and shaped by cells to form anisotropic inclusions of several (most often organized) layers of nanocrystals closely interacting with the intracellular membranes. This flattening increases over time as colloidosomes or fragments seem to condense into autophagosomes and lysosomes. This important reorganization of the nanocrystal assemblies upon contact with cell membrane suggests competing interactions between on the one hand adjacent nanocrystals coated with oleic acid and on the other hand nanocrystals and the hydrophobic parts of the cell membrane. Inside cells, nanocrystals tend to assemble again through an interdigitation process of oleic acid molecules. This indicates that interparticle interactions remain stronger than those of nanocrystals with the

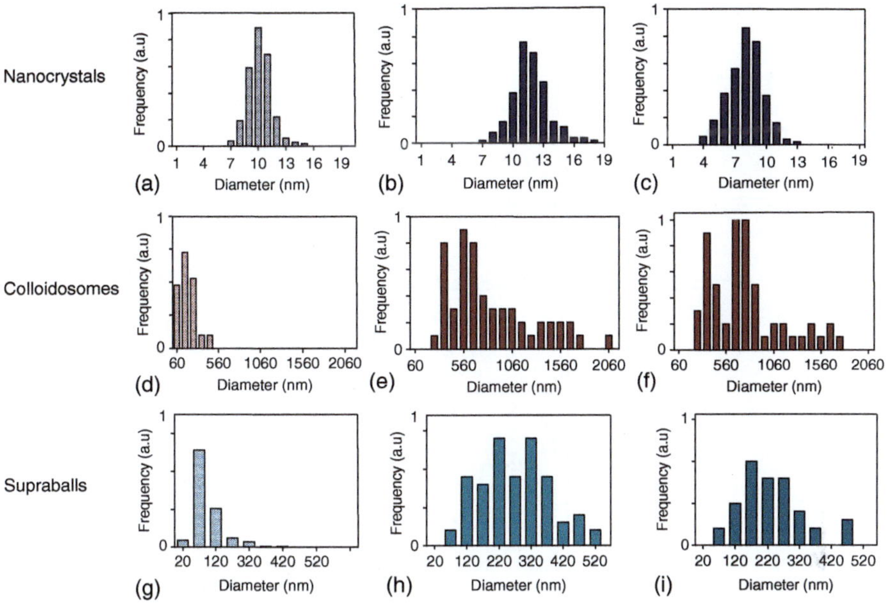

Figure 15.4 Histogram of dispersed nanocrystal (a, b, c), colloidosome (d, e, f), and supraball (g, h, i) diameters detected by cryo-TEM and from the 2D TEM slices of A341 cells at one day (a, d, g), two days (b, e, h), and eight days (c, f, i) following incubation with 28 $\mu g\,ml^{-1}$ of dispersed NCs (dark blue), colloidosomes (red), and supraballs (blue). To quantify the change in the colloidosome size in cells, we assume, even though all colloidosomes are not spherical, that they remain spherical. *Source:* Based on Nicolás-Boluda et al. [4].

biolipids. This might be explained by the fact that nanocrystals coated with oleic acid selectively interact. However, the presence of biolipids breaks the crystalline order, and reorganization to amorphous assemblies takes place. After eight days, organized layered structures of nanocrystals reminiscent of the colloidosome shell structure are clearly seen, demonstrating the enduring self-organization of 3D superlattices within intracellular compartments. Similar to dispersed nanocrystals, we notice an electron-dense protein matrix and ferritin close to ferrite nanocrystals, suggesting that the lysosomal degradation process might be ongoing although the individual nanocrystals still look intact.

15.2.2.3 Supraballs

Supraballs collectively interact with the plasma membrane at short times, mostly conserving their spherical structure after entering intracellular compartments (Figure 15.6). Endosomes are found to frequently contain intact supraballs in their lumen.

These compartments fuse with lysosomes over time, showing, at eight days, an electron-dense matrix surrounding well-preserved self-organized nanocrystals within supraballs. Similar to colloidosomes, but to a lesser extent, the apparent size of supraballs is found to be larger in cells than that in water suspension, suggesting

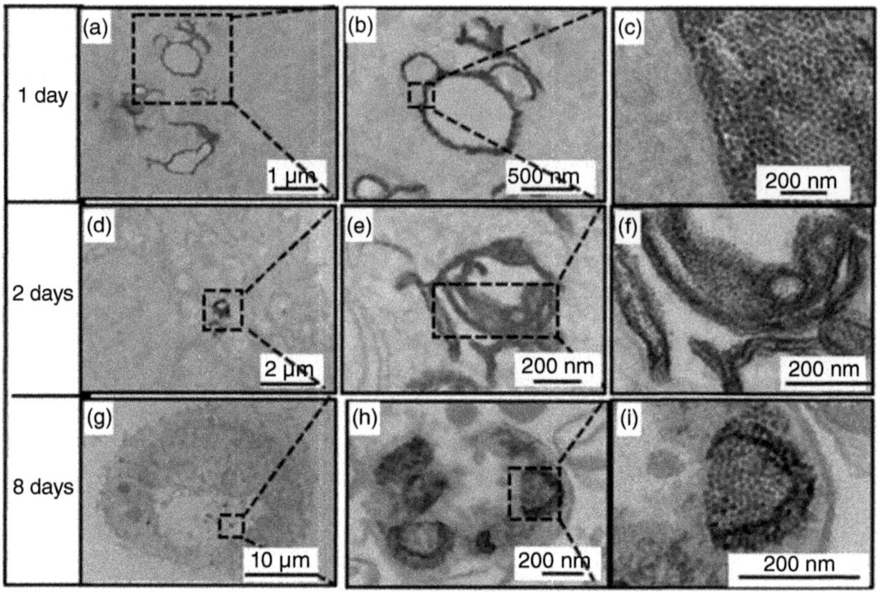

Figure 15.5 2D TEM slices images of A431 cells upon exposure to a magnetic field of 1 T during four hours following Day 1 incubation with colloidosomes: (a) overall pictures (b–d) intracellular and (e, f) extracellular internalizations. *Source:* Nicolás-Boluda et al. [2]/JOHN WILEY & SONS, INC.

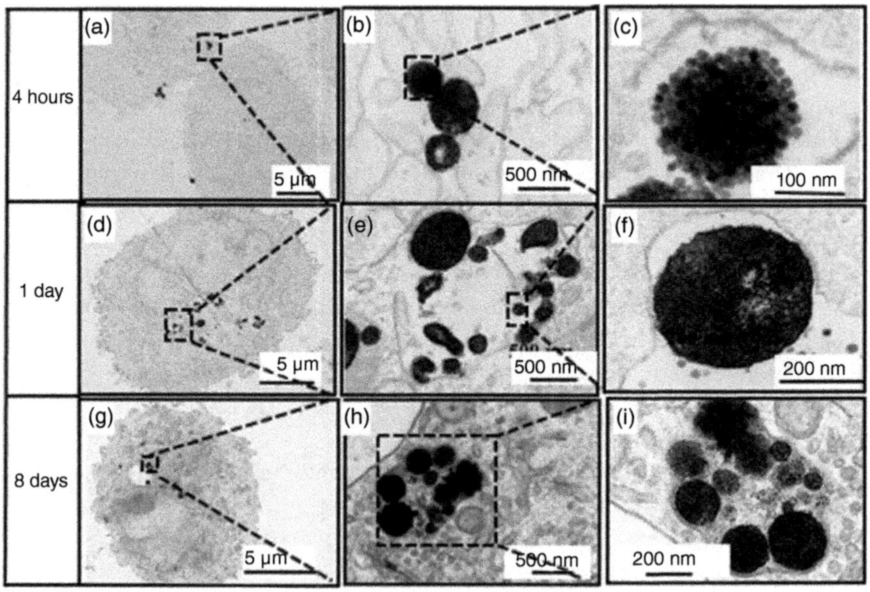

Figure 15.6 2D TEM slices images of A431 tumor cells incubated with supraballs for different points of time: 4 hours, (a–c) 1 day (d–f) and 8 days (g–i) with 28 μg l^{-1} iron concentration. *Source:* Nicolás-Boluda et al. [2]/JOHN WILEY & SONS, INC.

fusion and reorganization. However, after entering the cell, the apparent size does not significantly demonstrate that the reorganization mostly occurs at early times (Figure 15.5). Importantly, we continuously observe compact structures and spherical shapes of supraballs when they transit in the different intracellular compartments up to lysosomes. This is different compared with the colloidosomes that underwent significant shape deformations upon cell internalization. Clearly, this different fate in cells is due to the differences in flexibility, stiffness, and internal mobility of nanocrystals (see Chapter 14).

In conclusion, it is interesting to notice that whether nanocrystals have been self-organized into colloidosome or supraball structures before cell internalization, they conserve a locally ordered organization within the cell compartment for up to eight days, the final point of time of our experiment.

15.2.3 Structural Organization of Nanocrystals in Tumor Cells

To investigate the average structural organization and integrity of nanocrystals in cells with high statistics, small-angle X-ray scattering (SAXS) experiments are performed at different time points from 30 minutes to 48 hours during cell processing of dispersed nanocrystals, colloidosomes, and supraballs (Figures 15.7 and 15.8).

The scattered intensity of non-labeled cells is subtracted from the initial signal profile. This permits to isolate the contribution of nanocrystals to the scattering intensity (Figure 15.7). The signature of nanocrystals is detectable already after 30 minutes of incubation, and the signal-to-noise ratio improves with time. This shows that the cellular uptake is progressive and starts from the very beginning of the incubation period. The SAXS pattern at high q fits the form factor $P(q)$ of spheres with radius of around 5 nm in line with the cryo-TEM size and spherical shape of individual nanocrystals.

The structure factor, at low q, provides information on the local nanocrystal organization. $S(q)$ is obtained by dividing the scattered intensity by the form factor $P(q)$ of individual nanocrystals. For cells incubated with dispersed nanocrystals, $S(q)$ is characterized by a power law decay with an exponent of −1.7 (Figure 15.8a). This indicates

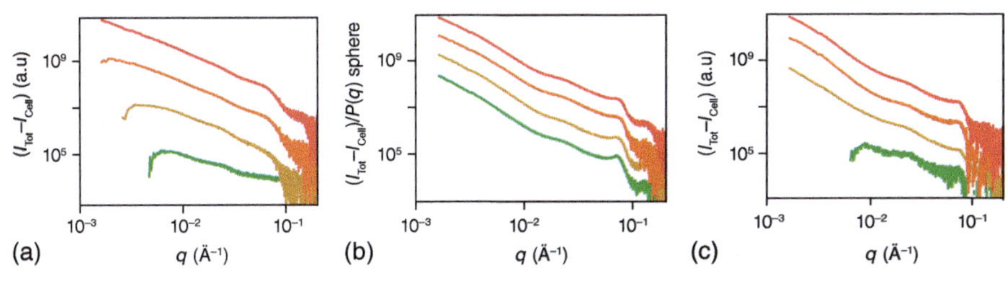

Figure 15.7 SAXS patterns of the total scattered intensity after subtraction of the cell contribution, $I_{total}(q) - I_{Cell}(q)$, (a, b, c) of cells after incubation for periods of 30 minutes (green), 4 hours (yellow), 1 day (orange), and 2 days (red) with (a) dispersed nanocrystals, (b) colloidosomes, and (c) supraballs. The curves presented are shifted by offsets from one to another for clarity. *Source:* (a–c) Based on Nicolás-Boluda et al. [4].

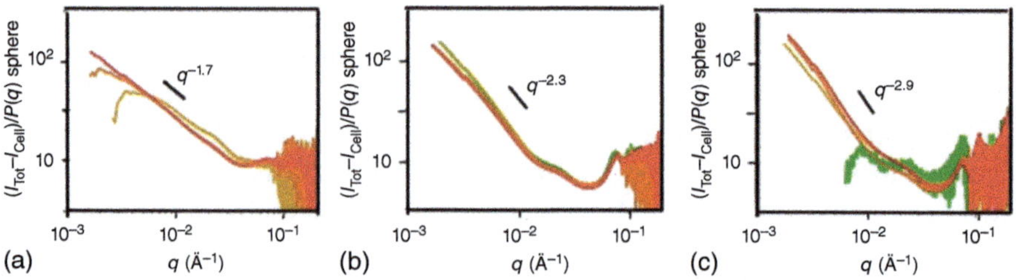

Figure 15.8 SAXS patterns of the total scattered intensity after subtraction of the cell contribution, $I_{total}(q) - I_{Cell}(q)$, divided by all the measured q range by the form factor $P(q)$ of 3.8 nm diameter sphere. We obtain the structure factor, $S(q)$, of NCs embedded in cells incubated with nanocrystals (a), colloidosomes (b), and supraballs (c) at the different time points. This is obtained after incubation for periods of 30 minutes (green), 4 hours (yellow), 1 day (orange), and 2 days (red). This reflects the rapid formation, after incubation, of loose fractal aggregates of nanocrystals (a) and denser structures of colloidosomes (b) and supraballs (c). *Source:* (a–c) Based on Nicolás-Boluda et al. [4].

the cell-induced formation of nanocrystal aggregates with an open structure and increasing aggregation number. This intracellular fractal aggregation of nanocrystals, similar to diffusion limited aggregation, is fully consistent with the TEM observations (Figure 15.3) and with previous small-angle neutron scattering results on endosomal clustering of iron oxide nanoparticles [8]. In contrast, for cells incubated with colloidosomes and supraballs, $S(q)$ is characterized by a power law decrease of the scattered intensity, with exponent of −2.3 and −2.9 for colloidosomes and supraballs, respectively (Figure 15.8b,c). This demonstrates a dense packing of nanocrystals with a further accentuated with supraballs. Importantly, these SAXS profiles do not change over 48 hours, confirming that the preassembly of nanocrystals enables maintaining of their ordered organization after cell internalization. Taken together, the SAXS and TEM structural analysis reveal drastic differences in cellular internalization of dispersed nanocrystals in comparison with their self-assemblies as deformable colloidosomes or as rigid supraballs. Dispersed nanocrystals are internalized in the form of loose fractal aggregates with rather low fractal dimension (1.7). In contrast, the local organization and interparticle distance of preassembled nanocrystals into suprastructures persist in cell organelles up to one-week post-labeling. Here, we demonstrate that self-organization of nanocrystals into the suprastructures outperforms current nanosystems in prescribing the intracellular organization of nanocrystals for at least one week. Note that the data obtained around $q = 0.15\,\text{Å}^{-1}$ are particularly noisy; consequently, it is impossible to observe the expected [111] peak of the fcc structure.

15.2.4 Interactions with Lysosomes

Dispersed nanocrystals, colloidosomes, and supraballs show very distinct organizational evolution in tumor cells, suggesting that they are not sensed and processed in the same way by endo/lysosomal compartments. As a consequence, we hypothesize that we could leverage their different structural and mechanical characteristics to modulate the intralysosomal distribution of nanocrystals in tumor cells.

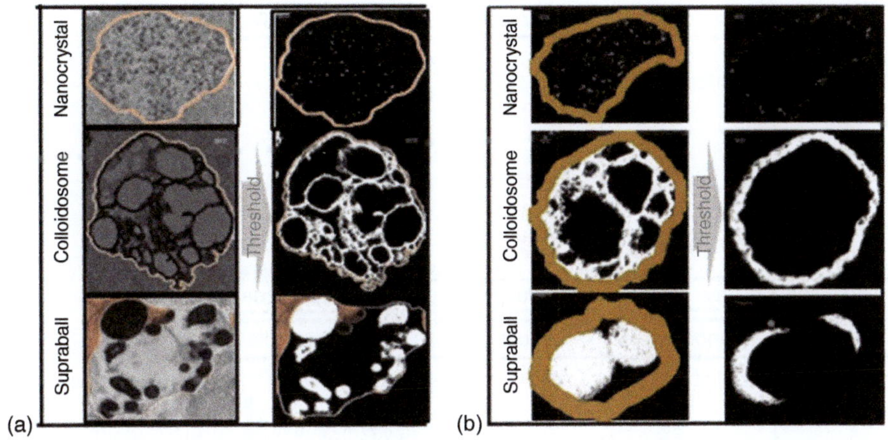

Figure 15.9 (a) Illustrative example of image analysis by manually delimiting the lysosomal membrane (pink line) and thresholding the area covered by nanocrystals (in white), hence analyzing the nanocrystal inside the lysosomes at day 1 and day 8 after internalization. (b) Proximity of nanocrystals to the lysosomal membrane. *Source:* Nicolás-Boluda et al. [2]/JOHN WILEY & SONS, INC.

Based on TEM image analysis (Figure 15.9), we have defined objective descriptors to characterize (i) the lysosome response to nanocrystals (lysosome size and circularity) and (ii) the intralysosomal distribution of nanocrystals (nanocrystal density and proximity to the lysosomal membrane). Electron-dense nanocrystals are identified as areas that are darker than biological structures and that can be detected by image analysis and thresholding (Figure 15.9a). We focus on lysosomes that contained nanocrystals by contouring lysosomal membranes at day 1 or day 8 post-labeling (Figure 15.9a). To evaluate whether lysosomes can adapt to the presence and organization of nanocrystals, we compare the apparent lysosome area at each condition and how it changed with time.

15.2.4.1 Lysosome Size

Surprisingly, at day 1, Figure 15.10a shows that the lysosomes containing colloidosomes are more than twofold larger than those containing initially dispersed nanocrystals (mean area of $4.1\,\mu m^2$ versus $1.8\,\mu m^2$) and 10-fold larger than lysosomes encapsulating supraballs ($4.1\,\mu m^2$ versus $0.4\,\mu m^2$). This suggests that the lysosomal compartment adapts not only to the carrier size but also to its ability to deform and to interact with the cell membrane dynamics. Deformable colloidosomes are internalized in much larger endosomes/lysosomes than the more rigid and less deformable supraballs. Supraballs are internalized in smaller compartments than dispersed nanocrystals. These differences arise from the early mechanisms of intracellular sequestration (day 1) as they tend to reduce at longer times. In particular, lysosomes containing colloidosomes drastically shrink in size between day 1 and day 8, whereas lysosomes containing dispersed nanocrystals or supraballs did not change their size significantly during the same period of time.

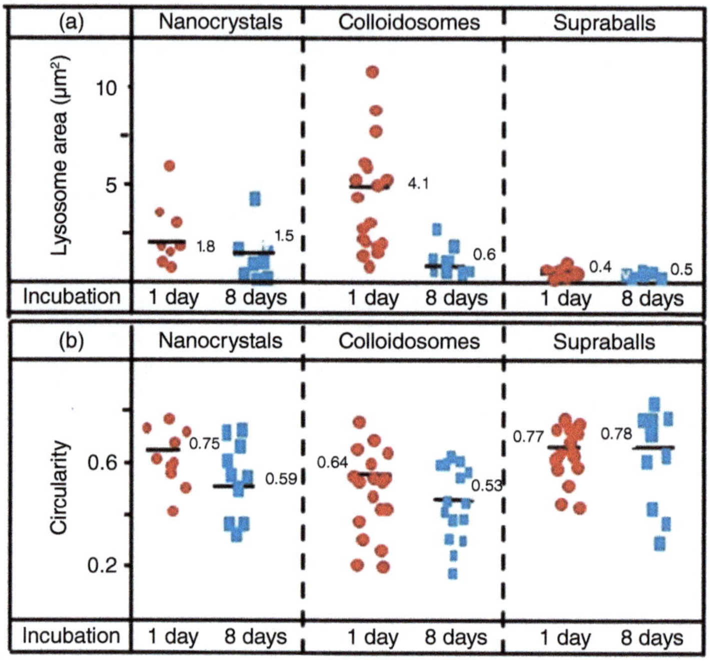

Figure 15.10 (a) Area of the lysosomes containing nanocrystals. (b) Circularity of lysosomes containing nanocrystals. *Source:* (a, b) Nicolás-Boluda et al. [4]/John Wiley & Sons.

15.2.4.2 Lysosome Shape

To further elucidate the lysosome sensing and adaptation to our nanostructures, the shape of the lysosome membrane is examined, and its circularity index is calculated (Figure 15.10b). Interestingly, lysosomes loaded with supraballs are the most circular, and their circularity, like their size, did not change from day 1 to day 8 (circularity of 0.77 and 0.78, respectively). This suggests that the rigid supraballs induce sustained membrane tension in the lysosomes. In contrast, colloidosomes are sequestered into lysosomes with much lower circularity (circularity of 0.64 at day 1), further decreasing at day 8 (0.53) and showing membrane bending undulations, analogous to flickering vesicles [9]. The large excess of membrane in colloidosome-loaded lysosomes accounts for the colloidosome flexibility, suggesting a reciprocal adaptation of host membranes to the colloidosome mechanical properties [10]. The coupling of membrane fluctuations to the colloidosome shell, giving rise to the so-called Helfrich entropic repulsion [11, 12], might favor colloidosome flattening and their concentration into the smaller, still more undulated, lysosomes observed at day 8. Lysosomes encapsulating dispersed nanocrystals that could be considered as the least perturbed system shows circularity of 0.75 at day 1, and it decreased to 0.59 at day 8. In conclusion, the host compartments clearly adapt to their cargo, with striking differences between rigid and deformable structures or dispersed nanocrystals. The high flexibility of colloidosomes favors their elastic deformation by cells, extended wrapping, and close interactions with cellular membrane with reciprocal adaptation and distortion

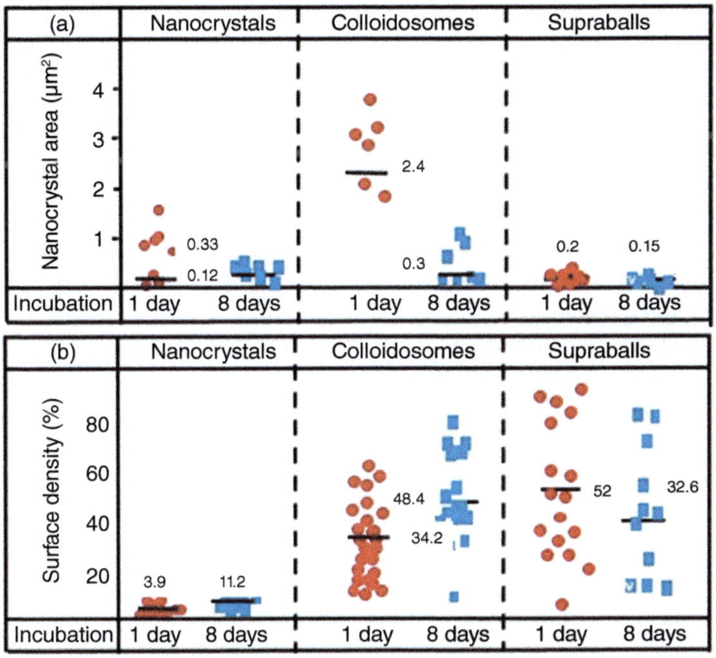

Figure 15.11 (a) Area covered by nanocrystals within the lysosome. (b) Percentage of area covered by nanocrystals in the lysosome surface. *Source:* (a, b) Nicolás-Boluda et al. [4]/John Wiley & Sons.

of the colloidosomes and of the host vesicle at short times that normalize at longer times [10]. It might be compared with cellular uptake and clearance of flexible erythrocytes, cell membrane fragments, or extracellular vesicles [13]. In contrast, the rigid and poorly deformable supraballs suffer few deformations over more than one-week post-internalization. Supraballs can be compared with viruses that adapt their mechanical properties to favor their entry into host cells and infectivity [14]. Overall, we observe dynamical changes in size, morphology, and membrane undulations of intracellular host compartments that reflect the mutual adaptation of intracellular membranes to the mechanical properties of self-assemblies.

15.2.4.3 Spatial Distribution and Density of Nanocrystals in Lysosomes

Lysosomal adaptation has important consequences for the density and spatial distribution of nanocrystals in these compartments. First, we evaluate the area covered by nanocrystals within the lysosome (Figure 15.11a), which reflects the amount of nanocrystals in each lysosome. The area covered by nanocrystals is found more than 20-fold higher and far more distributed for colloidosomes (mean of 2.4 μm [2]) in comparison with dispersed nanocrystals (0.12 μm^2) and supraballs (0.2 μm^2) at day 1. The area covered by nanocrystals in lysosomes evolves over time in different ways for dispersed nanocrystals and colloidosomes, mitigating at day 8 the initial differences found at day 1. For dispersed nanocrystal, at day 8, the nanocrystal area increases to 0.33 μm^2. This reflects accumulation of nanocrystal clusters in lysosomes. In contrast, the nanocrystal area drops to 0.3 μm^2 for colloidosomes and does not significantly change for supraballs. Colloidosomes are collectively engulfed at day 1 in very large and highly deformed compartments that became much smaller at day 8, suggesting intense reshaping and active membrane dynamics. On the other hand, supraball-containing lysosomes are much more stable over a week.

Considering the concomitant evolution of the lysosome size and nanocrystal load, the surface density of nanocrystals in lysosomes (ratio of surface occupied by nanocrystals over lysosome area) is deduced, as shown in Figure 15.11b. A first observation is that the use of supraballs maximizes the local density of nanocrystals, representing 52% of the lysosome area in average at day 1 and 32.6% at day 8. Hence, a compelling conclusion is that supraballs are particularly efficient to concentrate a large density of nanocrystals within small-sized lysosomes that do not change significantly in size and nanocrystal content over time. Remarkably, colloidosomes also optimize the local density of nanocrystals: it counts for 34.2% of the lysosome surface at day 1 and 52% at day 8. In comparison, dispersed nanocrystals only fill 3.9% of the lysosome surface area at day 1, increasing to 11.2% at day 8. The increase in the nanocrystal local density over time is due to the concurrent reduction of the size of host compartments that highlights lysosome accumulation and maturation in a way that depends on lysosomal cargo. Overall, the local density of nanocrystals increased from 5- to 12-fold by using suprastructures instead of dispersed nanocrystals. This is of crucial importance for controlling the physical response and biological outcome of intralysosomal nanocrystals. For example, cooperative magnetic properties that arise from supracrystalline organization of nanocrystals or ultrafast light to heat conversion demonstrated earlier with the same assemblies [15] could be easily transferred to cellular lysosomes, eliciting enhanced local effects. We anticipate that the optimized density of nanocrystals in the restricted area of lysosomes could amplify hot spot toxicity (local heating), mechanical stress, or radiosensitization for radiation therapy (see below).

15.2.4.4 Proximity of Nanocrystals to the Lysosome Membrane

The lysosomal membrane is recognized as a key target to induce cell death [16]. Indeed, lysosomal membrane permeabilization or breakdown causes the release of cathepsins and other hydrolases to the cytosol, which results in indiscriminate degradation of vital proteins. Lysosomal destabilization can be induced by the production of reactive oxygen species (ROS) through iron-catalyzed Fenton reactions [17], and it has been reported as a cytotoxic effect of local hyperthermia generated by magnetic nanoparticles [18]. Importantly, as ROS have a spatially limited range of activity, the localization of ROS mediators such as iron oxide nanoparticles with respect to the lysosomal membrane is of utmost importance. Therefore, we quantify the amount of nanocrystals that are localized at an arbitrary distance of 100 nm from the lysosome membrane (Figure 15.9b) to assess whether the proximity of nanocrystals to lysosome membrane can be modulated by the use of colloidosomes and supraballs.

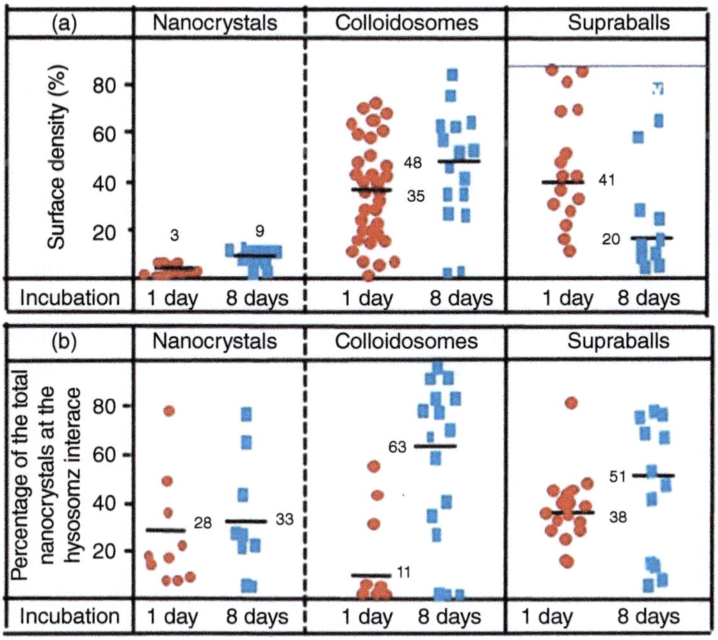

Figure 15.12 (a) Percentage of area occupied by nanocrystals within 100 nm of the lysosome membrane at 1 day and 8 days after internalization. (b) Percentage of the total number of nanocrystals in the lysosomes located within 100 nm from the lysosome membrane. *Source:* (a, b) Nicolás-Boluda et al. [4]/John Wiley & Sons.

At day 1, the surface density of nanocrystals in the vicinity of the membrane (percentage of area occupied by nanocrystals within 100 nm of the lysosome membrane) reaches a mean value of 35% for colloidosomes and 41% for supraballs being 10-fold and 12-fold larger than for dispersed nanocrystals (3.3%). This confirms membrane wrapping of the self-assemblies at the early stage of internalization. After eight days, the density of nanocrystals close to the membrane was 48% for colloidosomes, but decreased to 25% for supraballs, in comparison with 8% for dispersed nanocrystals (Figure 15.12a). This again reflects the durable interactions of colloidosomes with the lysosomal membrane while lysosome maturation tends to push aside the supraballs from the cell membrane.

Another relevant quantity is the fraction of the total nanocrystals in the lysosomes that lies close to the membrane (Figure 15.12b). Both colloidosomes and supraballs maximize this fraction after eight days with 63% and 51% of the total number of nanocrystals being located within 100 nm from the lysosome membrane. In contrast, dispersed nanocrystals distribute randomly as fractal clusters within the lysosome with a fraction close to the membrane of about 30%. We conclude that colloidosomes and supraballs are valuable nanostructures to optimize the density of nanocrystals close to the lysosomal membrane, which could enhance any biological effect on these vital cell components.

15.2.5 Magnetic Response of Internalized Nanocrystals

From the data described above, the nanocrystals keep their assemblies in compact network without keeping their crystalline structure. However, we know that the small distance between nanocrystals in internalized colloidosomes and supraballs facilitates collective properties due to magnetic dipolar interactions. The fact that the nanocrystal organization can endure biological processing by tumor cells is of upmost importance when considering that magnetic and optical properties of nanocrystals drastically depend on their local organization. For example, dipole–dipole interparticle interactions, magnetic ordering, and local arrangement of nanoparticles [18–20] strongly affect the magnetic to heat conversion of magnetic nanocrystals under alternating magnetic field. We and others have demonstrated that cell internalization can change the magnetic response and jeopardize the heating efficiency of different magnetic structures through uncontrolled magnetic interactions elicited by endosomal nanoparticle confinement and fractal clustering [21,22].

The response of intracellular nanocrystals to a uniform magnetic field is analyzed to assess whether the organization of nanocrystals as colloidosomes or supraballs could favor intracellular magnetic manipulation. After day 1 of incubation with dispersed nanocrystals, colloidosomes, and supraballs, followed by careful lavage, cells are exposed for four hours to a uniform magnetic field of 1 T.

With dispersed nanocrystals, the optical microscopy pattern does not show any effect. From 2D TEM observation of cells that are fixed in the presence of the magnetic field, the nanocrystals remain disordered in the cell and alignment of lysosomes containing dispersed nanocrystals (NCs) (Figure 15.13).

With colloidosomes, conspicuous dark rods (with length from 1 to 20 μm) in cells are all aligned along the direction of the applied magnetic field and follow with the direction of the field. This contrasts with no anisotropic structures for cells incubated with dispersed nanocrystals.

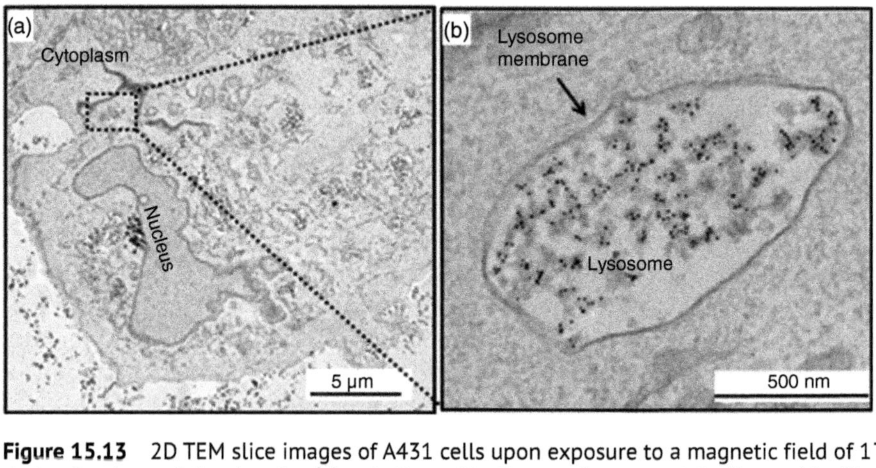

Figure 15.13 2D TEM slice images of A431 cells upon exposure to a magnetic field of 1 T during four hours following day 1 incubation with dispersed nanocrystals. *Source:* Nicolás-Boluda et al. [2]/JOHN WILEY & SONS, INC.

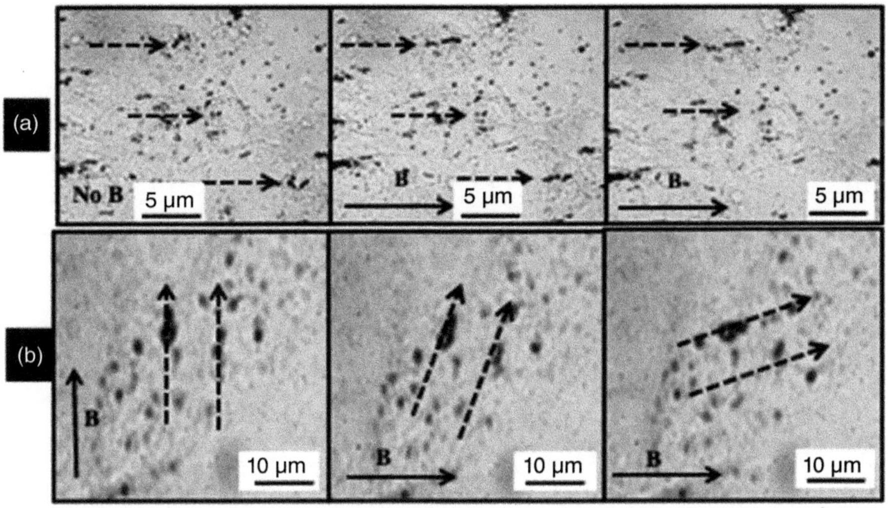

Figure 15.14 Bright-field images of colloidosomes, internalized in cells during one day. (a) Pictures are registered at various times from zero to one minute. Progressive alignment of internalized colloidosomes (subjected to a magnetic field). (b). Effect on pre-aligned internalized colloidosomes due to changing the orientation of the applied field by 90°. *Source:* Nicolás-Boluda et al. [2]/JOHN WILEY & SONS, INC.

The corresponding TEM image shows the formation of colloidosome chains aligned along the field direction in one minute of field exposure (Figure 15.14a). Once the colloidosomes are aligned, the magnetic field is rotated by 90°, and we observe reorientation of the chains (Figure 15.14b). A careful observation (see arrow in Figure 15.14b) shows that some of the chains are more easily manipulated than others.

High-magnification electron microscopy of colloidosome chains (Figure 15.15) reveals that some chains are fully intracellular consisting of lysosome alignments (Figure 15.15a–d), whereas others are surrounded by a cell membrane and anchored to the cell but are partly in the extracellular space (Figures 15.15a, e, f). This could explain their different dynamical response to the rotation of the magnetic field shown in Figure 15.14b.

By replacing colloidosomes with supraballs, dynamic alignments of supraparticles are also observed (Figure 15.16a). The rotation of the aligned supraballs (Figure 15.16b) seems slower than with colloidosomes under the same conditions (Figure 15.14b), which can be attributed to the presence of supraballs embedded in cytoplasm and entangled into membranes as shown in Figure 15.17 with intra- and extracellular alignments.

This markedly contrasts with the absence of any alignment of lysosomes containing dispersed nanocrystals (Figure 15.13). The formation of such long chains of colloidosomes (Figure 15.15) or supraballs (Figure 15.16) can be explained by the large magnetic dipolar interactions induced by the self-assemblies of nanocrystals both inside and close to the lysosome membrane.

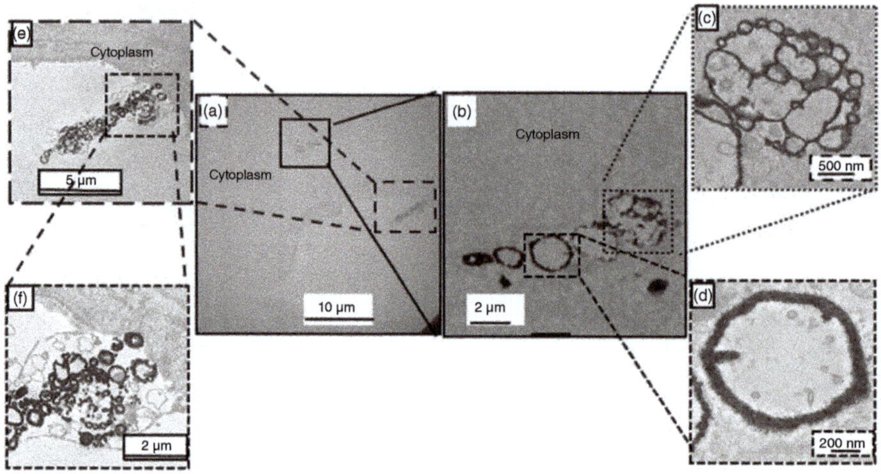

Figure 15.15 2D TEM slice images of A431 cells upon exposure to a magnetic field of 1 T during four hours following day 1 incubation with colloidosomes. *Source:* Nicolás-Boluda et al. [2]/JOHN WILEY & SONS, INC.

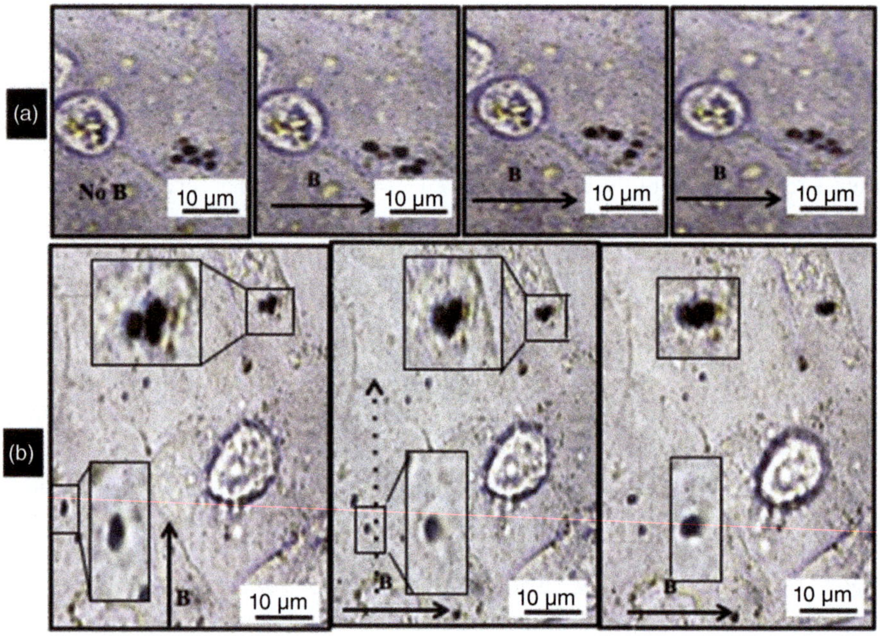

Figure 15.16 Bright-field images of supraballs, internalized in cells during one day. (a) Pictures are registered at various times from zero to one minute. Progressive alignment of internalized supraballs (subjected to a magnetic field). (b) Effect on pre-aligned internalized colloidosomes due to changing the orientation of the applied field by 90°. *Source:* Nicolás-Boluda et al. [2]/JOHN WILEY & SONS, INC.

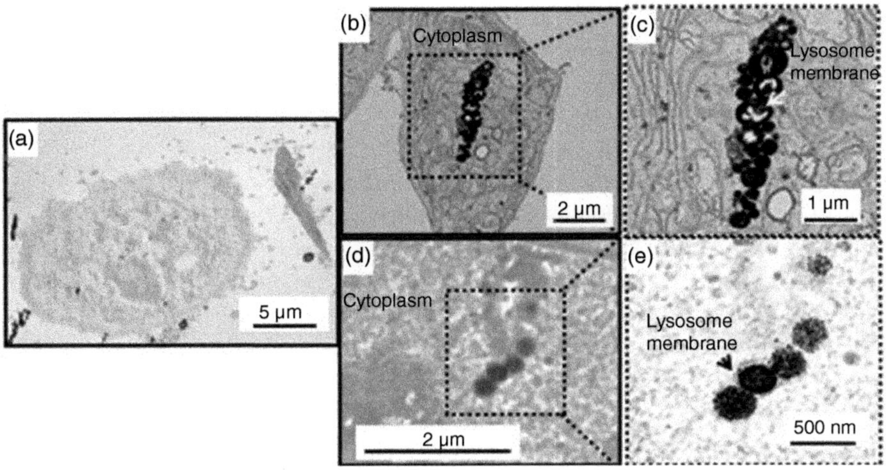

Figure 15.17 Various images obtained (a–e) from 2D TEM slice of A431 cells upon exposure to a magnetic field of 1T during 4h following Day 1 incubation with supraballs. *Source:* Nicolás-Boluda et al. [2]/JOHN WILEY & SONS, INC.

Overall, the field-induced arrangement of supraballs and colloidosomes can facilitate magnetic manipulations that are not possible with dispersed nanocrystals under our conditions. These observations, together with the sustained organization of nanocrystals within the cells, pave the way for the use of supracrystalline magnetic structures for biological or medical applications based on magnetic stimulation to activate specific molecular pathways or ion channels by magnetic force/torque-based methods.

Such anisotropic assembly of colloidosomes or supraballs could be rotated by changing the direction of the magnetic field, and the formation of chains could be monitored live as previously observed for magnetic endosomes [23], iron oxide nanoparticle-filled carbon nanotubes [24], or magnetic nanorods [11, 25].

15.3 Conclusion

We propose a new concept for modulating intracellular distribution, organization, fate, and manipulation of nanocrystals within intracellular organelles. The present study investigated for the first time the interactions of magnetic nanocrystals self-organized into colloidosomes or fcc supraballs with tumor cells, in comparison with their dispersed nanocrystal components. Colloidosomes are characterized as a flexible shell of ordered nanocrystals, while supraballs are described as rigid spherical nanocrystal assemblies. Remarkably, these supramolecular organizations of nanocrystals are differently sensed by tumor cells, triggering distinct intracellular fates mostly related to their different mechanical properties.

Structural and mechanical analogies of rigid supraballs with viruses could account for their full-membrane wrapping and long-lasting integrity within intracellular endosome/lysosome compartments for up to one week. In contrast, colloidosomes are rather seen by cells as "extracellular vesicles" exhibiting a deformable shell that complied with cellular membranes from the early stage of cell entry to the late stage of lysosomal processing. In turn, intracellular compartments strongly adapt to the uptake of colloidosomes by increasing first their size and then their membrane undulations at longer times.

Importantly, these differences in cellular sensing of nanostructures also influence their fate within lysosomes. These make it possible to modulate the density of magnetic nanocrystals in lysosomes and their proximity to the lysosome membrane. Even better than supraballs, colloidosomes could maximize the intralysosomal and close-to-membrane concentration of iron oxide nanocrystals particularly after eight days of lysosome maturation, transferring magnetic responsiveness to their host lysosomes.

While dispersed nanocrystals form uncontrolled fractal aggregates in lysosomes, the spatial ordering of nanocrystals into their self-assembled colloidosome and supraball structures could be maintained within the cell lysosomes up to at least eight days. This opens crucial opportunities for an enhanced control of physical properties of nanocrystals in the intracellular environment. Along with a tailored intracellular distribution of nanocrystals, a plethora of physical cooperative properties such as magnetic order or plasmon coupling could thus be translated to cellular components, allowing for more precise and tunable biological outcomes.

References

1 Yang, Z., Altantzis, T., Zanaga, D. et al. (2016). Supracrystalline colloidal eggs: epitaxial growth and freestanding three-dimensional supracrystals in nanoscaled colloidosome. *J. Am. Chem. Soc.* 138: 3493–3500.

2 Nicolás-Boluda, A., Vaquero, J., Laurent, G. et al. (2020). Photothermal depletion of cancer-associated fibroblasts normalizes tumor stiffness in desmoplastic cholangiocarcinoma. *ACS Nano* 14: 5738–5753.

3 Dobryden, I., Yang, Z., Claesson, P.M., and Pileni, M.P. (2020). Water dispersive suprastructures: an organizational impact on nanomechanical properties. *Adv. Mat. Interfaces* 7: 202001687. (1–10).

4 Nicolás-Boluda, A., Yang, Z., Dobryden, I. et al. (2020). Intracellular fate of hydrophobic nanocrystal self-assemblies in tumor cells. *Adv. Funct. Mater.* 30: 2004274.

5 Lartigue, L., Alloyeau, D., Kolosnjaj-Tabi, J. et al. (2013). Biodegradation of iron oxide nanocubes: high-resolution *in situ* monitoring. *ACS Nano* 7 (5): 3939–3952.

6 Kolosnjaj-Tabi, J., Di Corato, R., Lartigue, L. et al. (2014). Heat-generating iron oxide nanocubes: subtle "destructurators" of the tumoral microenvironment. *ACS Nano* 8: 4268–4283.
7 Volatron, J., Carn, F., Tabi, J.K. et al. (2017). Ferritin protein regulates the degradation of iron oxide nanoparticles. *Small* 13: 1602030.
8 Rivière, C., Wilhelm, C., Cousin, F. et al. (2007). Internal structure of magnetic endosomes. *Eur. Phys. J. E Soft Matter* 22 (1): 1–10.
9 Sackmann, E. (1994). Membrane bending energy concept of vesicle- and cell-shapes and shape-transitions. *FEBS Lett.* 346: 3–16.
10 Xiong, K., Zhao, J., Yang, D. et al. (2017). Cooperative wrapping of nanoparticles of various sizes and shapes by lipid membranes. *Soft Matter* 13: 4644–4652.
11 Mazuel, F., Mathieu, S., Di Corato, R. et al. (2017). Forced- and self-rotation of magnetic nanorods assembly at the cell membrane: a biomagnetic torsion pendulum. *Small* 13 (31): 1701274. Monzel, C. and Sengupta, K. (2016). *J. Phys. D: Appl. Phys.* 49 (24): 243002.
12 Ghochani, M., Nulton, J., Salamon, P. et al. (2010). Tensile forces and shape entropy explain observed crista structure in mitochondria. *Biophys. J.* 99 (10): 3244–3254.
13 Yi, X. and Gao, H. (2015). Cell membrane wrapping of a spherical thin elastic shell. *Soft Matter* 11: 1107–1115.
14 Greber, U.F. (2016). Virus and host mechanics support membrane penetration and cell entry. *J. Virol.* 90: 3802–3805.
15 Mazzanti, A., Yang, Z., Silva, M.G. et al. (2019). Light–heat conversion dynamics in highly diversified water-dispersed hydrophobic nanocrystal assemblies. *Proc. Natl. Acad. Sci. U.S.A.* 116 (17): 8161–8166.
16 Boya, P. and Kroemer, G. (2008). Lysosomal membrane permeabilization in cell death. *Oncogene* 27: 6434–6451.
17 Mak, I. and Weglicki, W. (1985). Characterization of iron-mediated peroxidative injury in isolated hepatic lysosomes. *J. Clin. Invest.* 75: 58–63.
18 Clerc, P., Jeanjean, P., Hallali, N. et al. (2018). Targeted magnetic intra-lysosomal hyperthermia produces lysosomal reactive oxygen species and causes Caspase-1 dependent cell death. *J. Controlled Release* 270: 120–134.
19 Tan, R., Carrey, J., and Respaud, M. (2014). Magnetic hyperthermia properties of nanoparticles inside lysosomes using kinetic Monte Carlo simulations: influence of key parameters and dipolar interactions, and evidence for strong spatial variation of heating power. *Phys. Rev. B* 90 (21): 214421.
20 Serantes, D., Simeonidis, K., Angelakeris, M. et al. (2014). Multiplying magnetic hyperthermia response by nanoparticle assembling. *J. Phys. Chem. C* 118 (11): 5927–5934.
21 Cabrera, D., Coene, A., Leliaert, J. et al. (2018). Dynamical magnetic response of iron oxide nanoparticles inside live cells. *ACS Nano* 12: 2741–2752.

22 Di Corato, R., Espinosa, A., Lartigue, L. et al. (2014). Magnetic hyperthermia efficiency in the cellular environment for different nanoparticle designs. *Biomaterials* 35 (24): 6400–6411.

23 Wilhelm, S., Tavares, A.J., Dai, Q. et al. (2016). Analysis of nanoparticle delivery to tumours. *Nat. Rev. Mater.* 1: 16014.

24 Liu, X., Marangon, I., Melinte, G. et al. (2014). Targeted labeling of early-stage tumor spheroid in chorioallantoic membrane model with upconversion nanoparticles. *ACS Nano* 8 (11): 11290–11304.

25 Zhu, L., Huang, W., Yang, F. et al. (2019). *Adv. Biosyst.* 3 (1): 1800246. Wilhelm, C., Gazeau, F., and Bacri, J.-C. (2003). *Phys. Rev. E* 67 (6): 061908.

16

Photothermal Effects in the Tumor Microenvironment

16.1 Colloidosomes and Supraballs

Spherical ferrite nanocrystals coated with oleic acid with 9.6-nm diameter and 8% polydispersity self-assemble to produce either shell and solid spherical assemblies called colloidosomes and supraballs respectively [1]. The same nanocrystals are also coated with dopamine to serve as a reference for isolated non-assembled positively charged nanocrystals dispersed in water. Independent of the coating agents, the structural study of the nanocrystals indicates a ferrite inverse spinel structure with the presence of wüstite. Consequently, we conclude that the nanocrystals used in this study are mainly Fe_3O_4. However, we cannot neglect the presence of γ-Fe_2O_3 and wüstite structure.

We design two types of water-dispersive self-assemblies of hydrophobic nanocrystals as illustrated in Figure 16.1. Colloidosome, empty vesicles with a crystalline colloidal shell of nanocrystals, are built via a template-confined chemical reaction (Figure 16.1a–e). The shell of the colloidosome is composed of an ordered self-assembly of nanocrystals in one to three layers. We underline that the colloidosomes are produced in the presence of oleic acid, dodecyltrimethylammonium bromide, and octadecene. Supraballs, homogeneous spherical 3D superlattices of nanocrystals, are generated via superlattice-matched epitaxial overgrowth (Figure 16.1f,g). The external layer of the oleic acid molecules is associated with dodecyltrimethylammonium bromide.

The size distributions of the self-assembly structures and their building blocks are obtained from transmission electron microscopy (TEM) images of dried samples and cryo-TEM images on hydrated samples. We find an average diameter of 10.1 ± 1.3 nm for dopamine-coated nanocrystals in water, 187 ± 93 nm for colloidosomes, and 91 ± 44 nm for fcc supraballs in their hydrated form. Note that the different size distribution of the supraballs and colloidosomes is imposed by the process of self-assembly and could not be varied by synthesis parameters in this

Metal Nano 3D Superlattices: Synthesis, Properties, and Applications, First Edition.

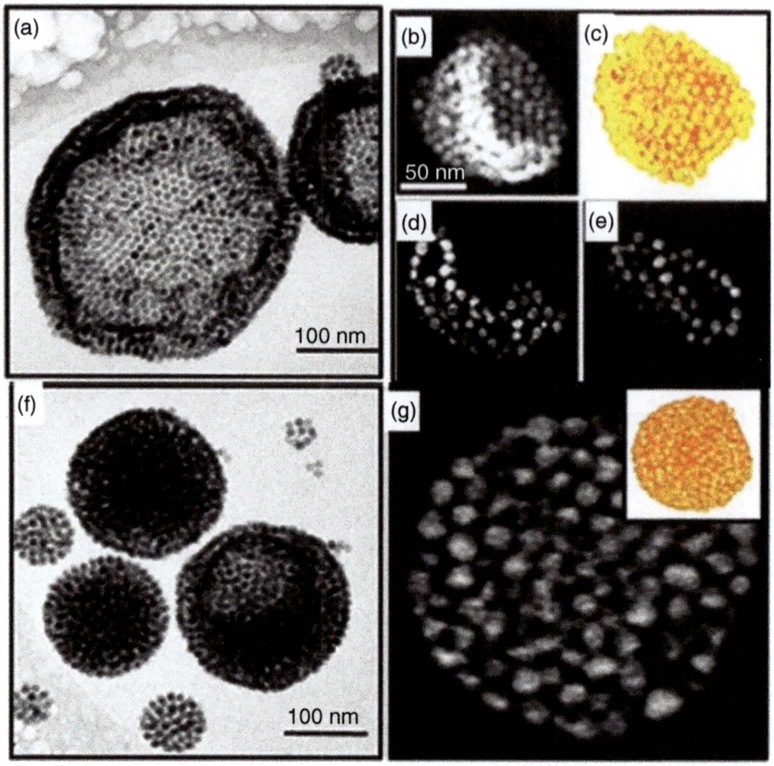

Figure 16.1 Self-assemblies of ferrite nanocrystals. Colloidosomes (a–e): (a) cryo-TEM image in water, (b) 2D high-angle annular dark field scanning TEM (HAADF-STEM) projection image of dried sample, and (c) 3D visualization of an electron tomogram (d, e) orthoslices through the 3D. Supraballs (f, g): (f) cryo-TEM image in water and (g) 2D HAADF-STEM orthoslice through the 3D tomogram shown in the inset. *Source:* Nicolas-Boluda et al. [2]/John Wiley & Sons, Inc.

study. A 3D reconstruction obtained by electron tomography on a dried sample shows the deformation and collapse of the colloidosome spherical structure (Figure 16.1b–e) while supraballs retain their shape (Figure 16.1f,g). Thus, the colloidosomes appear more deformable than the supraballs.

The zeta potential of dispersed nanocrystals (NCs) coated with dopamine and that of colloidosomes and supraballs dispersed in deionized water are +48, +43, and46 mV, respectively. The similarity between these various values clearly indicates that the zeta potential has a negligible influence on the data presented below.

16.2 Photothermal Properties: Apparent Contradiction Between the Global Heating and Cell Death

To evaluate the photoinduced heating properties of dispersed nanocrystals, colloidosomes, and supraballs, we have to compare the absorption spectra of the

Figure 16.2 Visible–NIR absorption spectra of Fe_3O_4 nanocrystals dispersed in solution (gray) and self-assembled in colloidosomes (red) and supraballs (blue) at equal iron concentrations.

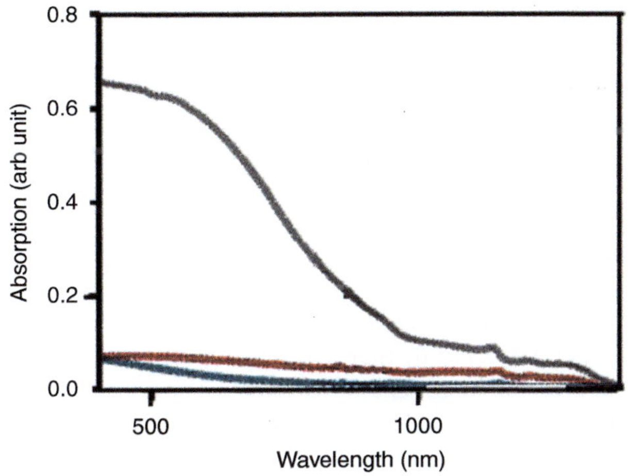

suprastructures and dispersed nanocrystals in aqueous solution. Unexpectedly, the suprastructure absorption spectra considerably dropped compared with their building blocks (Figure 16.2).

The photoactivated heating of the suprastructures and dispersed nanocrystals is obtained by subjecting 100 μl of suspension at fixed iron concentration (150 μg ml^{-1}) to an 808-nm laser power (1 W cm^{-2}) for five minutes. The temperature increment is the largest for supraballs (10.9 °C), followed by dispersed nanocrystals (8.8 °C), whereas the colloidosomes are heated by less than 2 °C. According to the drop of the suprastructure absorption spectra compared with that of dispersed nanocrystals, we could expect a lower temperature increment for both hybrids compared with dispersed nanocrystals. However, as shown in Chapter 14, nanocrystals self-assembled in superlattices act as nanoheaters embedded in a thermal reservoir formed by the coating agents [3]. This could explain the difference in the temperature increment between supraballs (spherical fcc 3D superlattices) and colloidosomes (closed structure of one or a few layers of nanocrystals). The rather large value obtained with dispersed nanocrystals compared with colloidosomes can be explained by the large difference in the number of absorbed photons at 808 nm.

Indeed, it has been shown that the temperature increase close to the heat nanosources is much higher than the macroscopic temperature in the surrounding medium [4, 5].

To be able to identify cell apoptosis and necrosis through the evaluation of the membrane permeability, three fluorescence dyes are used:

(i) Hoechst 33342 is well known for staining the nuclei of living cells. Bound to minor grooves of DNA, its fluorescence is markedly enhanced with blue/cyan emission around 461 nm. It allows to quantify all cells.
(ii) YO-PRO™ enters selectively in apoptosis cells with morphological features such as shrinkage and fragmentation. It is characterized by orange fluorescence.
(iii) Propidium iodide (PI) is specific for identifying dead (necrotic) cells. It green fluorescence centered around 617 nm is markedly enhanced upon binding to nucleic acids.

These dyes can be visualized by conventional fluorescence microscopes. Hence, the staining pattern resulting from the simultaneous use of these three dyes allows to distinguish alive, apoptotic, and necrotic cell populations as shown in Figure 16.3b,c.

To evaluate thermal damages on cells two different laser irradiation set-ups are used [6]:

16.2.1 Pellet of Nanocrystal-Loaded Cells

This combines local intracellular heating with global temperature increase in the pellet measured by thermal imaging. As previous studies demonstrated the detection of cell death apoptosis four to five hours post irradiation [2], cells that have been pelleted and irradiated are then put back in culture for four hours before cell death assessment (Figure 16.3a). Cells having internalized the different suprastructures without laser irradiation do not show any significant enhancement in apoptosis or necrosis level in comparison with the irradiated or uncontrolled cells. This confirms the data discussed above showing no significant toxicity in A431 cells up to concentration of 111 $\mu g\,ml^{-1}$ in the absence of irradiation. Furthermore, the suprastructures, nanocrystals, or the laser exposure did not affect cell viability itself.

After laser irradiation, cells having internalized dispersed nanocrystals show slightly enhanced level of apoptosis and necrosis. Figure 16.3b shows that the difference with non-irradiated cells is not statistically significant. In contrast, colloidosome and supraball-loaded cells feature a significant enhancement of the apoptosis (more than 15% of total cell number) and necrosis level (~15%) after irradiation (Figure 16.4a). These results are surprising if we consider the lower global temperature of the cell pellet for supraballs and colloidosomes in comparison with dispersed nanocrystals. This suggests that supraballs and to a larger extent colloidosomes could induce local thermal damages independently of the global temperature increment in the cell pellet. This has to be related to data shown in Chapter 14 [3] from which the light absorption in such nanostructures, dispersed in aqueous solution, initiates a chain energy transfer having various degrees of freedom coupled together. *in vitro*, the efficiency of supraballs is larger than colloidosomes. However, in cells, the colloidosome shape is not retained, and we can assume, but not quantify, that energy penetration is almost an even larger for colloidosomes than supraballs.

16.2.2 Monolayers of Cells Having Internalized the Fe_3O_4 Nanocrystals

To isolate thermal damages that arise from intracellular local heating, thermalized cell monolayers are exposed to the laser, and cell death is quantified in the same conditions as for the cell pellet (Figure 16.3c). As above, there is no effect of apoptosis or necrosis in the absence of laser irradiation. A significant increase of apoptosis/necrosis level after irradiation for the two suprastructures is observed compared with those of dispersed nanocrystals (Figure 16.3d). The level of cell damage induced by direct irradiation of cell monolayers is significantly higher (Figure 16.3d) than that

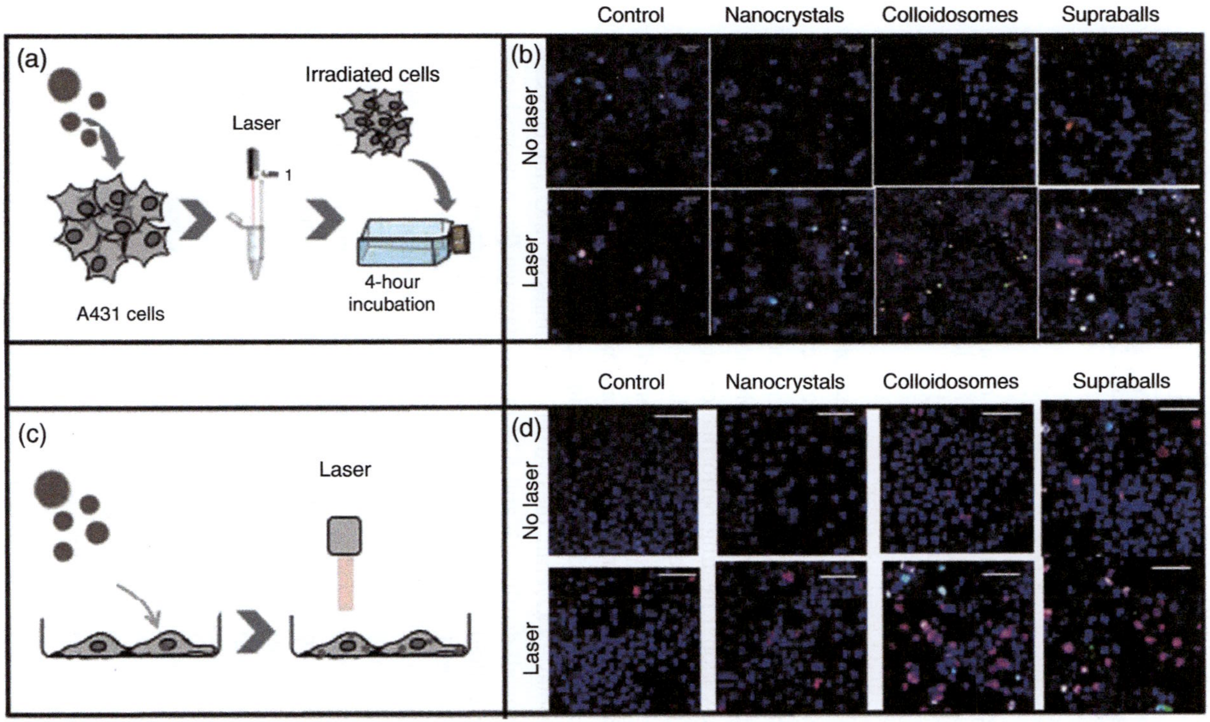

Figure 16.3 Evaluation of photothermal effects in A431 cells internalized with dispersed nanocrystals, colloidosomes, and supraballs: (a, b) heating capacity and (c, d) local cellular damages. (a, c) Scheme of the experimental setup to evaluate the global and local effect of laser irradiation: (a) After 10 minutes' laser irradiation of the nanoparticle-loaded cell pellet, cells are redeposited in cell culture flasks for four hours before evaluating the cell death. (c) A431 cells are incubated overnight with Fe_3O_4 nanocrystal, colloidosomes, and supraballs, and after rinsing, they were irradiated as monolayer with a 808-nm laser for 10 minutes. The cell damaged was evaluated four hours post-irradiation. (b, d) Percentage of cell necrosis and apoptosis upon laser exposition (or not) for 10 minutes of cell pellet (b) and cell monolayers (d). Cells were stained with Hoechst 33342 (cell nuclei, blue), YO-PRO (apoptotic cells, orange), and propidium iodide (necrotic cells, green). Scale bar = 50 μm. *Source:* (b, d) Nicolas-Boluda et al. [2]/John Wiley & Sons, Inc.

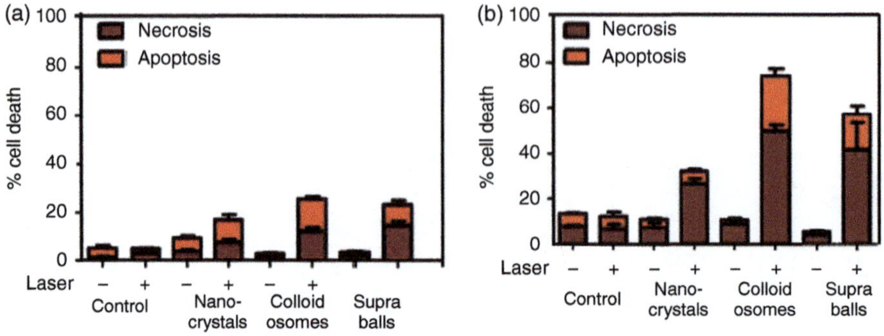

Figure 16.4 Percentage of cell necrosis and apoptosis upon laser exposition (or not) for 10 minutes of (a) cell pellet (heating capacity) and of (b) local cellular damages. *Source:* Nicolas-Boluda et al. [2]/John Wiley & Sons.

in pellet form (Figure 16.3b), although no macroscopic heating can be measured. Even though the laser irradiation experienced by each cell can be higher than in the cell pellet as there is no light screening by cells and nanocrystal absorption, we expect in this setup only local effect of intracellular nanoheaters. Remarkably, despite colloidosomes having low global heating performance in aqueous solution and even lower in cells, colloidosome-mediated intracellular local absorption induces the highest levels of cell damage compared with dispersed nanocrystals and supraballs (Figures 16.3d and 16.4b). Fe_3O_4 nanocrystals, despite exhibiting higher heating efficiency both in aqueous solution and in cell pellet, induce less cellular damage upon irradiation of cell monolayers. Overall, the two self-assemblies of Fe_3O_4 nanocrystals clearly induce massive cell death: 74% and 57% of cell damage (both necrosis and apoptosis) induced by colloidosome- and supraball-mediated photothermal therapy (PTT), respectively (Figure 16.4b), in comparison with less than 35% for dispersed nanocrystals. This experiment clearly demonstrates that self-assemblies of Fe_3O_4 nanocrystals, once internalized by cells, can elicit massive local damages that provoke cell apoptosis and necrosis without global heating of the environment.

At this point, we asked the following question: What is the origin of the apparent contradiction between the global heating efficiency and the biological outcome on cells? To answer this, we highlight the intracellular distribution of Fe_3O_4 nanocrystals. Major differences in the intracellular fate of Fe_3O_4 nanocrystals internalized as dispersed nanocrystals or as suprastructures up to eight days post labeling are shown in Figure 16.5 and Chapter 15 [2]. Dispersed nanocrystals tend to aggregate into endosomes that fuse with lysosomes, forming uncontrolled fractal structures, whereas supraballs and colloidosomes conserve their organized structures once internalized by tumor cells.

This demonstrate that self-assembled Fe_3O_4 nanocrystals could withstand intracellular trafficking and the harsh acidic environment of lysosomes. In addition, suprastructure self-assemblies maximize the intralysosomal density Fe_3O_4 nanocrystals with a high percentage close to the lysosomal membrane. Such local accumulation of Fe_3O_4 nanocrystals could explain that larger local damages could be obtained with

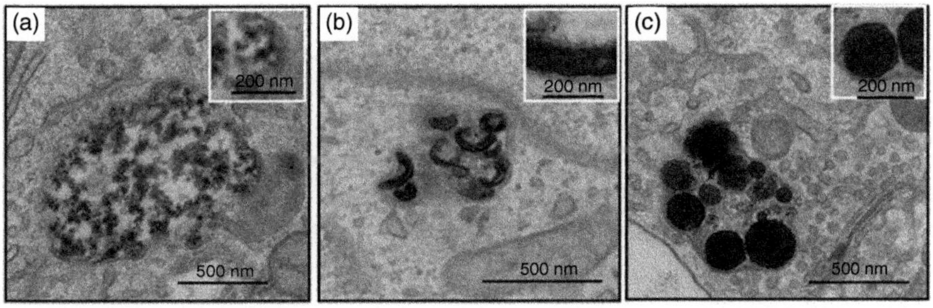

Figure 16.5 2D TEM slice images of dispersed nanocrystals (a), colloidosomes (b), and supraballs (c) in A431 tumor cells incubated for eight days with 28 μg l^{-1} iron concentration. Inset shows the fractal distribution of nanocrystals for dispersed nanocrystals and closed to the lysosomal membrane for both suprastructures. *Source:* Nicolas-Boluda et al. [7] Copyright Wiley-VCH GmbH. Reproduced with permission.

self-assembled nanocrystals in comparison with dispersed ones. In addition to the spatial density of Fe_3O_4 nanocrystals, their ordered structure into the lysosome, inherited from the self-organization of the assembly, might favor an accumulation of energy within these structures. Indeed, as shown in Chapter 14, the dilution of the metallic phase in the hybrid self-assemblies enables to increase the penetration depth of visible light in comparison with homogeneous metallic nanocrystals of similar size, thus enabling an energy load in the assembly. The same study revealed that coating agents of Fe_3O_4 nanocrystals act at short time as a reservoir of energy due to the tiny distance between nanocrystals and thermal transfer that propagates preferentially within the assembly. This reservoir effect and thermal resistance of the hydrophobic organic phase consequently induce an increase of the localized heating in the self-assemblies compared with the fractal aggregates of Fe_3O_4 nanocrystals. Hence, we hypothesize that the optimal accumulation and local density of Fe_3O_4 nanocrystals in lysosomes and their sustained compacity in lysosome and the affinity of the colloidosome shell for the lysosomal membrane can collectively explain the enhanced local damages that were observed upon irradiation, inducing massive cell death. The role of lipid bilayers and organic coating as a thermal barrier that favor local accumulation of energy has been demonstrated before using nanothermometry [8]. Here, we observe a direct biological consequence of the energy accumulation in the lysosome membrane due to the close proximity of the lysosome membrane to the absorbing Fe_3O_4 nanocrystals in colloidosomes and supraballs (insets in Figure 16.5). This accumulation may induce lysosome swelling and membrane permeabilization much more efficiently than for dispersed nanocrystals, without significant global heating. We thus conclude that the optimization of Fe_3O_4 nanocrystal density in lysosomes and proximity to the lysosome membrane together with the high-energy accumulation within the hydrophobic shell of colloidosomes or the hydrophobic core of the supraballs enhance local hot spots inducing cell death. Our study clearly demonstrates that cell damages are not directly related to the global heating of the environment but rather to the local energy accumulation that depends on ultrastructural distribution of Fe_3O_4 nanocrystals among themselves and within the lysosome. These data agree with the previous study showing that cancer cells can be particularly sensitive to agents that induce lysosomal cell death through lysosome membrane permeabilization or disruption [9]. It results in the activation of apoptosis signaling pathways and contributes to the amplification of death signals [10]. Lysosome membrane permeabilization can be triggered by a wide variety of agents, one of them being local heating [11, 12]. With colloidosomes, the local heating might trigger enhanced lysosome swelling, affecting more easily the lysosome integrity than supraballs or dispersed nanocrystals.

16.3 Suprastructures: Photothermal Properties in the *in vivo* Tumor Microenvironment

The tumors are stained by various compounds:

(a) Perls' Prussian blue pigments tissues or cells and permits to detect iron. This allows to localize Fe_3O_4 nanocrystals within the tissue.
(b) Hematoxylin eosin saffron (HES): hematoxylin stains the cell nuclei in blue, and eosin stains the extracellular matrix (ECM) and the cytoplasm in pink, while saffron allows the highlighting of collagen fibers (stained in yellow/orange). Fe_3O_4 nanocrystals appear in yellow/orange on HES staining. This trichromatic stain allows a general evaluation of a tissue sample and identification of anomalies such as the presence of necrotic regions.
(c) F4/80 detects macrophages in brown.

To test the feasibility of photothermal therapy mediated by self-assembled nanostructures in solid tumor, dispersed nanocrystals, colloidosomes, and supraballs are injected intratumorally (i.t.) in subcutaneous tumors of murine breast cancer model MET-1 [13]. The i.t. administration is chosen for two reasons:

(i) It is a clinically relevant option for tumors that should be shrink before surgery to improve local control (neoadjuvant treatment). Percutaneous or endoscopic administration of nanoparticles has been achieved in many solid tumors such as soft tissue sarcoma, breast, pancreatic, and gastrointestinal cancers, and endoscopic and intraluminal light illumination is also feasible through the same procedures [13–16].
(ii) The local administration of nanoparticles avoids the massive off-target accumulation of nanoparticles associated with intravenous or intra-arterial injection, minimizes systemic toxicity, and maximizes the tumor concentration of nanoheaters. It is also the best option for comparing nanostructures that may have very different biodistribution and access to the tumor when injected intravenously.

Here, the same mass of iron ([Fe] = 150 μg in 50 μl injected volume) is infused in tumors of approximately 150–200 mm^3. This allows a head-to-head comparison of the different nanostructure distributions in the tumor microenvironment and the outcome of light exposition. Injection of nanocrystals is performed in a unique injection point with a very slow infusion through 27G needle to maximize the distribution of nanostructures that are let for one day to diffuse and to be processed by cells of the tumor microenvironment before irradiation.

One day after i.t. administration, tumors are exposed for 15 minutes to an 808-nm laser at a power of 2 W cm^{-2}. Their surface temperature is monitored with a near-infrared camera (Figure 16.6a). A significant increase in the tumors' surface temperature is observed with all nanostructures (plateau temperature of T = 45, 44, and 43 °C for supraballs, dispersed nanocrystals, and colloidosomes) (Figure 16.6b) in comparison with the laser-exposed non-injected tumors.

One day after laser irradiation, tumors are harvested, and the local effect of dispersed nanocrystal-, colloidosome-, and supraball-mediated PTT is evaluated using histology. Tumors are stained with Perls' Prussian blue staining that allows the detection of iron in tissue samples to localize the Fe_3O_4 nanocrystals within the tissue. By using HES, a general evaluation of a tissue sample and identification of anomalies such as the presence of necrotic regions are done. The presence of necrotic regions is identified for each of the conditions (Figure 16.7).

The basal level of tumor necrosis without laser exposure is about 13% regardless of the nanostructures injected or not. However, we observe striking differences in the PTT-induced necrosis level depending on the nanostructures inside the tumors (Figure 16.8).

Even more surprisingly, the necrosis level does not correlate with the tumor temperature increase. Actually, the largest necrosis extension (64%) is observed for colloidosome-treated tumors that however heat at only 43 °C for less than 15 minutes under laser exposure. Supraball-mediated PTT at the same laser power induces a 50% necrosis extension for a surface temperature of 45 °C. Dispersed Fe_3O_4 nanocrystals on the other side induce a much lower level of tissue damage (24%) while the temperature reached to 44 °C. This evidences again that the level of global hyperthermia applied to a given tumor does not necessary translate to a higher level of cellular damage of the

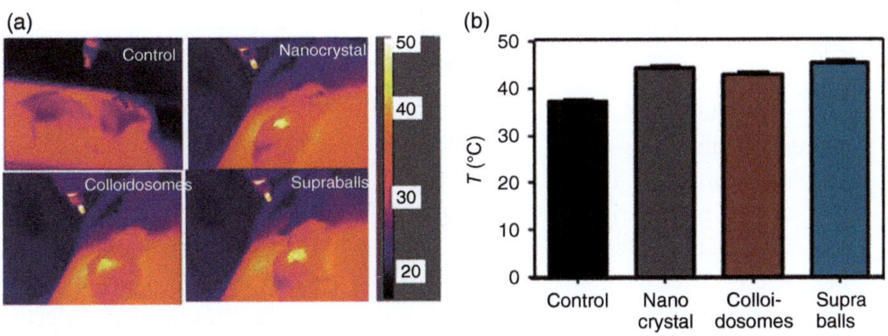

Figure 16.6 Colloidosome and supraball-mediated photothermal therapy *in vivo*. (a) *in vivo* infrared thermal images of tumors injected with 150 µg colloidosomes or supraballs and exposed 15 minutes to an 808-nm laser at 2 W cm^{-2}. (b) Average surface temperature of tumors exposed 15 minutes to an 808-nm laser at 2 W cm^{-2}. *Source:* (a) Nicolas-Boluda et al. [2]/John Wiley & Sons, Inc.

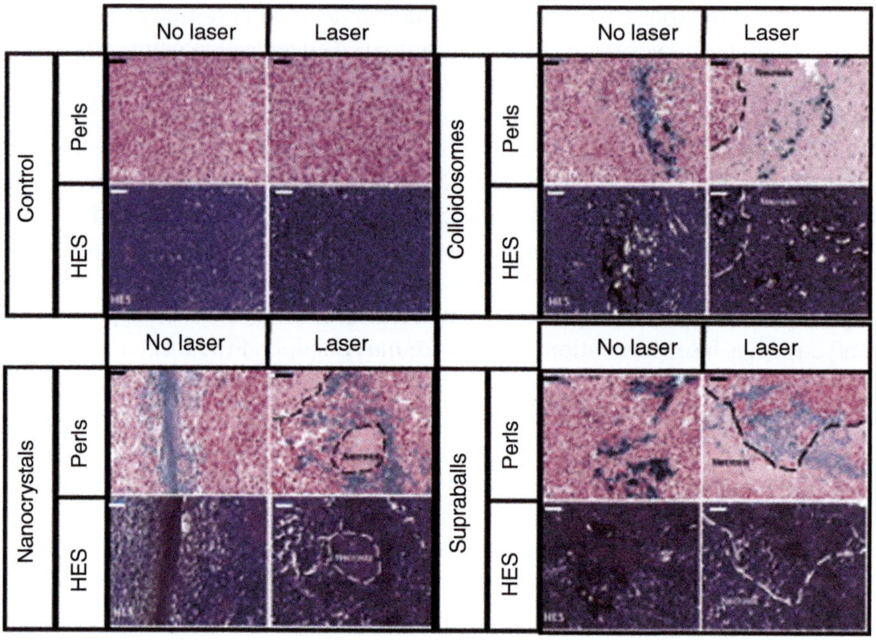

Figure 16.7 Histological (Perls and HES staining) analysis of MET-1 tumors injected with dispersed nanocrystals, colloidosomes, and supraballs 24 hours after 808-nm laser irradiation (15 minutes) or without laser exposure (at $2\,W\,cm^{-2}$). Scale bar = $50\,\mu m$. Nanocrystals appearing in blue with Perls staining are located in the tumor. Necrotic areas underlined by dashed line can be observed in tumors treated with laser close to nanocrystal localizations. *Source:* Nicolas-Boluda et al. [2]/John Wiley & Sons, Inc.

Figure 16.8 Quantification of the percentage of necrotic areas in MET-1 tumor exposed 15 minutes to an 808-nm laser at $2\,W\,cm^{-2}$. *Source:* Nicolas-Boluda et al. [2]/John Wiley & Sons, Inc.

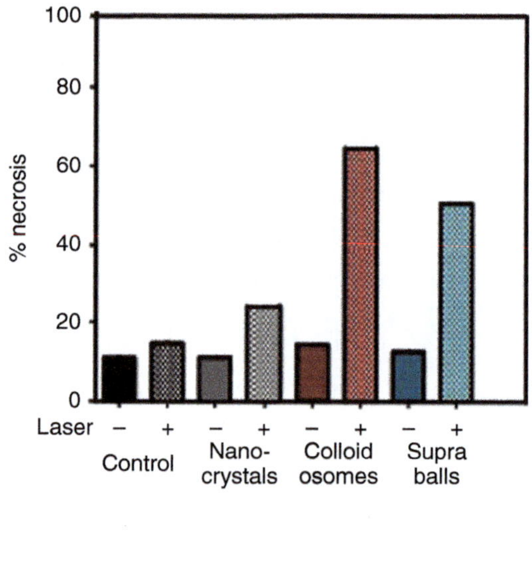

tumor tissue. Importantly, we observe that suprastructures induce a significant increase in tissue necrosis extension (>50% of the tissue) after a single-laser exposition and 15 minutes of mild hyperthermia while the biological outcome of their individual building blocks in similar experimental conditions and macroscopic temperature increase is not significant. These results at the tissue level in tumors fully correlate with the results obtained in the experiments described above at the cellular level, suggesting superior toxicity of intracellular heating mediated by self-assemblies in the tumor tissue.

16.4 Suprastructures Modulate the Distribution of Fe_3O_4 Nanocrystals in the Tumor Microenvironment

Tumors do not consist exclusively on tumor cells, but on a complex ecosystem including stromal cells (endothelial cells, fibroblasts, and immune cells) and ECM composed of a multitude of biomolecules such as collagen and hyaluronic acid, in what is called the tumor microenvironment [17]. The role of the heterogeneity of the tumor microenvironment in chemoresistance has been evidenced in the last decade [18]. However, the location and the local effects have been overlooked.

Let us first compare *in vitro* the internalization of dispersed nanocrystals, colloidosomes, and supraballs with separated monoculture of tumor cells (A431 – human epidermoid carcinoma cells), macrophages (RAW 264.7 – murine macrophages), and fibroblasts (hTERT-HSC – human immortalized liver fibroblasts). Iron quantification by inductively coupled plasma-atomic emission spectroscopy (ICP-AES) displayed in Figure 16.9 shows that supraballs could transfer more iron in macrophages than colloidosomes and dispersed nanocrystals. In contrast, iron capture in tumor cells was higher with colloidosomes than with supraballs and dispersed nanocrystals. Fibroblasts also present higher uptake of iron with colloidosomes and to a lesser extent with supraballs in comparison with dispersed nanocrystals that show more than 10 times smaller uptake.

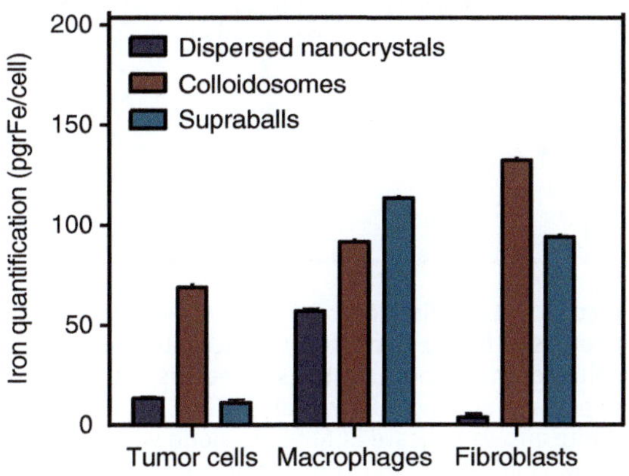

Figure 16.9 Iron quantification, in picogram per cell (pgrFe/cell), measured by ICP-AES in tumor cells (A431 cell line), macrophages (RAW 264.7 cell line), and fibroblasts (hTERT–HSC immortalized line) after day 1 of exposition to Fe_3O_4 nanocrystals, colloidosomes, or supraballs at [Fe] = 28 μg ml^{-1}. *Source:* Nicolas-Boluda et al. [2]/John Wiley & Sons, Inc.

Our comparative study emphasizes the poor efficacy of dispersed nanocrystals to accumulate into tumor cells or into fibroblasts, even if these cells are separately exposed to the Fe_3O_4 nanocrystals in the absence of ECM. Importantly, the only cell type that accumulates significant amount of dispersed nanocrystals (>50 pg) is the macrophages in line with the histological observations in tumors. Remarkably, the self-assembly of Fe_3O_4 nanocrystals within prescribed suprastructures can reverse this trend since colloidosomes are also substantially uptaken by tumor cells and fibroblasts.

Considering that the components of the tumor might interacts differently with the nanostructures, we investigate the differences in tissue distribution and local photothermal effects that could arise specifically from suprastructures. Importantly, we could observe striking differences one day after i.t. injection in the MET-1 breast tumor environment (Figure 16.10). As already mentioned, Perls staining detects iron in blue, the hematoxylin stains cell nuclei in blue, eosin stains the ECM and cytoplasm in pink, Fe_3O_4 nanocrystal appears in yellow/orange on HES staining, and F4/80 detects macrophages in brown. Let us consider the various structures:

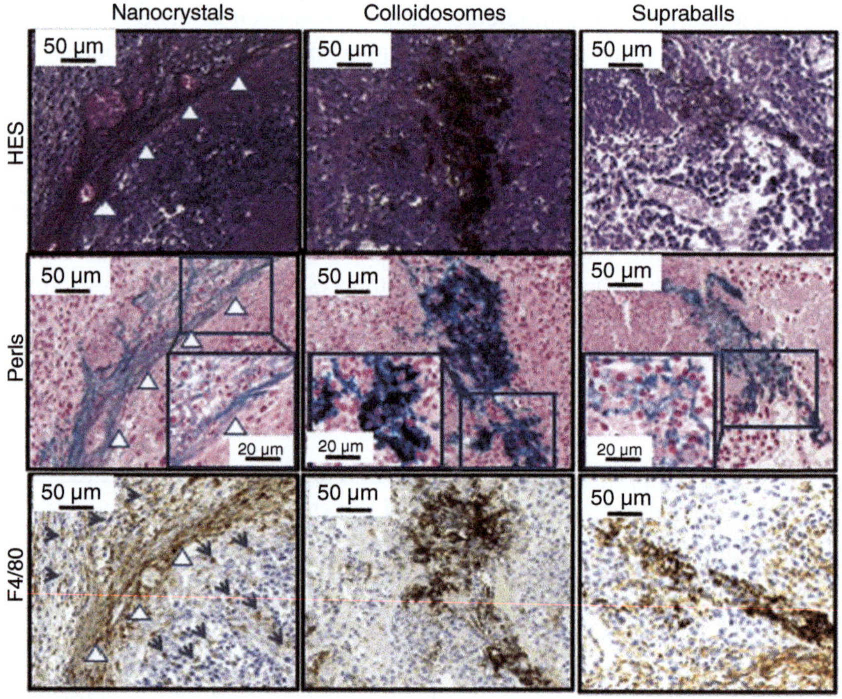

Figure 16.10 Histological (Perls and HES staining) and immunological (F4/80 labeling) analysis of MET-1 tumors day 1 after intratumoral injection of dispersed nanocrystals, colloidosomes, and supraballs. Insets display magnification of the gray square. Iron appears in blue with Perls staining and in yellow/orange in HES and F4/80 images. White arrows indicate ECM fibers along which most Fe_3O_4 nanocrystals are located. Gray arrows indicate the recruitment of macrophages (F4/80 brown labeling). Colloidosomes have mostly intracellular localization (see the inset on Perls staining), and supraballs have both intracellular and extracellular localizations. *Source:* Nicolas-Boluda et al. [2]/John Wiley & Sons, Inc.

(i) *Dispersed nanocrystals:* when dispersed nanocrystals are injected in tumors (Figure 16.10, left), they mainly localized in the extracellular space, in close contact with the ECM fibers. Such localization along collagen fibers was previously observed for iron oxide nanocubes of 20 nm [19]. In addition, a recruitment of F4/80 positive macrophages (indicated by blue arrows in Figure 16.10) is induced in the vicinities of Fe_3O_4 nanocrystals. In addition to a lower intracellular toxicity after irradiation, they are mostly located extracellularly with lesser impact on tumor cells.

(ii) *Colloidosomes:* these are mainly internalized by tumor cells (Figure 16.10, middle). They could effectively induce deleterious local heating within tumor cells due to their specific accumulation.

(iii) *Supraballs:* supraball location is both intracellular and extracellular (Figure 16.10, right). To a lesser extent than colloidosomes, supraballs could effectively induce deleterious local heating within tumor cells due to the specific accumulation.

This reveals that the different nanostructure architectures severely impact the nanostructure distribution in the heterogeneous tumor environment. Particularly, the suprastructures and more particularly colloidosomes have a more potent tropism for tumor cells than their dispersed building blocks (Fe_3O_4 nanocrystals) distributed mostly in between collagen fibers in this breast tumor model. This is an important finding to explain the extended necrosis damage inflicted by colloidosome- and supraball-mediated PTT in comparison with the insignificant effect of dispersed nanocrystals, despite having similar temperature increase.

Such differential cell internalization depending on both cell types and nanostructure architectures can be interpreted in view of their different mechanical properties, such as stiffness and deformability. Macrophages are key actors of immunity and first line of defense against pathogens, xenobiotics, or senescent cells. It has been reported that macrophages can sense the rigidity of phagocytic targets, such as red blood cells that become more rigid after senescence or in certain disease conditions (thalassemia, sickle cells) [20]. A number of studies have shown that stiffer particles are internalized to a higher extent than softer particles by macrophages and other immune cells [21, 22]. Our results show that stiffer supraballs are also more internalized by macrophages than colloidosomes with slightly lower Young's modulus and a more deformable structure [23]. One explanation for increased uptake of supraballs by phagocytic cells is that colloidosomes may undergo deformation in response to phagocytosis forces exerted by the cells that locally decreases their radius of curvature and makes phagocytosis more energetically costly and more difficult to achieve than with rigid undeformable supraballs [24]. In contrast to macrophages, the comparison of tumor cell uptake of soft/hard particles has shown complex findings without general trend emerging [25, 26]. In this case, epidermoid carcinoma cells show a preference for soft and deformable colloidosomes. This is also in line with our finding in the *in vivo* MET-1 breast tumor model confirming higher uptake of tumor cells for colloidosomes. To sum up, both in the heterogeneous microenvironment of tumors and in conditions in which only one cellular component is exposed to the nanostructures, the self-assembled colloidosomes have the unique potential to maximize their accumulation by tumor cells and fibroblasts in comparison with ECM and macrophages. This is the first demonstration that the self-organization of Fe_3O_4 nanocrystals and the resulting mechanical properties can be used to modulate the distribution of Fe_3O_4 nanocrystals in the tumor microenvironment.

To support the conclusions deduced from the *in vivo* experiments from which the location of Fe_3O_4 nanocrystals in tumors evolves with the way they are assembled or not, let us consider the ECM that is another important component of the tumor stroma differently expressed and organized depending on the origin and progression of the tumor. To investigate the interactions between the nanostructures and the ECM, we set up an *in vitro* model consisting of fibroblast sheets secreting ECM over stimulation with ascorbic acid [27]. The abundant ECM fibers surrounding fibroblasts can be observed on TEM image shown in Figure 16.11a. These fibroblast sheets are exposed to Fe_3O_4 nanocrystals, colloidosomes, or supraballs overnight, followed by careful washing. As in the *in vivo* tumor, Fe_3O_4 nanocrystals are mainly located along the ECM in between fibroblasts (Figure 16.11b). The colloidosomes are all internalized by fibroblasts and are absent from ECM (Figure 16.11c). Supraballs could be found in both compartments (Figure 16.11d). These results confirm that in the hybrid ECM/cellular environment, Fe_3O_4 nanocrystals are barely internalized by fibroblasts, while colloidosomes show the opposite behavior. Notably, supraballs are less internalized by fibroblasts than colloidosomes. The mechanical properties of colloidosomes featuring lower Young's modulus/stiffness and larger flexibility than supraballs may favor internalization by fibroblasts, while supraballs are more easily trapped in ECM fibers, as well as dispersed nanocrystals.

In conclusion, such observations on fibroblast sheets nicely correlate with those in the complex microenvironment of the tumor (Figure 16.10) from which dispersed Fe_3O_4 nanocrystals are mainly located in the extracellular compartments, colloidosomes internalized in the tumor cells, and supraballs both located in intracellular and extracellular compartments.

16.5 Suprastructures: Photothermal Effects on the Tumor Extracellular Matrix

Due to the important role of ECM in promoting tumor progression, chemoresistance, and immune exclusion, the ECM is recognized as a multicomponent target for advanced therapeutic strategies. Previous studies show that carbon nanotube-mediated PTT could induce the local denaturation of collagen fibers in tendon and in tumors [28]. We thus test the photothermal effects of the two hybrids and Fe_3O_4 nanocrystals on ECM in different organotypic models. To visualize the direct effect of PTT on the ECM of real tumors, we use *ex vivo* fresh 350-μm tumor slices incubated for two hours with dispersed nanocrystals, colloidosomes, and supraballs (Figure 16.12) and then exposed to 808-nm laser at $2\,W\,cm^{-2}$.

The EGI-1 tumor model, a human bile duct desmoplastic cancer, was chosen for its abundant ECM (unlike the MET-1 tumor model previously used) and linear organization of collagen fibers that can be observed and quantified using second

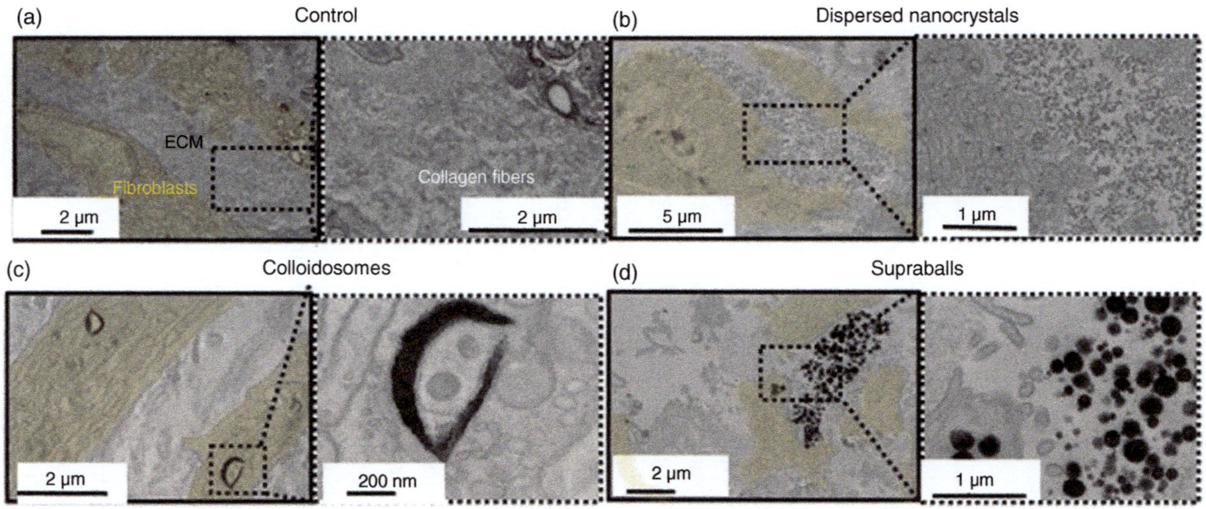

Figure 16.11 TEM images on 70-nm microtome sections of Fe_3O_4 nanocrystals, colloidosome, or supraballs distribution in the fibroblast/ECM *in vitro* model. The fibroblasts are highlighted in yellow in the lower magnification images. Images on the right are magnification of the dark quadrant on image on the left. (a) Control fibroblast sheet with dense ECM. (b) Dispersed Fe_3O_4 nanocrystals are mainly accumulate in the ECM compartment. (c) Colloidosomes are mainly internalized in fibroblasts. (d) Supraballs are distributed both within fibroblasts and within the ECM. *Source:* Nicolas-Boluda et al. [2]/ John Wiley & Sons, Inc.

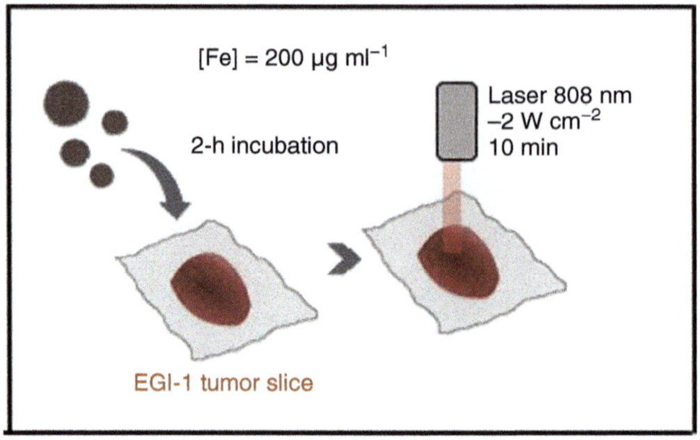

Figure 16.12 Colloidosome-mediated PTT also induces collagen denaturation in EGI-1 tumor slices. Schema of the experimental set-up. *Source:* Nicolas-Boluda et al. [2]/John Wiley & Sons, Inc.

harmonic generation (SHG) two-photon microscopy (Figure 16.13). In comparison with the nonirradiated tumors, the amount of collagen SHG signal was significantly reduced after irradiation of tumor slices exposed to colloidosomes but not to Fe_3O_4 nanocrystals or supraballs. Analyses of SHG images with the CT-FIRE software [28] also show a significant decrease in collagen fiber length of tumors exposed to colloidosomes and irradiated (Figure 16.14), which can be related to fiber denaturation induced by localized heating (Figure 16.13, red arrows).

This destructuration of collagen fibers on tumor slices by irradiated colloidosomes is confirmed in a tissue consisting of highly organized type-I collagen, a fresh sheep tendon. Using the same protocol as for the tumor slices, colloidosomes induce the apparition of SHG signal voids in the dense collagen matrix of the tendon. This means that the collagen fibers suffer a significant conformation change, leading to the loss of SHG properties, likely due to the local heat-induced denaturation of the collagen molecule. Figures 16.15 and 16.16 show a significant increase in the presence and size of these areas devoid of SHG signals observed when laser power is increased from 1 to 2 W cm^{-2}, whereas collagen fibers remain intact after laser irradiation in the presence of dispersed nanocrystals and supraballs.

The collagen structure is observed by high-resolution scanning electron microscopy (SEM) on the tendon before and after PTT (10 minutes at 2 W cm^{-2}). The flattened structure of colloidosomes and the more compact and rigid supraballs are easily recognizable in SEM images of nonirradiated tendon. After laser exposure, SEM observations confirm that colloidosome-mediated PTT induces the formation of "holes" on the dense collagen network (Figure 16.17a) close to colloidosomes. The same kind of holes is also seen in samples that had undergone supraball-mediated PTT (Figure 16.17b). This means that supraball-mediated PTT also induces collagen denaturation, but to a lesser extent probably not enough either to be detected with SHG microscopy in these conditions or to sit on both extracellular and intracellular compartments. These experiments focusing on collagen matrix on tumor and tendon slices clearly reveal the colloidosomes as the most efficient light-activated nanoheaters to induce local denaturation of the collagen fibers in our experimental conditions.

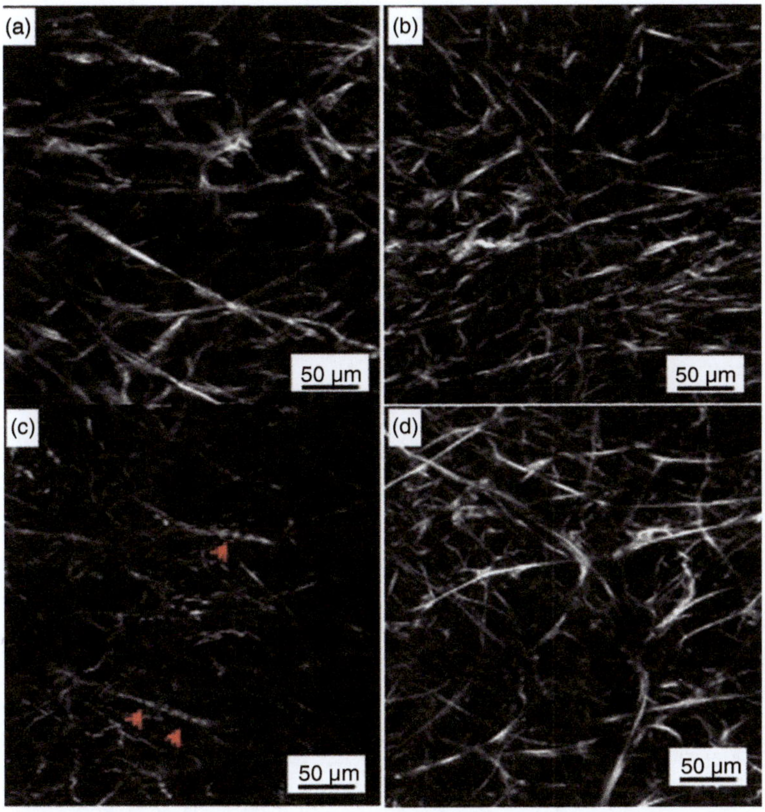

Figure 16.13 Colloidosome-mediated PTT also induces collagen denaturation in EGI-1 tumor slices: two photon microscopy SHG images of collagen network of EGI-1 (human bile duct) tumor slices incubated in the absence of nanocrystals (a), but with dispersed nanocrystals (b), colloidosomes (c), and supraballs (d) irradiated with an 808-nm laser. A clear reduction of collagen SHG signal can be observed after colloidosome-mediated PTT and decrease of fiber length (red arrows). *Source:* Nicolas-Boluda et al. [2]/John Wiley & Sons, Inc.

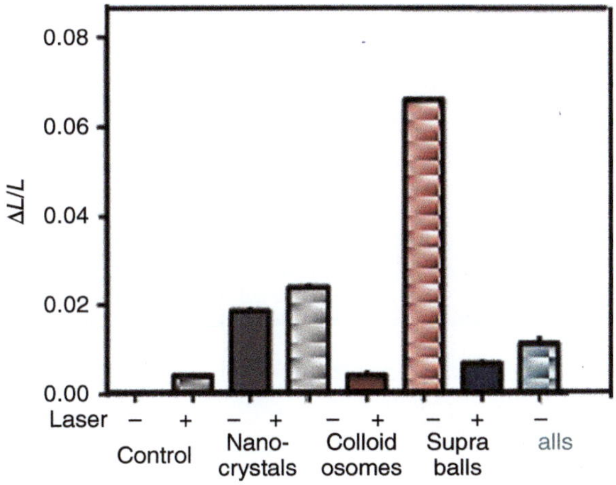

Figure 16.14 Colloidosome-mediated PTT also induces collagen denaturation in EGI-1 tumor slices. Percentage of two-photon microscopy SHG collagen signal in the different conditions. *Source:* Nicolas-Boluda et al. [2]/John Wiley & Sons, Inc.

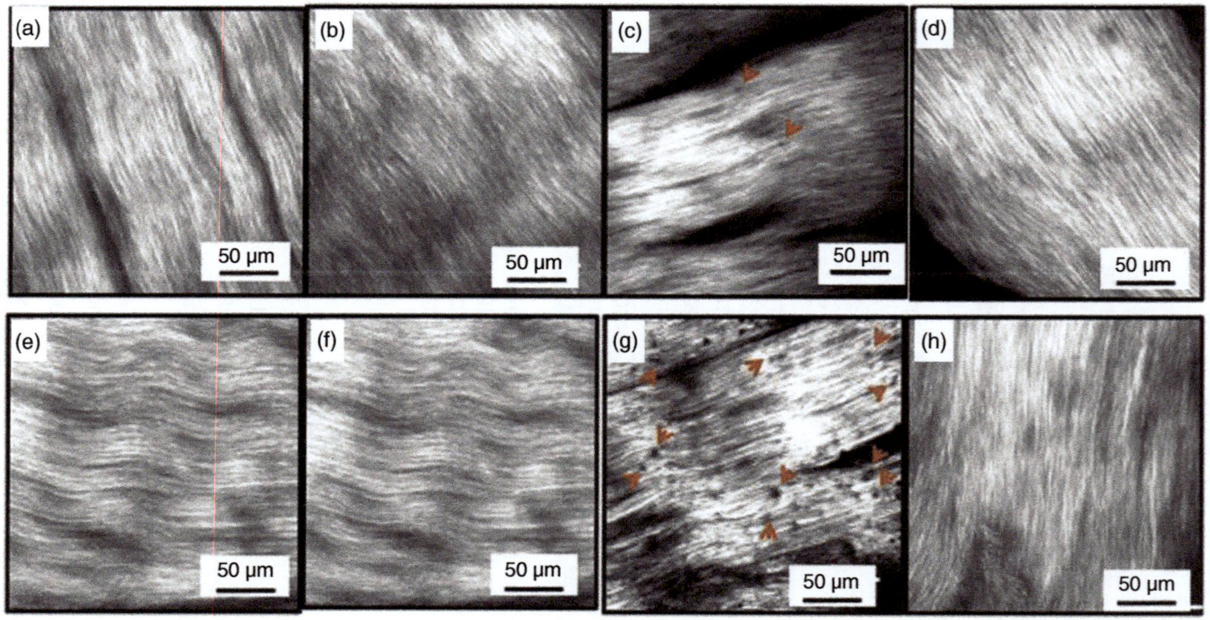

Figure 16.15 Colloidosome-mediated PTT induces collagen denaturation. Two photon second harmonic generation (SHG) microscopy images of sheep tendon collagen alone subjected (a) or not to light (e) and considered as controlled experiment or incubated with dispersed nanocrystals (b, f), colloidosomes (c, g), and supraballs (d, h) and subsequently irradiated with 808-nm laser with $1\,\mathrm{W\,cm^{-2}}$ (a–d) and $2\,\mathrm{W\,cm^{-2}}$ (e–h). Collagen denaturation is observed as areas devoid of collagen SHG signals within the dense collagen network (orange arrows). *Source:* Nicolas-Boluda et al. [2]/John Wiley & Sons, Inc.

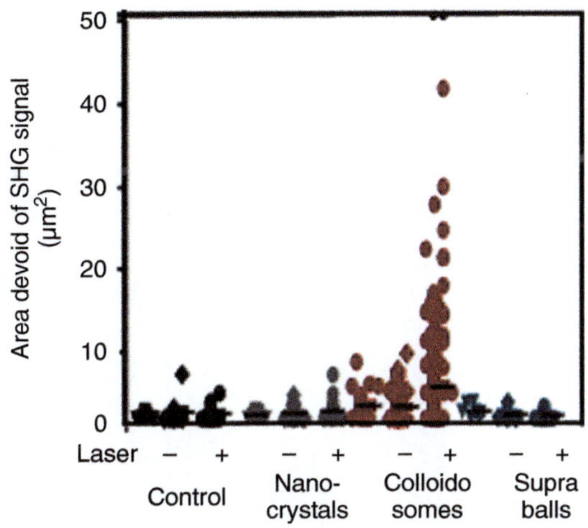

Figure 16.16 Quantification of the presence and size of the collagen denatured sites measured from Figure 16.15. *Source:* Nicolas-Boluda et al. [2]/John Wiley & Sons, Inc.

16.6 Conclusion

In this study, we focus on the photothermal and biological effects of Fe_3O_4 nanocrystal self-assembled into two different conformations: colloidosomes and supraballs. The different conformations confer them with unique properties that govern both their local light-to-heat conversion and their tumoral tissue distribution in comparison with dispersed nanocrystals. Remarkably, colloidosomes and supraballs are capable of inducing extended tumor necrosis after a 15 minutes' single near infrared (NIR) *in vivo* laser exposition and mild hyperthermia in a breast cancer model in mice. This impressive tissue damage correlates with enhanced cell death on tumor cell suspension and monolayers and collagen destructuration in tumor slices of a desmoplastic tumor and in a sheep tendon. However, these biological outcomes of PTT mediated by self-assemblies do not correlate with the global temperature increase that is measured by infrared camera. Indeed, dispersed Fe_3O_4 nanocrystals exhibit higher global heating capacity than colloidosomes, while their PTT biological effects are insignificant. In contrast, colloidosomes and to a lesser extent supraballs induced potent and localized thermal damages both in tumor cells and collagen matrix. This is a unique demonstration that global heating measurements is neither representative of local energy deposition nor predictive of the biological outcomes of nanostructure-mediated PTT. Our findings emphasize that nanoparticle distribution and light-to-heat energy conversion must be controlled at the subtissular and subcellular scales to optimize the biological outcomes of "nanohyperthermia." This represents a paradigm shift in the field of nanoparticle-mediated hyperthermia.

Our study also reveal different phenomena that explain the outstanding properties of self-assemblies, in comparison with dispersed nanocrystals, to optimize the NIR-induced local thermal damages in the biological environment: (i) an enhanced energy accumulation in the hybrid self-assemblies with the organic phase playing the role of localized thermal resistance [6], (ii) an optimized intralysosomal concentration of nanoheaters, and (iii) an optimized transfer of energy to the lysosome

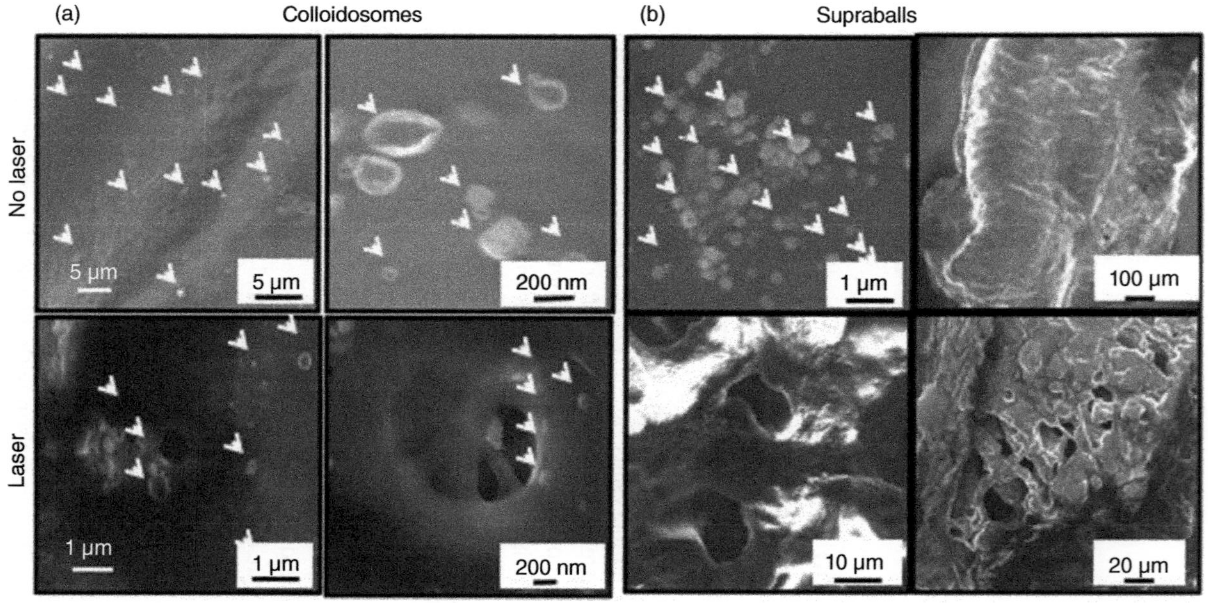

Figure 16.17 Colloidosome-mediated PTT induces collagen denaturation: SEM images of fresh sheep tendon incubated with suprastructure and irradiated or not with a NIR laser. (a) Colloidosomes: their location is indicated with the white arrows. (b) Supraballs: location of the supraballs is indicated with the white arrows. *Source:* Nicolas-Boluda et al. [2]/John Wiley & Sons, Inc.

membrane affording the affinity of colloidosomes to the cell membrane. All these features are direct consequences of the organization of the hybrid organic–inorganic self-assemblies and open new avenues for more precise control of photothermal conversion in the cellular and extracellular biological environments.

On top of these unique properties of hybrids as nanoradiators, we highlight very distinct distribution of the two hybrids and their dispersed building blocks (Fe_3O_4 nanocrystals) in the complex tumor microenvironment, taken as a whole (intratumoral administration) or by separating its different components (tumor cells, stromal cells, ECM). Indeed, Fe_3O_4 nanocrystals were mostly located in the ECM and macrophages, while supraballs were distributed partly in ECM and tumor cells. Owing to their different structure, flexibility, and affinity for cell membrane, colloidosomes were shown to be more efficiently internalized by tumor cells than stromal cells and to target the tumor cells even in the competitive multicomponent environment of the breast tumor model. This very efficient targeting of tumor cells by colloidosomes also correlates with enhanced PTT-induced necrosis in the tumor, while dispersed Fe_3O_4 nanocrystals have no significant effect. Here, we evidence that peculiar self-assemblies of nanocrystals allow for a better control of the tissular and intracellular distributions of the inorganic nanocrystals triggering different and more precise biological responses to the NIR irradiation. The complexity of the tumor microenvironment has been somehow overlooked in the research of nanomaterials for tumor therapy, which partly explain the low number of inorganic nanomedicines that has been translated into the clinics. Remarkably, using Fe_3O_4 nanocrystals or their self-assemblies as colloidosomes and supraballs allows us to target different components of the heterogeneous tumor ecosystem that is an important asset to envision new strategies of personalized therapy. One possible way of exploiting these differences for more personalized and precision medicine would be choosing the correct nanoassembly as function of the therapy target and the peculiarity of the tumor microenvironment of the patient. This offers the opportunity to adapt the choice of nanostructures to the particularity of the actual tumor (e.g. fibrosis level, macrophage infiltration, importance of fibroblasts) and to the intended biological outcome (tumor cell death, ECM destructuration, fibroblast depletion, macrophage depletion, etc.) for seeking a better control of therapy.

References

1 Yang, Z., Altantzis, T., Zanaga, D. et al. (2016). Supracrystalline colloidal eggs: epitaxial growth and freestanding three-dimensional supracrystals in nanoscaled colloidosome. *J. Am. Chem. Soc.* 138: 3493–3500.

2 Nicolas-Boluda, A., Yang, Z., Dobryden, I. et al. (2020). Intracellular fate of hydrophobic nanocrystal self-assemblies in tumor cells. *Adv. Funct. Mater.* 30 (40): 2004274.

3 Mazzanti, A., Yang, Z., Silva, M.G. et al. (2019). Light–heat conversion dynamics in highly diversified water-dispersed hydrophobic nanocrystal assemblies. *Proc. Natl. Acad. Sci. U.S.A.* 116: 8161–8166.

4 Riedinger, A., Guardia, P., Curcio, A. et al. (2013). Subnanometer local temperature probing and remotely controlled drug release based on azo-functionalized iron oxide nanoparticles. *Nano Lett.* 13: 2399–2406.

5 Dias, J.T., Mros, M., del Pino, P. et al. (2013). DNA as a molecular local thermal probe for the analysis of magnetic hyperthermia. *Angew. Chem. Int. Ed.* (in English), 52: 11526–11529.

6 Nicolás-Boluda, A., Vaquero, J., Laurent, G. et al. (2020). Photothermal depletion of cancer-associated fibroblasts normalizes tumor stiffness in desmoplastic cholangiocarcinoma. *ACS Nano* 14: 5738–5753.

7 Nicolas-Boluda, A., Yang, Z., Guilbert, T. et al. (2020). Self-assemblies of Fe3O4 nanocrystals: toward nanoscale precision of photothermal effects in the tumor microenvironment. *Adv. Funct. Mater.* 31 (4): 2006824. (2 of 17).

8 Bastos, A.R.N., Brites, C.D.S., Rojas-Gutierrez, P.A. et al. (2019). Thermal properties of lipid bilayers determined using upconversion nanothermometry. *Adv. Funct. Mater.* 29: 1905474.

9 Groth-Pedersen, L. and Jäättelä, M. (2013). Combating apoptosis and multidrug resistant cancers by targeting lysosomes. *Cancer Lett.* 332: 265–274.

10 Boya, P. and Kroemer, G. (2008). Lysosomal membrane permeabilization in cell death. *Oncogene* 27: 6434–6451.

11 Clerc, P., Jeanjean, P., Hallali, N. et al. (2018). Targeted magnetic intra-lysosomal hyperthermia produces lysosomal reactive oxygen species and causes Caspase-1 dependent cell death. *J. Controlled Release* 270: 120–134.

12 Fourmy, D., Carrey, J., and Gigoux, V. (2015). Targeted nanoscale magnetic hyperthermia: challenges and potentials of peptide-based targeting. *Nanomedicine* 10: 893–896.

13 Bonvalot, S., Rutkowski, P.L., Thariat, J. et al. (2019). NBTXR3, a first-in-class radioenhancer hafnium oxide nanoparticle, plus radiotherapy versus radiotherapy alone in patients with locally advanced soft-tissue sarcoma (Act.In.Sarc): a multicentre, phase 2–3, randomised, controlled trial [published correction appears in Lancet Oncol. 2019 Sep; 20(9): e468]. *Lancet Oncol.* 20: 1148–1159.

14 Casal, R.F., Schwalk, A.J., Fowlkes, N. et al. (2020). Endobronchial ultrasound-guided injection of NBTXR3 radio-enhancing nanoparticles into mediastinal and hilar lymph nodes: a swine model to evaluate feasibility, injection technique, safety, nanoparticle retention and dispersion. *J. Thoracic Disease* 12: 2317–2324.

15 Satiya, J., Schwartz, I., Tabibian, J.H. et al. (2020). Ablative therapies for hepatic and biliary tumors: endohepatology coming of age. *Trans. Gastroenterol. Hepatol.* 5: 15–18.

16 Alvarez-Sańchez, M.V. and Napoleón, B. (2016). Review of endoscopic radiofrequency in biliopancreatic tumours with emphasis on clinical benefits, controversies and safety. *World J. Gastroenterol.* 22: 8257–8270.

17 Balkwill, F.R., Capasso, M., and Hagemann, T. (2012). The tumor microenvironment at a glance. *J. Cell Sci.* 125: 5591.

18 Son, B., Lee, S., Youn, H. et al. (2017). The role of tumor microenvironment in therapeutic resistance. *Oncotarget* 8: 3933–3945.

19 Kolosnjaj-Tabi, J., Di Corato, R., Lartigue, L. et al. (2014). Heat-generating iron oxide nanocubes: subtle "destructurators" of the tumoral microenvironment. *ACS Nano* 8: 4268–4283.

20 Sosale, N.G., Rouhiparkouhi, T., Bradshaw, A.M. et al. (2015). Cell rigidity and shape override CD47's "self"-signaling in phagocytosis by hyperactivating myosin-II. *Blood* 125: 542–552.

21 Anselmo, A.C. and Mitragotri, S. (2017). Impact of particle elasticity on particle-based drug delivery systems. *Adv. Drug Delivery Rev.* 108: 51–67.

22 Anselmo, A.C., Zhang, M., Kumar, S. et al. (2015). Elasticity of nanoparticles influences their blood circulation, phagocytosis, endocytosis, and targeting. *ACS Nano* 9: 3169–3177.

23 Dobryden, I., Yang, Z., Claesson, P.M., and Pileni, M.P. (2020). Water dispersive suprastructures: an organizational impact on nanomechanical properties. *Adv. Mater. Interfaces* 8 (3): 2001687.

24 Yi, X. and Gao, H. (2015). Cell membrane wrapping of a spherical thin elastic shell. *Soft Matter* 11: 1107–1115.

25 Sun, H., Wong, E.H., Yan, Y. et al. (2015). The role of capsule stiffness on cellular processing. *Chem. Sci.* 6: 3505–3514.

26 Hartmann, R., Weidenbach, M., Neubauer, M. et al. (2015). Stiffness-dependent in vitro uptake and lysosomal acidification of colloidal particles. *Angew. Chem Int. Ed.* 54: 1365–1368.

27 Pillet, F., Gibot, L., Madi, M. et al. (2017). Importance of endogenous extracellular matrix in biomechanical properties of human skin model. *Biofabrication* 9: 025017.

28 Marangon, I., Silva, A.A.K., Guilbert, T. et al. (2017). Tumor stiffening, a key determinant of tumor progression, is reversed by nanomaterial-induced photothermal therapy. *Theranostics* 7: 329–343.

Index

Metal Nano 3D Superlattices: Synthesis, Properties, and Applications, First Edition.